Springer Water

The book series Springer Water comprises a broad portfolio of multi- and interdisciplinary scientific books, aiming at researchers, students, and everyone interested in water-related science. The series includes peer-reviewed monographs, edited volumes, textbooks, and conference proceedings. Its volumes combine all kinds of water-related research areas, such as: the movement, distribution and quality of freshwater; water resources; the quality and pollution of water and its influence on health; the water industry including drinking water, wastewater, and desalination services and technologies; water history; as well as water management and the governmental, political, developmental, and ethical aspects of water.

Anindita Nath · Bappaditya Koley ·
Biswajeet Pradhan · Arkoprovo Biswas ·
Tanupriya Choudhury · Jung-Sup Um
Editors

Geospatial Application of Coastal Hazards

Monitoring, Mitigation, Adaptation and Prediction

Springer

Editors
Anindita Nath
Department of Geography
Bankim Sardar College
Tangrakhali, West Bengal, India

Bappaditya Koley
Department of Geography
Bankim Sardar College
Tangrakhali, West Bengal, India

Biswajeet Pradhan
Centre for Advanced Modelling
and Geospatial Information Systems
University of Technology Sydney
Ultimo, NSW, Australia

Arkoprovo Biswas
Geological Studies Unit
Physics and Earth Science Division
Indian Statistical Institute
Kolkata, West Bengal, India

Tanupriya Choudhury
School of Computer Sciences
University of Petroleum and Energy Studies
Dehradun, Uttarakhand, India

Jung-Sup Um
Department of Geography
Kyungpook National University
Daegu, Korea (Republic of)

ISSN 2364-6934 ISSN 2364-8198 (electronic)
Springer Water
ISBN 978-3-032-20075-4 ISBN 978-3-032-20076-1 (eBook)
https://doi.org/10.1007/978-3-032-20076-1

This Springer imprint is published by the registered company Springer Nature Switzerland AG
The registered company address is: Gewerbestrasse 11, 6330 Cham, Switzerland

Dedicated with reverence at the lotus feet of Lord Shree Jagannatha, the eternal Lord of the Universe.

Foreword

The study of coastal hazards and disasters is crucial and immensely important for saving lives, protecting multi-billion-dollar economies, and building resilient communities in the 21st century. It equips scientists, planners and policymakers with the data needed to map vulnerabilities, optimize urban planning, and deploy nature-based defences against climate-driven threats like sea-level rise, storm surges, GLOFs, flash floods, drought etc. Coasts host major economic hubs, ports, and critical energy grids all over the world. In India every year about 400 hectares of coastal land, 75000 hectares of crop areas, 34000 residential houses and industrial establishments are affected by coastal hazards and disasters. In order to protect the coastline for future generation, a comprehensive understanding and holistic estimation of these coastal changes, impact assessment of extreme events on coastline is primary requirement for evolving sustainable coastal development strategies. The unplanned infrastructure combined with poor implementation of planning, relief and rehabilitation measures increases the human, physical and economic losses sustained during and after a disaster. Research on hydrodynamic forces helps engineers to design coastal structures and buildings that can withstand extreme events. High-resolution mapping and hydrodynamic models help to identify hazard zones, allowing local governments to establish efficient evacuation routes and reduce casualties. The real understanding of the mechanics of tropical cyclones, tsunamis, and storm surges enables the development of accurate Multi-Hazard Early Warning Systems (MHEWS). It also supports the sustainable management of marine sectors such as fishing, aquaculture, and tourism, minimizing the widespread loss of income and property. The study of coastal hazards incorporates the importance of coastal ecosystems—like mangroves, sea beaches, sand dunes, and coral reefs—as the first line of defence against flooding and erosion. It finally addresses disaster management beyond just emergency response into long-term disaster resilience—the ability to adapt, recover, and develop sustainably despite environmental fragility.

This edited book will enlighten the applications of geospatial tools and techniques in addressing coastal risks and disasters as well as the significance of resilience building and the promotion of sustainable practices in the coastal areas of the World. The book is made up with 25 chapters mostly dealt with remote sensing and GIS, coastal dynamics, Climate crisis, coastal hazards and vulnerability, resilience to coastal hazards and risks, coastal risk reduction measures, and sustainability. All the authors of each chapter used sophisticated techniques and methods for addressing several contemporary coastal issues and challenges. I strongly believe that it is one of the very few books in coastal geography and coastal hazards and disasters in India sufficiently meeting the demands of coastal researchers and post graduate students in specialized course. It is a major contribution of the authors and editors mostly to the promotion of the subject coastal geography at the College and University level. It is my pleasure in offering my felicitation to Anindita Nath, Bappaditya Koley, Biswajeet Pradhan, Arkoprovo Biswas, Tanupriya Choudhury, Jung-Sup Um for bringing out such a commendable publication. I also strongly recommend the edited book entitled **"Geospatial Application of Coastal Hazards: Monitoring, Mitigation, Adaptation and Prediction"** to students, research scholars, and teachers working on the coastal region across the World.

Sujit Mandal
[Former Professor and Head,
Department of Geography, University
of Gour Banga]
Professor of Geography and Former
Dean, Faculty of Science
Diamond Harbour Women's University
Diamond Harbour, West Bengal, India

Preface

Coastal regions across the globe are increasingly confronted by the intertwined challenges of climate change, environmental degradation, rapid urbanization, and anthropogenic pressures. Rising sea levels, shoreline erosion, tropical cyclones, flooding, salinity intrusion, and ecosystem degradation have transformed coastal landscapes into some of the most vulnerable yet critically important zones on Earth. In this context, geospatial technologies and Earth observation systems have emerged as indispensable tools for understanding, monitoring, and manag-ing the dynamic processes shaping coastal environments.

This edited volume, **Geospatial Application of Coastal Hazards: Monitoring, Mitigation, Adaptation and Prediction**, brings together a diverse collection of scholarly contributions that explore the application of Geographic Information Systems (GIS), Remote Sensing (RS), Earth Observation (EO), spatial modelling, machine learning, and decision-support frameworks in addressing contemporary coastal challenges. The chapters collectively provide an interdisciplinary platform integrating physical geography, environmental science, disaster management, climate studies, coastal engineering, social vulnerability assessment, and sustain-ability science.

The book begins by examining the foundational role of geospatial technologies in coastal hazard assessment, shoreline monitoring, and coastal vulnerability map-ping. Several chapters demonstrate the application of advanced tools such as DSAS, LiDAR, UAVs, SAR, MCDM approaches, and AI-driven spatial analytics to quantify shoreline change, coastal erosion, flooding risk, and sea-level rise impacts across diverse coastal environments in India, Bangladesh, Nigeria, Mexico, and the Caribbean. Through detailed case studies, the volume highlights the significance of integrating satellite observations with GIS-based analytical frame-works for evidence-based coastal governance and disaster risk reduction.

A major strength of this volume lies in its emphasis on climate resilience and eco-system-based adaptation. The chapters underscore the critical role of mangroves, coastal green belts, wetlands, and other nature-based systems in mitigating storm surges, shoreline retreat, and coastal flooding. Contributions focusing on the Sundarbans, Pichavaram, Chandipur, and other ecologically sensitive regions re-veal how ecosystem services, blue carbon valuation, and ecological restoration can support sustainable coastal management while strengthening local livelihoods and biodiversity conservation.

Equally important is the book's focus on the human dimensions of coastal vul-nerability. Several chapters address the socio-economic and health impacts of flooding and climate-related hazards on marginalized coastal populations, especially women and fishing communities. These studies illuminate the intersections of poverty, governance, gender inequality, livelihood insecurity, and environmen-tal change, thereby advocating for inclusive, community-centered adaptation strategies. The integration of participatory approaches, traditional knowledge sys-tems, and anticipatory climate action frameworks further enriches the discourse on resilience planning.

Another notable contribution of this volume is its exploration of emerging technologies and analytical advancements. The integration of machine learning, artificial intelligence, geostatistics, and Earth Observation science into coastal monitoring represents a transformative direction for future coastal management practices. By combining high-resolution spatial datasets with predictive modelling and decision-support systems, researchers and policymakers can improve hazard forecasting, enhance environmental monitoring, and design adaptive strategies for long-term sustainability. The geographic diversity of the case studies included in this book—from the Indian Sundarbans and Odisha coast to the Gulf of Cali-fornia, Nigeria, Bangladesh, and the Caribbean—provides a comparative under-standing of coastal processes and adaptation responses across varying socio-ecological contexts. This global perspective reinforces the idea that while coastal hazards manifest differently across regions, the need for integrated, science-driven, and socially inclusive management approaches remains universal.

This volume is intended for researchers, academicians, policymakers, planners, environmental managers, students, and practitioners working in the fields of coastal geography, environmental management, climate science, geospatial tech-nology, and disaster risk reduction. We hope that the insights presented in this book will contribute to strengthening interdisciplinary research, promoting sus-tainable coastal governance, and fostering resilient coastal futures in an era of accelerating environmental change.

Finally, we express our sincere gratitude to all contributing authors, reviewers, and collaborators whose expertise and dedication made this volume possible. Their scholarly contributions reflect a collective commitment to advancing knowledge and developing practical solutions for some of the most pressing coastal challeng-es of our time. We also acknowledge the support of institutions, research organi-zations, and local communities whose efforts continue to inspire innovation and resilience in coastal studies and management.

We trust that this book will serve as a valuable reference and encourage further research and dialogue toward sustainable and resilient coastal development worldwide.

West Bengal, India — Anindita Nath
West Bengal, India — Bappaditya Koley
Sydney, Ultimo, Australia — Biswajeet Pradhan
Kolkata, India — Arkoprovo Biswas
Dehradun, India — Tanupriya Choudhury
Daegu, Korea (Republic of) — Jung-Sup Um

Contents

Chapter 1
Geospatial Application in Coastal Hazards

Vedprakash Maralapalle, Jayatheja Muktinutalapati, Tammineni Gnananandarao, Abdullah Ansari, and Vinay Rangari

1.1 Introduction

Coastal areas are one of the most important and active territories on the face of the planet. They inhabit the slender sliver between the land and the sea, not only supporting these very special and delicate ecologies but also providing the focus point of human presence and economic prosperity. However, it is precisely due to this setting that coastlines are significantly exposed to various types of danger (both natural and man-made). -Flooding, erosion, tropical cyclones, tsunamis and the ever-present danger of sea-level rise all present a very serious danger to coastal communities, infrastructure and natural habitats. Management of such hazards has

V. Maralapalle (✉)
L.S. Raheja School of Architecture, Mumbai, India
e-mail: civilved@gmail.com

J. Muktinutalapati
RICS School of Built Environment, Amity University, Mumbai, India
e-mail: jayathejam@gmail.com

T. Gnananandarao
Department of Civil Engineering, Aditya University, Kakinada, Andhra Pradesh, India
e-mail: anandrcwing@gmail.com

A. Ansari
Earthquake Monitoring Center, Sultan Qaboos University, Muscat, Oman

Department of Civil Engineering, Inha University, Incheon, Republic of Korea

Department of Smart City Engineering, Inha University, Incheon, Republic of Korea

A. Ansari
e-mail: a.ansari@squ.edu.om

V. Rangari
Civil Engineering Department, Pimpri Chinchwad College of Engineering, Pune, India
e-mail: vinay.rangari@pccoepune.org

A. Nath et al. (eds.), *Geospatial Application of Coastal Hazards*, Springer Water,
https://doi.org/10.1007/978-3-032-20076-1_1

proved to be now of paramount essence in achieving the objective of disaster risk reduction, achieving sustainable development and protecting the livelihoods (Islam et al. 2017).

The development and growth of geospatial technologies over the past decades have put new prospects into perspective regarding understanding, monitoring, and mitigation of coastal hazards. Remote sensing, Geographic Information Systems (GIS), Global Navigation Satellite Systems (GNSS), and sophisticated spatial modeling tools have transformed the way hazards impacting coastal areas are managed by making management proactive and guided by data rather than reactive and based on post-disaster evaluation. Geospatial technologies enable communities and authorities to foresee risks, carry out specific interventions, and promote long-term resilience through their ability to combine tremendous quantities of geophysical data, enabling near-real-time-monitoring, enable dynamic modeling, and improve decision-making (Khan 2022).

This chapter gives us an in-depth discussion on how geospatial technologies can play a vital part in coastal management. We start by exploring the reasons why we need proactive management of coastal hazards based on the vulnerability of the coasts and the various threats they are exposed to. We then address the life-altering effects of geospatial technologies, which demonstrate how geospatial tools are now being employed in gauging, planning, and acting upon coastal risk.

1.1.1 Importance of Coastal Hazard Management

The presence of coastlines is having an overweight importance to the rest of the world: its population and economy. This is despite the fact that they barely represent a small percentage of the total amount of surface area on the Earth where almost 40 percent of the whole population of the world resides within 100 kilometers of the coast which also keeps rising to higher levels with increased urbanization and concentration of economic activities in these areas. Coastal areas harbor large population, urban centers and industries, tourist attraction sites, exclusive habitats, as well as critical infrastructure thus they are highly susceptible to a myriad of hazards (Zhu et al. 2018).

Coastal hazards are wide in spectrum. Tropical cyclones often cause storm surges and flooding, which over submerge low-lying regions with little warning costs lives and causes immense properties to be destroyed. Human modifications and sea level rise increasingly enhance the process of coastal erosion, which is generally a natural process and at times lead to loss of beaches, property and even communities. Even less common are tsunamis, which have the potential to cause destruction over huge distances and destroy infrastructure and kill people in large numbers. Unlike our other hazards which rise and fall quickly, sea-level rise is a long slow process that is insidious and threatens to entrap land, salinize freshwater aquifers, and exaggerate the effects of other dangers. In the meantime, saltwater intrusion can make the farmland unusable and poison the freshwater sources, destroying health and food security.

Historically, coastal hazard management methods have been in the reactive or response/recovery mode. Yet, the growing severity and frequency of risks on the coasts require proactive approach. The new paradigm focuses on the overall risk assessment and it includes the systematic identification of hazards, risk vulnerabilities and exposure to this risk. Being prepared takes center stage such as establishment of early warning systems, education of the community, and planning of evacuation. Such terms like adaptation and mitigation, which would be achieved with help of resilient infrastructure, restoration of ecological buffers, and planned strict rules, are key. An informed and swift response to the disaster is efficient in distributing resources and rapid recovery (Rajasree, Deo 2020). Figure 1.1 highlights the significance of coastal hazard management in protecting vulnerable coastal communities and ecosystems. It emphasizes the need for proactive planning and mitigation strategies to reduce risk and enhance resilience.

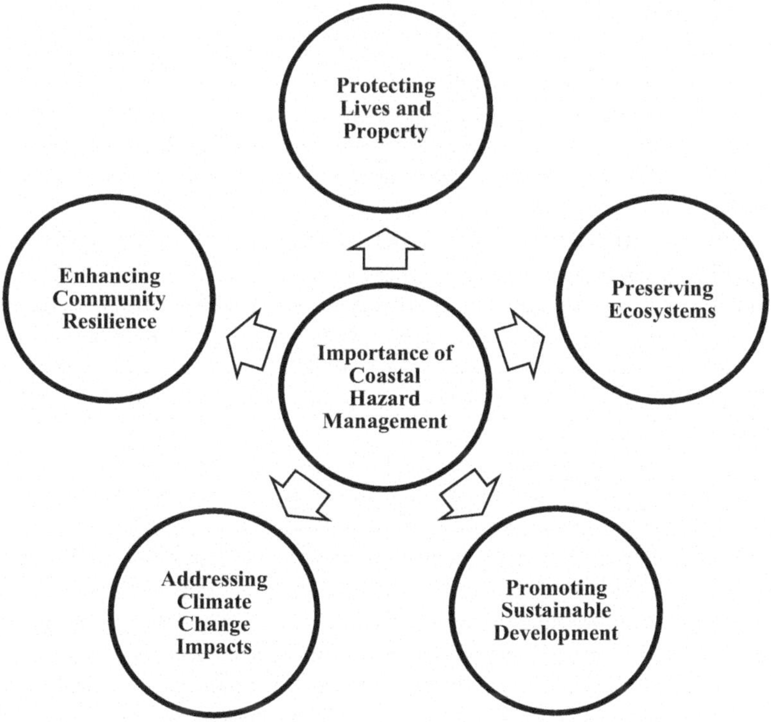

Fig. 1.1 Importance of coastal hazard management

1.1.2 Role of Geospatial Technologies

Geospatial technologies are a broad set of skills, methods and devices to take, process, analyze and display geo-referenced information. The flagship products in this suite are remote sensing, geographical information system (GIS), navigation satellite system (GNSS), and spatial model which can be used to simulate disasters and impacts (Benassai et al. 2014).

Such technologies are transformational in the manner they deal with coastal hazards management. Remote sensing technology has the capability of providing high-resolution satellite images that indicate current details of coastal topography, dangerous areas and land-use changes. GIS systems integrate physical, environmental and socio-economic data, thereby making it possible to map and analyze the hazardous areas. This integration can provide the strong basis of risk and vulnerability assessment, which will ultimately lead to the development of better planning and resources deployment (Waqas et al. 2023).

As far as monitoring is concerned, geospatial technologies support the continuous tracking of coastal processes. Satellite and sensor networks can monitor weather patterns, ocean conditions, shoreline modifications, ecosystem health and many more. GNS systems are used to measure changes in the coastal subsidence and sea-level changes with high precision issues which are used to determine the changing risks.

Another major application of geospatial technology is the use of early warning systems. When the meteorological, oceanographic, and even seismic data are incorporated into GIS platforms, authorities will create advanced warning systems against cyclones, tsunamis, and inundation of the coastal strip. Real-time spatial information presents the possibility of evacuation orders, precise response activities, and flow of important messages at the appropriate time hence minimizing losses of life and property to a great extent.

Due to geospatial modeling and simulation capabilities, a visualization or prediction of the evolution of hazards events like storm surges or both as well as the inundation of the under certain situations, and the coastal erosion can be developed. The models are used in both short-term disaster management operations and in the longer range adaptation planning because they identify the most vulnerable areas, suggest location of infrastructure, and prioritizing resources. The benefits of the geospatial approaches are numerous. They provide never before seen accuracy and timeliness, transmitting and sharing real-time, high-precision information across and over large sections and sometimes inaccessible regions. They render risk information understandable by visualizing the integration of diverse and multidimensional data to make it actionable by both planners and the masses. Moreover, they, as a rule, diminish the necessity to conduct expensive and risky field investigations, which makes it possible to monitor a matter more efficiently throughout a wide scope (Arun Kumar, Kunte 2012).

1.2 Types of Coastal Hazards

Coastal zones are critically important for their economic, ecological, and social value, but they also face a persistent and often intensifying suite of hazards. Understanding the nature and complexity of these coastal hazards is essential for building resilient coastal communities. Among the wide range of threats, the most impactful include storm surges and flooding, coastal erosion, and the twin phenomena of tsunamis and sea-level rise (Coelho et al. 2006). Figure 1.2 illustrates the various types of coastal hazards, including erosion, storm surges, tsunamis, and sea-level rise. These hazards pose significant threats to coastal infrastructure, communities, and natural habitats.

1.2.1 Storm Surges and Coastal Flooding

Storm surges are among the most dramatic and destructive hazards affecting coastal regions. Technically, a storm surge refers to the abnormal and often sudden rise in seawater level generated by intense wind systems, typically cyclones, hurricanes, or

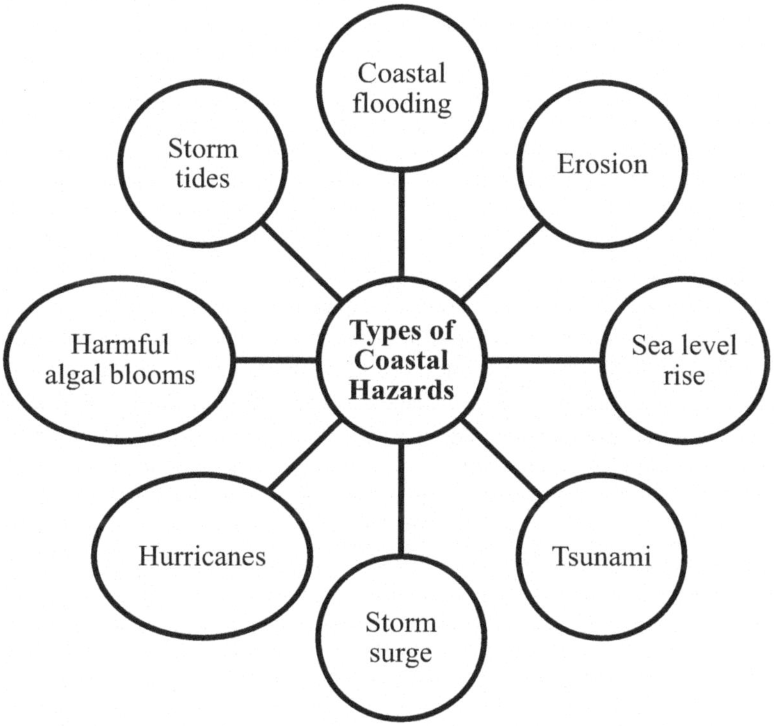

Fig. 1.2 Types of coastal hazards

strong tropical storms, combined with low atmospheric pressure. As these powerful storms approach the coastline, their fierce winds drive vast volumes of water onto the shore, raising the sea level significantly higher than normal tidal ranges. In severe situations, a storm surge can overwhelm coastal defenses, flood low-lying areas, and devastate communities within a matter of hours.

The impacts of storm surges are intensely magnified during high tides, sometimes giving rise to "storm tides," where flood waters penetrate farther inland, inundating roads, residential areas, and farmlands. Notable historical examples underscore their destructive potential: during Hurricane Katrina in 2005, the Gulf Coast of the United States witnessed catastrophic flooding with storm surges reaching more than six meters in height, leading to widespread devastation and loss of life. Similarly, the 2008 Cyclone Nargis produced storm surges in Myanmar that resulted in extensive damage and fatalities.

In the face of intensifying climate change, both storm surges and coastal flooding are expected to become more frequent and severe, underpinned by rising sea levels and shifting weather patterns. The compounded effect of these hazards underscores the need for rigorous early warning systems, sustainable land use planning, and continued investment in natural and engineered defenses (Fitton et al. 2018).

1.2.2 Coastal Erosion

Coastal erosion is the process by which coastlines are gradually worn away by the action of natural forces such as waves, currents, and tides, as well as occasional events like storms. Erosion is a fundamental and natural feature of dynamic coastlines, but human activities and climate-induced changes have dramatically accelerated this process in recent decades. Erosion can occur either as a rapid-onset event—such as during intense storms, when entire beaches or dunes may be stripped away overnight—or as a slow, chronic retreat of the shoreline over years or decades.

The major erosion agents are continuous waves, tidal movement of water and, increasingly, sea-level rise. The strengthening effect of storms is a part of the overall process as they pummel the coasts with huge waves and storm surges thus sweeping away sediments and destabilizing dunes and cliffs. Unless replenishment of the sand and sediment available to them continues, coastlines do not have the substrate to hold together. Development along the coastline, vegetations clearing, and the destruction of the natural barrier like mangroves and dunes makes the assets and structures built along the coastline more vulnerable. The effects of erosion are not only short term but also long term: land properties along the coast are lost, buildings that are meant to sustain life loss their foundations, the habitat is swept away, and the land that could be used is now permanently lost in the sea. For instance, in the United States, significant stretches of the Gulf and Atlantic coasts are eroding at average rates of up to six feet per year, with some areas experiencing far greater losses during severe weather events. In small island nations, entire communities are at risk of losing their homes and heritage to the relentless advance of the ocean.

To counteract coastal erosion, a multifaceted approach is necessary. This can include hard engineering solutions like seawalls and groynes, soft engineering such as beach nourishment; and, perhaps most sustainably, restoration of natural barriers like dunes, wetlands, and mangroves which absorb wave energy and stabilize sediments. Every intervention should be tailored to local geography, sediment budgets, and the balance between human needs and ecosystem preservation (Sarwar, Woodroffe, 2013).

1.2.3 Tsunamis and Sea-Level Rise

Among the most catastrophic coastal hazards, tsunamis are infamous for their devastating speed and overwhelming power. A tsunami consists of one or more massive ocean waves generated by the sudden displacement of the seafloor, usually from undersea earthquakes, but also from volcanic eruptions or underwater landslides. When a tsunami wave approaches shallow water near the coast, its speed slows and its height increases drastically, creating a wall of water that can inundate coastal settlements, infrastructure, and anything in its path with little or no warning.

While tsunamis are rare relative to other hazards, their impacts are profound. The Indian Ocean tsunami of 2004 remains one of the deadliest natural disasters in recorded history, causing over 220,000 deaths in Indonesia, Sri Lanka, Thailand, India, and other nations. Even well-prepared regions such as Japan have suffered severe losses, as seen in the 2011 Tōhoku earthquake and tsunami. The suddenness of these events, combined with the difficulty of predicting precise location and time, makes tsunami readiness—through education, warning systems, and clearly marked evacuation routes—an absolute priority for vulnerable regions.

The slow but inexorable upward creep of sea level threatens to permanently submerge coastal land, salinate freshwater reserves, and render certain islands and delta regions—such as the Maldives, Tuvalu, and the Sundarbans in South Asia—uninhabitable in coming decades. Impacts on human systems range from loss of housing and productive land to the displacement of whole communities and the loss of cultural sites (Deb et al. 2024).

1.3 Geospatial Technologies in Coastal Hazard Monitoring

Geospatial technologies have dramatically transformed how coastal hazards are monitored and managed, enabling scientists and decision-makers to detect, assess, and respond to dynamic coastal risks with unprecedented precision and efficiency. Within this sphere, the integration of remote sensing and satellite imagery, Geographic Information Systems (GIS), and cutting-edge sensor platforms including UAVs (drones) and LiDAR, has created a powerful toolkit for comprehensive coastal hazard analysis (Thomas et al. 2014). Figure 1.3 showcases the role of geospatial

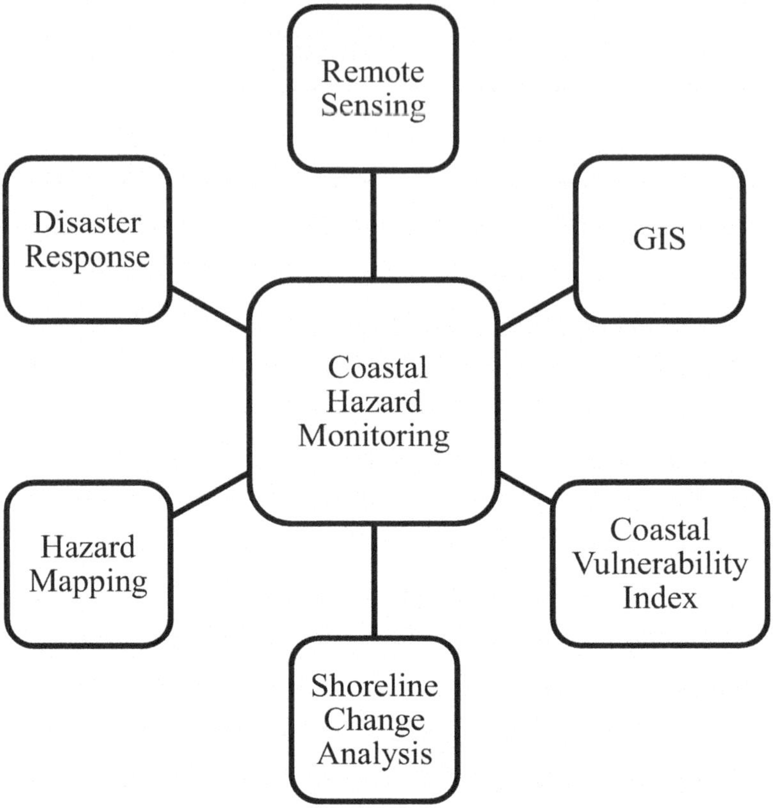

Fig. 1.3 Geospatial technologies in coastal hazard monitoring

technologies, such as GIS, remote sensing, and GPS in monitoring and assessing coastal hazards.

1.3.1 Remote Sensing and Satellite Imagery

Remote sensing is foundational to modern coastal hazard monitoring. It entails the acquisition of information about the Earth's surface from a distance, predominantly through satellites and, to a growing extent, airborne sensors. The value of remote sensing in coastal hazard management lies in its capacity to provide repetitive, synoptic, and multi-spectral observations of extensive coastal zones—capabilities that ground-based measurements cannot match. Contemporary satellites such as Landsat, Sentinel, IRS, and specialized ocean-observing missions collect vast arrays of imagery that can be analyzed for various coastal applications. For example, satellite-derived imagery enables the detection of shoreline changes, erosion hotspots, flood extents, and sediment movement. When cyclonic storms or tsunamis impact coastal zones, remote sensing provides critical post-event imagery for damage assessment, helping authorities identify inundated areas and prioritize emergency response.

Moreover, remote sensing supports the monitoring of persistent changes such as sea-level rise, wetland dynamics, and land-use changes driving increased vulnerability. Through time-series analysis, historical trends can be compared with current observations to reveal areas of accelerating erosion or urban expansion into hazardous floodplains. The strengths of remote sensing are further amplified by its ability to survey otherwise inaccessible or hazardous areas, contributing to consistent data acquisition regardless of local logistical constraints. The partnership between agencies (for example, India's ISRO and the Ministry of Environment, or the joint initiatives in the United States and Europe) ensures the ongoing development of satellite systems tailored for coastal management, with applications ranging from cyclone prediction to monitoring of marine pollution (Cokou et al. 2024).

1.3.2 GIS for Spatial Analysis

Geographic Information Systems (GIS) serve as the analytical engine behind the integration, visualization, and interpretation of geospatial data from multiple sources—including remote sensing, field surveys, and historical archives. In the context of coastal hazard monitoring, GIS is indispensable for synthesizing diverse datasets to model risks, track changes, and inform policy.

In practice, GIS is used to compile layers of information such as shoreline position, land use, population density, elevation, natural habitats, and infrastructure networks. This layered approach enables comprehensive hazard mapping and risk assessments. For example, coastal flood hazard mapping uses GIS to overlay data on elevation, hydrology, and socio-economic exposure, thereby identifying zones at risk from storm surge or sea-level rise. The same platforms are employed to chart erosion rates, simulate the movement of oil spills, and project urban encroachment onto sensitive or dangerous terrain.

The analytical power of GIS lies in its capacity for spatial modeling. Tools within GIS software allow for the simulation of disaster scenarios—such as the inundation extent of a hypothetical storm tide or the consequences of a further meter of sea-level rise. By animating and visualizing these scenarios, GIS enhances public understanding and supports the formulation of evidence-based response strategies (Davis et al. 2021).

1.3.3 UAVs and LiDAR Applications

The past ten years have seen the rise and development of UAVs as drones or Light Detection and Ranging (LiDAR) as a game-changing monitor of coastal hazards. UAVs that are fitted with high resolution cameras and sensors can create unmatched opportunities to survey and map coastal environments at high rates and in very small granularity. They are invaluably useful after calamities when ground transportation is blocked thus making it possible to determine the damages of a storm, the level of flooding or any rubbled infrastructure on a real-time basis. Out of crisis scenarios, a routine use of drones can be evaluating shoreline changes, already known dune or wetland health and even giving up to date imagery of an existing GIS database.

LiDAR, however, emits laser pulses that have centimeter-level accuracy by uses an aerially or terrestrially based platform to measure their elevation. The technology cannot be ignored when designing accurate digital elevation models (DEMs) of coastal terrain that are essential in realizing the flood risk in the area, locating erosion-prone sites and designing coastal defense constructions. LiDAR data can also unveil fine-resolution topographic variations and thus identify exposure to flooding, which is essential at the scenario modeling level as well as designing direct physical protection like seawalls or rebuilt dunes. LiDAR in association with UAV platforms also increases the opportunity to gather information in fast and repeated manners in more dynamic environments and records the changes over time caused by storms, rising seas, or anthropogenic changes.

Both UAV and LiDAR data are natively supported in the GIS environment, in which the resolution of their spatial data enriches all kinds of analysis, such as detecting change, or mapping areas of hazard at high resolution. With these technologies increasingly cost-effective and accessible in nature, their regular implementation is also being expanded, allowing even low-resource agencies or developing states to increase their monitoring and management capability of coastal hazards (Seenipandi et al. 2021).

1.4 Risk Assessment and Vulnerability Mapping

Useful geospatial technologies and advanced computation techniques can be of great essence in the risk assessments and vulnerability mapping exercises that are key in effective coastal hazard management. Such tools allow integrating various spatial data sets and applying advanced analytical methods, including machine learning, to improving the knowledge of coastal risks and thus making informed decisions. The chapter is divided around two topics; integration of spatial data and machine learning application in the coastal hazard study (Ganguli et al. 2018).

1.4.1 Spatial Data Integration

Risk evaluation of coastal hazards is in essence the apprehension of where the hazard can occur, what is accessible to the hazard i.e. assets or populations, and how susceptible the exposed aspects can be damaged. In order to attain such an understanding, it is necessary to integrate a wide variety of spatial data that can record these various dimensions, hazard, exposure, and vulnerability, into one cohesive, analyzable structure, mainly with the use of Geographic Information Systems (GIS). The integration of spatial data starts with the collection of geographic data of all sorts, e.g. satellite images, topographic maps, bathymetric data, weather, and meteorological data and socio-economic data. The different datasets have distinct, but complementary information which is required in the complete assessment of risk. Two examples are: The LiDAR or satellite altimetry elevation models indicate where flooding/inundation is likely to occur, and maps of lands uses and infrastructure can point out critical facilities in danger. Human exposure can be evaluated with the help of population distribution data that can be based on the census or remote sensed nighttime lights.

Nevertheless, these datasets cannot be integrated easily. Differences in spatial resolution, coordinate system, the temporal coverage, and the quality of the data can make their harmonization problematic. Strict data preprocessing procedure is required: data should be brought into the same coordinate and reference system, resampled to similar spatial scales and dates should be synchronized where required. It is also important to clean data and eliminate gaps and errors. Data quality and relevancy to local conditions also depends on the geomorphology of the coast, protective infrastructure in existence, and the trend of urban development.

These datasets after harmonization are overlapped in GIS to produce composite risk maps. As an example, probabilistically modeled or historically developed flood hazard maps get overlapped with building footprints and demographic data in order to define dangerous neighborhoods. In a similar manner, environmental hazards of erosion prone areas can be overlaid on the ecological habitat charts to determine the environmental vulnerability and human exposure. This integration mapping gives a clear geographic view of the hot spots of risk and targeted mitigation and emergency preparedness are enabled (Iyer et al. 2025).

1.4.2 Machine Learning in Coastal Hazard Analysis

A subfield of artificial intelligence is machine learning (ML) which deals with the learning of patterns in data through algorithms, and in the recent past, it has become quite a potent instrument in coastal hazard analysis. ML models, unlike traditional, deterministic models based on direct equations of physics, can find non-linear, high dimension relationships in large spatial and time dependent datasets and can be more precise in prediction and provide new information on hazard behavior, exposure and vulnerability. Predictive modeling is one of the well-known applications of ML in coastal hazard risk assessment. To give an example, the random forest, support vector machine, and neural network methods of supervised learning can then be trained using the existing data on hazards and impacts and environmental parameters to estimate the probability and magnitude of their occurrence in future. This comes in handy especially in predicting flood levels in storm surges, rate of erosion in vulnerable coastlines or even places prone to saltwater inundation.

ML is particularly useful when processing various types of data such as satellite images, weather data, sensor observational statistics, and socio-economic performance indications. The automation of the process of extracting a coastal feature using high-resolution remote sensing images (e.g. detecting changes in the shoreline, mapping the limits of a mangrove area, or depicting the destruction of infrastructure following a storm) has been automated using deep learning methods and, in particular, convolutional neural networks (CNNs). Such automation of feature extraction will result in speeding up spatial dataset update that is important in the risk assessment. Vulnerability classification is another important usage of ML. The vulnerability of coasts varies on different aspects such as the nature of construction material, the land use, the demographic character, and the ecology. The ML algorithms have the potential to incorporate these variables to group areas or assets into categories of vulnerability that are easier to oversimplify with conventional statistical methods. Clustering algorithms and unsupervised learning can also be used to reveal latent patterns and groupings in vulnerability data, highlighting previously unrecognized hotspots.

Challenges remain for the effective adoption of ML in this field. ML models require large, high-quality datasets for training; data scarcity, especially in developing regions, can limit performance. Interpretability of models is critical, as stakeholders must trust and understand predictions for policy uptake. Hybrid approaches that integrate physical process-based modeling with ML are increasingly advocated, combining the strengths of both for improved hazard forecasting.

In practice, the integration of ML into coastal hazard management is supported by advances in data availability and computational infrastructure. Cloud-based platforms and open-source ML libraries facilitate the processing of Big Data and model deployment in real time, enhancing early warning systems and adaptive management strategies (Iyer et al. 2024). Figure 1.4 demonstrates the application of machine learning in coastal hazard analysis, enabling the prediction of risk zones and pattern recognition from complex datasets.

1.5 Mitigation and Adaptation Strategies

Mitigation and adaptation strategies are essential components of coastal hazard management aimed at reducing risk, protecting lives and property, and enhancing the resilience of coastal communities. Such strategies imply the preventative types of measures in reducing the effect of hazards prior to an event, and the adaptation types of management to meet the current and future changes in the coastal environment.

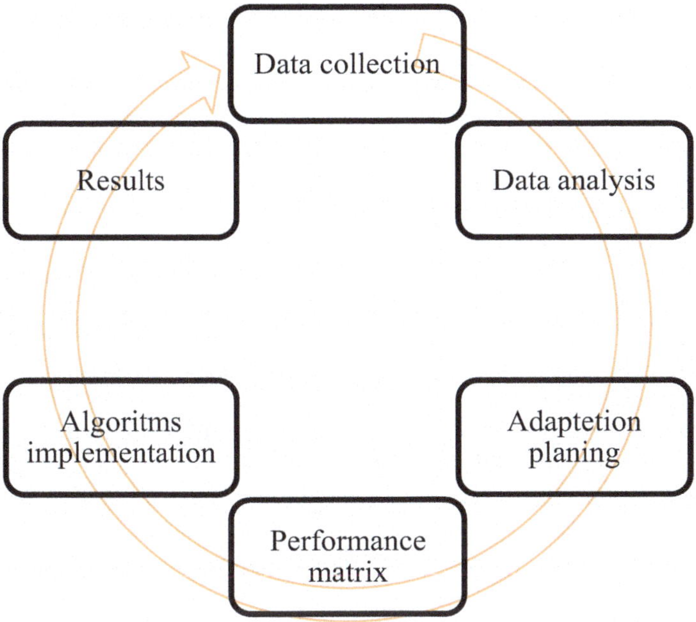

Fig. 1.4 Machine learning in coastal hazard analysis

Early warning systems and management policies in the coastal zones are two of the most important areas in this realm.

1.5.1 Early Warning Systems

Early warning systems (EWS) are considered to be one of the most efficient measures to mitigate the loss of human and economic resources due to the potential occurrences of coastal hazards best represented by storms, tsunami, or floodings. Collectively, these systems include monitoring technologies, data processing, risk estimation, and informed communication to create prospective warnings of approaching threats in order to permit people, communities, and authorities to implement shielding measures in a timely manner. A strong coastal hazard early warning system is based on the cumulative character of incorporating various technological and analytical segments. These comprise met and ocean sensors, space-based remote sensing, tide stations, seismic instrumentation, and ever more powerful weather and hydrodynamic numerical models that have the potential to forecast surge, wave heights, tsunami travel times and flood documentation with the growing accuracy. Geographic Information Systems (GIS) and computational platforms require real time data streams to render hazard forecasts and risk maps that give visual representations of the region's most prone to receive the impact.

The early warning systems have experienced profound enhancement due to changes in technology in terms of resolution and speed. An example would be that storm surge and inundation models of high-resolutions are currently integrating real-time data of weather stations, buoys, and satellites in order to produce regional forecast of floods hours or days ahead. Tsunami warning centers monitor seismic data and ocean-bottom pressure sensors to measure the source and possible areas of the seismic sea waves to alert in minutes after a wave meets the area. No less important than the systems of early warning is the communication infrastructure alerting the population and first responders. The current method of delivering warnings comes through various mediums that include text messaging, sirens, radio communications, television, social media, and applications on smartphones. All these different forms are able to provide wider coverage, as well as redundancy in the event of any of these channels malfunctioning in time of emergency.

In addition, the focus of early warning systems is shifting towards impact-based forecasting rather than the issuing of warnings. The method has more focus on making a forecast of not only the parameters of the physical hazard but also the probable impact people, infrastructure, and ecosystems. As an example, predictions on how deep the flood should reach in the neighborhoods or how long some essential services should be interrupted could be posted so as to enable more specific evacuation and other resource management schemes (Iyer et al. 2024).

1.5.2 Coastal Zone Management and Policy

Coastal zone management (CZM) and policies provide long term approaches to manage the balance between people using the coast, economically developing the coast, and reduction of hazard risks. Effective CZM combines scientific, risk, social, and regulatory tools to determine ways to facilitate sustainable development and hazard mitigation. Regulation of land use and development in risk zones is one of the leading objectives of CZM since it serves to diminish exposure and vulnerability. It is done by using a combination of zoning rules, building rules, set back lines and permitting systems which hinder or qualify development in flood-prone, erosive, or storm surge-prone areas. An example of this is setback lines where the idea is a minimum distance needed off the shore to consturct a new structure and ensure that the natural processes go on at the coast and also limit damages to the buildings.

Codes also usually have specifications on elevation of new constructions, flood resistant materials and constructions capable of standing up to coastal stresses. Such laws enhance structural strength and cuts down the cost of recovery following calamities. Coastal planners also embrace relocation or managed retreat in places that are high-risk and where defense and adaptation are not viable either technically or economically.

Already, CZM policies are focusing more on the use of so-called green solutions or nature-based designs as well as conventional gray infrastructure. The natural buffers dunes, mangrove forests, wetlands, and coral reefs can be replenished and maintained to perform several functions which are as follows: it provides protection to the shorelines by mitigating the wave energy, it can absorb the floodwater, it permits sediment deposition, and it also serves as the habitat to the biodiversity. These nature-based solutions offer economical and long-term protection with improved value of having healthy coastal ecosystems and community well-being. Adaptive management regards as one of the main concepts in modern CZM, as the coastal hazards and conditions change significantly with the climate change, rising of the sea level and anthropogenic factors. Policy frameworks are flexible in that, monitoring and iterative assessment are a part of policy framework design and they can be used to adapt to changing information by altering strategies. Such an adaptive strategy advocates the embrace of new technologies, evolving social requirements, and risks.

Another key component of a good coastal zone management is participatory governance. Involving local people, indigenous communities, business and other stakeholders in the decision-making process will enhance social acceptance and facilitate local knowledge application as well as more efficient regulatory compliance. Other alternatives to bottom-up approaches are volunteer surveillance programs, citizen science initiatives, and community adaption planning. The medium of policy usually involves coordination of numerous sectors and levels of governance to include local municipal and regional levels to national authorities in coastal hazard management. ICZM structures encourage cross-sector partnerships, and combine environmental protection with the reduction of disaster, economic growth and social fairness. Cumulative impacts could include land-use change, which impacts the availability of sediment upstream, or the additivity of coastal and river flood hazard (Akash et al. 2024).

1.6 Case Studies of Geospatial Applications

GIS systems have formed an inalienable part in the progressive improvement of coastal hazard control through the provision of high-quality, timely, and detailed spatial information that forms the basis of quality risk assessment, planning, and responsive action. The abundance of global and regional case studies available demonstrates how the incorporation of geospatial toolboxes together with remote sensing, GIS, UAVs, and spatial modeling techniques has changed the ability of humanity to know and to prevent risks in the coastal domain across various ecological and socio-economic setup (Miah et al. 2010).

1.6.1 Global and Regional Examples of Geospatial Applications in Coastal Hazard Management

Australia: Integrated Coastal Adaptation Planning Using CoastAdapt Resources.

Australia provides an excellent range of geospatial app development due to efforts organized by CoastAdapt, one of the most far-reaching resources hubs that facilitate climate change adaptation not only in Australia but worldwide. CoastAdapt has provided geospatial mapping tools, and climate scenario analysis functionality in small coastal towns of South Australia to conduct integrated coastal hazard assessment paying attention to vulnerabilities related to tourism. These evaluations integrate satellite imagery with GIS data layers that include the shoreline dynamics, land use, and socio-economic indicators to determine vulnerable infrastructure and ecosystems and coordinate the priorities setting adaptive responses.

In New South Wales, the CoastAdapt tools-based workshops assist the oyster farmers in the Far South Coast region to comprehend climate changes that happened in their area and how it had affected their region through the use of spatial information on flooding, rise in salinity, and temperature change. This geospatially enabled interaction encouraged stakeholder-led adaptation initiatives that increased resilience of the industries.

The airports in Cairns and Mackays, Queensland made use of CoastAdapt by applying an existing data set and risk screening tool to gain a better understanding of their vulnerability to sea-level rise and storm surge. Such geospatial evaluation informed structural improvement and operation-based planning, which were part of proactive climate risk management efforts.

Community focused spatial data integration and frameworks (based on GIS) were used by local councils, including Noosa and Kingborough Tasmania, to provide community-led climate change adaptation strategies that take into account the parameters of rising sea-levels, coastal erosions, and wild weather forecasts. These councils have also identified areas in which a managed retreat, nature-based defenses, and infrastructure hardening should be prioritized by reference to time-series satellite imagery and elevation data, and this is a good example of the successful policy integration of geospatial data. The level of partnership in the Victorian state who involved all 79 local councils demonstrates how coherence in adaptation action within geospatial data sharing and spatial planning at the regional level may support coherence in actions across levels.

United States: Flood Insurance and Coastal Hazards Mapping.

An example of a massive government-driven consolidation of geospatial information in coastal hazard mitigation can be seen through the establishment of the NFIP in the 1960s in the United States. The NFIP identifies flood prone areas using comprehensive GIS floodplain maps created by hydrological modeling, topographic data, and records of past storms and directs how the floodplain should be managed and how flood insurance should be provided. The program requires that the practice against the reduction of risk is incentivized by controlling the construction activities and the use of land in the mapped areas that experience hazard flooding.

Recent innovations include incorporating high-resolution LiDAR-derived digital elevation models (DEMs) into GIS platforms to refine flood risk predictions at local scales. This enhancement enables more precise delineation of flood extents from storm surge and heavy rainfall events, improving emergency planning and insurance risk assessments. Post-Hurricane Katrina and Superstorm Sandy, remote sensing-based damage assessments have been increasingly used to validate and update flood hazard maps and inform rebuilding strategies, demonstrating how geospatial technologies are embedded in disaster recovery cycles.

Additionally, the Tweed River Entrance Sand Bypass Project in the US integrates geospatial modeling to harmonize engineering with coastal sediment dynamics for beach stabilization, balancing environmental and human priorities.

United Kingdom: Coastal Erosion and Managed Realignment.

The United Kingdom has leveraged geospatial tools extensively to address coastal erosion and develop managed realignment policies. On the east of England, the government started the projects where the remote sensing, historical analysis of shoreline changes, and the vulnerability mapping based on GIS were used to determine the regions where the coastal defense system was insufficient and the residential areas were at the threat.

An excellent example is at Happisburgh, where the eroding sea fortifications have led to close geospatial assessments of hazards, such as elevation maps and predictions of erosion. These informed a rollback plan that would enable the coast to naturally retreat at the same time protecting any widespread infrastructure where possible. Multi-criteria analyses using GIS supported by weighting factors to consider ecological advantages and social and economic expenses resulted in planned realignment packages that were to restore intertidal habitats and resulted in an alleviation of future flooding risk.

The UK government agencies have enhanced coordination among the local councils and other stakeholders by helping to share raw data using the regional platforms that facilitate the process of adaptive governance provided through effective geospatial data-informed decision-making.

Asia-Pacific: Coastal Vulnerability and Community Engagement.

In the Asia-Pacific region, geospatial applications in coastal hazard management extend across both scientific assessment and participatory adaptation. In the Solomon Islands' Choiseul Bay, integrated climate risk and adaptation strategies were developed employing GIS and remote sensing to map flood-prone zones, erosion hotspots, and vulnerable settlements. This spatial information underpinned community consultations that incorporated indigenous knowledge alongside scientific data, ensuring culturally appropriate adaptation pathways including potential relocation of the provincial capital.

Similarly, the Torres Strait Regional Authority collaborated with researchers and state agencies in Australia to assess coastal risks through spatial data integration that combined habitat mapping, storm surge simulations, and sea-level rise projections. These assessments guided local adaptation measures while respecting traditional marine and land management practices.

In Tonga, geospatial capacity building efforts enhanced professional knowledge of coastal imaging and hazard mapping, enabling local coastal managers to apply near-real-time spatial data for improved decision-making. Community photo-monitoring programs in Western Australia offer another example of involving citizens in data collection, enriching geospatial databases while raising public awareness of coastal changes.

Small Island Developing States (SIDS): Monitoring and Adaptation Support.

Small island nations such as the Maldives, Tuvalu, and parts of the Caribbean have increasingly utilized geospatial technologies to confront the existential threats posed by sea-level rise and storm events. Satellite imagery and UAV surveys provide critical data on shoreline erosion, land loss, and damage from tropical cyclones, compensating for limited ground-based data collection capabilities.

GIS-based vulnerability assessments have identified critical infrastructure, freshwater lens locations, and population clusters at risk, informing adaptation policies including elevation of settlements, construction of seawalls, and ecosystem restoration. The integration of geospatial data with climate models supports scenario planning that underpins national resilience strategies and international funding proposals.

Furthermore, geospatial data supports early warning systems tailored to local contexts, enhancing community preparedness for cyclones and tsunamis.

Europe: Coastal Flooding Risk Management and Urban Adaptation.

European countries, particularly the Netherlands and Germany, are leaders in deploying state-of-the-art geospatial technologies for coastal flood defenses and urban coastal adaptation. The Netherlands combines LiDAR elevation data with hydrodynamic models within GIS frameworks to continuously monitor dike integrity, storm surge risks, and inundation extents. This integrated approach supports dynamic risk maps used in flood risk communication and spatial planning.

Urban adaptation initiatives in cities such as Hamburg employ geospatial analysis to map vulnerable neighborhoods, critical infrastructure, and green spaces that can serve as flood water retention areas. These spatial insights guide investments in resilient urban design, including permeable surfaces, floodable parks, and adaptive building codes, illustrating integrated human-environmental system management.

Collaborative and Cross-Border Initiatives.

Many coastal regions facing shared challenges have benefited from cross-border geospatial cooperation. For example, the Great Barrier Reef Marine Park Authority, Queensland Government, and environmental partners collectively apply GIS and remote sensing to monitor reef health and coastal vulnerabilities, aligning conservation with hazard mitigation.

Similarly, regional climate modeling projects in south-eastern Australia combine geospatial datasets to support adaptation planning across multiple jurisdictions, businesses, and communities (Ahmed et al. 2021; Ramakrishnan et al. 2023; Jana, Hegde 2016; Kantamaneni et al. 2018).

1.7 Future Trends and Challenges

Advances in geospatial technologies and computational capabilities are pivotal in shaping the future of coastal hazard management. The integration of artificial intelligence (AI) and big data analytics is revolutionizing how risks are assessed, hazards are monitored, and adaptive strategies are designed. Meanwhile, emerging policy and research priorities focus on fostering resilience amid accelerating climate change impacts, population growth, and environmental degradation. Below is a detailed examination of these future trends and challenges in coastal hazard management, highlighting advances in AI and big data analytics and outlining critical policy and research recommendations (Rufat et al. 2015).

1.7.1 Advances in AI and Big Data Analytics

Artificial Intelligence (AI) and big data analytics are reshaping the landscape of coastal hazard management by enabling more precise, rapid, and comprehensive analysis of complex and voluminous spatial and temporal datasets. These technologies leverage advances in machine learning, deep learning, cloud computing, and sensor technologies to enhance early warning systems, risk assessment, vulnerability mapping, and adaptive planning.

Big Data Integration: Coastal hazard monitoring now involves massive streams of heterogeneous data—from high-resolution satellite and LiDAR imagery to real-time sensor networks measuring oceanographic, meteorological, and geophysical parameters. AI-powered big data platforms facilitate the integration, storage, and processing of these diverse datasets, overcoming traditional limitations related to volume, velocity, and variety. For example, cloud computing services enable scalable, on-demand computational power, allowing coastal managers to analyze data at unprecedented speeds and scales, from local coastline changes to regional hazard hotspots (Ariffin et al. 2023).

Improved Hazard Prediction and Scenario Modeling: Machine learning algorithms excel at detecting patterns in large datasets, enabling the prediction of coastal hazard events with enhanced accuracy. Deep learning models, such as convolutional neural networks (CNNs), automate the classification of satellite imagery to identify shoreline erosion, flood extents, or damaged infrastructure post-disaster. These models improve over time through training on historical datasets, thus refining forecasts for storm surges, tsunami inundation, and coastal flooding.

Moreover, AI-based approaches are increasingly used to develop dynamic scenario models that can simulate multiple future conditions under different climate change trajectories, land-use patterns, and socio-economic developments. This capacity supports proactive planning by highlighting areas of emerging risk and testing the effectiveness of mitigation measures, thus supporting adaptive management.

Enhancing Early Warning Systems and Real-Time Monitoring: AI develops early warning systems by combining multi-source of information including seismic sensors, tide gauges, weather stations and social media feed. The Natural language processing algorithms descreen the social media and the emergency reports to give situational awareness in case of coastal hazards. As an example, warning systems that are AI-powered may provide higher priorities and spread warnings tailored to the projected severity of the impact and the susceptibility of a population, streamlining the emergency response plans and evacuation processes.

Vulnerability and Risk Assessment: AI facilitates the more sophisticated vulnerability mapping based on the intricate correlation between environmental, infrastructural, and social factors. The classification in machine learning may divide communities along socio-economic lines, infrastructure vulnerability, ecological sensitivity, thus representing the opportunity to target the risk communication, as well as resource distribution. AI is also used in quantification of uncertainties where probabilistic estimates of risk are offered, which make the stakeholders more confident in the decision-making process.

Challenges in AI Adoption: As much as these have taken place, there are some difficulties that have been taken place. Such limitations may restrict the definition of AI models because of data shortage and inequality in data quality, particularly in developing coastal areas. Algorithmic bias and lack of transparency in "black-box" models require explainability techniques to ensure trustworthiness. Ethical considerations concerning data privacy and surveillance also demand careful governance.

Finally, human capacity building remains vital. Training coastal managers in AI and big data analytics is essential to maximize benefits, complemented by ongoing research integrating AI with traditional physical modeling for a hybrid, robust risk assessment framework.

1.7.2 Policy and Research Recommendations

To harness the full potential of emerging technologies and confront persistent and future coastal threats, coordinated policy and research efforts are critical. The following recommendations provide a strategic pathway toward resilient, adaptive, and equitable coastal hazard management (Mourin et al. 2018).

1. **Promote Integrated and Multi-Scale Governance**: Coastal hazards traverse administrative boundaries, requiring governance models that integrate local, regional, and national efforts. Policies should incentivize cross-jurisdictional collaboration and data sharing, supported by interoperable geospatial platforms

and shared early warning systems. Adaptive governance frameworks that incorporate stakeholder participation, including indigenous and local communities, will ensure context-sensitive and inclusive solutions.
2. **Invest in Data Infrastructure and Open Access**: Sustainable coastal hazard management depends on accessible, high-quality spatial and socio-economic data. Governments and international agencies should prioritize investments in data acquisition (e.g., satellite missions, sensor networks), standardization, and maintenance. Open data policies promote transparency, innovation, and community engagement, facilitating the co-production of knowledge and locally informed adaptation strategies.
3. **Advance AI and Big Data Research Toward Explainability and Ethics**: Research must focus on developing AI models that are interpretable, robust to biases, and respectful of privacy. Methodologies integrating physical, social, and ecological aspects of coastal hazards should be emphasized. Collaborative research involving computer scientists, coastal engineers, social scientists, and policymakers will foster the development of holistic AI tools tailored to real-world decision-making contexts.
4. **Bridge Gaps in Capacity and Equity**: Building technical expertise and institutional capacity in vulnerable and resource-limited regions must be a global priority. Tailored training programs, technology transfer initiatives, and inclusive knowledge networks will empower local communities to utilize geospatial and AI tools effectively. Addressing digital divides ensures that advances in coastal hazard management benefit all, mitigating socio-economic disparities.
5. **Integrate Climate Change Projections and Scenario Planning**: Policy frameworks should mandate the incorporation of the latest climate models and sea-level rise scenarios into coastal hazard assessments and planning. Dynamic risk mapping and scenario analysis provide foresight into emerging vulnerabilities and opportunity spaces for adaptive investment. This anticipatory approach supports sustainable development and ecosystem-based adaptation efforts.
6. **Foster Nature-Based Solutions and Hybrid Approaches**: Research on scientific and policy sides, should uncover collaborations between the engineered defense, and the natural coastal features, like wetlands, dunes and reefs. These hybrid measures will increase community, resilience, and biodiversity. The technologies of geospatial information are essential monitoring and evaluation mechanisms to maximize design, execution, and long-term effectiveness.

1.8 Conclusion

The coastal areas in the globe are one of the most worthwhile and therefore at-risk environments. The recent surge of coastal hazards in coastal areas due to climate change, population, and environmental degradation has heightened the needs of proper approaches to managing the hazard. In this chapter of the book, we have seen how geospatial technologies transform the process of coastal hazard management in terms of identification and monitoring of hazards to the assessment of risks, mitigation and adaptation to risks. This final part summarizes the main findings, provides easy-to-determine guidelines on the up-and-coming research to avail more resilient coastal nation residents flowing with changing problems.

1.8.1 Summary of Key Findings

The exposure and vulnerability of coastal areas which are rising require integrated solutions. The coastal areas are home to an outsized percentage of the globe and the economic activity compared to a negligible portion of the planet. Storm surges, flooding, erosion, tsunamis, and sea-level rise impose their threat to human lives, structures, natural ecologies as well as livelihood along the coastal areas. These hazards are made worse by the consequences of climate change with increasing sea levels and more severe storms and human demands in the form of urban sprawl and ecosystem disturbance. As a complex and dynamic phenomenon, coastal risks need comprehensive solutions and approaches occurring on the frontiers of environment, social and economic aspects. The geospatial technologies are the core of contemporary coastal hazard management. The advantage of remote sensing and satellite imagery is that it offer wide, frequent, and fine-scale observations that are necessary in revealing shoreline changes, flood reach, disaster effects, and the long-term coastal processes.

Geographic Information Systems (GIS) can be used as a highly effective framework to combine various types of any spatial data- physical topography and socio-economic indicators, etc. and apply them in terms of risk mapping as well as scenario analysis. The new technologies like Unmanned Aerial Vehicles (UAVs), Light Detection and Ranging (LiDAR) provide high precision, localized data that is important to detailed maps of elevation as well as quick post disaster assessment. In combination with one another, these technologies enable the real-time monitoring, early warning and informed decision-making.

Predictive and vulnerability assessment are taken to the next level with the help of advanced analytics, in particular, integrating machine learning and big data. ML algorithms have been used to effectively eliminate the need to treat nonlinear associations in a spatial and temporal dataset, and thus will result in better predictive models of hazards, automated interpretation of features and more subtle classifications of vulnerability. Big data platforms and cloud computing infrastructure allow the processing of massive, heterogeneous data streams in near real-time, drastically improving early warning systems and impact forecasting. These analytical advances represent a paradigm shift from reactive risk management towards proactive, anticipatory planning.

Effective mitigation and adaptation strategies root in early warning systems, regulatory policies, and nature-based solutions. Early warning systems that leverage geospatial data and AI-driven analytics save lives by providing timely and localized hazard alerts. Coastal zone management policies, including land-use zoning, building regulations, and managed retreat, help reduce exposure and promote resilient development patterns. Importantly, ecosystem-based adaptation through the restoration and conservation of natural buffers offers sustainable, multi-benefit protections to shorelines. Adaptive and participatory governance ensures that mitigation and adaptation measures are socially equitable, contextually appropriate, and flexible in the face of changing conditions.

Global and regional case studies demonstrate diverse successful applications of geospatial technologies. From Australia's CoastAdapt initiatives and the US National Flood Insurance Program's advanced floodplain mapping to the UK's managed realignment programs and small island developing states' vulnerability assessments—these examples underscore the vital role of spatial tools in guiding effective, evidence-based coastal resilience measures. Collaborations across sectors, governance scales, and borders are essential for maximizing the benefits of geospatial data and fostering inclusive risk management.

1.8.2 Directions for Future Research

While significant strides have been made, several research avenues warrant attention to strengthen coastal hazard management in the coming decades (Evelpidou et al. 2018):

(i) Enhanced Integration of Physical, Social, and Ecological Data: Current geospatial analyses often emphasize physical hazard modeling or socio-economic vulnerability independently. Future research should focus on integrated socio-ecological modeling frameworks that couple physical coastal processes with human systems and ecosystem dynamics. This integration will improve holistic assessments of risk and resilience, capturing feedbacks and trade-offs between natural and built environments.

(ii) Development and Validation of Hybrid Modeling Approaches: Combining physics-based models with AI-driven data analytics holds promise for improving hazard prediction accuracy and interpretability. Research should explore hybrid methodologies that leverage the strengths of mechanistic simulations and data-informed pattern recognition, particularly for complex events like compound flooding or multi-hazard interactions. Validation with empirical data and transparent reporting of uncertainties will enhance credibility and usability.

(iii) Expansion and Standardization of Data Collection and Sharing: Addressing data gaps in underserved regions, especially in developing countries and small island states, is critical. Future efforts must prioritize the deployment of affordable sensors, community-based monitoring, and open-access data platforms to democratize coastal hazard information. Standard protocols for data integration will facilitate interoperability among diverse datasets and across jurisdictions.

(iv) Advances in Real-Time Monitoring and Early Warning Systems: Research should continue to enhance sensor technologies, data assimilation methods, and AI-driven alert systems that integrate multi-hazard data streams. Innovations in social media analytics, crowd-sourced reporting, and networked communication systems can improve situational awareness and public engagement during hazard events, boosting responsiveness and minimizing losses.

(v) Focus on Socio-Technical Systems and Community Resilience: Understanding how technological solutions interact with social structures and community behavior remains under-explored. Interdisciplinary research combining geospatial sciences with social sciences can provide insights on risk perception, communication effectiveness, and adaptive capacities. Participatory GIS and citizen science initiatives should be evaluated for their role in empowering vulnerable populations.

(vi) Addressing Climate Change Uncertainty in Coastal Planning: Sea-level rise and climate projections contain inherent uncertainties that complicate decision-making. Research should develop robust, flexible planning frameworks and scenario tools that accommodate uncertainty, enabling adaptive management and phased interventions. Integration of socio-economic scenarios will further refine future risk landscapes.

(vii) Ecosystem-Based Adaptation and Hybrid Infrastructure Research: Empirical and modeling studies assessing the long-term efficacy, co-benefits, and trade-offs of green-gray infrastructure combinations are needed. Spatial monitoring of ecological responses to restoration and protective interventions will provide critical feedback for optimization and scaling.

References

Ahmed N, Howlader N, Hoque MAA, Pradhan B (2021) Coastal erosion vulnerability assessment along the eastern coast of Bangladesh using geospatial techniques. Ocean Coast Manag 199:105408. https://doi.org/10.1016/j.ocecoaman.2020.105408

Akash SH, Sarkar SK, Bindajam AA, Kumari R, Talukdar S, Mallick J (2024) Assessment of coastal vulnerability using integrated fuzzy analytical hierarchy process and geospatial technology for effective coastal management. Environ Sci Pollut Res 31(41):53749–53766. https://doi.org/10.1007/s11356-024-31896-5

Ariffin EH, Mathew MJ, Roslee A, Ismailluddin A, Yun LS, Putra AB, Yusof KMKK, Menhat M, Ismail I, Shamsul HA (2023) A multi-hazards coastal vulnerability index of the east coast of peninsular Malaysia. Int J Disaster Risk Reduct 84:103484. https://doi.org/10.1016/j.ijdrr.2022.103484

Arun Kumar A, Kunte PD (2012) Coastal vulnerability assessment for Chennai, east coast of India using geospatial techniques. Nat Hazards 64:853–872. https://doi.org/10.1007/s11069-012-0274-4

Benassai G, Di Paola G, Aucelli P, Passarella M, Mucerino L (2014) An inter-comparison of coastal vulnerability assessment methods. In: Engineering geology for society and territory–volume 4: marine and coastal processes. Springer, pp. 45–49
Coelho C, Silva R, Veloso-Gomes F, Taveira Pinto F (2006) A vulnerability analysis approach for the Portuguese west coast. In: Risk analysis V: simulation and hazard mitigation, vol 1. WIT Press, pp 251–262
Cokou PK, Tamini LD, Pepin S, Khasa DP, Abbas Y, Lamhamedi MS (2024) Evaluating multi-criteria decision-making methods for sustainable management of forest ecosystems: a systematic review. Forests 15(10):1728. https://doi.org/10.3390/f15101728
Davis CR, Berke P, Holloman MDE, Griffard MR, Haynes S, Johnson E, Warraich Z, Crisostomo-Morales L, Benissan DGG, Gillespy C (2021) Support strategies for socially marginalized neighborhoods likely impacted by natural hazards. Coastal Resilience Center, The University of North Carolina, Chapel Hill
Deb D, Uddin MM, Mahbub-E-Kibria ASM, Das MK, Hasan M (2024) Coastal vulnerability assessment to multi-hazards in the exposed coast of southeastern coastal region of Bangladesh. Reg Stud Mar Sci 73:103484. https://doi.org/10.1016/j.rsma.2024.103484
Evelpidou N, Kampolis I, Karkani A (2018) Geomorphic features associated with erosion. In: Natural hazards. CRC Press, pp 205–232
Fitton JM, Hansom JD, Rennie AF (2018) A method for modelling coastal erosion risk: the example of Scotland. Nat Hazards 91:931–961. https://doi.org/10.1007/s11069-017-3161-0
Ganguli S, Islam S, Garai J (2018) Physico-chemical assessment of water bodies and socio-economic analysis from the coastal belt of Chittagong. Indones J Environ Manag Sustain 2(4):107–117
Islam MA, Murshed S, Kabir SMM, Farazi AH, Gazi MY, Jahan I, Akhter SH (2017) Utilization of open source spatial data for landslide susceptibility mapping at Chittagong District of Bangladesh—an appraisal for disaster risk reduction and mitigation approach. Int J Geosci 8(4):577–598. https://doi.org/10.4236/ijg.2017.84031
Iyer R, Maralapalle V, Mahesh P, Patil D (2024) Generative AI and LLM: case study in e-commerce. In: Generative AI and LLMs: natural language processing and generative adversarial networks. De Gruyter, pp 253–272. https://doi.org/10.1515/9783111425078-013
Iyer R, Saranya A, Maralapalle V, Raskar D (2025) Reconfigurable intelligent surface-assisted backscatter communication for enabling 6G IoT networks. In: NMM S, Dhanaraj R (eds) Applications and challenges of reconfigurable intelligent surfaces in 6G. IGI Global Scientific Publishing, pp 123–146. https://doi.org/10.4018/979-8-3693-8099-4.ch005
Jana AB, Hegde AV (2016) GIS based approach for vulnerability assessment of the Karnataka coast, India. Adv Civ Eng:5642523. https://doi.org/10.1155/2016/5642523
Kantamaneni K, Phillips M, Thomas T, Jenkins R (2018) Assessing coastal vulnerability: development of a combined physical and economic index. Ocean Coast Manag 158:164–175. https://doi.org/10.1016/j.ocecoaman.2018.03.031
Khan SR (2022) Resource potentials of the marine areas of Bangladesh. In: Bangladesh geosciences and resources potential. CRC Press, pp 317–353
Miah G, Bari N, Rahman A (2010) Resource degradation and livelihood in the coastal region of Bangladesh. Front Earth Sci China 4:427–437. https://doi.org/10.1007/s11707-010-0119-6
Mourin MM, Ferdaus AA, Hossain MJ (2018) Landslide susceptibility mapping in Chittagong district of Bangladesh using support vector machine integrated with GIS. In: International Conference on Innovation in Engineering and Technology (ICIET), pp. 1–5
Rajasree BR, Deo MC (2020) Assessment of coastal vulnerability considering the future climate: a case study along the central west coast of India. J Waterw Port Coast Ocean Eng 146(2):05019005. https://doi.org/10.1061/(ASCE)WW.1943-5460.0000536
Ramakrishnan R, Shaw P, Rajput P (2023) Coastal vulnerability map of Jagatsinghpur district, Odisha, India: a satellite based approach to develop two-dimensional vulnerability maps. Reg Stud Mar Sci 57:102747. https://doi.org/10.1016/j.rsma.2022.102747

Rufat S, Tate E, Burton CG, Maroof AS (2015) Social vulnerability to floods: review of case studies and implications for measurement. Int J Disaster Risk Reduct 14:470–486. https://doi.org/10.1016/j.ijdrr.2015.09.013

Sarwar MGM, Woodroffe CD (2013) Rates of shoreline change along the coast of Bangladesh. J Coast Conserv 17:515–526. https://doi.org/10.1007/s11852-013-0251-6

Seenipandi K, Ramachandran KK, Chandrasekar N (2021) Modeling of coastal vulnerability to sea-level rise and shoreline erosion using modified CVI model. In: Remote sensing of ocean and coastal environments. Elsevier, pp 315–340

Thomas JV, Arunachalam A, Jaiswal RK, Diwakar PG, Kiran B (2014) Dynamic land use and coastline changes in active estuarine regions–a study of Sundarban delta. In: The international archives of the photogrammetry, remote sensing and spatial information sciences, vol 40, pp 133–139

Waqas M, Humphries UW, Wangwongchai A, Dechpichai P, Ahmad S (2023) Potential of artificial intelligence-based techniques for rainfall forecasting in Thailand: a comprehensive review. Water 15(16):2979. https://doi.org/10.3390/w15162979

Zhu ZT, Cai F, Chen SL, Gu DQ, Feng AP, Cao C, Qi HS, Lei G (2018) Coastal vulnerability to erosion using a multi-criteria index: a case study of the Xiamen coast. Sustainability 11(1):93. https://doi.org/10.3390/su11010093

Chapter 2
Coastal Changes Determination Using GIS and Remote Sensing

V. R. Mageshen, S. Janarth, Udhayakumar Sekaran, and Dhayalan Vaithiyanathan

2.1 Introduction

Coastal zones, which support vast populations and a variety of ecosystems, are some of the Earth's most sensitive and dynamic interfaces. Understanding the effects of anthropogenic factors like pollution, land use change, and coastal development, as well as natural processes like tides, wave action, and sea level rise, requires monitoring these areas. Frequent evaluation facilitates coastal planning, enables early environmental degradation detection, and aids in the development of robust adaptation plans. Monitoring the coast protects natural areas and biodiversity in addition to human settlements.

Natural forces like storms, waves, and tides constantly change coastlines. However, these changes have been hastened by human activities such as urban growth, sand mining, harbour construction, and aquaculture, which has led to increased loss of habitat, erosion, and decreased coastal stability Adam et al. (2010). By causing extreme weather events and rising sea levels, climate change exacerbates

V. R. Mageshen
Amrita School of Agricultural Sciences, Coimbatore, India

S. Janarth (✉)
Vanavarayar Institute of Agriculture, Pollachi, India
e-mail: janadpm@gmail.com

U. Sekaran
Irrigation and Soil Fertility Management, Department of Crop and Soil Science, Oregon State University, Oregon, USA
e-mail: udayakumar.sekaran@oregonstate.edu

D. Vaithiyanathan
Department of Agricultural Engineering, Hindustan College of Engineering and Technology, Coimbatore, India
e-mail: dhayalan.agri@hicet.ac.in

A. Nath et al. (eds.), *Geospatial Application of Coastal Hazards*, Springer Water,
https://doi.org/10.1007/978-3-032-20076-1_2

these consequences and increases the hazards to coastal ecosystems and communities. Anthropogenic changes are increasingly a major factor in determining the patterns of coastal erosion and accretion, rather than just natural processes Baig et al. (2020). For the purpose of planning for strategy and hazard mitigation, it is crucial to quantify these changes.

Remote sensing (RS) and geographic information systems (GIS) have become essential instruments for coastal monitoring. Repetitive, synoptic observations across wide spatial extents are provided by RS, which records dynamic changes in vegetation, built-up areas, and land-water interfaces. In order to evaluate coastal changes, pinpoint susceptible areas, and facilitate evidence-based decision making, GIS combines these geographical datasets with models. When combined, they make it possible to manage coastal zones sustainably and economically Bhatt et al. (2007). GIS-based decision support tools improve readiness for disasters and encourage data-driven planning for coastal belt growth, infrastructure, and conservation.

2.2 Remote Sensing for Coastal Change Detection

Satellite-Based Coastal Monitoring.

The observation and study of coastal changes has relied heavily on satellite-based remote sensing because of its cost-effectiveness, high temporal resolution, and synoptic coverage. Consistent observation of shoreline dynamics, changes in land use, and changes in coastal ecosystems has been made possible by a number of satellite missions, including Landsat, Sentinel, MODIS, and IRS. Time-series imaging from these databases aids in the mapping of vegetation changes, accretion, erosion, and urban growth. Depending on the goals of the study, the spatial resolution can be selected from moderate (e.g., MODIS: 250–1000 m) to high-resolution (e.g., Sentinel-2: 10 m). Since1972, Landsat imagery has been available for a long time, making it especially useful for detecting coastal change Camfield & Morang (1996). This allows researchers to track changes in shoreline position and land cover classes over time. Satellite Datasets Used for Coastal Studies is given in (Table 2.1).

2.2.1 Multi-Spectral and Hyperspectral Imaging for Coastal Studies

Differentiating between distinct forms of land cover, such as vegetation, bare terrain, and water bodies, is made easier by multispectral remote sensing, which gathers data in distinct spectral bands. Contrarily, hyperspectral imaging records data in hundreds of small, continuous bands, making it possible to distinguish between coastal elements like sand, silt, and species of vegetation with greater accuracy Guneroglu (2015). These methods are essential for mapping mangrove forests, tracking the

Table 2.1 Summary of satellite datasets used for coastal monitoring

Sl. no.	Satellite/Sensor	Spatial resolution	Temporal resolution	Application
1.	Landsat-8 OLI	30 m	16 days	Shoreline mapping, LULC change
2.	Sentinel-2 MSI	10 m	5 days	Mangrove, wetland classification
3.	Sentinel-1 SAR	10 m	6–12 days	Flood mapping, coastline extraction
4.	ALOS PALSAR	10–25 m	~46 days	Terrain analysis, vegetation structure
5.	LiDAR (airborne)	0.5–2 m	On-demand	DEM generation, coastal elevation

quality of the water along the coast, and identifying algal blooms. Interpretation of complicated coastal settings is improved through integration with spectral indices as SAVI (Soil Adjusted Vegetation Index), NDVI (Normalised Difference Vegetation Index), and NDWI (Normalised Difference Water Index). Hyperspectral sensors' great spectral fidelity makes it possible to assess biodiversity in estuarine and intertidal zones and identify subtle ecological changes JRC (2020).

2.2.2 Synthetic Aperture Radar (SAR) Applications in Coastal Analysis

During monsoon or cyclone-prone seasons, Synthetic Aperture Radar (SAR), which functions in the microwave portion of the electromagnetic spectrum and can operate in all weather conditions and through cloud cover, is perfect for monitoring the coast. Sentinel-1, RADARSAT, and ALOS PALSAR are examples of SAR sensors that are used to map flooded areas, detect oil spills, track coastline positions, and determine fluctuations in surface roughness. SAR's utility for monitoring intertidal regions and mangroves is increased by its capacity to quantify backscatter from wetland plants and waterlogged areas Himmelstoss et al. (2018). In deltaic areas and low-lying coastal plains, advanced SAR techniques like interferometric SAR (InSAR) enable vertical displacement monitoring, which is helpful in detecting land subsidence.

2.2.3 LiDAR for Shoreline Mapping and Elevation Modeling

An active remote sensing method called Light Detection and Ranging (LiDAR) creates high-resolution elevation data by using laser pulses. In coastal research, airborne LiDAR sensors are particularly useful for producing Digital Elevation Models (DEMs), that are crucial for defining shorelines, identifying topographic changes, and identifying areas that are vulnerable to flooding Kuenzer & Voigt (2013). LiDAR is appropriate for tracking sand dynamics, barrier islands, and estuary systems because it can precisely map minute topography differences even in heavily vegetated coastal zones. Elevation data is essential for modelling the effects of sea level rise and creating evacuation plans for vulnerable coastal communities, making its use in climate change adaptations more and more significant Luo et al. (2013).

2.2.4 *Change Detection Techniques in Remote Sensing*

Remote sensing change detection techniques are used to detect and measure changes in land use patterns and coastal landforms over time. Principal component analysis (PCA), post-classification comparison, image differencing, and based on machine learning classification are some of the methods. These methods can show urban encroachment, wetland deterioration, coastline retreat, and land reclamation. Long-term trend analysis is supported by multi-temporal information from platforms such as Sentinel and Landsat. By considering the shape and spatial context of land features, object-based classification techniques have significantly increased the accuracy of coastal detection of changes while lowering the classification errors typical of pixel-based techniques Zheng et al (2023). Comparison of Change Detection Techniques in Coastal Remote Sensing is given in (Table 2.2).

Table 2.2 Comparison of change detection techniques in coastal remote sensing

Sl. no.	Technique	Advantages	Limitations
1.	Image differencing	Simple to implement, fast	Sensitive to atmospheric variations
2.	Post-classification comparison	Provides thematic clarity	Classification errors propagate
3.	Principal component analysis	Reduces dimensionality, highlights changes	May not directly identify land cover categories
4.	Object-based classification	Contextual accuracy, handles high-res imagery well	Computationally intensive
5.	Machine learning (RF, SVM)	High accuracy, adaptable to complex data	Requires labeled training data
6.	Deep learning (CNNs)	Excellent for feature recognition	Requires large datasets and computation power

Table 2.3 Shoreline erosion and accretion statistics using DSAS (1990–2020)

Sl. No	Coastal stretch	Net shoreline movement (NSM, m)	End point rate (EPR, m/year)	Change type
1.	Nagapattinam	−60	−2.0	Erosion
2.	Cuddalore	−45	−1.5	Erosion
3.	Thoothukudi	+30	+1.0	Accretion
4.	Kanyakumari	+15	+0.5	Accretion
5.	Chennai (North coast)	−20	−0.7	Erosion
6.	Pichavaram (Mangrove belt)	−10	−0.3	Minor erosion

2.3 GIS Applications in Coastal Management

2.3.1 Shoreline Mapping Using GIS

By combining satellite images, aerial photos, and field data to define coastal boundaries, GIS plays a crucial part in coastline mapping. GIS makes it possible to calculate net shoreline movement, identify erosion hotspots, and track coastline movements through vectorisation and overlay analysis Maiti & Bhattacharya (2009). Predictive modelling and intervention area prioritisation are supported by the historical examination of shorelines taken from multi-date images Selemat et al (2019). Coastal regulatory zone (CRZ) enforcement and the planning of protective infrastructure, such as groynes and seawalls, depend heavily on GIS-based shoreline mapping. Shoreline Erosion and Accretion Statistics Using DSAS (1990–2020) is given (Table 2.3).

2.3.2 Spatial Analysis for Coastal Erosion and Accretion

Both accretion and erosion are natural processes that are impacted by human activity, sediment transport, and wave energy. Zones undergoing erosion or deposition are mapped using GIS technologies such buffer analysis, spatial interpolation, and hotspot detection Mullick et al. (2020). Planning coastline stabilisation measures, modelling the sediment budget, and identifying sensitive areas are all made easier by these insights. For example, the creation of erosion risk indices which inform coastal zone management strategies and policy choices is made possible by the integration of GIS with wave and tide data.

2.3.3 *Digital Elevation Models (DEM) for Coastal Vulnerability Assessment*

For topographic investigation in coastal areas, DEMs are crucial. GIS may model elevation profiles and simulate inundation scenarios under different sea-level rise projections using data from LiDAR, SRTM, or photogrammetry Zhou et al (2022). Analyses based on DEMs aid in determining the degree of population exposure, flood risk, and possible financial damages. Targeted preparedness for emergencies and adaptive land-use planning are made possible by vulnerability indices that integrate elevation, density of population, and land use data in GIS contexts Singh (1989).

2.3.4 *Integration of GIS with Hydrodynamic Models for Flood Forecasting*

Using data on tides, waves, and storm surges, hydrodynamic models model the flow of water. These models offer spatially explicit predictions for floods when used with GIS. The accuracy of the model is improved with GIS inputs including drainage networks, slope, and land cover Rajawat et al. (2005). Mapping output flood levels and extents helps with infrastructure design and evacuation planning by identifying at-risk areas. For flood impact evaluations in deltas and estuaries, for instance, GIS coupling using MIKE21, HEC-RAS, and Delft3D is used.

2.4 Methodologies for Coastal Change Detection

2.4.1 *Digital Shoreline Analysis System (DSAS) for Shoreline Change Assessment*

A popular GIS extension for determining shoreline change rates is called DSAS, which was created by the USGS. It measures the separation between many historical coastline locations along transects that are created perpendicular to a baseline. Important indicators that measure erosion or accretion trends include End Point Rate (EPR), Linear Regression Rate (LRR), and Net Shoreline Movement (NSM). Decision-makers can prioritise sites for action, evaluate the efficacy of coastal constructions, and track coastline retreat via DSAS. Work flow of GIS and remote sensing in coastal monitoring is shown in (Fig. 2.1).

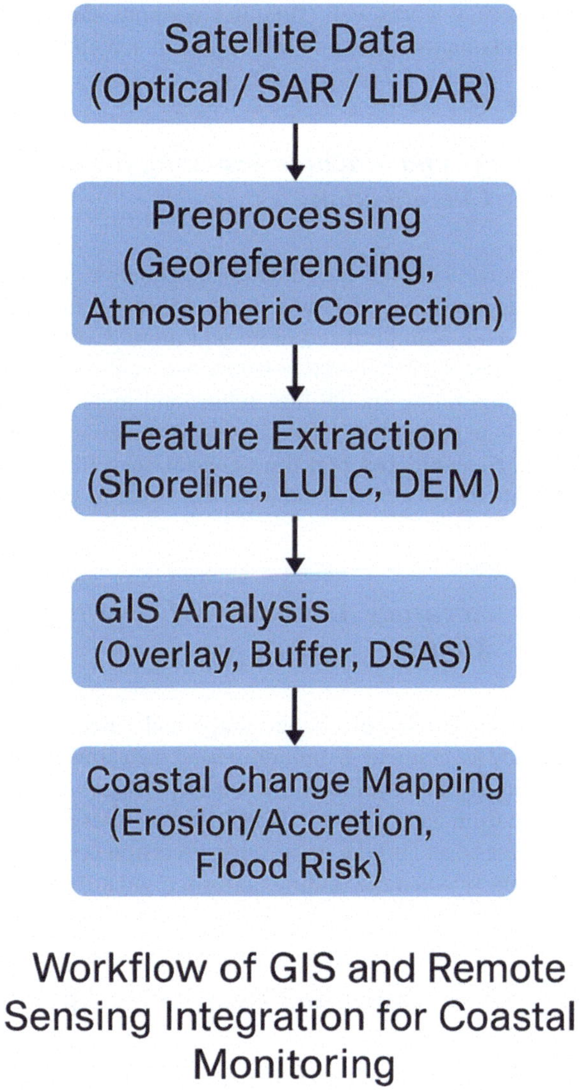

Fig. 2.1 Work flow of GIS and remote sensing integration for coastal monitoring

2.4.2 Object-Based Image Analysis (OBIA) for Coastal Feature Extraction

OBIA accurately classifies coastal features including beaches, mangroves, and urban patches by dividing satellite imagery into significant items based on spectral and spatial properties. OBIA minimises noise and misclassification in contrast to pixel-based methods, particularly in high-resolution images. When mapping diverse coastal habitats, where various land cover classes may be represented by comparable spectral signatures, it works well. Tidal flat mapping, salt marsh monitoring, and small-scale land reclamation detection have all been accomplished with OBIA.

2.4.3 AI and Machine Learning for Automated Coastal Classification

In automated coastal classification and change detection, machine learning methods like Random Forest, Support Vector Machines, and Convolutional Neural Networks (CNNs) are being utilised more and more. Even in intricate coastal environments, these models can accurately categorise land cover thanks to labelled training data. By incorporating spatial hierarchies and textural information, deep learning techniques—in particular, CNNs—can enhance coastline extraction and flood zone delineation. Scalable and real-time coastal surveillance is made possible through the integration of AI with cloud platforms such as Google Earth Engine.

2.4.4 Accuracy Assessment of Coastal Change Detection Models

The dependability of change maps and categorisation findings is guaranteed by accuracy assessment. Commonly used measures include confusion matrices, overall accuracy, Kappa coefficient, and user/producer accuracy. High-resolution images or ground truth data are used for validation. Assessing temporal consistency, positional correctness, and border delineation precision is very crucial in coastal investigations, particularly when the findings inform regulatory zoning and policy decisions.

2.5 Case Studies and Real-World Applications

2.5.1 Case Study on Shoreline Erosion Monitoring Using GIS and Remote Sensing

Because of cyclonic activity and human encroachment, coastline erosion has been a significant worry in Tamil Nadu, India. Significant erosion was found in the districts of Nagapattinam and Cuddalore, according to a study that used Landsat imagery from 1990 to 2020 combined with DSAS in GIS. The findings revealed that in a few of transects, EPRs exceeded-2 m/year. Government efforts to build groynes and encourage mangrove regeneration in places vulnerable to erosion were bolstered by these revelations. Multi-temporal shoreline changes in Tamil Nadu coast using Digital Shoreline Analysis System (DSAS) are shown in (Fig. 2.2a and b).

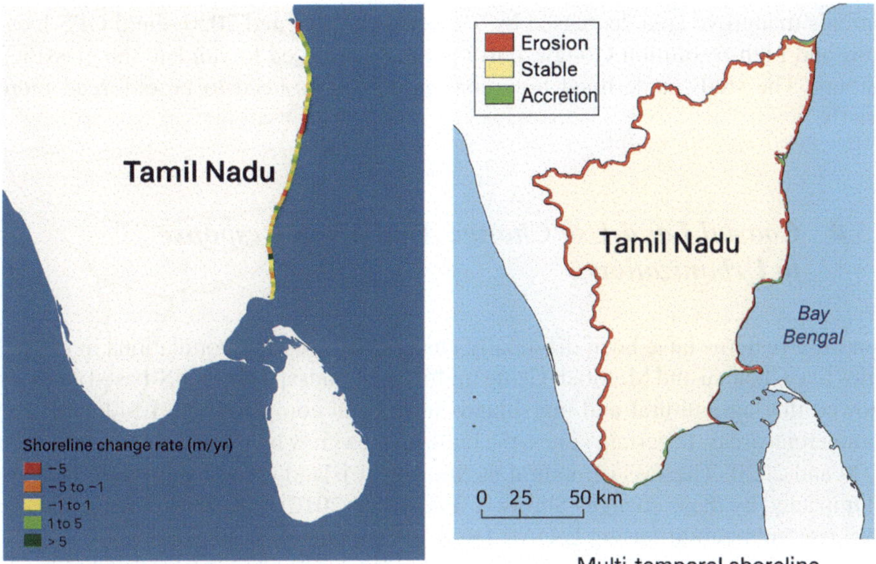

Multi-temporal shoreline change in Tamil Nadu coast using DSAS

(a)

Multi-temporal shoreline change in Tamil Nadu coast using DSAS

(b)

Fig. 2.2 Multi-temporal shoreline change in Tamil Nadu coast using DSAS **a** change rate **b** area location

2.5.2 Application of Satellite-Based Flood Mapping in Coastal Areas

Sentinel-1 SAR images were utilised to map flooding in real time during the 2015 Chennai floods. Even under cloudy conditions, SAR's all-weather capability aided in determining the amount of flooding Wernette et al (2017). Authorities were able to identify areas that were severely impacted and effectively mobilise relief efforts by combining GIS analysis with municipal boundary data. Planning for rehabilitation and processing insurance claims also made use of the flood maps.

2.5.3 Use of Machine Learning for Mangrove Forest Change Detection

Sea level rise, urbanisation, and aquaculture have all contributed to the destruction of mangrove habitats along India's east coast. According to a study that used Google Earth Engine's Random Forest classification and Sentinel-2 data, the Muthupet region's mangrove area decreased by 7% between 2010 and 2020. Field GPS locations and high-resolution Google Earth photos were used to validate the classified outputs. The study underlined how conservation laws need to be enforced more strictly.

2.5.4 *Coastal Land-Use Change Analysis in Response to Urbanization*

Land-use patterns have been drastically changed by growing populations in coastal cities like Chennai and Mumbai. Using multi-date Landsat data, a GIS-based analysis showed that agricultural and marshlands have been converted into residential and commercial areas. In certain zones, the built-up area grew by more than 50% between 2000 and 2020. The loss of natural streams, heat island effects, and flood risk are all impacted by these changes Shibly & Takewaka (2012). Coastal elevation and sea level rise vulnerability using LiDAR DEM is shown in (Fig. 2.3a and b).

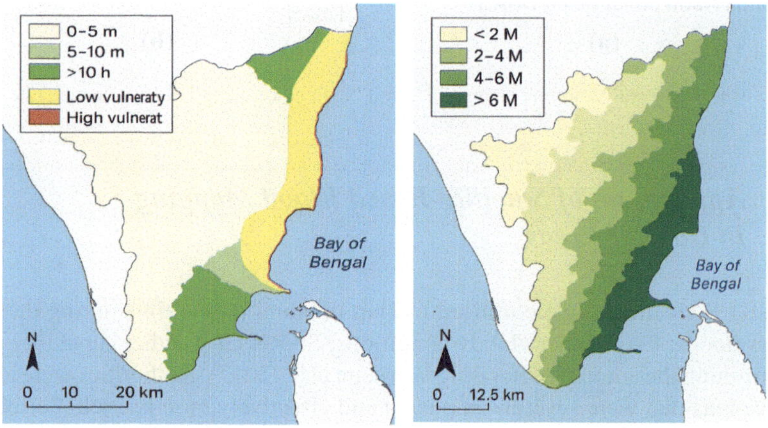

Coastal elevation and sea-level rise vulnerability using LiDAR DEM

Coastal elevation and sea-level rise vulnerability using LiDAR DEM

Fig. 2.3 Coastal elevation and sea level rise vulnerability using LiDAR DEM

2.6 Challenges and Future Directions

2.6.1 Limitations of Current GIS and Remote Sensing Techniques

Several limits still exist despite technical progress. These include limited ground truth available for validation, data gaps caused by the cloud cover in optical data, and insufficient spatial resolution of publicly available photography. Disparities in geographical, spectral, and temporal resolution make it difficult to integrate data from several sources. Significant computational resources are also needed to process massive datasets.

2.6.2 Need for High-Resolution Datasets and Real-Time Monitoring

UAV photography and high-resolution satellite data (such as WorldView and PlanetScope) provide finer spatial detail, which is essential for identifying small-scale coastal changes. In order to facilitate early warning and quick response during coastal hazards, there is also an increasing need for near real-time monitoring systems. Building resilient coastal towns requires initiatives that combine AI-driven alarm systems with near real-time satellite feeds.

2.6.3 Future Advancements in Geospatial Technology for Coastal Applications

Coastal monitoring is being revolutionised by emerging technologies like artificial intelligence, UAV-based photogrammetry, and cloud-based GIS systems like Google Earth Engine. Furthermore, decentralised, localised, and participative coastal management frameworks will be made possible by open access to high-frequency images and developments in edge computing.

2.6.4 Policy Recommendations for Sustainable Coastal Management

Geospatial data must be included into environmental impact assessments, disaster mitigation plans, and coastal regulation zone mapping for effective coastal government. Coastal managers' capacity growth, interagency cooperation, and data sharing

Table 2.4 Policy recommendations based on coastal vulnerability index

S.no	Vulnerability category	Recommendation
1.	High vulnerability (e.g., low elevation, erosion zones)	Construct bio-shields, enforce CRZ rules, implement early warning systems
2.	Medium vulnerability (e.g., urbanizing zones)	Regulate construction, promote eco-engineering solutions
3.	Low vulnerability (e.g., stable, vegetated regions)	Conserve natural buffers, encourage sustainable tourism
4.	Deltaic and estuarine zones	Integrate river basin and coastal planning, restore wetland ecosystems
5.	Island ecosystems	Restrict development, invest in relocation and climate resilience

should all be encouraged by policy frameworks. Using mobile GIS tools for community-based monitoring can guarantee bottom-up planning methods and increase involvement. Policy Recommendations Based on Coastal Vulnerability Index were given in (Table 2.4).

2.7 Conclusion

Strong, data-driven monitoring systems are required since the world's coastal zones are becoming more and more susceptible to changing environmental and human stressors. Assessing, evaluating, and monitoring coastal changes with greater accuracy, spatial coverage, and temporal frequency has been made possible by the revolutionary combination of Geographic Information Systems (GIS) and Remote Sensing (RS).

An extensive review of the methods and instruments utilised in the detection of coastal change has been given in this chapter. While GIS makes geographical analysis, modelling, and decision support easier, remote sensing platforms—such as optical, SAR, and LiDAR sensors—offer strong capabilities for multi-temporal and multi-spectral observations. The accuracy of extracting coastal features and quantifying change has been significantly improved by methods like the Digital Shoreline Analysis System (DSAS), Object-Based Image Analysis (OBIA), and Machine Learning (ML).

The case studies presented demonstrate the practical application of these geospatial technologies in various coastal contexts, ranging from flood mapping in the Sundarbans to mangrove monitoring using AI to erosion of shorelines assessments in Tamil Nadu. Emerging innovations including real-time data capture, UAV deployment, AI-driven analytics, and cloud-based services are gradually overcoming certain limits pertaining to data resolution, access, and integration

GIS and remote sensing applications must be not only embraced but also mainstreamed into community-based planning, coastal legislation, and policy frameworks in order to guarantee sustainable coastal development. Interdisciplinary cooperation, technological advancement, and evidence-based governance—supported by the potential of geospatial science—are the ways forward.

References

Adam E, Mutanga O, Rugege D (2010) Multispectral and hyperspectral remote sensing for identification and mapping of wetland vegetation: a review. Wetl Ecol Manag 18(3):281–296

Baig MRI, Ahmad IA, Shahfahad TM, Rahman A (2020) Analysis of shoreline changes in Vishakhapatnam coastal tract of Andhra Pradesh, India: an application of digital shoreline analysis system (DSAS). Ann GIS 26(4):361–376. https://doi.org/10.1080/19475683.2020.1815839

Bhatt CM, Rao GS, Manjunath KR, Dadhwal VK (2007) Analysis of land use/land cover dynamics of Indian coastal districts using remote sensing. J Indian Soc Remote Sens 35(4):281–290

Camfield FE, Morang A (1996) Defining and interpreting shoreline change. Ocean Coast Manag 32(3):129–151

Güneroğlu A (2015) Coastal changes and land use alteration on northeastern part of Turkey. Ocean Coast Manag 118:225–233

Himmelstoss EA, Farris AS, Henderson RE, Kratzmann MG, Ergul A, Zhang O, Zichichi JL, Thieler ER (2018) Digital shoreline analysis system (version 5.0). US Geol Surv. https://code.usgs.gov/cch/dsas/

JRC (2020) Digital shoreline analysis system (DSAS) user guide. U.S, Geological Survey

Kuenzer C, Voigt S (2013) Remote sensing of coastal environments. CRC Press

Luo S, Wang H, Cai F (2013) An integrated risk assessment of coastal erosion based on fuzzy set theory along Fujian coast, Southeast China. Ocean Coast Manag 84:68–76. https://doi.org/10.1016/j.ocecoaman.2013.07.007

Maiti S, Bhattacharya AK (2009) Shoreline change analysis and its application to prediction: a remote sensing and statistics based approach. Mar Geol 257(1–4):11–23

Mullick MRA, Islam KA, Tanim AH (2020) Shoreline change assessment using geospatial tools: a study on the Ganges deltaic coast of Bangladesh. Earth Sci Inform 13:299–316. https://doi.org/10.1007/s12145-019-00423-x

Rajawat AS, Chauhan HB, Nayak SR (2005) Shoreline change analysis for Dahanu taluka, Maharashtra, using remote sensing and GIS. J Indian Soc Remote Sens 33(4):463–468

Selamat SN, Maulud KNA, Mohd FA, Rahman AAA, Zainal MK, Wahid MAA et al (2019) Multimethod analysis for identifying the shoreline erosion during northeast monsoon season. J Sustain Sci Manag 14:43–54

Shibly AM, Takewaka S (2012) Morphological changes along Bangladesh coast derived from satellite images. Proc Coast Eng 3:41–45

Singh A (1989) Review article digital change detection techniques using remotely-sensed data. Int J Remote Sens 10(6):989–1003

Wernette P, Shortridge A, Lusch DP, Arbogast AF (2017) Accounting for positional uncertainty in historical shoreline change analysis without ground reference information. Int J Remote Sens 38(13):3906–3922

Zheng H, Li X, Wan J, Xu M, Liu S, Yasir M (2023) Automatic coastline extraction based on the improved instantaneous waterline extraction method and correction criteria using SAR imagery. Sustainability 15(9):7199. https://doi.org/10.3390/su15097199

Zhou Y, Li X, Gong P (2022) Deep learning for coastal wetland mapping using multi-sensor data. Remote Sens Environ 270:112847

Chapter 3
Monitoring Coastal Transformations Globally: Short- and Long-Term Insights from Multi-Resolution Remote Sensing

Saroj Senapati and Gouri Sankar Bhunia

3.1 Introduction

As a result of both natural and man-made forces, coastal zones are one of the planet's most vulnerable and dynamic habitats, always changing. Nayak (2002) points out that these places are constantly changing and that regular, systematic evaluation of coastal areas is crucial to comprehending and controlling the rapid alterations that influence them. The changing patterns of the shoreline, especially sedimentation and erosion, are important markers of the health of the coast. Wave action, tides, currents, sediment supply, and sea-level variability all interact to influence these naturally long-term geomorphic phenomena (Kannan et al. 2016; Ranasinghe 2016). However, human activities are rapidly overshadowing or speeding up natural coastal dynamics. Significant changes in shoreline positions have been observed worldwide as a result of extensive changes to the natural sediment budget and hydrodynamic framework brought about by accelerating urbanization, port and harbor improvement, land reclamation, excavation for sand, and the enlargement of hard coastal structures (such as seawalls, groynes, and breakwaters) (Thampanya et al. 2006; Jha et al. 2012).

The evaluation of land use and land cover change (LULCC) has been transformed by remote sensing (RS) and geographic information systems (GIS), which provide effective and affordable substitutes for conventional ground-based techniques. These tools are especially useful for monitoring landscape processes because they make it possible to systematically gather, store, analyze, and visualize geospatial data across time and place. According to Kasereka et al. (2010), RS and GIS technologies provide high-resolution, timely, and reliable data—all of which are essential for

S. Senapati
Annapurna Memorial College of Education, Kashinagar, West Bengal, India

G. S. Bhunia (✉)
Kharagpur, West Bengal, India
e-mail: rsgis_gouri@rediff.com

A. Nath et al. (eds.), *Geospatial Application of Coastal Hazards*, Springer Water,
https://doi.org/10.1007/978-3-032-20076-1_3

comprehending the swift changes in land use, particularly in highly volatile and environmentally sensitive regions like coastal zones. They are more accurate and scalable than conventional field surveys, but they also take less time and money.

The Landsat satellite series, in particular Landsat 5 TM (Thematic Mapper), Landsat 7 ETM+ (Enhanced Thematic Mapper Plus), Landsat 8 OLI (Operational Land Imager), and Landsat 9, offers one of the most extensive and reliable long-term datasets among the different RS platforms for examining medium-resolution changes in shoreline dynamics and land use. These publicly accessible records, which span the 1980s to the present, are crucial for multitemporal evaluations of coastal change (Apostolopoulos and Nikolakopoulos 2020). The importance of satellite images in researching multi-decadal patterns of environmental phenomena, such as coastline evolution, beach erosion, sediment transport, and wetland changes, is further highlighted by Apostolopoulos and Nikolakopoulos (2021). They are crucial for macro-scale coastal zone management because of their capacity to examine vast spatial expanses and extended time periods.

However, high-resolution and precise local assessments of beach and coastal morphology often rely on in-situ techniques such as, Global Navigation Satellite Systems (GNSS) beach surveys, which provide accurate elevation and positional data (Baptista et al. 2008); Terrestrial laser scanning (TLS) or LiDAR, which offers dense and highly accurate topographic data (Kaehler et al. 2022); Structure-from-Motion (SfM) photogrammetry, a technique that reconstructs 3D models from 2D photographs, often used with drone imagery for beach profiling (Talavera et al. 2020). These techniques have limited temporal and spatial coverage because of logistical, financial, and labor limitations, despite providing excellent positional accuracy and vertical resolution. As such, their efficacy in analyzing coastal morphodynamics at the regional or global scale is diminished. The accuracy and dependability of coastal LULCC assessments are nevertheless being improved by continuous developments in satellite technology, image correction techniques, and machine learning categorization.

In light of the aforementioned discussion, the current study uses satellite-based remote sensing data at various spatial and temporal resolutions to examine and measure the dynamic changes that appear along coastlines around the world.

3.2 Methods

Since the early 2000s, when satellite data became more readily available, there has been an increase in scholarly interest in monitoring coastal alteration, according to the bibliometric analysis. The goal of the study is to comprehend how multi-resolution remote sensing, with a focus on geospatial technologies and satellite data, has been used to monitor both short-term and long-term coastal changes worldwide. "Coastal transformation," "short-term change," "long-term change," "geospatial technology," and "satellite data" are some of the thematic keywords that were employed. Dimensions AI, an extensive research database, provided the data. This query was refined

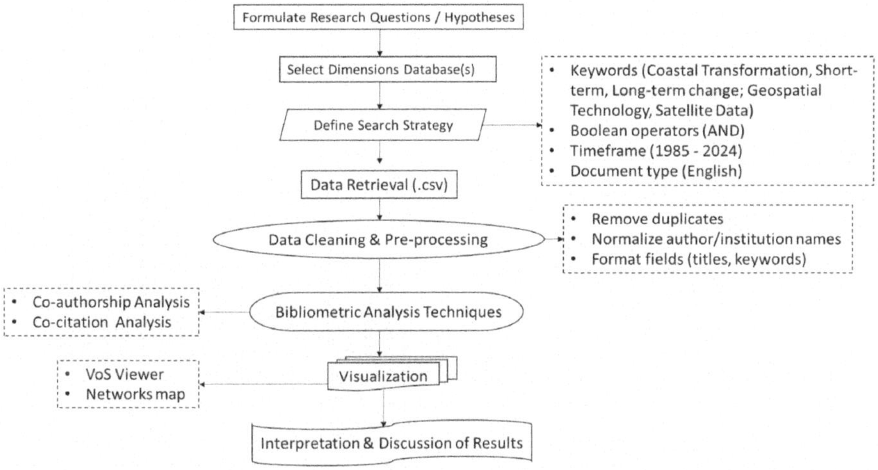

Fig. 3.1 Overall methodology of the bibliometric analysis

using the Boolean operator "AND" to make sure that every document that was retrieved had all of the pertinent keywords. The search was limited to the English language and encompassed the years 1985–2024 in order to capture both historical and current research. Articles, conference papers, reviews, and other peer-reviewed publications of various kinds were included in the filtered search results.

In order to eliminate duplicate entries and standardize fields like author names, universities, and keywords, the data was cleaned before analysis. This stage guaranteed correctness and consistency in subsequent statistical and network analysis. Measures like publications, citations, publications with citations (i.e., the proportion of publications that have at least one citation), the Field Citation Ratio (FCR), and the Relative Citation Ratio (RCR) have all been examined. Evaluation of research production, impact, and field-normalized performance across time is made possible by these metrics. In order to conduct network visualizations for co-authorship and co-citation analysis, the cleaned dataset had been imported into VOSviewer. Figure 3.1 provides an illustration of the analysis's overall process.

3.3 Results and Discussion

There was extremely little publishing activity in the subject between 1985 and 1999; typically, there were less than 20 articles annually. Citation counts stayed low, but in 1998 there was a noticeable increase (62 publications), which signaled the start of academic attention. Due to restricted satellite access and data availability, this time frame represents the early stages of remote sensing applications in coastal studies. Both publications and citations have steadily increased since 2000. The number of citations rose from 56 in 2000 to 1724 in 2009, demonstrating the increasing

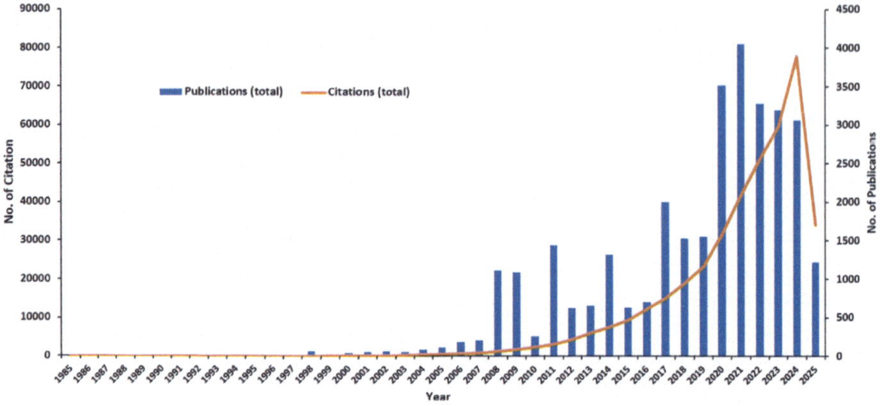

Fig. 3.2 Analysis of publications and citations published information on costal transformation during the period between 1985 and 2025

scholarly influence. The globalization of geospatial technology, higher-resolution satellite images (such as Landsat 7 and MODIS), and growing worries about coastal vulnerability are all associated with this phase (Fig. 3.2). The number of publications and scholarly influence in the field have skyrocketed since 2017. 2021 had a peak of 4059 publications, closely followed by 2024 at 3063. The number of citations has dramatically increased from 15,064 in 2017 to an all-time high of 78,021 in 2024, demonstrating the growing urgency on a global scale to comprehend and address coastal changes brought on by urbanization, climate change, and sea level rise. Research quality and quantity have increased thanks to technologies like artificial intelligence (AI), machine learning, multi-resolution data fusion, and improved satellite platforms (like Sentinel-2 and PlanetScope).

As remote sensing technologies and global attention to coastal change were still in their infancy, both FCR and RCR values remained 0 over the first 15 years (1985–1999), suggesting either a dearth of research papers in this area or articles that had not yet been referenced. High FCR values for a number of years (e.g., 2005: 3.29, 2010: 4.52, 2014: 4.50) show that articles received far more citations than the norm for their discipline. RCR values rise sharply, reaching 1.64 in 2010 and 1.36 in 2016, suggesting that these studies had greater sway than the NIH standard. In the majority of years, such as 2017 (4.03), 2019 (3.44), and 2020 (3.90), the FCR stays high, indicating ongoing intellectual achievement (Fig. 3.3). According to RCR, which peaked in 2021 (1.79), papers published in that year had almost twice the impact of the median publication in their field. In line with international initiatives to use geospatial data for sustainable coastal management, the data shows excellent and significant research, particularly after 2010.

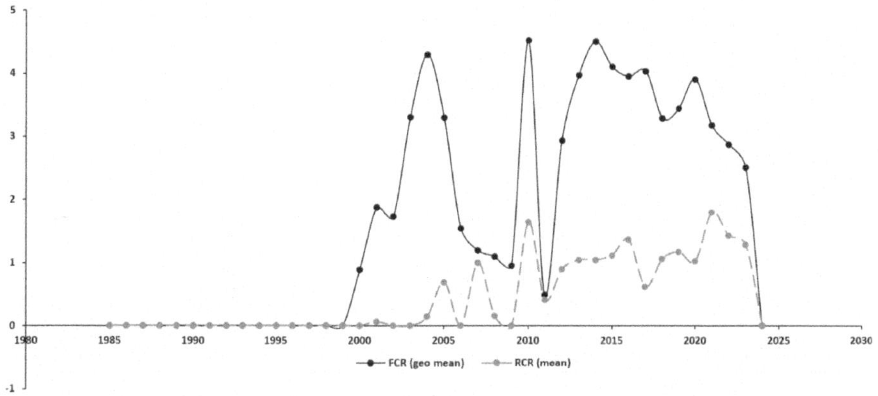

Fig. 3.3 Estimation of field citation ratio (FCR) and relative citation ratio (RCR) of published information on costal transformation

3.3.1 Analysis of Research Publications

Based on performance criteria and the number of scientific publications, the current analysis identifies the ten most productive years from 1985 to 2025. These metrics—number of publications, proportion of articles with citations (≥1 citation), Field Citation Ratio (FCR), and Relative Citation Ratio (RCR)—provide a thorough picture of the quality, influence, and effect of research across time. With 4059 papers, 2021 saw the most publications and the peak of research activity in this field. Its high citation rate (100 percent with at least one reference) indicates the importance, prominence, and direct influence of these works even if it is not the oldest year. Furthermore, papers published in 2010 were cited more than four times as frequently as the norm for their field, according to the FCR, which peaked at 4.52 in 2010. These publications outperformed the NIH standard by a large margin, as evidenced by their 2012 RCR of 1.79, demonstrating their ongoing scientific importance (Table 3.1).

90.41% of the 3524 papers published in 2020 had at least one citation, demonstrating a strong citation performance. Both historical depth and recent recognition are highlighted by the FCR of 4.50 (2014) and RCR of 1.64 (2010), which imply that fundamental ideas and techniques from earlier decades were revived in light of contemporary coastal monitoring issues. The year 2022 continued the upward trend in research output with 3286 articles. Of these papers, a significant 86.54% were cited at least once. A convergence of heritage citations and contemporary relevance is suggested by the FCR of 4.29 (2004) and RCR of 1.42 (2022), which are probably affected by developments in satellite technologies such as Sentinel-2 and the growing application of AI-driven geospatial analytics. 3202 publications surfaced in 2023, demonstrating the field's ongoing vitality. Despite a minor decline to 81.82%, the citation rate is typical for new articles that haven't had much time to amass

Table 3.1 Top ten research publications and analytics during the period between 1985 and 2025

Rank	Publications	Publication with citations (%, ≥1 citation	FCR (mean)	RCR (mean)
1	4059 (2021)	100 (1991)	4.52 (2010)	1.79 (2012)
2	3524 (2020)	90.41 (2016)	4.50 (2014)	1.64 (2010)
3	3286 (2022)	86.54 (2013)	4.29 (2004)	1.42 (2022)
4	3202 (2023)	81.82 (2004)	4.11 (2015)	1.36 (2016)
5	3063 (2024)	78.63 (2010)	4.04 (2017)	1.28 (2023)
6	2006 (2017)	76.83 (2013)	3.97 (2013)	1.16 (2019)
7	1555 (2019)	75.89 (2005)	3.95 (2016)	1.11 (2015)
8	1535 (2018)	75.75 (2015)	3.90 (2020)	1.05 (2018)
9	1440 (2011)	74.88 (2012)	3.44 (2019)	1.04 (2014)
10	1318 (2014)	74.19 (2002)	3.31 (2003)	1.04(2013)

citations. The RCR of 1.36 (2016) and the FCR of 4.11 (2015) highlight how techniques and information from previous eras continue to greatly influence contemporary coastal monitoring strategies. This trend of output was maintained in 2024 with 3063 publications. An FCR of 4.04 (2017) and a citation rate of 78.63% demonstrate a developing research ecosystem. Further demonstrating that new work is already being cited frequently—which is remarkable considering its recentness—is the RCR of 1.28 (2023). The remaining five years in the top ten—2017, 2019, 2018, 2011, and 2014—account for 1318–2006 articles each year. According to field standards, their FCR values (3.31–3.97) demonstrate influential research, and their citation rates (74–77%) are still high. Even these older articles appear to have substantial scholarly significance today, as indicated by the above-average performance indicated by the RCRs, which range from 1.04 to 1.16.

3.3.2 Co-Authorship Analysis

Co-authorship analysis is essential for determining the composition and robustness of cooperative networks and for comprehending the ways in which academic communities have developed around this multidisciplinary field of study. 61 active researchers were included in the dataset, which was screened for pertinent articles from 1985 to 2024. These writers were mapped using VOSviewer according to their co-authorship links, creating an interconnected network with each link signifying at least one joint publication with another author and each node representing a separate researcher. There were 164 direct co-authorship ties in the network, which represented the official scientific partnerships formed over time. With a total co-authorship frequency of 617, which is a measure of the total number of co-authored documents across all links, the field demonstrated a comparatively high level of collaborative participation (Fig. 3.4). Interestingly, the research identified ten different clusters, each of which

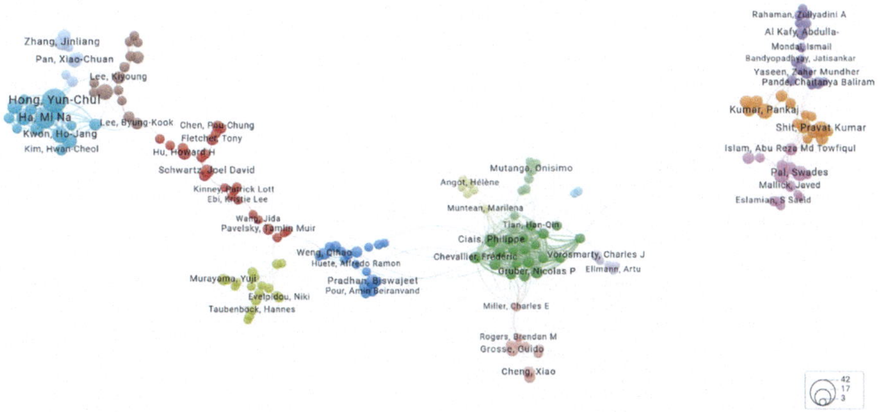

Fig. 3.4 Co-authorship analysis of published research article on costal transformation

corresponds to a group of authors who work together more frequently than those in other clusters. These clusters could represent research teams, organizations, or regional partnerships concentrating on particular facets of coastal monitoring, like coastline erosion, shifting land cover, or using remote sensing technology to track sea level rise.

3.3.3 *Co-Citation Analysis*

A co-citation analysis was carried out using VOSviewer in order to identify the conceptual framework and underlying literature supporting studies on global coastal alterations using multi-resolution remote sensing. Between 1985 and 2024, 234 scholars whose work was regularly co-cited in pertinent literature made up the co-citation network. There were 1001 co-citation relationships found in this network, each of which represented a pair of researchers whose work was referenced jointly in at least one citing article. There were 2474 co-citation instances overall, which suggests a solid academic background and regular cross-referencing of significant scientific findings. These researchers were arranged into 17 different clusters by the study, which was shown using VOSviewer. In the VOSviewer map, the thickness of the links indicates the strength of co-citation relationships, whilst the node size indicates how many times a specific author has been co-cited (Fig. 3.5). In order to indicate particular schools of thought or multidisciplinary convergence, clusters are frequently color-coded, signifying close-knit co-citation communities.

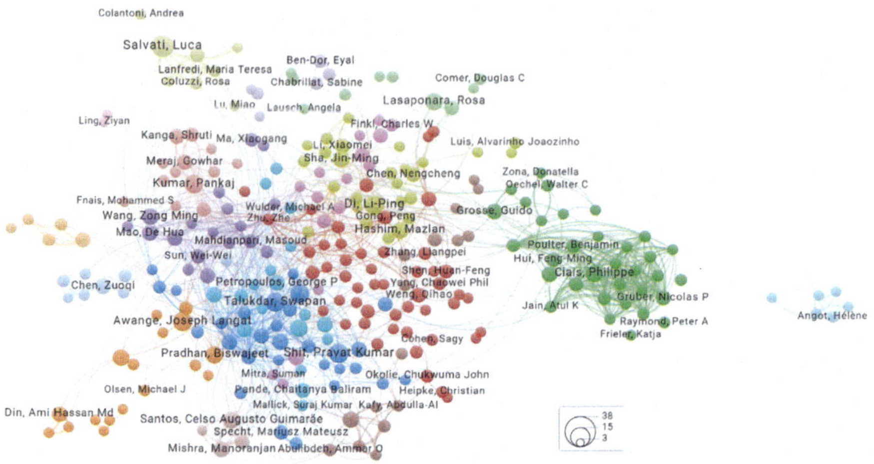

Fig. 3.5 Co-citation analysis on costal transformation

3.3.4 *Remote Sensing for Coastal Monitoring*

Over the past 50 years, the development of satellite remote sensing technology has fundamentally changed our capacity to evaluate and comprehend coastal processes at the local, regional, and global levels (Table 3.2). Significant advancements in spatial, spectral, and temporal resolution have been made in the field since the launch of Landsat-1 in 1972, which signaled the beginning of Earth resource observation from space (Roy et al. 2014). Researchers can now more precisely evaluate shoreline shifts, sediment transport, land use/land cover changes (LULCC), and coastal erosion/accretion thanks to satellite missions like MODIS, Sentinel-1 and 2, and the Landsat series (5, 7, 8, and 9) (Apostolopoulos and Nikolakopoulos, 2020; Wulder et al. 2019). The collection and analysis of satellite imagery with different geographical and temporal resolutions, or multi-resolution satellite data, has proven crucial in documenting both long-term trends and short-term dynamic occurrences in coastal regions. While moderate-resolution platforms such as Landsat (30 m) and MODIS (250–1000 m) are appropriate for measuring large-scale patterns and multi-decadal changes, high-resolution data (e.g., from WorldView or Sentinel-2 at 10–30 m) enables comprehensive shoreline mapping and infrastructure impact evaluations (Gens, 2010; Kuleli et al. 2011).

Because of its 16-day return period and uninterrupted archive since the 1970s, Landsat is still one of the most used remote sensing platforms for long-term coastal monitoring. For coastline delineation and vegetation change studies, Landsat 8 and 9 provide enhanced radiometric resolution thanks to the Operational Land Imager (OLI) and Thermal Infrared Sensor (TIRS) (Roy et al. 2014). Because of its 16-day return period and uninterrupted archive since the 1970s, Landsat is still one of the most used remote sensing platforms for long-term coastal monitoring. For coastline

Table 3.2 Key Remote sensing studies in coastal monitoring during the period between 1985 and 2024

Geography	Objectives	Data used	Methods/ techniques	Results	References
Arabian gulf	Monitor coastal urbanization and environmental changes.	MODIS; DubaiSat; Landsat OLI	Chlorophyll extraction algorithms; atmospheric correction models.	Detected red tide outbreaks; improved chlorophyll estimation in dusty atmospheres.	Dahy et al. 2024
Sundarban Delta, India	Analyze decadal changes in shoreline and mangrove cover.	Landsat series (MSS, TM, ETM+, OLI)	NDVI analysis; GIS-based change detection.	Noted significant reduction in dense vegetation cover; shoreline retreat observed in sea-facing islands.	Biswas et al. 2024
Africa	Dynamic coastal mapping using sentinel data.	Sentinel-1 (SAR); Sentinel-2 (optical)	Supervised machine learning; integration of radar and optical data.	Enhanced coastline mapping in cloud-prone regions; improved detection of coastal changes.	Liu et al. 2024
Andhra Pradesh, India	Assess and predict shoreline changes.	Landsat MSS, TM, ETM+, OLI	Geostatistical analysis; DSAS (digital shoreline analysis system).	Identified erosion and accretion patterns; predicted future shoreline positions.	Vaila and Ilango, 2024
Global	Evaluate shoreline changes due to sea-level rise	Landsat, Sentinel-2	Multi-temporal image analysis	Projected significant shoreline retreat by 2100 under high emission scenarios	Vousdoukas et al. 2020a, b
Taiwan	Detect and monitor long-term landslides in urbanized areas.	Landsat TM, ETM+, OLI; nighttime light data; ASTER DEM	Random forest classification; integration of multi-seasonal imagery.	Achieved high accuracy in landslide detection; identified persistent and reoccurring landslide areas.	Chen et al. 2019

(continued)

Table 3.2 (continued)

Geography	Objectives	Data used	Methods/ techniques	Results	References
Texas, USA	Assess long-term coastline changes and their association with sea level variation, hurricanes, and sediment transport.	Landsat TM, ETM+, OLI (30 m)	Time series analysis of land-water maps; validation with LIDAR data.	Identified spatial-temporal changes in coastline; correlated changes with sea level rise and hurricane events.	Tu et al. 2018
Global	Assess coastal erosion and accretion patterns	Landsat series (5, 7, 8)	Time-series analysis using Google earth engine	Identified global hotspots of beach erosion and accretion	Luijendijk et al. 2018a, b
Australia	Assess coastal vegetation changes	Sentinel-2	NDVI analysis	Observed significant mangrove dieback linked to climatic factors	Duke et al. 2017
China (Bohai Bay)	Monitor coastal urbanization impacts	Landsat series	Land use classification and change detection	Rapid urban expansion leading to loss of coastal wetlands	Wang et al. 2014
Japan (Tohoku region)	Evaluate tsunami-induced coastal changes	ALOS PALSAR, Landsat	Pre-and post-event image comparison	Documented extensive coastal geomorphological changes post-tsunami	Goto et al. 2011
Caribbean	Monitor coral reef health	MODIS	Ocean color analysis	Detected coral bleaching events correlated with SST anomalies	Mumby et al. 2004

delineation and vegetation change studies, Landsat 8 and 9 provide enhanced radiometric resolution thanks to the Operational Land Imager (OLI) and Thermal Infrared Sensor (TIRS) (Roy et al. 2014).

3.3.5 Global Monitoring Using Google Earth Engine and Big Earth Data

The availability and processing of remote sensing data have been completely transformed by the introduction of cloud-based geospatial technologies, especially Google Earth Engine (GEE). GEE, created by Gorelick et al. (2017), makes it possible for academics to quickly and affordably access and examine petabytes of recent and historical satellite imagery. An effective instrument for researching coastal morphodynamics at various temporal and spatial scales is made possible by the democratization of Earth observation data, which has lasted for more than thirty years. The application of GEE has resulted in historic research evaluating long-term shoreline changes on a worldwide scale. Luijendijk et al. (2018a, b) produced a global-scale evaluation of beach erosion and accretion patterns by analyzing nearly 36 years of coastline data using Landsat imagery within the GEE platform. Similar to this, Vousdoukas et al. (2020a, b) used modeling techniques in conjunction with remote sensing to forecast future shoreline changes under different climatic scenarios. The resulting consequences for coastal resilience and climate adaptation sparked a great deal of public and policy interest. Using satellite-derived shorelines in Europe, Castelle et al. (2021a, b, c, d) examined shoreline response and beach rotation at sub-annual to decadal scales at the local level. Similar methods were used by Vieira da Silva et al. (2016) and Vos et al. (2019a) to gain a better understanding of the effects of short-term events like storms and localized beach morphodynamics. The value of remote sensing in tracking the efficacy of coastal fortifications was further emphasized by Miranda et al. (2019).

Global analyses provide important insights, but they also have unique benefits and drawbacks. On the plus side, platforms such as GEE provide unparalleled scalability, enabling researchers to process and access reliable long-term satellite records (e.g., MODIS, Sentinel-1/2, and Landsat) without the need for costly gear. This accessibility democratizes science by allowing scholars from underdeveloped nations to participate in globally significant research projects (Table 3.3). Additionally, reproducibility and transparency in methodologies are made possible by the standardized workflows, which is essential for policy-making and scientific validation. Nevertheless, there are drawbacks to these worldwide evaluations. Small-scale erosion hotspots, thin mangrove fringes, and narrow beaches are examples of fine-scale coastal characteristics that may be overlooked by their frequent reliance on medium-resolution images (e.g., 30 m for Landsat). Furthermore, global assessments usually demand for generalizations or simplifications (such as uniform thresholding or fixed buffer distances) that could not accurately represent regional geomorphological or socioecological heterogeneity. Estimates of coastline position are further complicated by cloud cover, tidal fluctuations, and discrepancies in sensor calibration. According to Castelle et al. (2021a, b, c, d), water turbidity, wet sand/dry sand distinction, and fluctuating wave run-up make it difficult to derive precise shorelines from optical footage, even with sophisticated automation.

3.3.6 Local-Scale Coastal Monitoring Techniques

In order to capture fine-resolution topography and morphologic characteristics that are not visible in medium-resolution satellite datasets, local-scale coastal monitoring techniques are essential. Photogrammetry, Light Detection and Ranging (LiDAR),

Table 3.3 Global monitoring using google earth engine and big earth data

Objectives	Geography	Data used	Methods/ techniques	Key outcomes	References
To assess long-term shoreline, change of sandy beaches worldwide	Global (33,700 km of sandy coastline)	Landsat 5, 7, 8 (1984–2016); GEE	Automatic shoreline detection; statistical trend analysis	24% of sandy beaches eroding >0.5 m/ year; major hotspots in West Africa, south and Southeast Asia	Luijendijk et al. 2018a, b
To model future shoreline, retreat due to sea level rise and wave climate change	Global (sandy coasts)	CMIP5 projections, historical shoreline data (satellite-derived), DSAS	Integrated numerical modeling with satellite data	Estimated loss of ~50% of sandy beaches globally by 2100 under RCP8.5	Vousdoukas et al. 2020a, b
To create global-scale monthly surface water maps and dynamics	Global	Landsat archive (1984–2015); GEE	Classification of water bodies using spectral indices; temporal frequency analysis	Mapped over 30 years of surface water changes; major loss in Central Asia, gains in Africa	Pekel et al. 2016
To derive near-global coastline dynamics and beach rotation patterns	10,000+ sites globally	Landsat (via GEE), shoreline proxies, tide corrections	Wavelet transform analysis; interannual variability detection	Detected rotation and seasonal behavior of sandy coasts globally	Vos et al. 2019a, b, c
To assess global mangrove forest change	Global	Landsat (1984–2014); GEE	Change detection using supervised classification	2.1% net loss globally; Southeast Asia highest loss rate	Hamilton & Casey, 2016

(continued)

Table 3.3 (continued)

Objectives	Geography	Data used	Methods/ techniques	Key outcomes	References
To map changes in artificial surfaces over time	Global	Landsat TM, ETM+, OLI (1985–2015); GEE	Global human settlement layer; pixel-based classification	Rapid urban expansion detected in Asia and Africa	Liu et al. 2020
To analyze climate-driven vegetation changes globally	Global terrestrial ecosystems	MODIS NDVI, Landsat; GEE	Trend analysis using Mann-Kendall; correlation with temperature & precipitation	Greening in high latitudes; browning in tropics	Chen et al., 2019
To monitor global crop phenology	Global croplands	MODIS, Sentinel-2; GEE	Time series smoothing; phenological markers	Detected regional shifts in planting/ harvesting cycles	Becker-Reshef et al., 2020
To quantify anthropogenic impacts on deltas	48 major global deltas	Landsat (1984–2015); GEE	Sediment budget modeling; land change classification	Net land loss in ~70% of deltas; highest in Asia	Higgins et al. 2018
To track flood extent and exposure globally	Global flood-prone areas	MODIS, Landsat; GEE + UNOSAT data	Automated flood detection; exposure overlay with population	255 M people exposed to floods 2000–2015	Tellman et al. 2021

Global Navigation Satellite Systems (GNSS), and high-resolution Unmanned Aerial Vehicles (UAVs) are already commonplace instruments for local coastal assessments. UAVs can study coastal dynamics, dune migration, beach volume changes, and vegetation cover in great detail since they can give ultra-high-resolution photography down to the sub-decimeter level. Digital surface models (DSMs) and orthomosaics—which are crucial for volumetric and morphological analysis—can be produced by UAV photogrammetry employing Structure from Motion (SfM) approaches (Gonçalves and Henriques, 2015; Casella et al. 2016). These tools are especially helpful for tracking erosion/accretion patterns at the beach scale and for post-storm assessments.

In order to validate satellite-derived coastline positions and calibrate elevation data, GNSS-based surveying—which frequently employs Real-Time Kinematic (RTK) systems—improves the positional accuracy of coastal profiles to the centimeter level (Baptista et al. 2008). Likewise, high-precision elevation models that facilitate dune and flood modeling, sediment budgeting, and habitat mapping are provided by terrestrial or aerial LiDAR. To evaluate beach morphology and flood resistance infrastructure, for instance, Kaehler et al. (2022) showed how LiDAR data can be combined with UAV outputs (Table 3.4).

A multiscale approach to coastal monitoring is made possible by the combination of these high-resolution technologies with satellite data, particularly medium-to high-resolution datasets from Landsat, Sentinel-2, or WorldView. Satellites provide regional and temporal context, but LiDAR and UAVs provide site-specific accuracy. This collaboration is especially helpful for enhancing the classification accuracy of land cover maps, identifying subpixel shoreline characteristics, and calibrating remote sensing algorithms (Talavera et al. 2020; Casella et al. 2017).

3.3.7 Challenges in Satellite-Based Shoreline Detection

Although satellite-based shoreline detection is becoming more and more popular for coastal monitoring because of its synoptic and temporal coverage, its accuracy and dependability are limited by a number of technological and environmental issues. The impact of water level fluctuations and tidal effects, which can cause notable differences in coastal position, is a major worry. Particularly in low-gradient coastal zones, tidal oscillations change the location of the land-water interface. Shorelines taken from satellite images may represent several tidal stages if tidal phase and water level are not corrected for during image acquisition, which could complicate evaluations of long-term development. For instance, the difference in tidal stage across scenes can add up to tens of meters of apparent coastline fluctuation in studies that use Landsat data with 16-day revisit intervals (Pardo-Pascual et al. 2012; Luijendijk et al. 2018a, b). This can be lessened by adding tidal models or tide-gauge corrections to shoreline recognition algorithms, however many beaches across the world lack high-resolution access to this data.

Other recurring issues are cloud cover, sensor limitations, and spatial resolution. The great susceptibility of optical satellite sensors, such as MODIS, Sentinel-2, and Landsat, to cloud blocking might cause data gaps or delay time series studies. For example, Sentinel-2 has a 10 m resolution but is frequently covered by clouds in tropical and temperate regions. This means that either cloud masking methods (like Fmask) or cloud-free composites are required, which may result in a reduction in temporal fidelity. Shoreline demarcation may also be impacted by sensor problems including calibration drift, spectral noise, or inaccurate geolocation. Furthermore, the scale of identifiable features is determined by resolution constraints (Table 3.5). Sub-meter resolution satellites, such as WorldView or PlanetScope, offer limited spatial or temporal coverage and are not cheap, but they can record delicate coastline details, such as minor embayments and jetties. Conversely, coarser resolution sensors (like MODIS at 250–500 m) are insufficient for narrow beaches or urban coastlines and are only appropriate for large-scale shoreline alterations.

Table 3.4 Local-scale coastal monitoring using high-resolution coastal monitoring techniques

Objectives	Geography	Data used	Methods	Outcomes	References
Assess UAV-SfM photogrammetry accuracy for beach monitoring	NW Portugal	UAV imagery, GCPs, Agisoft Photoscan	Structure-from-motion (SfM), DSM, orthomosaic generation	Achieved 10 cm resolution; vertical accuracy of ~3.5–5 cm	Gonçalves & Henriques, 2015
Monitor beach morphological changes seasonally	Ligurian Sea, NW Mediterranean	UAV, GPS, GCPs	SfM photogrammetry; pre/post-winter comparison	Detected sand erosion/accretion zones; validated UAV repeatability	Casella et al. 2016
Compare LiDAR vs UAV-SfM for topography	Riccione, Italy	UAV-LiDAR, UAV imagery	Co-registration; accuracy validation using checkpoints	Found LiDAR more consistent in vegetated areas; SfM had higher detail in exposed terrain	Duo et al. 2021
Evaluate UAV-GNSS integration for beach profile surveys	Lido di Spina, Italy	UAV (phantom 4 RTK), RTK-GNSS	DSM generation; volumetric change analysis	RTK UAV provided ~5 cm vertical accuracy; cost-effective vs GNSS-only	Vecchi et al. 2021
Monitor foredune volume changes	Truc Vert Beach, France	Airborne LiDAR, UAV	Multi-temporal LiDAR; UAV photogrammetry comparison	Dune migration and sediment gain observed; integration improved temporal resolution	Robin et al. 2020
Measure post-storm beach recovery	SE England	UAV (DJI), RTK-GNSS, Sentinel-2	SfM from UAV; validation with GNSS; time-series sentinel analysis	UAV captured small-scale changes post-storm; sentinel used for broader trend	Fairley et al. 2021
Assess shoreline variability with stereo photogrammetry	Texas Gulf Coast, USA	ZED 2i stereo camera, UAV imagery	Close-range photogrammetry; DSMs; UAV validation	Achieved high temporal resolution at sub-meter accuracy	Pilartes-Congo et al. 2025

(continued)

Table 3.4 (continued)

Objectives	Geography	Data used	Methods	Outcomes	References
Analyze sediment transport & beach nourishment	Lord Howe Island, Australia	UAV imagery (multi-date)	SfM photogrammetry; DSM differencing	Detected artificial sand deposition effects; suitable for remote coastlines	Lee et al. 2022
Evaluate UAV-LiDAR for dune habitat monitoring	Coast of the Netherlands	UAV-LiDAR, field RTK-GNSS	3D point cloud processing; vegetation analysis	High-resolution dune morphology and plant distribution mapping	Lin et al. 2019
Build low-cost coastal monitoring solution	Baltic Sea, Poland	UAV, RTK-GNSS, TLS	UAV-SfM vs TLS; accuracy comparison	UAV more efficient, TLS more accurate for micro-topography	Mazur-Marzec et al. 2025
Use TLS and UAV for sand dune dynamics	Poland (Sopot Beach)	UAV imagery, TLS, RTK-GNSS	DSMs from TLS and UAVs; cross-validation	UAV-derived DSMs valid, though noisier than TLS; vegetation affected accuracy	Gracia et al., 2025
Track sediment budget and dune changes	Truc Vert Beach, France	UAV surveys, airborne LiDAR	Volumetric change analysis; beach-dune interaction	Dunes accreted landward over time; UAV useful for event-based change	Castelle et al. 2021a, b, c, d
Evaluate UAV performance on rocky coasts	Galicia, Spain	UAV RGB imagery	SfM; DEM and orthomosaic production	Successfully mapped complex rocky terrain; image overlap critical	González-Guillén et al. 2022
Monitor coastal nourishment using UAV	Lord Howe Island, Australia	UAV photogrammetry	BACI design; shoreline extraction	Verified artificial nourishment effectiveness; UAV effective in remote areas	Melnyk & Harris, 2026
Real-time shoreline tracking using stereo cameras	Texas Gulf Coast, USA	Stereo video data	Photogrammetry + point cloud from stereo vision	Provided high-frequency shoreline updates; low cost, high utility	Pilartes-Congo et al. 2025

Table 3.5 Challenges in satellite-based shoreline detection using remote sensing and GIS

Objectives	Geography	Data used	Methods	Outcomes	Reference
Assess global shoreline change over 30+ years	Global (all continents)	Landsat 5, 7, 8 (1984–2016); tidal data; OpenStreetMap	Automated shoreline extraction using Landsat; correction for tidal effects using hydrodynamic models	Found 24% of world's sandy beaches are eroding; highlighted need to account for tidal stage in multi-date imagery	Luijendijk et al. (2018a, b)
Develop sub-pixel precision method for shoreline extraction	Western Mediterranean (Spain)	Landsat TM/ ETM+; LIDAR; GNSS control data	Developed SHOREX algorithm; tide-corrected shoreline mapping; used LiDAR validation	Achieved ±2 m accuracy; emphasized limitations of single-band thresholding and importance of water level correction	Pardo-Pascual et al. (2012)
Global shoreline dynamics mapping and analysis	Global	Landsat imagery (1984–2016); MODIS for context; DEMs; validation datasets	Global shoreline evolution mapped using automated detection; waterline detection at MSL; incorporated SRTM DEMs	Identified accretion and erosion trends; highlighted importance of resolution and validation using independent datasets	Mentaschi et al. (2018)
Validate automatic shoreline detection techniques	Spain (Valencia coast)	Sentinel-2, Landsat-8, UAV images; GNSS surveys	Validated satellite-derived shorelines against UAV and field data; assessed uncertainties	Confirmed necessity of robust reference data; Sentinel-2 showed ±3 m RMSE when validated	Ribas et al. (2020)

(continued)

Table 3.5 (continued)

Objectives	Geography	Data used	Methods	Outcomes	Reference
Assess geolocation accuracy of satellite sensors	Various U.S. test sites	Landsat 7 ETM+, SPOT, IRS, ASTER	Quantified geolocation errors; evaluated misregistration between sensors	Geolocation errors up to 50 m for uncorrected data; emphasized importance of geometric correction in shoreline work	Di et al. (2003)
Develop automated multi-sensor shoreline detection approach	Netherlands	Sentinel-2, Landsat-8, UAV imagery, tide data	Combined optical and SAR data; used NDWI and machine learning classifiers; validated with UAV data	Emphasized tidal and seasonal effects; suggested sensor fusion improves detection consistency	Vos et al. (2019a, b, c)

Shorelines determined by satellite still need to be accurate and validated. For remotely sensed shorelines to be validated, ground-truth data—usually acquired using GNSS surveys, UAV mapping, or LiDAR—is fundamental. However, particularly in developing nations, such in-situ statistics are frequently out-of-date or unavailable. The significance of rigorous validation frameworks utilizing high-resolution ground data, error budgets, and uncertainty quantification is emphasized by studies such as Mentaschi et al. (2018) and Pardo-Pascual et al. (2019). Auxiliary datasets including tide models, wave data, and DEMs must be integrated because of the potential for interpretation ambiguities caused by elements like wave run-up, seasonal vegetation, and sand wetness. Furthermore, shoreline recognition systems (such as automatic classifiers or thresholding of spectral indices like NDWI) differ in their sensitivity to terrain, land cover, and sensor type, which affects consistency and comparability over time and place.

3.3.8 Comparative Analysis: Short-Versus Long-Term Transformations

In order to comprehend the dynamics of coastal transformation, it is necessary to distinguish between short-term (such as seasonal to interannual fluctuations) and long-term (such as decadal to multi-decadal trends) changes in the coastline, morphology, and sediment budget. The detection of patterns in coastal transformation is mostly dependent on time-series analysis. Theil-Sen Estimator, Kalman Filtering, and Linear Regression are tools commonly used to measure the pace of long-term shoreline change (e.g., Luijendijk et al. 2018a, b; Mentaschi et al. 2018). For short-term analysis, change-point detection, wavelet transform, and seasonal decomposition methods help isolate seasonal variations, storm-related erosion episodes, or rapid anthropogenic modifications (Vos et al. 2019a, b, c; Castelle et al. 2021a, b, c, d). Google Earth Engine (GEE) has enabled automated and large-scale time-series analysis of multi-decadal satellite imagery, including from the Landsat and Sentinel missions.

High-energy events like storm surges, flash floods, or cyclones frequently cause short-term alterations that lead to sudden erosion or accretion (e.g., Harley et al. 2017). For example, near the eastern Indian coast, the 2004 Indian Ocean tsunami resulted in an immediate and significant shoreline retreat. On the other hand, as noted by Vousdoukas et al. (2020a, b) and Mentaschi et al. (2018), long-term changes appear as slow trends driven by sea-level rise, a decline in the sediment supply, and ongoing human activity (such as building ports or damming rivers). While gradual changes can have regional to global ramifications, rapid changes are usually confined but have a larger influence. Both tendencies were highlighted throughout decades by studies such as Bishop-Taylor et al. (2021) in Australia and Almeida et al. (2021a, b) in Brazil using GEE-based shoreline extraction techniques.

By contrasting short- and long-term coastal changes, scientists may determine vulnerable regions, evaluate how resilient natural features (such dunes and mangroves) are, and guide climate adaptation plans (Table 3.6). This comparative understanding is used by tools like the Integrated Coastal Zone Management (ICZM) frameworks and the Coastal Vulnerability Index (CVI). For example, to inform coastal protection measures, Vousdoukas et al. (2020a, b) created worldwide coastline retreat predictions under sea-level rise scenarios. By combining remote sensing data with field measurements, socioeconomic statistics, and hydrodynamic models, risk zonation maps and resilience indices that are tailored for particular regions (such as urban coasts, deltas, and small island nations) can be created.

3.4 Policy Implications and Future Directions

The incorporation of cutting-edge technologies into policy frameworks is crucial as coastal regions confront increasing challenges due to anthropogenic pressures, sea level rise, and climate change. Cloud computing, open data, artificial intelligence (AI), machine learning (ML), and remote sensing are all combining to transform coastal zone management (CZM) and bring it closer to the Sustainable Development Goals (SDGs) and climate adaptation plans.

Adaptive planning techniques that tackle present and future coastal issues are now a part of Integrated Coastal Zone Management (ICZM). Adaptive planning techniques, like the Adaptation Pathways approach, provide frameworks for creating

Table 3.6 comparative analysis of studies focusing on short-and long-term coastal transformations

Objectives	Geography	Data used	Methods	Outcomes	Reference
Assess global shoreline change over 30+ years	Global	Landsat 5, 7, 8 (1984–2016); tidal data; OpenStreetMap	Automated shoreline extraction using Landsat; correction for tidal effects using hydrodynamic models	Found 24% of world's sandy beaches are eroding; highlighted need to account for tidal stage in multi-date imagery	Luijendijk et al. (2018a, b)
Develop a toolkit for global shoreline mapping	Global	Landsat 5, 7, 8; Sentinel-2; UAV imagery	Developed CoastSat: a python toolkit leveraging Google earth engine for shoreline extraction	Enabled global-scale, high-resolution shoreline mapping; facilitated detection of both rapid and gradual changes	Vos et al. (2019a, b, c)
Analyze competing morphodynamic patterns at the coast	Various global sites	Multi-source datasets including satellite imagery and in-situ measurements	Comparative analysis of morphodynamic patterns; assessment of natural vs. anthropogenic influences	Identified complex interactions between natural processes and human activities influencing coastal changes	Castelle et al. (2021a, b, c, d)
Evaluate global coastal morphodynamics over 32 years	Global	Landsat imagery (1984–2015); global surface water explorer dataset	Analyzed changes in water presence along over two million virtual transects	Found that eroded land area (~28,000 km^2) is twice the gained land area; anthropogenic factors are dominant drivers	Mentaschi et al. (2018)

(continued)

Table 3.6 (continued)

Objectives	Geography	Data used	Methods	Outcomes	Reference
Project future shoreline retreat under sea-level rise scenarios	Global	Satellite-derived shoreline data; sea-level rise projections	Modeled shoreline changes under various climate scenarios	Predicted significant erosion of sandy coastlines by 2100; emphasized need for coastal adaptation strategies	Vousdoukas et al. (2020a, b)
Map annual shoreline positions across Australia	Australia	Landsat archive (1988–2019); tide data; field measurements	Developed sub-pixel shoreline extraction method using modified NDWI; validated with extensive ground data	Produced high-accuracy annual shoreline maps; detected changes as small as 0.31 m/year	Bishop-Taylor et al. (2021)
Compare shoreline changes in two Brazilian deltas (1986–2021)	Brazil (São Francisco and Paraíba do Sul deltas)	Landsat 5 and 8 imageries; Google earth engine	Applied modified NDWI for shoreline detection; analyzed spectral trajectories over time	Identified areas of accretion and erosion; linked changes to anthropogenic activities like dam construction	Almeida et al. (2021a, b)

strong plans to deal with unforeseen future changes, especially those brought on by the effects of climate change. These approaches make sure that management techniques can adapt to changing environmental conditions by highlighting the value of adaptability and long-term vision in coastal planning. Additionally, the use of GIS and satellite remote sensing technology has improved the ability to track changes along the shore, enabling better decision-making in ICZM procedures. Policymakers may more effectively evaluate risks, plan for sustainable development, and put effective adaptation measures into place by combining these technologies.

The monitoring and accomplishment of SDGs, especially those pertaining to life below the water (SDG 14) and climate action (SDG 13), are greatly aided by remote sensing technology. Remote sensing makes it possible to evaluate biodiversity, ecosystem health, and the effects of climate change by supplying reliable and thorough data on coastal and marine ecosystems. Furthermore, by incorporating data from remote sensing into frameworks for climate adaptation, susceptible areas can be identified, which helps guide resource allocation and focused interventions. This collaboration between policy and technology promotes the sustainable management of coastal areas and increases the efficacy of climate resilience solutions.

The analysis and interpretation of intricate coastal data sets has changed with the introduction of AI and ML. Proactive risk management is made possible by these technologies, which make it easier to foresee coastal risks like erosion and flooding. AI-powered models, for example, are able to examine enormous volumes of satellite footage in order to identify trends and predict possible dangers to ecosystems and coastal infrastructure. The democratization of access to high-resolution geospatial data through cloud computing platforms such as Google Earth Engine has made it possible to conduct extensive environmental monitoring and research. This ease of access encourages cooperative research and well-informed decisions, especially in underresourced areas. Transparency and community involvement in coastal management are further improved by open data efforts. Publicly accessible data allows stakeholders to take part in decision-making, promoting inclusive and equitable adaptation tactics. This cooperative strategy guarantees that a range of viewpoints are considered when developing and implementing coastal regulations.

3.5 Way Forward

By highlighting important trends, obstacles, and gaps in the existing research landscape, the current study provides a solid basis for future developments. The network and bibliometric studies yield a number of strategic initiatives and recommendations for improving scientific research and real-world coastal monitoring applications. Promoting data harmonization and interoperability across different remote sensing technologies is urgently needed. To enable more precise, comparable, and long-term evaluations of coastal change, consistent processing and analysis frameworks must be created, given that several satellite missions (such as Landsat, Sentinel, and MODIS) produce a variety of datasets. The democratization of geospatial information for all stakeholders necessitates international cooperation in the development of open-access platforms, uniform data repositories, and AI-powered analytics.

The focus of this study is on increasing the temporal and spatial resolution of coastal monitoring, particularly in vulnerable and underrepresented areas such polar coastal zones, low-lying deltas, and small island developing states (SIDS). Localized insights and context-specific policy solutions would be made possible by supporting focused finance and capacity-building initiatives in these regions. Furthermore, encouraging interdisciplinary cooperation is crucial. Experts in remote sensing, coastal engineers, urban planners, climate scientists, and local governance organizations need to better integrate, according to the co-authorship and co-citation networks. Academic research and policy implementation can be more closely aligned with the establishment of transdisciplinary coastal observatories and innovation laboratories.

Additionally, by utilizing developments in machine learning, cloud computing, and IoT-enabled sensor networks, the study suggests a move toward predictive and real-time coastal monitoring systems. Planning for disaster risk reduction, shoreline change modeling, and early warning capabilities can all be greatly improved by combining these technologies with satellite data. The use of community-based participatory methodologies in coastal transformation research is becoming more and more popular. It is possible to create monitoring frameworks that are more inclusive and socially responsive by coordinating high-resolution satellite data with ground-truthing initiatives and indigenous knowledge.

3.6 Conclusion

The current study provides a thorough scientific and bibliometric examination of the changing field of coastal studies, especially as it relates to satellite-based geospatial technology. This study emphasizes how multi-resolution remote sensing is becoming increasingly relevant both academically and practically for tracking coastal transformation. It demonstrates a move away from discrete case studies and toward worldwide, data-driven, and methodologically complex research that tackle both short-term and long-term changes in coastal habitats. By combining environmental studies, climate change adaptation, and geospatial sciences, the discipline has developed into a strong multidisciplinary area. The study covers a wide time period (1985–2025) and explores the amount, significance, and cooperative trends of worldwide research on tracking changes along the coast caused by both natural and man-made factors. According to the bibliometric results, the study shows a notable increase in publishing output after 2000, with a sharp increase after 2010, which corresponded with improvements in Earth observation system technology and rising worries about urban encroachment, sea level rise, and coastal vulnerability. The 2010–2023 decade saw a boom in citation measures, such as the Field Citation Ratio (FCR) and Relative Citation Ratio (RCR), which demonstrated the strong impact and pertinence of recent research contributions in this area in addition to a spike in volume. According to the data, the most significant years in terms of scholarly output and citations were 2021 and 2022.

The study measures the influence of articles using Dimensions AI, displaying metrics that indicate rising citation rates and a high proportion of referenced papers, especially from 2015 to 2024. Dense co-authorship links between 61 researchers and 10 different clusters are shown by the VOSviewer-based network analysis, suggesting an increase in transdisciplinary and global collaboration. With 234 scholars and 17 clusters included, the co-citation study highlights the growth of important knowledge domains and influential figures influencing the conversation on coastal monitoring.

In addition to revealing a developing field, the bibliometric insights show that there is a strong push for scientists, politicians, and coastal managers to use satellite data and analytics for both long-term planning and real-time decision-making. Finally, this study effectively charts the course, patterns, and effects of this change over the past forty years. Remote sensing has evolved from a supplementary instrument to a key component in the global response to coastal concerns.

Conflict of Interest No any conflict of interest.

References

Almeida LP, Granja HM, Ferreira Ó, Dias JA (2021a) National shoreline changes in Brazil derived from satellite data and their relation to coastal processes. J Coast Res 37(3):453–465. https://doi.org/10.2112/JCOASTRES-D-20-00105.1

Almeida et al (2021b) Historical shoreline mapping of the Brazilian coast using earth engine. Remote Sens Appl: Soc Environ 23:100548

Apostolopoulos D, Nikolakopoulos K (2020) Evaluation of shoreline change using remote sensing and GIS: case study of a coastal area in Western Greece. Eur J Remote Sens 53(1):99–109. https://doi.org/10.1080/22797254.2019.1706314

Apostolopoulos D, Nikolakopoulos K (2021) Monitoring of coastal areas using multi-temporal satellite data: methodologies and case studies. Geocarto Int 38(1):1–21. https://doi.org/10.1080/10106049.2021.1875369

Baptista P et al (2008) Beach morphodynamics and GNSS data accuracy: a case study at Faro Beach. Port J Coast Res 24(2B):96–104. https://doi.org/10.2112/05-0608.1

Becker-Reshef I et al (2020) Strengthening agricultural decisions in developing countries with earth observation data. Nat Food 1:200–205. https://doi.org/10.1038/s43016-020-0056-y

Bishop-Taylor R, Sagar S, Lymburner L, Beaman RJ, Brooke BP (2021) Deploying digital earth Australia's intertidal extents model to map Australia's intertidal zone. Estuar Coast Shelf Sci 249:107094. https://doi.org/10.1016/j.ecss.2020.107094

Biswas S, Sarkar K, Das, (2024) Decadal changes (1980–2021) of shoreline and mangrove cover in sundarban delta, India, Using Remote Sensing and GIS J Phys Oceanogr 54:1655–1674. https://doi.org/10.1175/JPO-D-23-0019.1

Castelle B, Marieu V, Bujan S, Splinter KD, Robinet A, Senechal N, Ferreira S (2021a) Impact of climate change on coastal sediment budget: evidence from satellite-derived shorelines and numerical modelling. Earth Sci Rev 215:103575. https://doi.org/10.1016/j.earscirev.2021.103575

Casella E et al (2016) Mapping sand beach morphology changes using UAVs. Remote Sens 8(1):60. https://doi.org/10.3390/rs8010060

Castelle B et al (2021b) Challenges in long-term shoreline monitoring using satellite-derived data. Remote Sens Environ 264:112603. https://doi.org/10.1016/j.rse.2021.112603

Casella E et al (2017) Accuracy assessment of UAV-derived orthomosaics and surface models in coastal environments. Remote Sens 9(1):20. https://doi.org/10.3390/rs9010020

Castelle B et al (2021c) Coast sat: a google earth engine-enabled python toolkit to extract shorelines from publicly available satellite imagery. Environ Model Softw 139:105041. https://doi.org/10.1016/j.envsoft.2021.105041

Castelle et al (2021d) Competing morphodynamic patterns at the coast: beyond simple cause-effect paradigms. Earth Sci Rev 221:103777

Chen C et al (2019) Global land surface greening due to rising CO_2 concentrations. Nat Sustainability 2:122–129. https://doi.org/10.1038/s41893-019-0220-7

Chen THK, Prishchepov AV, Fensholt R, Sabel CE (2019) Detecting and monitoring long-term landslides in urbanized areas with nighttime light data and multiseasonal Landsat imagery across Taiwan from 1998 to 2017. Remote Sens Environ 225:317–327

Dahy B, Al-Memari M, Al-Gergawi A, Burt JA (2024) Remote sensing of 50 years of coastal urbanization and environmental change in the Arabian Gulf: a systematic review. Front Remote Sens 5:1422910. https://doi.org/10.3389/frsen.2024.1422910

Di K, Ma R, Li R (2003) Geometric comparison and evaluation of IKONOS, QuickBird, and OrbView-3 high-resolution satellite imagery. Photogramm Eng Remote Sens 69(5):469–479. https://doi.org/10.14358/PERS.69.5.469

Duke NC et al (2017) Large-scale dieback of mangroves in Australia. Mar Freshw Res 68(10):1816–1829

Duo J et al (2021) High-accuracy co-registration of UAV-LiDAR and photogrammetric data. Drones 9(1):49. https://doi.org/10.3390/drones9010049

Fairley I et al (2021) Integrating UAV and satellite observations for storm impact analysis. Remote Sens 13(4):687. https://doi.org/10.3390/rs13040687

Gens R (2010) Remote sensing of coastlines: detection, extraction and monitoring. Int J Remote Sens 31(7):1819–1836

Gracia V, Dietrich MM, Sierra JP, Valero F, Espanya A, Mösso C, Sánchez-Arcilla A. (2025) Using UAVs to monitor the evolution of restored coastal dunes. Remote Sens 17(19):3263. https://doi.org/10.3390/rs17193263

Gonçalves JA, Henriques R (2015) UAV photogrammetry for topographic monitoring of coastal areas. ISPRS J Photogramm Remote Sens 104:101–111. https://doi.org/10.1016/j.isprsjprs.2015.02.009

Gorelick N, Hancher M, Dixon M, Ilyushchenko S, Thau D, Moore R (2017) Google earth engine: planetary-scale geospatial analysis for everyone. Remote Sens Environ 202:18–27. https://doi.org/10.1016/j.rse.2017.06.031

Goto K et al (2011) Tsunami-induced boulder transport and coastal erosion in Tohoku. Japan Marine Geology 290(1–4):1–9

Hamilton SE, Casey D (2016) Creation of a high spatio-temporal resolution global database of continuous mangrove forest cover for the 21st century (CGMFC-21). Glob Ecol Biogeogr 25(6):729–738. https://doi.org/10.1111/geb.12449

Harley MD, Turner IL, Kinsela MA et al. (2017) Extreme coastal erosion enhanced by anomalous extratropical storm wave direction. Sci Rep 7:6033. https://doi.org/10.1038/s41598-017-05792-1

Higgins SA et al (2018) Human and climate impacts on sediment delivery to deltas. Glob Planet Chang 159:29–46. https://doi.org/10.1016/j.gloplacha.2017.10.007

Jha SK, Nayak S, Roy PS (2012) Long-term shoreline changes along the Indian coast. J Coast Res 28(5):1142–1146. https://doi.org/10.2112/JCOASTRES-D-10-00129.1

Kaehler CA et al (2022) Terrestrial laser scanning for monitoring coastal erosion and beach morphology: a systematic review. Remote Sens Environ 267:112745. https://doi.org/10.1016/j.rse.2021.112745

Kasereka MM et al (2010) Application of GIS and remote sensing techniques in monitoring urban sprawl in a third world city: a case study of Kisangani, DR Congo. J Sustain Dev Afr 12(3):193–207

Kannan R, Prabhu R, Ramalingam M, Chandrasekar N (2016) Assessment of shoreline changes and coastal vulnerability using geospatial techniques for the Tamil Nadu coast. India Nat Hazards 84(1):417–438. https://doi.org/10.1007/s11069-016-2429-z

Kuleli T et al (2011) Automatic detection of shoreline change on coastal Ramsar wetlands of Turkey. Ocean Eng 38(10):1141–1149

Lee JR et al (2022) Remote beach nourishment assessment using UAVs. Coast Eng J 64(2):205–221. https://doi.org/10.1080/21664250.2022.2032340

Liu X et al (2020) High-resolution urban expansion dynamics from Landsat data using Google earth engine. Remote Sens Environ 240:111691. https://doi.org/10.1016/j.rse.2020.111691

Lin Y-C, Cheng Y-T, Zhou T, Ravi R, Hasheminasab SM, Flatt JE, Troy C, Habib A. (2019) Evaluation of UAV LiDAR for mapping coastal environments. Remote Sens 11(24):2893. https://doi.org/10.3390/rs11242893

Liu Q, Yuan F, Adams C, Rebelo L-M, Wellington M (2024) Dynamic coastal mapping using sentinel-1 and sentinel-2 data through digital earth africa. Int Arch Photogramm Remote

Sens Spatial Inf Sci, XLVIII-4-2024:325–330. https://doi.org/10.5194/isprs-archives-XLVIII-4-2024-325-2024
Luijendijk A, Hagenaars G, Baart F, Donchyts G, Aarninkhof S (2018a) The state of the world's beaches. Sci Rep 8(1):6641. https://doi.org/10.1038/s41598-018-24630-6
Luijendijk A et al (2018b) The state of the world's beaches. Sci Rep 8:6641
Mazur-Marzec H, Grabowski Ł, Węgrzyn A, Błaszczyk A, Cegłowska M, Dąbek P, Farooq M, Górecka E, Jurczak-Kurek A, Kaczorowska AK, Kaczorowski T, Kataržytė M, Konkel R, Kotlarska E, Overlingė D, Surosz W, Toruńska-Sitarz A, Walat S, Zielenkiewicz M, Węgrzyn G. (2025 Oct 28) Unlocking the biotechnological potential of Baltic microorganisms. Front Microbiol 16:1682611. https://doi.org/10.3389/fmicb.2025.1682611. PMID: 41229690; PMCID: PMC12602537
Mentaschi L, Vousdoukas MI, Pekel JF, Voukouvalas E, Feyen L (2018) Global long-term observations of coastal erosion and accretion. Sci Rep 8(1):12876. https://doi.org/10.1038/s41598-018-30904-w
Melnyk S, Harris BS (2026) Droning on: neoichnology in the age of UAV photogrammetry. Ichnos 33(1):1–9. https://doi.org/10.1080/10420940.2025.2609269
Miranda P, Leão ZMAN, Kikuchi RKP (2019) Evaluation of coastal protection structures using satellite images and digital shoreline analysis. Ocean Coast Manag 174:18–27. https://doi.org/10.1016/j.ocecoaman.2019.03.006
Mumby PJ et al (2004) Remote sensing of coral reefs and their physical environment. Mar Pollut Bull 48(3–4):219–228
Nayak SR (2002) Application of remote sensing to coastal zone management in India. Indian J Mar Sci 31(1):1–10
Pardo-Pascual JE, Almonacid-Caballer J, Balaguer-Puig M, Palomar-Vázquez J (2012) Automatic extraction of shorelines from Landsat TM and ETM+ multi-temporal images with subpixel precision. Remote Sens Environ 123:1–11. https://doi.org/10.1016/j.rse.2012.02.024
Pekel JF et al (2016) High-resolution mapping of global surface water and its long-term changes. Nature 540:418–422. https://doi.org/10.1038/nature20584
Pilartes-Congo J et al (2025) High-frequency shoreline monitoring using stereo photogrammetry. https://arxiv.org/abs/2504.06464
Ranasinghe R (2016) Assessing climate change impacts on open sandy coasts: a review. Earth Sci Rev 160:320–332. https://doi.org/10.1016/j.earscirev.2016.07.011
Ribas F, Simarro G, Arriaga J, Luque P. (2020) Automatic shoreline detection from video images by combining information from different methods. Remote Sens 12(22):3717. https://doi.org/10.3390/rs12223717
Robin N et al (2020) Long-term coastal dune evolution monitored by LiDAR and UAV. J Coast Res 95(sp1):160–169. https://doi.org/10.2112/SI95-032.1
Roy DP et al (2014) Landsat-8: science and product vision for terrestrial global change research. Remote Sens Environ 145:154–172
Talavera A et al (2020) Monitoring beach morphology using low-cost UAVs and photogrammetry: a case study in the mediterranean. Remote Sens 12(8):1314. https://doi.org/10.3390/rs12081314
Tellman B et al (2021) Satellite imaging reveals increased proportion of population exposed to floods. Nature 596:80–86. https://doi.org/10.1038/s41586-021-03695-w
Tu M-C, Smith P, Filippi AM (2018) Hybrid forward-selection method-based water-quality estimation via combining Landsat TM, ETM+, and OLI/TIRS images and ancillary environmental data. PLoS ONE 13(7): e0201255. https://doi.org/10.1371/journal.pone.0201255
Thampanya U, Vermaat JE, Sinsakul S, Panapitukkul N (2006) Coastal erosion and mangrove progradation of southern Thailand. Estuar Coast Shelf Sci 68(1–2):75–85. https://doi.org/10.1016/j.ecss.2006.01.011
Vecchi M et al (2021) Evaluation of UAV RTK and GNSS data for coastal monitoring. J Mar Sci Eng 9(11):1194. https://doi.org/10.3390/jmse9111194

Vos K, Splinter KD, Harley MD, Simmons JA, Turner IL (2019a) CoastSat: a google earth engine-enabled python toolkit to extract shorelines from publicly available satellite imagery. Remote Sens 11(17):2045

Vousdoukas MI, Mentaschi L, Hinkel J, Ward PJ, Feyen L (2020a) Assessing the effect of anthropogenic climate change on long-term coastal erosion. Nat Clim Chang 10(9):859–863. https://doi.org/10.1038/s41558-020-0697-0

Vousdoukas MI et al (2020b) Sandy coastlines under threat of erosion. Nat Clim Chang 10:260–263

Vieira da Silva GM, Martínez ML, Yoskowitz D (2016) Satellite-derived monitoring of coastal erosion at multiple spatial and temporal scales. Remote Sens Environ 180:15–27. https://doi.org/10.1016/j.rse.2016.02.031

Vos K, Harley MD, Splinter KD, Simmons JA, Turner IL (2019c) Sub-annual to multi-decadal shoreline variability from publicly available satellite imagery. Coast Eng 150:160–174. https://doi.org/10.1016/j.coastaleng.2019.04.004

Vos K et al (2019b) The global variability of beach seasonality. Nat Commun 10:1–8. https://doi.org/10.1038/s41467-019-10754-5

Wang XH et al (2014) Coastal wetland loss and its impact on ecosystem services in Bohai Bay, China. Estuar Coast Shelf Sci 139:1–9

Wulder MA et al (2019) Current status of Landsat program, science, and applications. Remote Sens Environ 225:127–147

Chapter 4
Geospatial Analysis of Coastal Erosion and Accretion Trends at Chandipur Coast, India Using Digital Shoreline Analysis System (DSAS)

Rupsa Mondal, Abhisek Santra, Ananya Sarkar, Shaikh Abu Sufiyan Ali, Fathma Kamalludin Mallik, Archisman Saha, Sinchan Banerjee, Suman Chatterjee, Manisha Baral, Supratim Karmakar, Suman Rout, and Debasmita Mukherjee

4.1 Introduction

The shoreline marks the boundary where land meets the ocean (Kabir et al. 2020), and its location is constantly adjusted by the physical forces and human land use/cover (LULC) changes (Ahmed 2011; Carter and Woodroffe 1994; Islam et al. 2021). LULC transformations are intricately linked to shifts in shoreline positions, making it critical to map and monitor these changes with precision (Anselin and Griffith 1988). Shoreline change is a highly dynamic and multifactorial geophysical process influenced by a combination of natural forces, internal environmental dynamics, and human interventions (Thoai et al. 2019). Internal factors such as wave action, tidal inundation, near-shore currents, and sediment budget fluctuations contribute significantly to the shifting of shoreline positions (Williams et al. 2018). Simultaneously, external forces such as storm surges, coastal flooding, intense rainfall, and catastrophic events like tsunamis add episodic stress to the coastal landscape (Chowdhury et al. 2022; Zzaman et al. 2021). These natural drivers are increasingly compounded by human activities, including dredging, deforestation, slope destabilization, port construction, and infrastructure development, which often accelerate

R. Mondal · A. Santra (✉) · A. Sarkar · S. A. S. Ali · F. K. Mallik · A. Saha · S. Banerjee · M. Baral · S. Karmakar · S. Rout · D. Mukherjee
Department of Geography, School of Basic and Applied Sciences (SoBAS), Adamas University, Barasat, Kolkata, West Bengal, India
e-mail: abhisek1.santra@adamasuniversity.ac.in; avisek.santra@gmail.com

S. Chatterjee
Department of Geography, Loreto College, Kolkata, India

A. Nath et al. (eds.), *Geospatial Application of Coastal Hazards*, Springer Water,
https://doi.org/10.1007/978-3-032-20076-1_4

shoreline retreat and, in many cases, lead to inland migration (Islam et al. 2023; Kuleli et al. 2011). The intricate interplay between natural geophysical forces and anthropogenic interventions further complicates the process of shoreline movement. For instance, coastal protection structures and complex infrastructure may serve immediate needs but often alter sediment dynamics, leading to imbalanced erosion and deposition patterns..

Monitoring shoreline movement over time has become essential in understanding these complex processes and their implications for coastal management (Jana et al. 2017; Jana et al. 2016a; Jana et al. 2014). Accurate temporal assessment is critical because the shoreline characterizes a transition zone between land and water, a zone continuously shaped by climatic variability, hydrodynamic conditions, and human interventions (Jana et al. 2016b; Kale et al. 2019). Coastal change processes are driven by geomorphic, hydrological, tectonic, and atmospheric influences, including sea-level fluctuations, longshore sediment transport, tidal currents, storm surges, erosion-deposition cycles, and land subsidence (Jana et al. 2012; Jana et al. 2016b; Nassar et al. 2019). At the same time, growing urbanization, particularly in ecologically sensitive coastal belts such as Brazil and South Asia, has triggered the rapid development of tourism infrastructure, housing, and real estate projects, significantly altering natural shoreline stability (Santos et al. 2021). Consequently, both short and long-term monitoring is critical for recognizing evolving spatial patterns and managing coastal risk (Baral et al. 2018; Roy et al. 2018).

In this framework, the advancement of geoinformatics has revolutionized the field of shoreline monitoring. These technologies offer multiple advantages over traditional field-based approaches, including broader spatial coverage, time efficiency, cost effectiveness, and the ability to provide consistent and unbiased data (Esmail et al. 2019; Santos et al. 2021). Semi-automated methods are increasingly used to delineate shorelines and analyze their temporal dynamics, allowing researchers to assess rates of change and underlying patterns more effectively (Yamano et al. 2006). Furthermore, advanced spatial analysis has enabled the abstraction of physiographic and bathymetric datasets, as well as the development of integrated coastal assessments (Emam and Soliman 2020; Gibbs and Richmond 2017; Joseph et al. 2016; Lafon et al. 2002; Saravanan et al. 2019). These methods are widely used to quantify erosion and accretion, enhancing understanding of shoreline dynamics across coastal environments (Jana et al. 2017; Jana et al. 2016a; Jana et al. 2014; Ahmed et al. 2021; Hwang et al. 2014; Natesan et al. 2015).

Historically, shoreline change detection relied on ground surveys, paper maps, and aerial photographs (Hapke et al. 2006). Although accurate, these methods were time-consuming and limited in spatial coverage. Technological innovations, since introduction, have more precise tools such as the GPS & GNSS, LiDAR and Drone sensors, which offer high-resolution shoreline positioning (Tak et al. 2020). Despite their high accuracy, these tools often come with high costs, limited accessibility, and need for trained personnel (Moussaid et al. 2015). This has made satellite remote sensing an increasingly attractive alternative. These images offer extensive temporal coverage, the ability to monitor inaccessible regions, and the availability of free data sources like the Landsat programme (Kuenzer et al. 2014; Toure et al. 2019). Though satellite data may not match the spatial resolution of LiDAR and UAVs, it enables repeatable, wide-scale shoreline mapping, especially when integrated with GIS.

A robust shoreline monitoring framework typically includes satellite-based shoreline digitization and rate-of-change analysis using GIS tools (Ford 2013). One widely adopted ArcGIS extension in this domain is the "Digital Shoreline Analysis System (DSAS)", established by the USGS (Himmelstoss et al. 2018). It has been successfully applied in various case studies worldwide to analyze temporal shoreline shifts. Studies by Matin and Hasan (2021), Sharma and Patnaik (2022), and Qiao et al. (2018) have applied DSAS in combination with satellite imagery to debt significant long-term shoreline changes. These findings provide critical insights for coastal planners, disaster managers, and environmental policymakers. They provide a scientific basis for preparing coastal vulnerability assessments, formulating zoning regulations, and designing coastal adaptation and protection strategies (Ngowo et al. 2021). Therefore, in this research, the shoreline change was assessed along the part of the Chandipur coast using DSAS tool considering the bi-temporal LISS-III images of 2014 and 2024. The study aimed to estimate the shoreline using a semi-automated method and assess erosion and accretion rates along the Chandipur coast. A "decision matrix" was developed to classify the coastal area into four sections, allowing policymakers to monitor the urgency of protection interventions. As climate change continues to intensify, sea level rise and extreme weather events will only grow in importance, underscoring the need for accurate shoreline monitoring. The integration of geospatial technologies into coastal zone management ensures not only more precise and timely data but also supports informed decision-making for shoreline development.

4.2 Study Area

The study site was considered for the current research, a part of Chandipur coastal stretch within the Balasore district of Orissa, India (Fig. 4.1). The coast represents physiographically diverse and environmentally sensitive stretch along the eastern Indian coastline. The area is characterized as a tidal or marshy flat, forming part of a dynamic intertidal zone, positioned above the "high-water line", and the subtidal zone, lying below the low-water line, are part of this coastal continuum. However, no clear differentiation exists between the supratidal and "intertidal areas" (Chakrabarti 1991).

A prominent feature of the Chandipur coast is its wide tidal flat, stretching approximately 3.8 Km in width, fronted by a thin sandy seashore in the northern part. The coast is backed by "aeolian dunes", which transition inland into areas of recent alluvium depositions. To the east, the Buribalam River forms a key fluvial input, shaping an emerged terrace interspersed with marshes. The southern segment of the sandy area, beyond the "tidal flat", includes the "Balasore shelf", which exhibits a two-dimensional shoreface. Accelerated coastal erosion has been reported in this region, primarily due to sea encroachment and mismatches between wave and landform

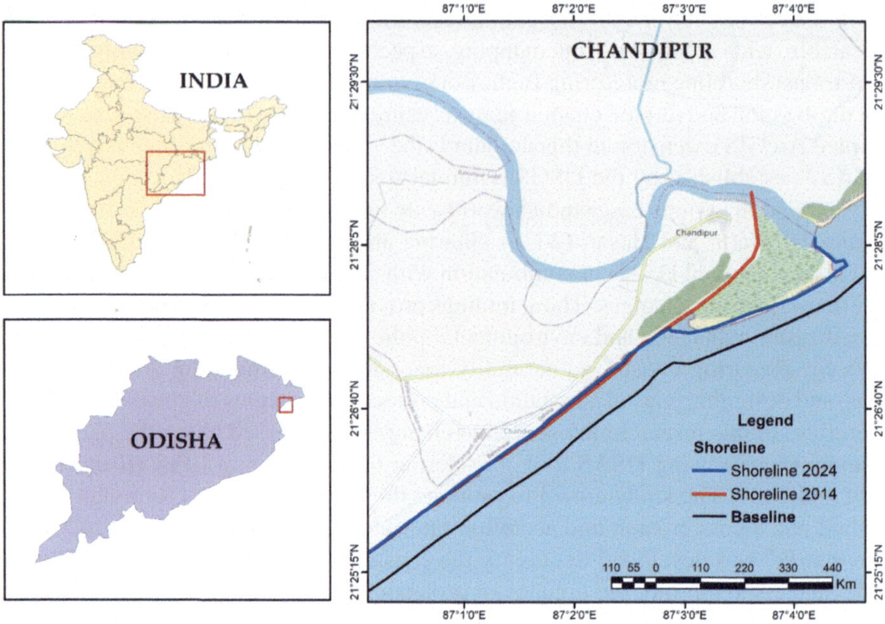

Fig. 4.1 Study area

processes, resulting in geomorphic disequilibrium (Mukhopadhyay et al. 2011). The shoreline trends NE-SW, aligned obliquely to prevailing wave directions, further contributing to sediment redistribution.

From a geological perspective, the coastal plain comprises "semi-consolidated to unconsolidated Quaternary sediments and recent alluvium", including fluvial silt, sand, clay, iron nodules, pebbles, marine clay, and calcareous concretions. These represent both fluvial and fluvio-marine depositional profiles. The area falls within a tropical monsoon climatic zone, with a mean annual temperature of 27 °C, ranging from 13 °C in winter to 37 °C in summer. The average monthly rainfall is 120 mm, concentrated predominantly throughout the monsoon season (June to October) (Mukhopadhyay et al. 2011).

Hydrologically, the coast is influenced by a "semi-diurnal tidal cycle" with a tidal range between 1.87 m (neap tide) and 4.89 m (spring tide) (Sarkar et al. 1991), and storm gushes rarely surpassing 1 m (Mukherjee et al. 1987). The low-energy, micro- to meso-tidal environment fosters the development of barrier bars, shore-parallel ridges, and tidal channels. The Buribalam estuary, located at the northeastern margin, is separated from the tidal flat by a spit-transformed barrier bar, producing a quite swampy hinterland (Saha et al. 2023). Sediments from the river, reworked by marine forces, create minor swales and dune systems that actively migrate inland, forming a complex mosaic of depositional landforms. Overall, the Chandipur coast exemplifies a transitional zone between fluvial, tidal, and marine influences, undergoing rapid morphological changes due to both physical processes and human-caused pressures, making it a critical region for coastal geomorphological and vulnerability studies.

4.3 Database

The shoreline change analysis of the Chandipur coast was assessed based on bi-temporal cloud-free LISS-III data of the years 2014 and 2024 for the month of August. The images contain four bands, starting from green to SWIR, and were acquired from the Bhoonidhi portal. The reason behind selecting the month was for validation purpose. Field collected GPS data was applied to validate the extracted shoreline of the year 2024. Atmospheric correction was done using the FLAASH module of the ENVI software. The correction algorithm was based on the "MODTRAN 4 radiative transfer model" (Santra et al. 2021; Santra et al. 2019). The corrected images were co-registered with "WGS 84 UTM Zone 45" projection system.

4.4 Methodology

4.4.1 Shoreline Extraction

Reliable shoreline mapping is crucial for informed policy-making, effective infrastructure planning, and effective environmental protection, particularly in mitigating risks associated with shoreline retreat. Various methods are used for detecting coastline positions, including traditional coastal maps, historical photographs, ground surveys, GPS-based field measurements, aerial imagery, and satellite data processed through digital image classification or elevation modelling (Dolan et al. 1980; Islam et al. 2023). Although conventional field surveys yield high precision, they are time-intensive and impractical for monitoring large or inaccessible coastal regions. On the other hand, modern technologies such as LiDAR and coastal video imaging systems improve spatial accuracy and temporal coverage, but they are financially demanding (Turner et al. 2016). Using multi-temporal satellite images, Mitra et al. (2017) applied a combination of spectral indices, ISODATA classification, thresholding, and ISH transformation techniques to identify the shorelines. Following the guidelines provided by different researchers, in this study shorelines of 2014 and 2024 were extracted using "Modified Normalized Water Index (MNDWI)" (1) (Xu 2006).

$$MNDWI = \frac{\rho_G - \rho_{SWIR}}{\rho_G + \rho_{SWIR}} \tag{1}$$

where ρ_G and ρ_{SWIR} are reflectance values in green and short-wave infrared bands respectively.

The MNDWI values range from -1 to $+1$, with positive values indicating water bodies and negative values showing land surface. To delineate the coastline, only land and water elements are needed, enabling a binary classification (0 for water and 1 for land) (Ji et al. 2009). The shoreline of 2024 was validated using the field-collected ground truths during satellite pass.

4.4.2 Shoreline Change Analysis Using DSAS

The "Digital Shoreline Analysis System (DSAS)", developed by the "United States Geological Survey (USGS)", is a powerful open-source tool widely adopted for analyzing and monitoring shoreline change. Designed as a user-friendly extension within ArcGIS, DSAS facilitates the assessment of long-term coastal dynamics by leveraging geospatial data (Oyedotun 2014). Developed with support from Environmental Systems Research Institute (ESRI), DSAS calculates shoreline change statistics by comparing multiple shoreline positions captured across different time periods using satellite images. The standard workflow involves defining a baseline, generating transects perpendicular to this baseline, and calculating change rates, such as loss or deposition, based on how the coastline shifts over time (Raj et al. 2020). In the present research, transects were generated from the baseline to the shorelines of 2014 and 2024. These transects intersect the shoreline at various points, enabling detailed spatial analysis of coastal dynamics (Islam et al. 2023). Given the growing importance of coastal zone management, monitoring changes has become an area of sustained scientific attention, with several studies focusing on the spatial and temporal variability of shorelines (Ottman et al. 2019; Samanta and Paul 2016; Vickers 2017).

Among the tools available for analyzing shoreline displacement, DSAS is regarded as particularly effective due to its ability to accurately track changes in shoreline position across multiple time frames (Raj et al. 2020). The "Federal Emergency Management Agency (FEMA)" has endorsed the use of transects projected from a baseline as a standard method for assessing long-term coastal change and erosion in vulnerable zones (Crowell et al. 1997). A baseline is a reference line aligned with the general direction of past shorelines (Jackson et al. 2012; Topah Apandi et al. 2022). Based on user needs, it can be a linear or curved line placed landward or seaward of past shorelines (Gómez-Pazo et al. 2022). Shoreline change estimates can vary depending on the transect orientation along the baseline (Jackson et al. 2012; Khallaghi and Pontius Jr. 2022). In this study, a manually digitized baseline was created as a new feature class, with erosion (negative) and accretion (positive) defined by its orientation, placed landward, seaward, or along the shoreline center (Himmelstoss et al. 2018). Here, the seaward baseline was constructed. After digitizing the baseline and extracting shorelines, some attribute fields such as "OBJECTID, SHAPE, SHAPE LENGTH, DATE, UNCERTAINTY", etc. were created for calculation in the DSAS tool. These attributes informed DSAS about transect order and baseline position, aiding accurate rate-of-change calculations. The negative values denote erosion, while the positive values highlight accretion.

DSAS employs a series of pre-defined algorithms to quantify shoreline change rates by analyzing multiple time-series shorelines along transects (Thieler et al. 2009). In this study, two shoreline change statistical parameters were applied. The "Net Shoreline Movement (NSM)" was used to measure the total distance of shoreline movement over the entire study period. "End Point Rate (EPR)"—a rate-based statistic was used to compute the coastline variation rates (Awad and El-Sayed 2021; Thieler et al. 2009). The NSM parameter estimates the linear gap between the initial and the current coastline positions along each transect drawn perpendicular to the coast (2) (Himmelstoss et al. 2018).

$$NSM = (d_{2024} - d_{2014}) \tag{2}$$

where, d_{2024} and d_{2014} are displacements in 2024 and 2014, respectively.

This measurement reflects the cumulative displacement of the shoreline but does not account for the rate of change over time. The EPR is calculated by dividing the shoreline movement distance (between the earliest and latest shoreline data) by the time elapsed between those two observations (3) (Dolan et al. 1991; Himmelstoss et al. 2018; Santra et al. 2011).

$$EPR = \frac{d_{2024} - d_{2014}}{t_{2024} - t_{2014}} \tag{3}$$

where d_{2024} and d_{2014} are the shoreline displacements in 2024 and 2014 respectively and t_{2024} and t_{2014} are the time frames.

Unlike NSM, EPR provides a temporal rate of coastline change, typically expressed in meters per year. It is widely used due to its simplicity and applicability when only two shoreline positions are available (Ozturk and Sesli 2015). Finally, DSAS-generated change statistics were linked to the transects, and NSM data were extracted. The resulting output included intersection coordinates and shoreline distances relative to the baseline, which were exported into Microsoft Excel for further analysis. The total extent of erosion and accretion was quantified through graphical and tabular interpretation, which was subsequently transformed into spatial maps.

4.5 Results and Discussion

4.5.1 Decadal Change along Chandipur Coast

The present research presents the decadal change of shoreline along the Chandipur coast from 2014 to 2024 using EPR and NSM models. The EPR values across the coast exhibit considerable spatial variability, reflecting a combination of erosion and accretion zones. Figure 4.2 classifies shoreline segments into nine classes ranging from severe erosion (<= −5.1 m/year) to very high accretion (> 3.0 m/year). Zone 'a' located in the SW part of the study area, is predominantly marked by erosional trends, with change rates from −5.1 m/year to −0.5 m/year. This consistent erosion can be attributed to wave-induced sediment transport, a reduced sediment supply, and the lack of natural or artificial coastal buffers. On the other hand, zone 'b' is located northeast of the estuarine outlet of Buribalam, showing a contrasting accretional pattern, with accretion rates ranging from +0.5 m/year to > + 3.0 m/year. The presence of green-shaded transects indicates stable and rapidly advancing shorelines, possibly resulting from sediment deposition by tidal inflow or anthropogenic modifications such as beach nourishment or port-related sediment trapping. This clear spatial demarcation suggests differential coastal processes operating along the two segments of the coast. These processes may be influenced by local hydrodynamic conditions, land use patterns, and geomorphological features.

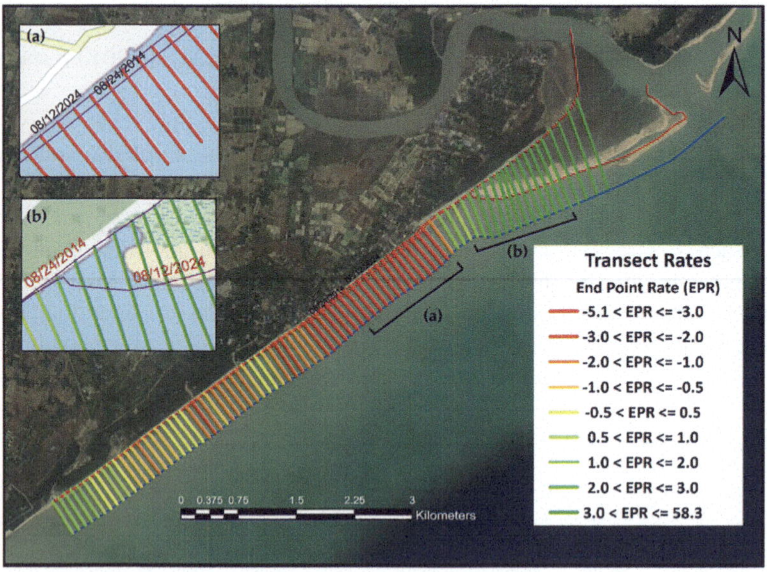

Fig. 4.2 Shoreline change rate using EPR

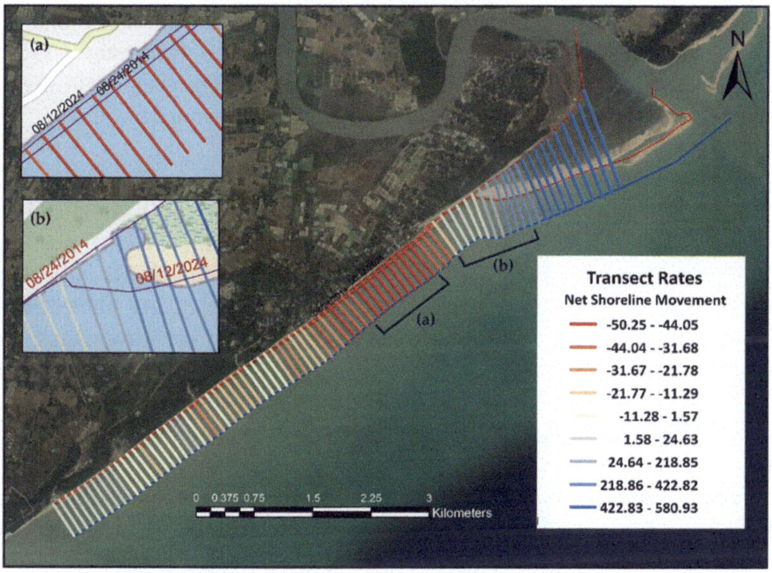

Fig. 4.3 Net shoreline movement along Chandipur coast

The NSM values—representing linear distance of shoreline change between 2014 and 2024—further corroborate the trends seen in EPR results (Fig. 4.3). The transects are colored based on absolute change values, ranging from −50.25 m (erosion) to + 580.93 m (accretion). In zone ‘a’, most of the net change values are negative, with extensive sections falling within the −50.25 m to −11.29 m range. These values signify substantial landward migration of the shoreline. The erosional hotspots coincide with the areas of high negative EPR values, reinforcing the pattern of progressive coastal retreat. Conversely, zone ‘b’ displays highly positive net shoreline shift values, particularly in the +218.86 m to +580.93 m range, indicating intense shoreline advancement. This region corresponds to high positive EPR values, suggesting either natural sediment accretion or successful human intervention. The high positive net shoreline changes values in this segment highlight the potential for significant land gain in less wave-exposed environments or estuarine zones. A transition zone between segments ‘a’ and ‘b’ exhibits mixed signals, with NSM values near zero (−11.28 m to +1.57 m), suggesting relative shoreline stability.

4.5.2 Statistical Summary of Shoreline Changes

The analysis using NSM and EPR reveals that a distinct spatial heterogeneity exists in shoreline behavior. From Table 4.1, out of 86 transects erosion was observed in 58.1% of transects, while accretion occurred in 41.9%. Total eroded area was observed 5 sq. Km, while accretion accounted for 3.6 sq. Km. The NSM values range from −50.2 m to +581 m, with a mean of 78.3 m, indicating net land gain. EPR values range from −5.04 m/year to +58.3 m/year, averaging 7.89 m/year, again suggesting dominant accretional trends, although localized erosion is more frequent.

The coast is divided into four segments (sections I to IV) (Fig. 4.4), which correspond spatially to transect-based segments A-D (Fig. 4.5) in the EPR and NSM plots. In the southwestern segment within transects 1–12, the NSM varies between −6.86 m to +17.3 m with an average of +7.49 m. EPR varies from −0.69 m/year to 1.74 m/year, averaging +0.75 m/year (Table 4.2). Some transects indicate slight accretion (Fig. 4.5). This clearly implies that this zone is relatively stable, like influenced by balanced sediment dynamics or shielding from direct wave attack. In sect. II/B, within transects 13 to 24, NSM ranges from −21.8 m to +3.71 m, averaging −7.37 m. EPR ranges here from −2.18 m/year to +0.37 m/year with a mean of −0.74 m/year. The majority of the transects in this segment show slight to moderate erosion.

This area exhibits incipient retreat, possibly due to reduced sediment supply, longshore drift, or minor geomorphic changes. NSM varies between −23.6 m and + 1.57 m with an average − 11.2 m in sect. III/C in between transects 25 to 48. EPR ranges here from −2.37 m/year to +0.16 m/year, averaging −1.13 m/year. Therefore, all transects in this region show clear erosional trends, with moderate retreat distances. This is a critical erosion zone, possibly influenced by estuarine sediment trapping, wave dynamics, or human interventions such as upstream construction or embankment redirection. Finally, in sect. IV/D in the northeastern section, between transects 49 and 85, NSM reaches a maximum of +581 m, with an average of + 140 m. EPR peaks at +58.3 m/year, with an average of +14.1 m/year. This shows a dramatic and consistent increase in both NSM and EPR, indicating rapid accretion. This segment is experiencing significant shoreline advancement, likely due to sediment input from estuarine mouths, river systems, or engineered modifications, such as breakwaters. However, localized erosion (minimum NSM −50.2 m) persists in selected transects, reflecting complex coastal dynamics.

The statistical overview and sectional transect analysis collectively point toward a bipolar shoreline behavior. Sections I to III, i.e., from southwest to central Chandipur, represent low to moderate erosion zones with varying intensities. Particularly, section III is the most vulnerable, with sustained landward shoreline retreat. Section IV, by contrast, is undergoing massive shoreline accretion, both in rate and extent. Its average NSM (140 m) and EPR (14.1 m/year) are notably higher than those of the remaining parts of the coast (Table 4.3). These patterns are indicative of localized forcing mechanisms, such as hydrodynamic influences like tidal

Table 4.1 Descriptive statistics of the changes in shoreline (2014–2024)

Parameter	Mean	Median	Min	Max	SD
NSM	78.3	−6.41	−50.2	581	192
EPR	7.89	−0.645	−5.04	58.3	19.4

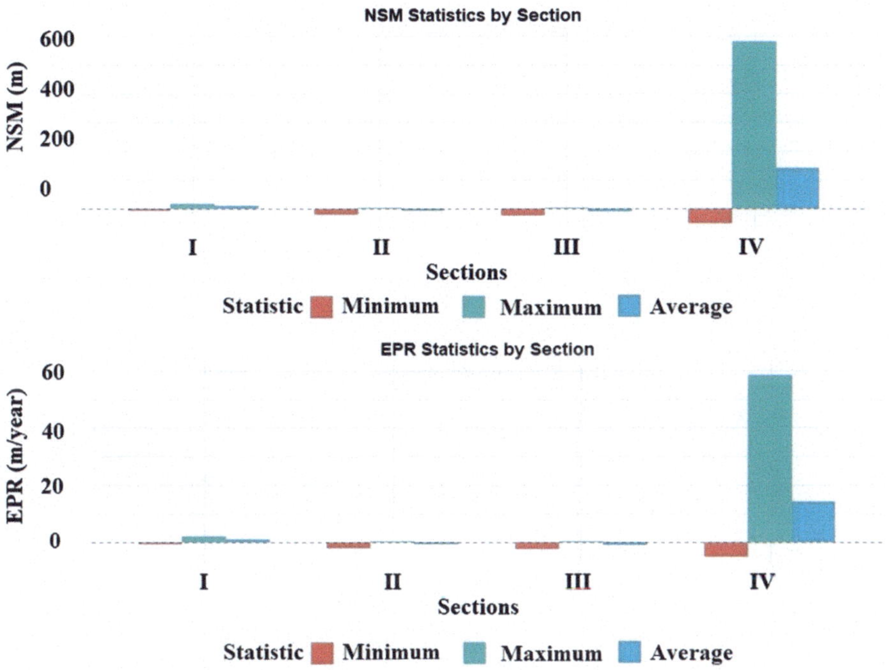

Fig. 4.4 Sectional shoreline shift statistics

influx. Estuarine plume sedimentation, and longshore sediment transport; anthropogenic impacts including port construction, embankments, and dredging activities; and natural factors like riverine discharge, monsoon sedimentation, and shoreline geomorphology.

The results highlight the importance of location-specific coastal management strategies. Section III requires urgent erosion control measures, soft engineering (e.g., beach nourishment, dune restoration) or hybrid interventions. Sections I and II should be monitored regularly to prevent further degradation, section IV, while experiencing accretion, must be evaluated for potential over-development, shoreline instability, or sediment redistribution impacts downstream. Moreover, the dominant erosion frequency (58.1%) despite overall net accretion (mean NSM 78.3 m) reveals the importance of combining rate-based (EPR) and distance-based (NSM) indicators to inform both short-term planning and long-term resilience strategies.

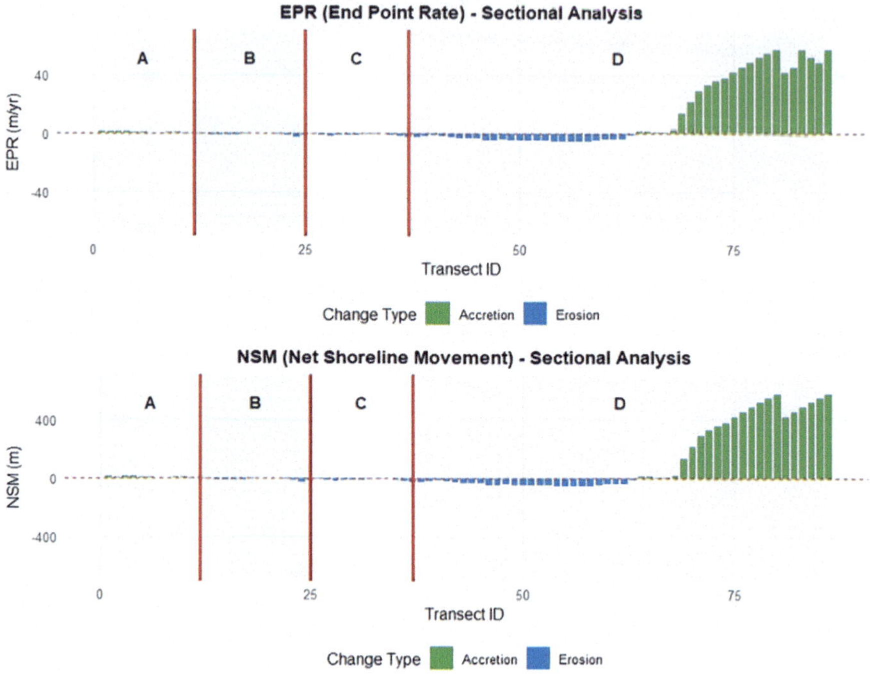

Fig. 4.5 Transect-wise sectional statistics

Table 4.2 Erosion and accretion statistics of the shoreline over the years (2014–2024)

Parameter	Accretion count	Erosion count	Accretion percent	Erosion percent	Erosion in Sq km	Accretion in Sq km
NSM	36	50	41.9	58.1	5	3.6
EPR	36	50	41.9	58.1	5	3.6

Table 4.3 Erosion and accretion statistics of the shoreline over the years (2014–2024)

Section	NSM min	NSM max	NSM average	EPR min	EPR max	EPR average
I	−6.86	17.3	7.49	−0.69	1.74	0.75
II	−21.8	3.71	−7.37	−2.18	0.37	−0.74
III	−23.6	1.57	−11.2	−2.37	0.16	−1.13
IV	−50.2	581	140	−5.04	58.3	14.1

4.6 Conclusion

The ten-year shoreline change assessment, integrating both EPR and NSM datasets, highlights an apparent dichotomy in coastal dynamics along the Chandipur coast. The southwestern stretch is undergoing consistent erosion, both in terms of rate and absolute magnitude, indicating a critical need for coastal protection measures, such as dune restoration, vegetative buffers, or complex engineering solutions. In contrast, the northwestern zone exhibits strong accretion, which could be either naturally beneficial or indicative of sediment deposition imbalances that may result in downstream erosion over time. Though DSAS-derived transect-level insights and statistical interpretation, it is evident that coastal interventions must be nuanced and location-specific. Particularly, erosion mitigation in central sectors and sediment balance assessment in the northeastern zone are critical for sustainable coastal zone management. This spatial disparity highlights the importance of localized management strategies over a uniform coastal policy. The study analyzed decadal shoreline change from 2014 to 2024 using LISS III images. It could be extended in the future to assess long-term historical change and prediction analysis. However, the results emphasize the utility of DSAS-based long-term shoreline monitoring in identifying vulnerable zones, supporting predictive modelling, and planning sustainable coastal interventions.

Acknowledgement The authors acknowledge ISRO and USGS for providing the data and model, respectively, for this research.

References

Ahmed A (2011) Some of the major environmental problems relating to land use changes in the coastal areas of Bangladesh. J Geogr RegNal Plan 4:1–8

Ahmed N, Howlader N, Hoque MA-A, Pradhan B (2021) Coastal erosion vulnerability assessment along the eastern coast of Bangladesh using geospatial techniques. Ocean Coast Manag 199:105408. https://doi.org/10.1016/j.ocecoaman.2020.105408

Anselin L, Griffith DA (1988) Do spatial effects really matter in regression analysis? Pap Reg Sci 65(1):11–34. https://doi.org/10.1111/j.1435-5597.1988.tb01155.x

Awad M, El-Sayed HM (2021) The analysis of shoreline change dynamics and future predictions using automated spatial techniques: case of El-OMAYED on the Mediterranean coast of Egypt. Ocean Coast Manag 205:105568. https://doi.org/10.1016/j.ocecoaman.2021.105568

Baral R, Pradhan S, Samal RN, Mishra SK (2018) Shoreline change analysis at chilika lagoon coast, India using digital shoreline analysis system. J Indian Soc Remote Sens 46(10):1637–1644. https://doi.org/10.1007/s12524-018-0818-7

Carter RWG, Woodroffe CD (eds) (1994) Coastal evolution: late quarter shoreline morph dynamics. Cambridge University Press

Chakrabarti P (1991) Morphostratigraphy of coastal quaternaries of West Bengal and Subarnarekha delta. Indian J Earth Sci 15:219–225

Chowdhury MA, Zzaman RU, Tarin NJ, Hossain MJ (2022) Spatial variability of climatic hazards in Bangladesh. Nat Hazards 110(3):2329–2351. https://doi.org/10.1007/s11069-021-05039-3

Crowell M, Douglas BC, Leatherman SP (1997) On forecasting future U.S. shoreline positions: a test of algorithms. J Coast Res 13(4):1245–1255

Dolan R, Fenster MS, Holme SJ (1991) Temporal analysis of shoreline recession and accretion. J Coast Res 7(3):723–744

Dolan R, Hayden BP, May P, May S (1980) Reliability of shoreline change measurements from aerial photographs. Shore Beach 48(4):22–29. https://www.scopus.com/inward/record.uri?eid=2-s2.0-0019068233&partnerID=40&md5=0f2e40791b5cbcb21240cb236cc360e4

Emam WWM, Soliman KM (2020) Quantitative analysis of shoreline dynamics along the Mediterranean coastal strip of Egypt. Case study: Marina El-Alamein resort. In: Elbeih SF, Negm AM, Kostianoy A (eds) Environmental remote sensing in Egypt. Springer International Publishing, pp 575–594. https://doi.org/10.1007/978-3-030-39593-3_18

Esmail M, Mahmod WE, Fath H (2019) Assessment and prediction of shoreline change using multi-temporal satellite images and statistics: case study of Damietta coast. Egypt Appl Ocean Res 82:274–282. https://doi.org/10.1016/j.apor.2018.11.009

Ford M (2013) Shoreline changes interpreted from multi-temporal aerial photographs and high resolution satellite images: wotje atoll, Marshall Islands. Remote Sens Environ 135:130–140. https://doi.org/10.1016/j.rse.2013.03.027

Gibbs AE, Richmond BM (2017) National Assessment of shoreline change—summary statistics for updated vector shorelines and associated shoreline change data for the north coast of Alaska, U.S.-Canadian border to icy cape. (Open-File Report 2017–1107)

Gómez-Pazo A, Payo A, Paz-Delgado MV, Delgadillo-Calzadilla MA (2022) Open digital shoreline analysis system: ODSAS v1.0. J Mar Sci Eng 10(1)

Hapke CJ, Reid D, Richmond BM, Ruggiero P, List J (2006) National Assessment of shoreline change part 3: historical shoreline change and associated coastal land loss along Sandy shorelines of the California coast (Open File Report 2006-1219)

Himmelstoss EA, Henderson RE, Kratzmann MG, Farris AS (2018) Digital shoreline analysis system (DSAS) version 5.1 user guide (Open-File Report 2021–1091)

Hwang CS, Choi CU, Choi JS (2014) Shoreline changes interpreted from multitemporal aerial photographs and high resolution satellite images. A case study in Jinha beach. Korean J Remote Sens 30(5):607–616. https://doi.org/10.7780/KJRS.2014.30.5.6

Islam MK, Fahad MNH, Chowdhury MA, Islam SLU (2023) Shoreline change rate estimation: impact on salt production in Kutubdia Island using multi-temporal satellite data and geo-statistics. Remote Sens Appl Soc Environ 30:100957. https://doi.org/10.1016/j.rsase.2023.100957

Islam MS, Uddin MA, Hossain MA (2021) Assessing the dynamics of land cover and shoreline changes of Nijhum Dwip (Island) of Bangladesh using remote sensing and GIS techniques. Reg Stud Mar Sci 41:101578. https://doi.org/10.1016/j.rsma.2020.101578

Jackson CW, Alexander CR, Bush DM (2012) Application of the AMBUR R package for spatio-temporal analysis of shoreline change: Jekyll Island, Georgia, USA. Comput Geosci 41:199–207. https://doi.org/10.1016/j.cageo.2011.08.009

Jana A, Maiti S, Biswas A (2017) Appraisal of long-term shoreline oscillations from a part of coastal zones of Sundarban Delta, eastern India–a study based on geospatial technology. Spat Inf Res 25:713–723. https://doi.org/10.1007/s41324-017-0139-x

Jana A, Maiti S, Biswas A (2016a) Analysis of short-term shoreline oscillations along midnapur-Balasore coast, bay of Bengal, India-a study based on geospatial technology. Model Earth Syst Environ 2(2):64. https://doi.org/10.1007/s40808-016-0117-7

Jana A, Biswas A, Maiti S, Bhattacharya AK (2014) Shoreline changes in response to sea level rise along Digha coast, eastern India: an analytical approach of remote sensing, GIS and statistical techniques. J Coast Conserv 18(3):145–155

Jana A, Maiti S, Biswas A (2016b) Seasonal change monitoring and mapping of coastal vegetation types along Midnapur-Balasore coast, bay of Bengal using multi-temporal Landsat data. Model Earth Syst Environ 2(1):7

Jana A, Sheena S, Biswas A (2012) Morphological change study of Ghoramara Island, eastern India using multi-temporal satellite data. Res J Recent Sci 1(10):72–81
Ji L, Zhang L, Wylie B (2009) Analysis of dynamic thresholds for the normalized difference water index. Photogramm Eng Remote Sens 11:1307–1317
Joseph MB, Preston DL, Johnson PTJ (2016) Integrating occupancy models and structural equation models to understand species occurrence. Ecology 97(3):765–775. https://doi.org/10.1890/15-0833.1
Kabir MA, Salauddin M, Hossain KT, Tanim IA, Saddam MMH, Ahmad AU (2020) Assessing the shoreline dynamics of Hatiya Island of Meghna estuary in Bangladesh using multiband satellite imageries and hydro-meteorological data. Reg Stud Mar Sci 35:101167. https://doi.org/10.1016/j.rsma.2020.101167
Kale MM, Ataol M, Tekkanat İS (2019) Assessment of shoreline alterations using a digital shoreline analysis system: a case study of changes in the Yeşilırmak Delta in northern Turkey from 1953 to 2017. Environ Monit Assess 191(6):398. https://doi.org/10.1007/s10661-019-7535-8
Khallaghi S, Pontius RG Jr (2022) Area method compared with transect method to measure shoreline movement. Geocarto Int 37(20):5963–5984. https://doi.org/10.1080/10106049.2021.1926556
Kuenzer C, Ottinger M, Liu G, Sun B, Baumhauer R, Dech S (2014) Earth observation-based coastal zone monitoring of the Yellow River Delta: dynamics in China's second largest oil producing region over four decades. Appl Geogr 55:92–107. https://doi.org/10.1016/j.apgeog.2014.08.015
Kuleli T, Guneroglu A, Karsli F, Dihkan M (2011) Automatic detection of shoreline change on coastal Ramsar wetlands of Turkey. Ocean Eng 38(10):1141–1149. https://doi.org/10.1016/j.oceaneng.2011.05.006
Lafon V, Froidefond JM, Lahet F, Castaing P (2002) SPOT shallow water bathymetry of a moderately turbid tidal inlet based on field measurements. Remote Sens Environ 81(1):136–148. https://doi.org/10.1016/S0034-4257(01)00340-6
Matin N, Hasan GMJ (2021) A quantitative analysis of shoreline changes along the coast of Bangladesh using remote sensing and GIS techniques. Catena 201:105185. https://doi.org/10.1016/j.catena.2021.105185
Mitra SS, Mitra D, Santra A (2017) Performance testing of selected automated coastline detection techniques applied on multispectral satellite imageries. Earth Sci Inf 10(3):321–330. https://doi.org/10.1007/s12145-017-0289-3
Moussaid J, Fora AA, Zourarah B, Maanan M, Maanan M (2015) Using automatic computation to analyze the rate of shoreline change on the Kenitra coast, Morocco. Ocean Eng 102:71–77. https://doi.org/10.1016/j.oceaneng.2015.04.044
Mukherjee KK, Das S, Chakrabarti A (1987) Common physical sedimentary structures in a beach-related Open Sea Siliciclastic tropical tidal flat at Chandipur, Orissa and evaluation of weather conditions through discriminate analysis. Senckenb Marit 19:261–293
Mukhopadhyay A, Mukherjee S, Hazra S, Mitra D (2011) Sea level rise and shoreline changes: a geoinformatic appraisal of Chandipur coast, Orissa. Int J Geol Earth Environ Sci 1(1):9–17
Nassar K, Mahmod WE, Fath H, Masria A, Nadaoka K, Negm A (2019) Shoreline change detection using DSAS technique: case of North Sinai coast. Egypt Mar Georesources Geotechnol 37(1):81–95. https://doi.org/10.1080/1064119X.2018.1448912
Natesan U, Parthasarathy A, Vishnunath R, Kumar GEJ, Ferrer VA (2015) Monitoring longterm shoreline changes along Tamil Nadu, India using geospatial techniques. Aquat Procedia 4:325–332. https://doi.org/10.1016/j.aqpro.2015.02.044
Ngowo RG, Ribeiro MC, Pereira MJ (2021) Quantifying 28-year (1991–2019) shoreline change trends along the Mnazi Bay–Ruvuma estuary Marine Park, Tanzania. Remote Sens Appl Soc Environ 23:100607. https://doi.org/10.1016/j.rsase.2021.100607
Ottman N, Ruokolainen L, Suomalainen A, Sinkko H, Karisola P, Lehtimäki J et al (2019) Soil exposure modifies the gut microbiota and supports immune tolerance in a mouse model. J Allergy Clin Immunol 143(3):1198–1206.e1112. https://doi.org/10.1016/j.jaci.2018.06.024

Oyedotun TDT (2014) Shoreline geometry: DSAS as a tool for historical trend analysis. Geomorphol Tech 2:1–12. https://www.scopus.com/inward/record.uri?eid=2-s2.0-85044030902&partnerID=40&md5=7a99e3e078ed4345e67d8cb232587f2b
Ozturk D, Sesli FA (2015) Shoreline change analysis of the kizilirmak lagoon series. Ocean Coast Manag 118:290–308. https://doi.org/10.1016/j.ocecoaman.2015.03.009
Qiao G, Mi H, Wang W, Tong X, Li Z, Li T et al (2018) 55-year (1960–2015) spatiotemporal shoreline change analysis using historical DISP and Landsat time series data in Shanghai. Int J Appl Earth Obs Geoinf 68:238–251. https://doi.org/10.1016/j.jag.2018.02.009
Raj N, Rejin Nishkalank RA, Chrisben Sam S (2020) Coastal shoreline changes in Chennai: environment impacts and control strategies of southeast coast, Tamil Nadu. In: Hussain CM (ed) Handbook of environmental materials management. Springer International Publishing, pp 1–14. https://doi.org/10.1007/978-3-319-58538-3_223-1
Roy S, Mahapatra M, Chakraborty A (2018) Shoreline change detection along the coast of Odisha, India using digital shoreline analysis system. Spat Inf Res 26(5):563–571. https://doi.org/10.1007/s41324-018-0199-6
Saha B, Samanta P, Mukhopadhyay S, Eriksson PG (2023) Effects of transgression on sedimentation system Vis-à-Vis coastal erosion on the Chandipur coast, Odisha, India. J Indian Assoc Sedimentol 40(1):5–19. https://doi.org/10.51710/jias.v40iI.276
Samanta S, Paul SK (2016) Geospatial analysis of shoreline and land use/land cover changes through remote sensing and GIS techniques. Model Earth Syst Environ 2(3):108. https://doi.org/10.1007/s40808-016-0180-0
Santos CAG, Nascimento TVM, Mishra M, Silva RM (2021) Analysis of long-and short-term shoreline change dynamics: a study case of João Pessoa city in Brazil. Sci Total Environ 769:144889. https://doi.org/10.1016/j.scitotenv.2020.144889
Santra A, Mitra D, Mitra S (2011) Spatial modeling using high resolution image for future shoreline prediction along Junput coast, West Bengal. India Geo-Spat Inf Sci 14(3):157–163. https://doi.org/10.1007/s11806-011-0522-z
Santra A, Mitra D, Mitra SS (2021) Effect of statistical relative radiometric normalization on spectral response of mangrove vegetation of Indian sundarbans–a comparative performance evaluation on sentinel 2A multi-spectral data. In: Mukhopadhyay A, Mitra D, Hazra S (eds) Sundarbans Mangrove Systems-A Geo-Informatics Approach. CRC Press
Santra A, Santra Mitra S, Mitra D, Sarkar A (2019) Relative radiometric normalisation—performance testing of selected techniques and impact analysis on vegetation and water bodies. Geocarto Int 34(1):98–113. https://doi.org/10.1080/10106049.2017.1367424
Saravanan S, Parthasarathy KSS, Vishnuprasath SR (2019) Chapter 4-monitoring spatial and temporal scales of shoreline changes in the Cuddalore region, India. In: Ramkumar M, James RA, Menier D, Kumaraswamy K (eds). Elsevier, Coastal Zone Management, pp 99–112. https://doi.org/10.1016/B978-0-12-814350-6.00004-5
Sarkar S, Bose PK, Bandhyopadhayay S (1991) Intertidal occurrence of mesoscale scours in the Bay of Bengal, India, and their implications. Sediment Geol 75(1):29–37. https://doi.org/10.1016/0037-0738(91)90048-I
Sharma G, Patnaik KVKRK (2022) Assessment of Coringa mangrove shoreline migration using geospatial techniques. J Oper Oceanogr 15(2):126–135. https://doi.org/10.1080/1755876X.2020.1840245
Tak WJ, Jun KW, Kim SD, Lee HJ (2020) Using drone and LiDAR to assess coastal erosion and shoreline change due to the construction of coastal structures. J Coast Res 95(sp1):674–678. https://doi.org/10.2112/SI95-131.1
Thieler ER, Himmelstoss EA, Zichichi JL, Ergul A (2009) The digital shoreline analysis system (DSAS) version 4.0—an ArcGIS extension for calculating shoreline change
Thoai DT, Dang AN, Kim Oanh NT (2019) Analysis of coastline change in relation to meteorological conditions and human activities in Ca mau cape, Viet Nam. Ocean Coast Manag 171:56–65. https://doi.org/10.1016/j.ocecoaman.2019.01.007

Topah Apandi EB, Salleh SA, Rahim HA, Adnan NA (2022) Mapping of coastline changes in mangrove forest using digital shoreline analyst system (DSAS). IOP Conf Ser: Earth Environ Sci 1067(1):012036. https://doi.org/10.1088/1755-1315/1067/1/012036

Toure S, Diop O, Kpalma K, Maiga AS (2019) Shoreline detection using optical remote sensing: a review. ISPRS Int J Geo Inf 8(2)

Turner IL, Harley MD, Short AD, Simmons JA, Bracs MA, Phillips MS, Splinter KD (2016) A multi-decade dataset of monthly beach profile surveys and inshore wave forcing at Narrabeen. Aust Sci Data 3(1):160024. https://doi.org/10.1038/sdata.2016.24

Vickers NJ (2017) Animal communication: when I'm calling you, will you answer too? Curr Biol 27(14):R713–R715. https://doi.org/10.1016/j.cub.2017.05.064

Williams AT, Rangel-Buitrago N, Pranzini E, Anfuso G (2018) The management of coastal erosion. Ocean Coast Manag 156:4–20. https://doi.org/10.1016/j.ocecoaman.2017.03.022

Xu H (2006) Modification of normalised difference water index (NDWI) to enhance open water features in remotely sensed imagery. Int J Remote Sens 27(14):3025–3033. https://doi.org/10.1080/01431160600589179

Yamano H, Shimazaki H, Matsunaga T, Ishoda A, McClennen C, Yokoki H et al (2006) Evaluation of various satellite sensors for waterline extraction in a coral reef environment: majuro atoll. Marshall Isl Geomorphol 82(3):398–411. https://doi.org/10.1016/j.geomorph.2006.06.003

Zzaman RU, Nowreen S, Billah M, Islam AS (2021) Flood hazard mapping of Sangu River basin in Bangladesh using multi-criteria analysis of hydro-geomorphological factors. J Flood Risk Manag 14(3):e12715. https://doi.org/10.1111/jfr3.12715

Chapter 5
Spatiotemporal Dynamics and Forecasting of the Shoreline in Sinaloa, Gulf of California: 40 Years of Analysis with Landsat Data and DSAS

Yedid Guadalupe Zambrano-Medina, Marco A. Mora-Ramírez, Evangelina Avila-Aceves, Francisco Roberto Estrada-Valdez, Wenseslao Plata-Rocha, Lidia Yadira Perez-Aguilar, Enrique Morales-Acuña, and Rosendo Romero-Andrade

Y. G. Zambrano-Medina (✉) · M. A. Mora-Ramírez
Laboratorio Universitario de Calidad del Aire, Facultad de Ciencias Químicas, Benemérita Universidad Autónoma de Puebla, Puebla de Zaragoza 72570, Puebla, México
e-mail: yedidzambrano@uas.edu.mx

Posgrado en Tecnologías Agrícolas Limpias, Instituto de Ciencias, Benemérita Universidad Autónoma de Puebla, Puebla de Zaragoza 72570, Puebla, México

Y. G. Zambrano-Medina · E. Avila-Aceves · W. Plata-Rocha · R. Romero-Andrade
Facultad de Ciencias de la Tierra y el Espacio, Universidad Autónoma de Sinaloa, Culiacán 80013, Sinaloa, México

E. Avila-Aceves · W. Plata-Rocha · L. Y. Perez-Aguilar
Laboratorio Nacional CONAHCYT de Tecnologías de la Información Geoespacial para los Sistemas Socioecológicos Resilientes (LaNCTIGeSSR), Toluca de Lerdo 50110, Mexico

F. R. Estrada-Valdez
Facultad de Ciencias Económicas y Sociales, Universidad Autónoma de Sinaloa, Culiacán 80013, Sinaloa, México

L. Y. Perez-Aguilar
Facultad de Informática Culiacán, Universidad Autónoma de Sinaloa, Culiacán 80013, Sinaloa, México

E. Morales-Acuña
Laboratorio de Oceanografía Satelital y Clima (LOSyC), Instituto Politécnico Nacional, Centro Interdisciplinario para el Desarrollo Integral Regional, Unidad Sinaloa, Guasave 81049, Sinaloa, México

A. Nath et al. (eds.), *Geospatial Application of Coastal Hazards*, Springer Water,
https://doi.org/10.1007/978-3-032-20076-1_5

5.1 Introduction

Shoreline analysis is crucial for monitoring and sustainable management of coastal areas, given their ecological importance, high productivity, and rich biodiversity (Jana et al. 2017, 2016a, 2014). These spaces are fragile and are subject to both natural and anthropogenic pressures. Globally, approximately 40% of the population lives within 100 km of the coast, which intensifies the impacts on coastal ecosystems (Sevilla et al. 2018). One of the most significant challenges in coastal areas is the rise in sea level, which exacerbates erosion and flooding problems, as well as promotes saline intrusion in aquifers and estuaries (Hamid et al. 2021; Islam 2025). IPCC climate change projections of sea level rise indicate these effects could intensify towards the end of the century, requiring more effective and sustainable monitoring strategies (Gupta et al. 2025; Jamal et al. 2025).

In this context, spatiotemporal monitoring of the shoreline is crucial for understanding the coastal system's response to environmental and anthropogenic pressures. The analysis of coastal dynamics using geospatial tools has become increasingly important, especially in the face of climate change and unplanned coastal development (Jana et al. 2012, 2016b; Astuti et al. 2023; Enríquez et al. 2016; Islam 2025). Remote sensing has proven to be an effective technique for identifying the position of the shoreline, using visible and infrared bands, to leverage the spectral differences between water, soil, and vegetation (Alesheikh et al. 2007; Gens 2010). Several studies have used satellite imagery and geospatial technologies to assess its evolution. Researchers widely accept the use of Landsat images due to their temporal and spatial coverage; Landsat images also represent an accessible alternative to more costly and laborious traditional methods (Baig et al. 2020; Williams et al. 2006).

One of the most widely used tools for studying coastal erosion and accretion processes is the Digital Shoreline Analysis System (DSAS), a statistical analysis extension developed by the United States Geological Survey (USGS). Over the past decades, DSAS has proven essential for analyzing the spatial and temporal dynamics of shorelines worldwide. The system enables the calculation of Net Shoreline Movement (NSM) and rates of change using methods such as End Point Rate (EPR), Linear Regression Rate (LRR), and Weighted Linear Regression Rate (WLR), by integrating multiple shoreline positions over time (Hossain et al. 2022). In addition, researchers have applied predictive models, such as the Kalman Filter, to forecast future shoreline positions based on time series data (Gümüş 2024; Yan et al. 2021).

Several studies have applied DSAS in diverse geographical and environmental contexts. In South Africa, Mshelia et al. (2025) analyzed 33 years of coastal change, estimating an error margin of ±0.32 m/year. In Thailand, Chawalit et al. (2025) incorporated machine learning algorithms to automate shoreline classification. In India, numerous studies have documented regional erosion and accretion trends using a range of spatial configurations, such as transects spaced at 50, 100, or 500 m (Baig et al. 2020; Giri et al. 2024). Research in polar regions like Svalbard (Nicu et al. 2020), and in South American countries such as Colombia (Fuentes et al. 2022)

DSAS has demonstrated the utility for evaluating the effects of human activity and natural phenomena on coastal retreat.

Despite these global advances in using tools such as DSAS for monitoring and analyzing coastal dynamics, the Mexican coast continues to lag in the systematic application of this type of methodology. This situation is worrisome considering that approximately a quarter of the country's population lives in coastal plains, where disorderly urban growth and the development of intensive economic activities, such as the tourism industry, fishing, and agriculture, have generated important environmental conflicts, directly affecting the stability and equilibrium of the coast (CONABIO 2022).

In this context, the state of Sinaloa represents a critical case, with more than 600 km of coastline and a distinctly dynamic morpho-sedimentary environment. This coastal region is subject to both natural and anthropogenic pressures, as well as growing vulnerability to extreme hydrometeorological events. Like many coastal regions worldwide, Sinaloa's coastline is shaped by complex interactions between physical processes and human activities. Natural drivers of coastal morphodynamics include wave action, tides, storms, and sea level rise, while human-related factors include expanding urbanization, the construction of hydraulic infrastructure, intensive agriculture, and the disruption of sediment transport (Franco-Ochoa et al. 2020; Özpolat and Demir 2019; Zambrano-Medina et al. 2023). These combined factors have led to significant episodes of both erosion and accretion along the Sinaloa coast, threatening the ecological integrity of coastal ecosystems and placing critical infrastructure and human settlements at risk. In this regard, the coast of Sinaloa has shown particularly dynamic behavior, with erosion rates more intense than in other regions of the country (Martínez et al. 2025), underscoring the need for continuous monitoring using specialized tools and robust geospatial approaches.

Therefore, this chapter aims to generate solid knowledge on shoreline displacement, erosion, and accretion rates, as well as future projections of the Sinaloa coastline, based on the analysis of Landsat imagery from 1981 to 2020 using geospatial tools such as DSAS. The goal is to provide foundational information that supports better coastal management in Sinaloa in the face of climate change, natural variability, and human pressures. This approach also contributes to achieving the Sustainable Development Goals (SDGs), particularly Goal 14, by providing continuous and precise data on coastal changes, critical for conservation efforts (Tsai 2022). The use of this tool with multi-temporal data and standardized methodological criteria would strengthen coastal management strategies and enhance Mexico's role in the global field of geospatially informed coastal analysis.

Given the economic, ecological, and social relevance of the region, where strategic sectors such as fisheries, agriculture, and tourism converge, accurate information on coastline evolution is essential to support land-use planning, risk adaptation, and the conservation of ecosystem services. This study also includes a comprehensive review of the application of DSAS in international coastal monitoring efforts, emphasizing its effectiveness in quantifying erosion and accretion rates and its integration into territorial planning, risk management, and climate change adaptation processes. By contextualizing the use of DSAS across various coastal regions worldwide, this review consolidates its role as a key tool for the analysis and sustainable management of coastal areas.

5.2 International Application of DSAS in Coastal Monitoring

The Digital Shoreline Analysis System (DSAS), developed by the United States Geological Survey (USGS), is a widely recognized tool for researching and managing shoreline change worldwide. Since the 1990s, it has been part of the Coastal Change Hazards Program and is currently available in version 5.1. This version enables the calculation of change statistics and the forecasting of future shoreline positions using a vector time series of georeferenced shorelines (Himmelstoss et al. 2021; Thieler et al. 2009). DSAS primarily calculates shoreline change rates using time-series data, applying metrics such as Net Shoreline Movement (NSM), End Point Rate (EPR), Linear Regression Rate (LRR), and Weighted Linear Regression Rate (WLR).

Over the past decades, DSAS has proven to be a key tool for the spatial and temporal analysis of shoreline dynamics at the international level (Table 5.1). Research in highly sensitive coastal environments, such as the Norwegian fjords (Nicu et al. 2020), and in densely urbanized and vulnerable areas like Visakhapatnam (Baig et al. 2020) it has demonstrated its effectiveness across diverse erosion and accretion scenarios. More recent studies—such as those by Mshelia et al. (2025) in South Africa and (Chawalit et al. 2025) In Thailand, they have further highlighted the capabilities of DSAS, remote sensing, and advanced techniques, such as machine learning algorithms, for extracting shoreline positions.

Obtaining coastlines over time requires data sources that ensure continuity, suitable resolution, and multi-temporal availability. Among the most widely used sources are the Landsat satellite image series, which have been extensively applied due to their long-term temporal coverage since the 1970s, moderate spatial resolution, and free accessibility. Numerous studies (Acharyya et al. 2023; Arjasakusuma et al. 2021; Astuti et al. 2023; Ateş et al. 2024; Aziz 2024; Baig et al. 2020; Basheer Ahammed and Pandey 2022; Bera and Maiti 2019; Das and Dhorde 2022; Giri et al. 2024; Harikrishna et al. 2024; Kassouk et al. 2024; Mann et al. 2023; Mshelia et al. 2025; Mukhopadhyay et al. 2018; Ozturk and Maras 2024; Singh et al. 2023; Song et al. 2021; Thasarathan et al. 2024; Yasir et al. 2020; Zhang et al. 2024; Jana et al. 2017, 2016a, 2014, 2012, 2016b) have employed this satellite series to analyze coastal dynamics, demonstrating its effectiveness in shoreline delineation and monitoring across diverse geographic contexts.

In recent years, the use of higher-resolution sensors for shoreline extraction has increased, including Sentinel-2 (Benhur et al. 2024; Lacerda Barros et al. 2024), as well as data from SPOT (Abdul Maulud et al. 2022), IKONOS, GeoEye (Radosavljevic et al. 2016), and PlanetScope (Hastuti et al. 2024), which allow for the detection of coastal changes with greater spatial detail. Shorelines have also been delineated

Table 5.1 International studies applying DSAS, statistical metrics used, analysis period, and reported erosion rates

Country	DSAS statistics	Time period	Erosion (–) and accretion (+) rate (m yr^{-1})	Source
Canada	EPR	1952–2011	–5.5 and 2.7	Radosavljevic et al. (2016)
India	EPR, LRR	1990–2015	–0.58 and 1.26	Mukhopadhyay et al. (2018)
India	EPR, NSM, LL/LG	1975–2017	–4.78 and 4.7	Bera and Maiti (2019)
Guinea	EPR, LRR, WRL	1948–2012	–3.8 and 3.9	Fotsi et al. (2019)
India	EPR, LRR	1991–2018	–1.16 and 1.62	Baig et al. (2020)
Noruega	EPR, LRR	1927–2019	–0.76	Nicu et al. (2020)
China	EPR, NSM, LRR	2000–2019	–2.08	Yasir et al. (2020)
Indonesia	EPR, LRR, NSM, WLR, SCE	2000–2019	4.12	Arjasakusuma et al. (2021)
India	LRR, EPR, NSM	2000–2018	–2.22	Nath et al. (2021)
EE. UU	EPR	1976–1993	–0.82 and 0.47	De Lima et al. (2021)
Grece	LRR, NSM, EPR, SCE	1945–2021	–0.25 and 0.34	Apostolopoulos et al. (2022)
India	EPR, JKR, LRR	2000–2019	–0.02	Das and Dhorde (2022)
Bangladesh	EPR, LRR, WLR	1985–2021	–22.57	Sultana et al. (2023)
Indonesia	LRR, EPR	2004–2019	–12.7	Astuti et al. (2023)
Ghana	EPR	2005–2022	–1	Angnuureng et al. (2023)
Spain	NSM, LRR, LCI	1930–2022	–1.16	Fernández-Hernández et al. (2023)
Ghana	EPR, NSM, WLR	1913–2016	–16.57	Mann et al. (2023)
Sri Lanka	LRR, NSM, EPR, SCE	1988–2022	–4.7	Thasarathan et al. (2024)
India	EPR, LRR, LMS	1991–2022	–0.45 and 1.05	Giri et al. (2024)
India	EPR, LRR, NSM	1992–2022	3.75	Harikrishna et al. (2024)
Turkey	EPR, NSM. SCE, LRR	1948–2021	–1.5	Ozturk and Maras (2024)

(continued)

Table 5.1 (continued)

Country	DSAS statistics	Time period	Erosion (–) and accretion (+) rate (m yr^{-1})	Source
Tunisia	NSM, EPR, LRR	1985–2019	–2.24 and 1.3	Kassouk et al. (2024)
India	LRR, NSM, SCE	2016–2022	–1.63	Benhur et al. (2024)
China	NSM, EPR, LRR	1985–2020	–7.62 and 13.54	Zhang et al. (2024)
Brazil	EPR, LRR	2002–2024	–1.1 and 0.5	(Lacerda Barros et al. (2024)
South Africa	EPR, NSM, LRR	1990–2023	–0.22	Mshelia et al., (2025)
Thailand	EPR, NSM	1988–2023	–13.55	Chawalit et al. (2025)

using aerial photogrammetry and the generation of high-resolution orthomosaics, which complement satellite-based analyses by validating or refining shoreline positions in critical or difficult-to-access areas (Angnuureng et al. 2023; Apostolopoulos et al. 2022; Fernández-Hernández et al. 2023; Fuentes et al. 2022); as well as by topographic surveys (Fotsi et al. 2019; Liang et al. 2022; Nicu et al. 2020). The choice of input data depends on the period of analysis, study scale, required precision, and the availability of historical data. Significant limitations remain that hinder methodological standardization and regional comparisons. On the one hand, the heterogeneity in data resolution (ranging from Landsat imagery to 3-m orthomosaics) and in the temporal frequency of observations makes a consistent evaluation of results across studies challenging.

On the other hand, studies vary in their configurations regarding transect spacing and applied metrics, resulting in differing levels of precision and comparability. For instance, some use transects every 2.5 m (Abdul Maulud et al. 2022) or every 50 m (Baig et al. 2020), while others apply intervals of up to 100 m (Mshelia et al. 2025), with perpendicular transect lengths ranging from 500 to 1,000 m, or fixed at 500 m (Harikrishna et al. 2024). This variability complicates the comparison of results and hinders the development of standardized protocols. Moreover, few studies report measurement uncertainty or margins of error; among them, Mshelia et al. (2025), report an annual error of ± 0.32 m. It should also be noted that factors such as geometric distortions, topographic variations, cartographic inconsistencies, and projection errors can affect accuracy, especially given the dynamic and uncertain nature of coastal environments (Kankara et al. 2015; Parthasarathy and Deka 2019).

The diverse applications of DSAS demonstrate its adaptability in studies of erosion, accretion, coastal vulnerability, land-use planning, and assessing the impact of climate change. Its adoption by scientific, governmental, and academic institutions underscores its value as an accessible, robust, and scientifically validated tool for analyzing coastal dynamics at both local and regional scales.

5.3 Study Area

The state of Sinaloa, located in northwestern Mexico, features an extensive coastal stretch of approximately 656 km along the Gulf of California (Fig. 5.1). Its coastline displays remarkable geomorphological heterogeneity, comprising bays, estuaries, sandy bars, marshes, beaches, and deltas, making it a highly dynamic coastal system subject to erosion, sedimentation, and extreme hydrometeorological events. Climatically and oceanographically, the Sinaloa coast lies within a transition zone between arid-subhumid and tropical-dry environments, characterized by seasonal summer rainfall influenced by the North American monsoon and an active tropical cyclone season from June to November. These factors intensify morphodynamic processes along the coastal strip. The coastal plain and deltaic systems cover a substantial portion of the area, including lagoons and salt marshes that account for 28.9% of the Sinaloa coastal plain. Additionally, the coastline hosts at least 12 bays and 15 significant estuaries distributed along its length.

In terms of population density, Sinaloa has a state average of 53 inhabitants per km^2, slightly below the national average of 64 inhabitants per km^2. Administratively, ten of the state's 20 municipalities have a coastline: Ahome, Guasave, Angostura, Navolato, Culiacán, Elota, San Ignacio, Mazatlán, El Rosario, and Escuinapa. However, population densities in these coastal municipalities are considerably higher. According to data from (INEGI's 2020) Census, the most densely populated coastal municipalities are Culiacán (808,416 inhabitants), Mazatlán (441,975 inhabitants), Ahome/Los Mochis (298,009 inhabitants), and Guasave (77,849 inhabitants), concentrating a significant portion of the regional population and economic activity along the coast (INEGI 2020).

Sinaloa's coastline features numerous beaches and beachfront settlements of significant natural, social, and tourist value. In the municipality of Escuinapa, the coastal nucleus of Teacapán and its beaches, such as La Tambora, Las Cabras, Las

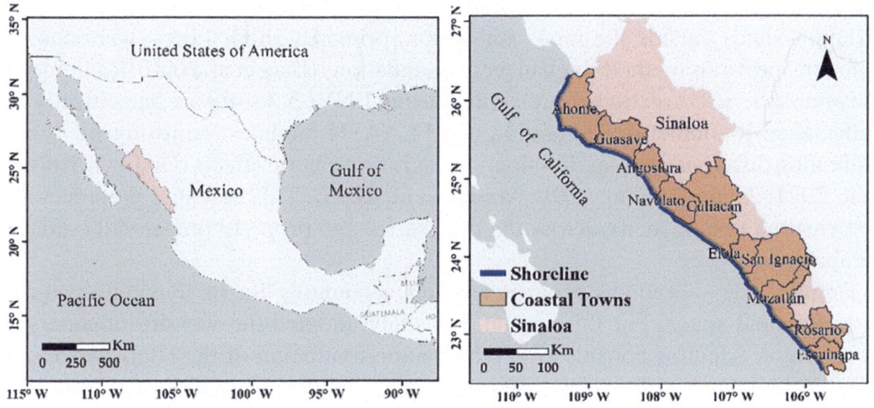

Fig. 5.1 Study area location in the Northwest of Mexico

Lupitas, and Cuatro Surcos, stand out as popular attractions, surrounded by estuaries, mangroves, and lagoon systems. Further north, Mazatlán's approximately 17 km of urban coastline hosts one of the state's main tourist centers, where residential neighborhoods and hotel, recreational, and commercial infrastructures directly border the waterfront. In Navolato, Altata Beach is a prominent tourist and recreational destination experiencing rapid urban growth, framed by the Altata-Ensenada del Pabellón estuarine-wetland system, which is part of the Ramsar site network. Further north in Guasave, Las Glorias Beach combines tourism, fishing, and agricultural activities, with housing and developments close to the shoreline. Lastly, in Ahome, the port of Topolobampo holds strategic importance due to its commercial activity and proximity to beaches and lagoon systems such as El Maviri, which also serves as a local tourism hub. The distribution of coastal communities and beaches with direct shoreline occupation underscores the urgent need for detailed assessments of coastline evolution, especially in areas with exposed infrastructure and sensitive ecosystems.

5.4 Methodology

5.4.1 Shoreline Extraction and Displacement Detection Methods

The study utilized satellite images from Landsat sensors (MSS, TM, ETM+, OLI), obtained from the United States Geological Survey's (USGS) Earth Explorer, and geometrically corrected to the WGS 84/UTM coordinate system. The analyzed period spans 1981–2020, aiming to assess the spatio-temporal evolution of the Sinaloa coastline. To achieve a robust representation of coastal variability, five specific years were selected (1981, 1993, 2004, 2010, and 2020), as multiple time points increase the accuracy of change rate calculations, according to Sekovski et al. (2014). To ensure image quality and reduce atmospheric interference, cloud cover, and tidal variation, dates outside the monsoon season, primarily in October—were chosen, following previous methodological recommendations (Baig et al. 2020). Radiometric and atmospheric corrections were applied using ENVI 5.3 software, specifically the Reflectance Radiometric Calibration and FLAASH modules, adjusting for sensor calibration differences, solar illumination, and atmospheric effects (Godwyn-Paulson et al. 2021; Jalaludin et al. 2020; Masria et al. 2015). This essential preprocessing step ensured homogeneity across the time series and properly prepared the images for analysis.

Defining the shoreline's position requires accounting for its dynamic nature in both time and space. For this reason, the study adopted the wet-dry boundary as an indicator, which is considered a reliable approximation of the High-Water Line (HWL). This method is widely used in coastal monitoring studies due to its clear visibility in satellite imagery and its correspondence with the most recent high tide (Barbaro et al. 2019; Boak and Turner 2005; Kermani et al. 2016; Oyedotun 2016).

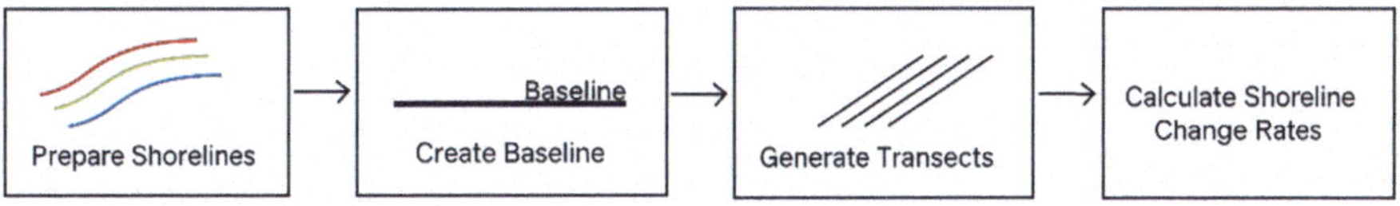

Fig. 5.2 Workflow of shoreline change analysis using DSAS

This approach enables a representative identification of the shoreline without the need for in situ tide gauge data.

For shoreline extraction, the single band threshold was used, and the Normalized Difference Water Index (NDWI) (McFeeters 1996), was calculated with bands 4 (Green) and 6 (Near InfraRed (NIR)) of the Multispectral Scanner System (MSS) sensor, and bands 3 (Green) and 5 (NIR) of the Operational Land Imager (OLI) sensor, using the following equation:

$$NDWI = \frac{\lambda_g - \lambda_n}{\lambda_g + \lambda_n}$$

where λ_g and λ_n represent the green and NIR bands of Landsat data, respectively.

This index enhances the contrast between water bodies and land surfaces, enabling a more accurate delineation of the coastal boundary as observed from satellite imagery (Ashtekar et al. 2019; Mondejar and Tongco 2019). The resulting images were binarized and processed in ArcGIS, where threshold segmentation was applied to distinguish land from water (Liu et al. 2017; Zorlu and Kusak 2025). Next, they converted the entities to vectors, extracted the shoreline, and saved it as a shapefile. Subsequently, the raster entities were converted to vectors, the shoreline was extracted, and the output was saved as a shapefile. In some cases, minor visual adjustments guided by the NDWI results were applied to refine the shoreline's definition.

Subsequently, once the shorelines were obtained, we applied DSAS (Fig. 5.2) to calculate the displacement and rates of change among the shorelines.

The workflow began with the preparation of shoreline data and the creation of a geodatabase within a GIS environment, which included the shoreline time series and their corresponding date attributes. A baseline was then constructed—typically derived from the earliest shoreline—smoothed and shifted offshore to serve as a reference line (Crowell et al. 1991; Himmelstoss et al. 2018). From this baseline, transects were generated at consistent 50 m intervals, oriented perpendicularly to intersect all shoreline positions throughout the time series. The DSAS tool was subsequently used to calculate various shoreline change metrics, including the Net Shoreline Movement (NSM), End Point Rate (EPR), and Linear Regression Rate (LRR). Definitions, formulas, and explanations for each of these metrics are presented in Table 5.2.

Table 5.2 Summary of shoreline change statistics used in DSAS

Statistic	Definition	Equation	Meaning of terms
NSM	Net distance between the oldest and most recent shoreline positions	$NSM = X_f - X_i$	X_f : final Shoreline position; X_i : initial, shoreline position
EPR	The rate of shoreline changes between two shoreline positions	$EPR = \frac{NSM}{T}$	T : time interval between shoreline dates (in years)
LRR	Estimates the average rate of shoreline change by fitting a least-squares regression line to all shoreline positions over time	$Y = a + bx$	Y : distance from the baseline (m); X : time (years); b : slope (rate of Shoreline change in m/year); a : y-intercept

5.4.2 *Shoreline Extraction and Displacement Detection Methods*

The Kalman Filter model (Kalman 1960), incorporated into DSAS as a predictive tool, was employed to forecast the future position of the coastline. This model integrates historical shoreline data with its own estimates to iteratively predict new shoreline positions (Himmelstoss et al. 2021; Long and Plant 2012). It operates by minimizing the error between observed and modeled positions through a sequence of temporal observations, continually adjusting its predictions. The algorithm begins with an initial linear regression rate derived from historical shoreline positions, which serves to initialize the model and generate sequential forecasts (Godwyn-Paulson et al. 2021; Long and Plant 2012). With each new observed shoreline position, the model compares its prediction with the actual observation, updating both the rate of change and associated uncertainties. This feedback mechanism progressively enhances the model's accuracy and enables the generation of confidence bands around the predicted shoreline (Himmelstoss et al. 2021). However, the approach assumes that the factors influencing past coastal dynamics will remain consistent in the future, an assumption that constitutes a fundamental limitation of the model.

In this study, the Kalman filter was applied to forecast shoreline positions for the years 2030 and 2040, using historical shoreline data from 1981 to 2020. As part of the model validation process, a predicted shoreline for the year 2020 was generated. This modeled shoreline was then compared with the actual 2020 coastline, which had been extracted through the interpretation of Landsat imagery. The accuracy of the prediction was evaluated by calculating the root-mean-square error (RMSE) using the following equation:

$$RMSE = \frac{\sqrt{(X_p - X_r)^2 + (Y_p - Y_r)^2}}{n}$$

where X_p and Y_p are the predicted model coordinates of the coast, and X_r and Y_r are the actual coordinates of the coast, and n are total number of evaluated points.

This value enables the estimation of the spatial accuracy of the prediction model. To evaluate the correspondence between modeled and observed rates of change, we used a scatter plot of the model's derived LRR and actual rates from historical data to calculate the linear regression coefficients and correlation coefficient. This additional comparison allows us to quantify how adequately the model represents coastal dynamics.

5.5 Results and Discussion

5.5.1 Analysis of Shoreline Change Between 1981 and 2020

The NSM results indicate that a total of 12,677 transects were analyzed along the Sinaloa shoreline (Fig. 5.3). The overall average shoreline displacement was – 43.45 m, revealing a predominance of landward movement (erosion). The finding further supports this trend that 74.54% of the transects exhibited negative displacement, with an average shoreline retreat of −109.71 m and a maximum recorded erosion of −995.97 m. In contrast, 25.46% of the transects showed shoreline progradation, with a maximum positive displacement of 1,382.39 m and an average advance of 140.50 m, indicating localized accretion in specific sectors of the coast (Table 5.3). These results suggest that, although accretion is present in specific areas, erosion has been the predominant process along the Sinaloa coastline during the study period.

In this study, shoreline change rates derived from both the End Point Rate (EPR) and the Linear Regression Rate (LRR) were classified into five categories according to their magnitude (m yr^{-1}), as summarized in Table 5.4.

The analysis based on End Point Rate (EPR) statistics revealed an average shoreline change rate of −1.17 m yr^{-1}, indicating a predominant erosional trend along the Sinaloa coast (Fig. 5.4). Among the evaluated transects, 74.54% exhibited erosion, with a maximum retreat rate of −31.12 m yr^{-1} and an average rate of −3.30 m yr^{-1} in these sectors. In contrast, 25.46% of the transects showed accretion, with a maximum rate of 69.61 m yr^{-1} and an average rate of 4.71 m yr^{-1} in accreting areas (Table 5.5).

For a more detailed spatial analysis, the coastline was divided into three zones: north, central, and south (Fig. 5.4). In the northern zone (Fig. 5.4a), the average EPR was −0.99 m yr^{-1}, with a maximum retreat rate of −31.12 m yr^{-1}. The central zone (Fig. 5.4b) exhibited a slightly lower average rate of −0.87 m yr^{-1}, with a maximum of −20.87 m yr^{-1}. In contrast, the southern zone (Fig. 5.4c) recorded the highest average retreat rate at −1.67 m yr^{-1}, although its maximum value (−12.51 m yr^{-1}) was lower than in the other two zones. These results indicate that, while erosion is a dominant process along the entire Sinaloa coastline, its magnitude and spatial distribution vary across sectors, with the southern zone emerging as the most vulnerable in terms of average shoreline retreat.

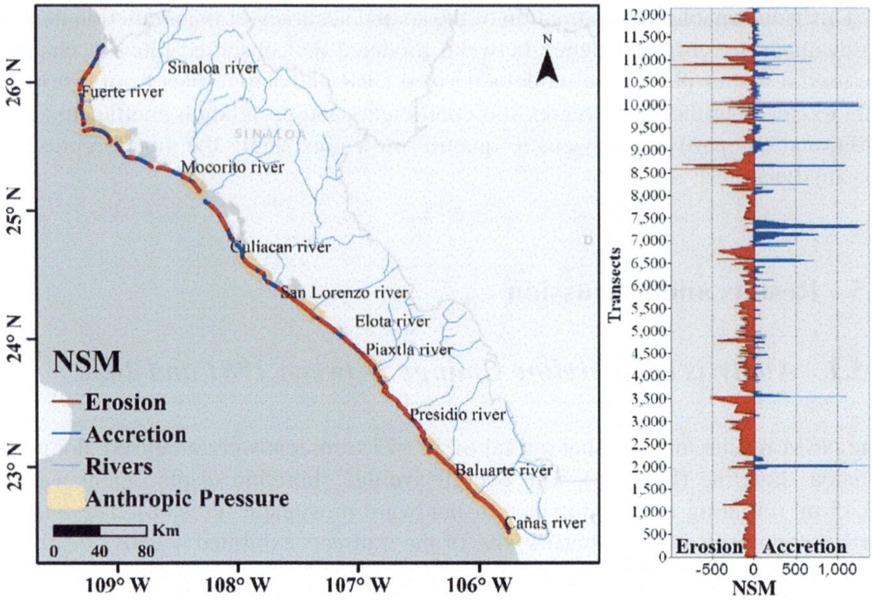

Fig. 5.3 Net shoreline movement (NSM) along the Sinaloa shoreline during 1981–2020. The red and blue colored line segments represent the erosion and accretion shoreline transects, respectively

Table 5.3 Summary of NSM results for the Sinaloa shoreline (1981–2020)

Statistics	NSM
Number of transects	12,677
Average distance (m)	–43.45
Number of transects with negative distance	9450
The percentage of all transects that have a negative distance	74.54
Maximum negative distance (m)	–995.97
Average of all negative distances (m)	–109.71
Number of transects with negative distance	3227
The percentage of all transects that have a positive distance	25.46
Maximum positive distance (m)	1382.39
Average of all positive distances (m)	140.50

Table 5.4 Classification of shoreline change rates based on EPR and LRR

Category	Range m yr^{-1}
High erosion	<–4
Moderate erosion	–4 to –2
Low erosion	–2 to –1
Stability	–1 to 0
Accretion	>0

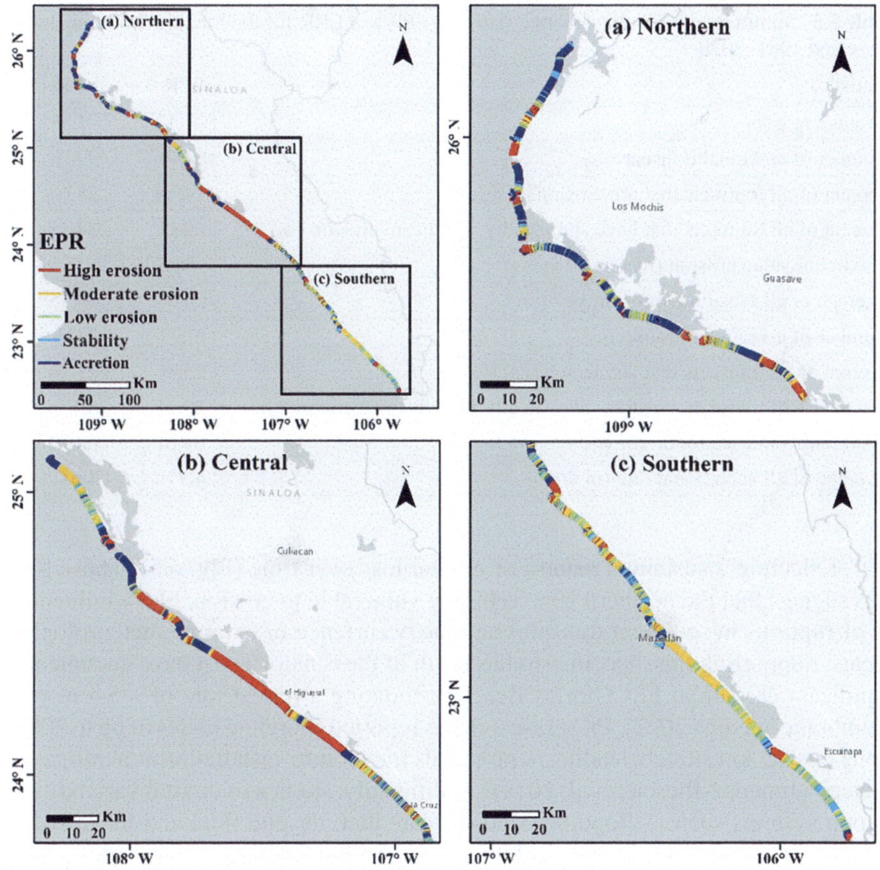

Fig. 5.4 Shoreline change rates estimated using EPR along the Sinaloa shoreline (1981–2020). Subdivisions: **a** Northern zone, **b** Central zone, **c** Southern zone

The Linear Regression Rate (LRR) statistic yielded an average erosion rate of −1.39 m yr^{-1}, slightly higher than the estimate obtained by the End Point Rate (EPR), indicating a more pronounced shoreline retreat when considering the overall trend across the entire study period. Erosional transects accounted for 78.64% of the total, with a maximum retreat rate of −35.83 m yr^{-1} and an average of −3.45 m yr^{-1} in these zones. Conversely, 21.74% of transects showed accretion, with a maximum advance of 68.73 m yr^{-1} and an average rate of 4.51 m yr^{-1} (Table 5.5). This pattern suggests that, despite localized sediment gains, the predominant historical trend is a sustained loss of shoreline.

In the northern coastal zone of Sinaloa, the DSAS analysis reveals a clear predominance of erosional processes. The Linear Regression Rate (LRR) statistics confirm this trend, with a mean rate of −1.34 m yr^{-1} and a maximum retreat of −35.83 m

Table 5.5 Summary of shoreline change statistics EPR and LRR for the Sinaloa shoreline during the period 1981–2020

Statistics	EPR	LRR
Average rate (m yr^{-1})	–1.17	–1.39
Number of erosional transects	9450	9921
Percent of all transects that are erosional (%)	74.54	78.64
Percent of all transects that have statistically significant erosion (%)	69	34.15
Maximum value erosion (m/year)	–31.12	–35.83
Average of all erosional rates (m yr^{-1})	–3.30	–3.45
Number of accretional transects	3227	2756
Percent of all transects that are accretional (%)	25.46	21.74
Percent of all transects that have statistically Significant accretion (%)	20.45	5.17
Maximum value accretion (m yr^{-1})	69.61	68.73
Average of all accretional rate (m yr^{-1})	4.71	4.51

yr^{-1}, reflecting a sustained pattern of coastal loss over time (Fig. 5.5). These findings suggest that the northern zone is highly vulnerable to erosion, likely influenced by disruptions in sediment transport and the occurrence of extreme meteorological events. Supporting this, specific studies north of the Sinaloa coast have documented significant erosion at Las Glorias Beach, estimating a retreat rate of −5.4 m yr^{-1} (Zambrano-medina 2022). Prior research has reported shoreline losses of up to 300 m along this 4.5 km stretch, leading to proposals for geotube installation as a mitigation strategy (Jiménez-Illescas et al. 2019). Additionally, studies of coastal variability in lagoon systems such as Topolobampo indicate that, despite tidal and local hydrodynamic influences, the system exhibits a natural tendency to export sediments to the Gulf of (Montaño-Ley et al. 2014), with potential implications for its internal stability and the adjacent beaches.

In the central zone (Fig. 5.6), the Linear Regression Rate (LRR) statistic indicated an average erosion rate of −1.13 m yr^{-1}, with a maximum retreat of −21.32 m yr^{-1}, reflecting a sustained erosive trend over time in this sector of the coast. Nuevo Altata Beach is among the most severely eroded areas, with an average shoreline retreat of 3.2 m yr^{-1}. Coastal erosion here has reached critical levels, posing a threat to the stability of homes and tourist facilities near the shore. Local media reports from 2018 document a sea advance of up to 60 m in just seven months, which led to the preventive evacuation of nearby buildings due to the imminent risk of collapse. By 2023, ongoing wave impacts, exacerbated by extreme weather and rising sea levels, had resulted in at least 80 abandoned or semi-ruined luxury homes, valued at approximately 2–3 million pesos each. These conditions were widely reported by the media and social networks, highlighting the physical and social decline of the coastal zone. Geological studies corroborate these observations, confirming that the marine barriers in the Altata-Pabellones lagoon system experience rebound erosion,

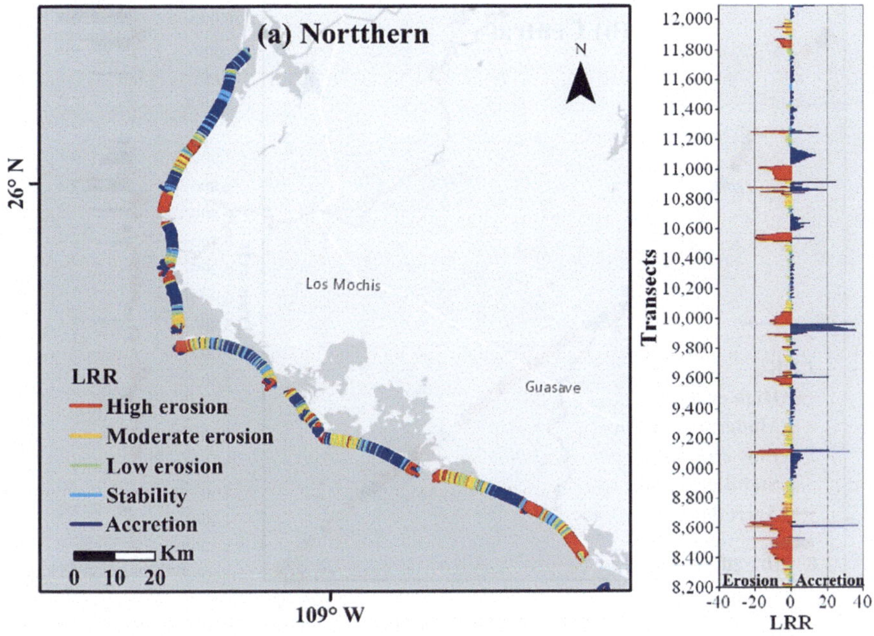

Fig. 5.5 Estimated shoreline change rates using LRR along the Northern zone of the Sinaloa coast

characterized by narrow beaches and frequent erosion escarpments (Carbajal et al. 2018).

Finally, in the southern zone (Fig. 5.7), the Linear Regression Rate (LRR) registered the highest average rate, with -1.70 m yr^{-1}, although with a lower negative maximum than in other regions (-12.78 m yr^{-1}). In the specific case of Mazatlán beaches, the results are in line with previous research that has also documented significant coastal erosion. For example, (Oyedotun et al. 2018), reported average rates of -1.9 ± 0.9 m yr^{-1}in various sectors of the Sinaloa coast, including Mazatlán. According to Valderrama-Landeros et al. (2022) the major causes of the historical erosion along the tourist city of Mazatlán are a combination of unplanned real estate construction facing the beach, the foredune and beach berm fragmentation, and the intensity and direction of the swell from tropical storms and hurricanes. Beaches south of Escuinapa such as Teacapán showed high erosion rates of up to 2.7 m yr^{-1}; this municipality has practically seen its beaches disappear in the last four decades, a problem that has left its beaches unprotected against waves and natural disasters such as Hurricane Willa in 2018. This problem has been present since the 70 s of the last centuries, with the dredging for the Cuautla channel in Nayarit, which has modified the water currents in Marismas Nacionales.

The analysis of the applied statistics indicates that a significant portion of the Sinaloa coastline is affected by erosive processes resulting from the interaction

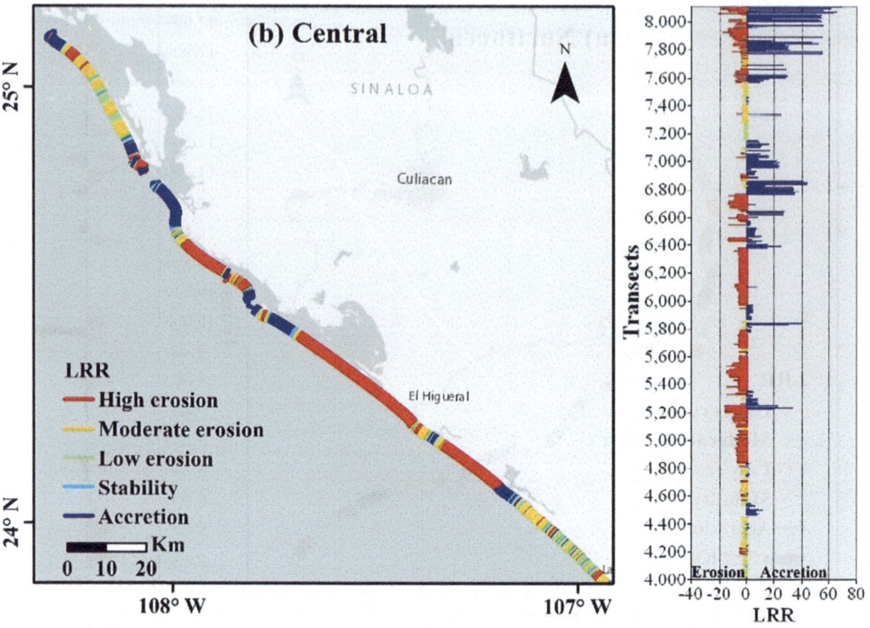

Fig. 5.6 Shoreline change rates estimated using LRR along the central zone of the Sinaloa coast

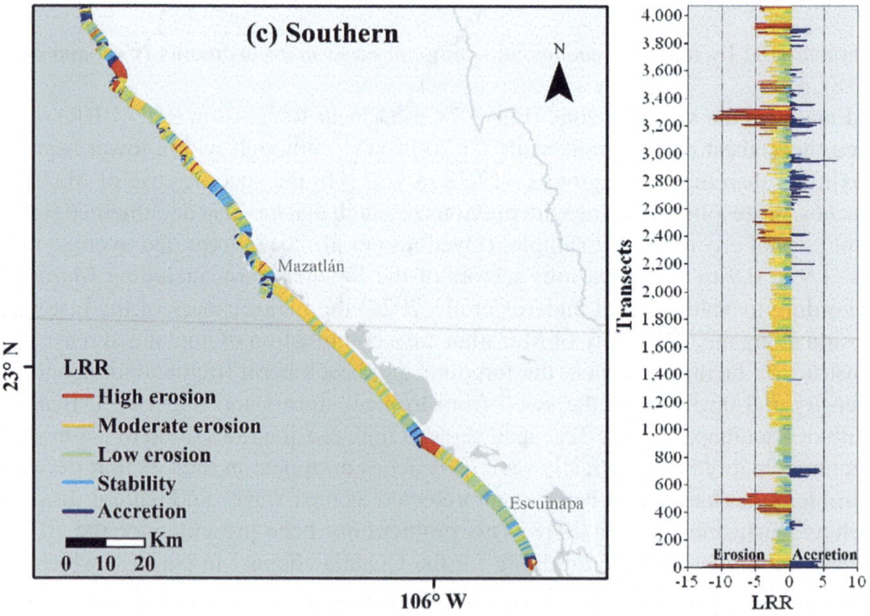

Fig. 5.7 Shoreline change rates using LRR along the southern zone of the Sinaloa coast

between natural factors and human activities. One of the leading causes is the disruption of fluvial sediment transport due to the construction of dams on the state's major rivers, which reduces sediment supply to coastal zones and alters the morphodynamic balance, especially near river mouths (Jiménez-Illescas Ángel et al. 2018). This situation is further compounded by extensive human interventions such as breakwaters, jetties, and other coastal protection structures, which modify natural coastal dynamics and lead to uneven sediment redistribution. The high anthropogenic pressure linked to population concentration exacerbates this scenario, as nearly 80% of the state's inhabitants reside in coastal areas, increasing the vulnerability of tourist, urban, and industrial infrastructure to erosion-related risks. Additionally, Sinaloa's location in a region highly exposed to tropical cyclones contributes to heightened morphodynamic variability. These extreme weather events trigger intense, short-term erosion episodes that cause lasting changes to the coastline, impacting both ecosystems and built assets.

5.5.2 Shoreline Forecast

Using LRR and the Kalman filter, shoreline projections for 2030 and 2040 were generated based on historical data from 1981, 1993, 2004, 2010, and 2020. The model outputs indicate a generalized retreat along the Sinaloa coast, with specific sectors particularly susceptible to erosional processes (Fig. 5.8). Areas near river mouths, coastal urban centers, and tourist beaches under high anthropogenic pressure are expected to experience the most significant changes. Specifically, the model forecasts that beaches such as Las Glorias, Nuevo Altata, Matatlán, and Escuinapa will be highly vulnerable to coastal erosion, likely facing substantial shoreline setbacks by 2030 and 2040.

Las Glorias Beach in Guasave is one of the most affected sites in the state, exhibiting a historical erosion rate exceeding 5 m $yr^{-1}$and a shoreline loss of up to 300 m. If this trend continues, an additional retreat of 110–130 m is projected by 2040 (Fig. 5.8a). Along the central coast, Nuevo Altata Beach in Navolato has experienced significant coastal front degradation linked to changes in the sediment dynamics of the Altata-Ensenada del Pabellón lagoon-wetland system. Projections estimate a retreat rate between 2 and 3 m yr^{-1} (Fig. 5.8b), which could result in a loss of up to 55 m by 2040, severely impacting terrestrial connectivity and tourism infrastructure.

On the southern coast of the state, several beaches in Mazatlán are showing concerning erosion patterns. At Isla de la Piedra, the coastline is currently retreating at approximately 4.7 m yr^{-1}, with projections indicating a cumulative loss of over 90 m by 2040 if current trends continue. Playa La Escondida and Cerritos experience annual erosion rates between 3.5 and 4.0 m, which could result in an estimated retreat of 70–80 m within the same period, threatening hotel and tourism infrastructure. Meanwhile, Olas Altas Beach has recorded historical retreats of up to 40 m in recent decades, with projections warning of intensified erosion without protective

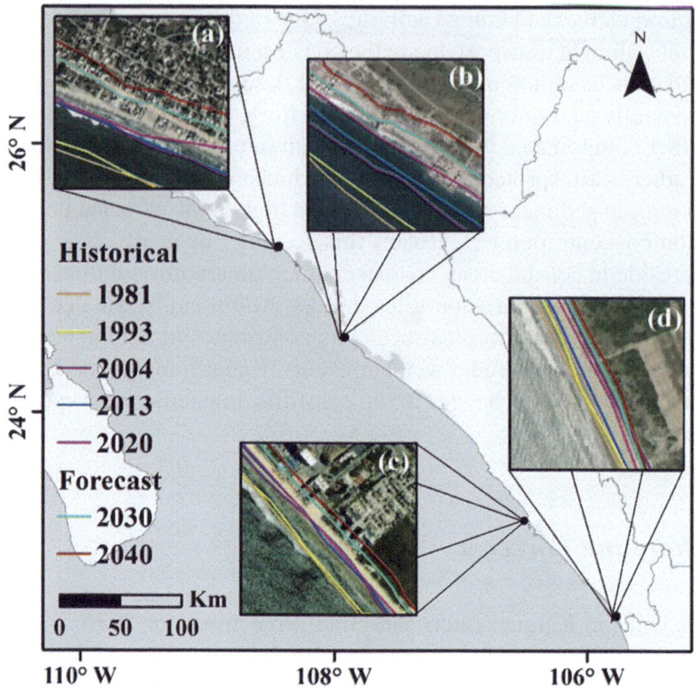

Fig. 5.8 Shoreline history and forecast for **a** Las Glorias, **b** Nuevo Altata, **c** Escondida, and **d** Teacapán beaches

or restorative interventions (Fig. 5.8c). In addition to Mazatlán, the coastal strip of Escuinapa exhibits unstable sedimentological characteristics, characterized by high erosion rates during extreme events, such as storms and cyclones. Forecasts suggest retreats of up to 60 m by 2040 (Fig. 5.8d), particularly near lagoon mouths and deltaic zones like Teacapán, where coastal morphodynamics are especially active.

In more than 90% of the transects, the projected lines for 2030 and 2040 remained within the uncertainty band generated by the Kalman model, aligning with similar results obtained in other coastal regions. The RMSE of the model was ±8.43 m, and the regression analysis between observed and modeled LRR values yielded $R^2 = 0.87$, supporting the reliability of the model for future projections.

5.6 Conclusions

The multitemporal analysis of coastal dynamics in the state of Sinaloa, conducted using DSAS with Landsat imagery from 1981 to 2020, revealed a dominant trend of erosion. Seventy-five percent of the transects analyzed showed significant shoreline retreat, with an average rate of -3.45 m yr^{-1} in strictly erosive areas. For the entire coastline, the average rate estimated by the LRR statistic was -1.39 m yr^{-1}, confirming a sustained loss over time. Values derived from the NSM and EPR statistics support this trend, showing an average net retreat of -43.45 m and an overall rate of -1.17 m yr^{-1}, respectively. Although specific areas exhibited accretion, with maximum rates exceeding 68 m/year, these remain isolated cases that do not alter the general erosive pattern.

The Kalman filter model enabled projections of the coastline's future position for 2030 and 2040, indicating that if current conditions persist, coastal retreat will continue and likely intensify—especially on beaches such as Las Glorias, Nuevo Altata, and sites in Mazatlán and south of Escuinapa. These projections underscore the high vulnerability of the Sinaloa coastline to a combination of natural factors—such as tropical cyclones, rising sea levels, and extreme waves—and anthropogenic pressures including rigid infrastructure construction, sediment retention by dams, and unplanned urban development.

These findings underscore the importance of integrating this information into coastal management and territorial planning to mitigate exposure and vulnerability in the most critical sectors of the Sinaloa coast. Furthermore, this study confirms that the use of tools like DSAS, combined with statistical and geospatial analysis, provides an effective, replicable, and cost-efficient methodology for monitoring coastal dynamics in regions facing significant anthropogenic and natural challenges.

5.7 Recommendations and Future Perspectives

This study's findings highlight that coastal erosion in Sinaloa extends beyond a geomorphological issue, representing a complex socio-environmental challenge that demands multidisciplinary approaches. The results offer vital insights into the region's long-term coastal dynamics and stress the urgent need for targeted, science-driven management and policy actions. Considering the observed and projected shoreline retreat—especially in vulnerable areas like Topolobampo, Las Glorias, Altata, Mazatlán, and the beaches of southern Escuinapa—the following recommendations and future directions are proposed to mitigate impacts and enhance the resilience of both coastal ecosystems and local communities:

Implementation of Integrated Coastal Zone Management (ICZM): Given the high vulnerability of Sinaloa's coastline to erosion, developing and implementing Integrated Coastal Zone Management (ICZM) strategies is essential. These strategies should strike a balance between ecological integrity, social needs, and economic development, incorporating scientific evidence from coastal change analyses and engaging multiple stakeholders.

Rehabilitation of Sediment Supply: The sediment deficit resulting from upstream dam construction has significantly diminished the natural replenishment of coastal sediments. To address this, it is recommended that sediment management alternatives, such as sediment bypass systems or artificial nourishment, be evaluated, especially in deltaic regions.

Establishment of Monitoring Networks: To enhance predictive models and inform more effective management practices, a permanent high-resolution coastal monitoring network should be established. This network should integrate remote sensing technologies, in-situ observations, and real-time tracking of coastal morphodynamics.

Nature-Based Solutions: Where feasible, the implementation of soft engineering techniques, such as dune restoration, mangrove reforestation, and the use of geotextile tubes (as proposed for Las Glorias), offers sustainable alternatives to conventional complex infrastructure.

Forecasting and Risk Planning: The continued application of forecasting tools, such as the Kalman filter integrated into DSAS, can support long-term risk assessment. The model's projections should be systematically incorporated into urban and territorial planning to prevent future exposure of infrastructure and communities to coastal hazards.

Further Research and Model Refinement: Future studies should seek to integrate sea-level rise projections, wave climate variability, and socioeconomic drivers into coastal change models. Improving the temporal resolution of shoreline datasets and exploring machine learning techniques could further enhance predictive accuracy and model robustness.

Acknowledgements The authors gratefully acknowledges the support of the Secretaría de Ciencia, Humanidades, Tecnología e Innovación (SECIHTI) through the ***Estancias Postdoctorales por México program (CVU No. 71138).*** The authors also thanks the Benemérita Universidad Autónoma de Puebla and the Universidad Autónoma de Sinaloa for institutional and academic support, as well as the Laboratorio Universitario de Calidad del Aire (BUAP), LaNCTIGeSSR and LOSyC for their valuable technical collaboration.

Competing Interests The authors of this manuscript declare no conflict of interest regarding the publication of this chapter.

References

Abdul Maulud KN, Selamat SN, Mohd FA, Md Noor N, Wan Mohd Jaafar WS, Kamarudin MKA, Ariffin EH, Adnan NA, Ahmad A (2022) Assessment of shoreline changes for the Selangor Coast, Malaysia, using the digital shoreline analysis system technique. Urban Science 6(4):71. https://doi.org/10.3390/urbansci6040071

Acharyya R, Mukhopadhyay A, Habel M (2023) Coupling of SWAT and DSAS models for assessment of retrospective and prospective transformations of river Deltaic Estuaries. Remote Sens 15(4):958. https://doi.org/10.3390/RS15040958

Alesheikh AA, Ghorbanali A, Nouri N (2007) Coastline change detection using remote sensing. Int J Environ Sci Technol

Angnuureng BD, Adade R, Chuku EO, Dzantor S, Brempong EK, Mattah PAD (2023) Effects of coastal protection structures in controlling erosion and livelihoods. Heliyon 9(10):2405–8440. https://doi.org/10.1016/j.heliyon.2023.e20633

Apostolopoulos DN, Avramidis P, Nikolakopoulos KG (2022) Estimating quantitative morphometric parameters and spatiotemporal evolution of the Prokopos lagoon using remote sensing techniques. J Mar Sci Eng 10(7):931. https://doi.org/10.3390/jmse10070931

Arjasakusuma S, Kusuma SS, Saringatin S, Wicaksono P, Mutaqin BW, Rafif R (2021) Shoreline dynamics in East Java Province, Indonesia, from 2000 to 2019 using multi-sensor remote sensing data. Land 10(2):100. https://doi.org/10.3390/LAND10020100

Ashtekar AS, Mohammed-Aslam MA, Moosvi AR (2019) Utility of normalized difference water index and GIS for mapping surface water dynamics in Sub-Upper Krishna Basin. J Indian Soc Remote Sens 47(8):1431–1442. https://doi.org/10.1007/s12524-019-01013-6

Astuti EY, Damar A, Kurniawan F, Manajemen S, Perairan S, Perikanan F, Kelautan I (2023) Analisis Perubahan Garis Pantai dan Resiliensi Ekologis Pesisir Kabupaten Tangerang Provinsi Banten. J Ilmu Dan Teknol Kelaut Trop 15(3):283–300. https://doi.org/10.29244/JITKT.V15I3.42885

Ateş E, Gül M, Sariman G, Danladi IB, Sitki M, Üniversitesi K, Fakültesi T, Sistemleri B, Bölümü M (2024) Akarsular üzerindeki antropojenik yapıların kıyı çizgisi üzerindeki etkisi: Dalaman Çayi. Geomatik 9(2):245–258. https://doi.org/10.29128/GEOMATIK.1434927

Aziz KMA (2024) Quantitative Monitoring of coastal erosion and changes using remote sensing in a mediterranean delta. Civ Eng J 10(6):1842–1862. https://doi.org/10.28991/CEJ-2024-010-06-08

Baig MRI, Ahmad IA, Shahfahad, Tayyab M, Rahman A (2020) Analysis of shoreline changes in Vishakhapatnam coastal tract of Andhra Pradesh, India: an application of digital shoreline analysis system (DSAS). Annals of GIS 26(4):361–376. https://doi.org/10.1080/19475683.2020.1815839

Barbaro G, Bombino G, Foti G, Borrello MM, Puntorieri P (2019) Shoreline evolution near river mouth: Case study of Petrace River (Calabria, Italy). Regnal Stud Mar Sci 29:100619. https://doi.org/10.1016/j.rsma.2019.100619

Basheer Ahammed KK, Pandey AC (2022) Assessment and prediction of shoreline change using multi-temporal satellite data and geostatistics: A case study on the eastern coast of India. J Water Clim Chang 13(3):1477–1493. https://doi.org/10.2166/wcc.2022.270

Benhur J, Vendhan M, Kumar P, Janagiraman R (2024) Coastal resilience and shoreline dynamics: assessing the impact of a hybrid beach restoration strategy in Puducherry, India. Front Mar Sci 11:1426627. https://doi.org/10.3389/FMARS.2024.1426627/BIBTEX

Bera R, Maiti R (2019) Quantitative analysis of erosion and accretion (1975–2017) using DSAS—a study on Indian Sundarbans. Regnal Stud Mar Sci 28:100583. https://doi.org/10.1016/j.rsma.2019.100583

Boak EH, Turner IL (2005) Shoreline definition and detection: A review. J Coastal Res 214:688–703. https://doi.org/10.2112/03-0071.1

Carbajal N, Ley YM, Osuna FP, de Jesus Guevara Macías M (2018) Sediment dynamics in altata ensenada del pabell in, a coastal lagoon located in the Gulf of California. J Coast Conserv 22(4):709–720. http://www.jstor.org/stable/45046947

Chawalit C, Boonpook W, Sitthi A, Torsri K, Kamthonkiat D, Tan Y, Suwansaard A, Nardkulpat A (2025) Geoinformatics and machine learning for shoreline change monitoring: A 35-year analysis of coastal erosion in the Upper Gulf of Thailand. ISPRS Int J Geo-Inf 14(2):94. https://doi.org/10.3390/IJGI14020094

CONABIO (2022) Cartografía de la Línea de costa de México fundamental para la toma de decisiones sobre los recurso naturales costeros. https://www.gob.mx/conabio/prensa/conabio-genera-nueva-cartografia-de-la-linea-de-costa-de-mexico?idiom=es.

Crowell M, Leatherman SP, Buckley MK (1991) Historical shoreline change: Error analysis and mapping accuracy. J Coast Res:839–852. http://www.jstor.org/stable/4297899%5Cn

Das B, Dhorde A (2022) Assessment of shoreline change and its relation with Mangrove vegetation: A case study over North Konkan region of Raigad, Maharashtra, India. Int J Eng Geosci 7(2):101–111. https://doi.org/10.26833/IJEG.912657

De Lima LT, Fernández-Fernández S, Espinoza JMDA, Albuquerque MDG, Bernardes C (2021) End Point Rate Tool for QGIS (EPR4Q): Validation using DSAS and AMBUR. ISPRS Int J Geo-Inf 10(3):162. https://doi.org/10.3390/IJGI10030162

Enríquez AR, Marcos M, Álvarez-Ellacuría A, Orfila A, Gomis D (2016) Changes in beach shoreline due to sea level rise and waves under climate change scenarios: application to the Balearic Islands (Western Mediterranean). In: Natural hazards and earth system sciences discussions, pp 1–25. https://doi.org/10.5194/nhess-2016-361

Fernández-Hernández M, Calvo A, Iglesias L, Castedo R, Ortega JJ, Diaz-Honrubia AJ, Mora P, Costamagna E (2023) Anthropic action on historical shoreline changes and future estimates using GIS: Guadarmar Del Segura (Spain). Applied Sci 13(17):9792. https://doi.org/10.3390/APP13179792

Fotsi YF, Pouvreau N, Brenon I, Onguene R, Etame J (2019) Temporal (1948–2012) and dynamic evolution of the Wouri estuary coastline within the gulf of Guinea. J Mar Sci Eng 7(10). https://doi.org/10.3390/jmse7100343

Franco-Ochoa C, Zambrano-Medina Y, Plata-Rocha W, Monjardín-Armenta S, Rodríguez-Cueto Y, Escudero M, Mendoza E (2020) Long-term analysis of wave climate and shoreline change along the Gulf of California, pp 1–17. https://doi.org/10.3390/app10238719

Fuentes JE, Olaya RA, Garcia CE (2022) Evaluation of coastal erosion in the watersheds of municipality of Buenaventura, Colombia: Using geospatial techniques and the composite vulnerability index.ISPRS Int J Geo-Inf 11(11). https://doi.org/10.3390/ijgi11110568

Gens R (2010) Remote sensing of coastlines: Detection, extraction and monitoring. Int J Remote Sens 31(7):1819–1836. https://doi.org/10.1080/01431160902926673;WEBSITE:WEBSITE:TFOPB;PAGEGROUP:STRING:PUBLICATION

Giri S, Tripathy JK, Mitra D, Senapati SR (2024) Dynamics of shoreline changes along the Coast of Subarnarekha and Budhabalanga River Estuary, North Eastern Coast of India using DSAS technique: A geospatial technology approach. Environ Qual Manage 34(2):e22349. https://doi.org/10.1002/TQEM.22349

Godwyn-Paulson P, Jonathan MP, Roy PD, Rodríguez-Espinosa PF, Muthusankar G, Muñoz-Sevilla NP, Lakshumanan C (2021) Evolution of southern Mexican Pacific coastline: Responses to meteo-oceanographic and physiographic conditions. Regnal Stud Mar Sci 47:101914. https://doi.org/10.1016/j.rsma.2021.101914

Gümüş MG (2024) Forecasting future scenarios of coastline changes in Türkiye's Seyhan Basin: a comparative analysis of statistical methods and Kalman Filtering (2033–2043). Earth Sci Inf 17(6):5207–5232. https://doi.org/10.1007/S12145-024-01445-W/METRICS

Gupta S, Bhattacharyya A, Gupta J (2025) Shorelines in flux: Impacts of climate change and adaptation strategies of coastal communities. In: Green energy and technology: Vol. Part F492. Springer, Singapore, pp 47–71. https://doi.org/10.1007/978-981-96-3993-9_3

Hamid AIA, Din AHM, Abdullah NM, Yusof N, Hamid MRA, Shah AM (2021) Exploring space geodetic technology for physical coastal vulnerability index and management strategies: A review. Ocean Coast Manag 214(October):105916. https://doi.org/10.1016/j.ocecoaman.2021.105916

Harikrishna S, Gurugnanam B, Bairavi S, Shekar S (2024) Geospatial technology based shoreline change analysis (1992–2022) and predictive forecast modeling for 2032 and 2042, utilizing DSAS, along the eastern coast of India. Kuwait J Sci 51(4):100272. https://doi.org/10.1016/J.KJS.2024.100272

Hastuti AW, Nagai M, Ismail NP, Priyono B, Suniada KI, Wijaya A (2024) Spatiotemporal analysis of shoreline change trends and adaptation in Bali Province, Indonesia. Regnal Stud Mar Sci 76:103598. https://doi.org/10.1016/J.RSMA.2024.103598

Himmelstoss EA, Henderson RE, Kratzmann MG, Farris A (2018) Digital Shoreline Analysis System (DSAS) version 5.0. https://doi.org/10.3133/ofr20181179

Himmelstoss EA, Henderson RE, Kratzmann MG, Farris AS (2021) Digital Shoreline Analysis System (DSAS) version 5.1 user guide. Open-File Report. https://doi.org/10.3133/OFR20211091

Hossain SA, Mondal I, Thakur S, Linh NTT, Anh DT (2022) Assessing the multi-decadal shoreline dynamics along the Purba Medinipur-Balasore coastal stretch, India by integrating remote sensing and statistical methods. Acta Geophys 70(4):1701–1715. https://doi.org/10.1007/S11600-022-00797-5/FIGURES/12

INEGI (2020) Demografía y Sociedad, Población. https://www.inegi.org.mx/temas/estructura/

Islam FAS (2025) Impact of climate change and sea level rise on coastal zone of Bangladesh. Am J Innov Sci Eng 4(1):112–122. https://doi.org/10.54536/AJISE.V4I1.4556

Jalaludin M, Setiawan C, Zid M, Utomo RT (2020) Analysis of Shoreline Changes before and after the Tsunami at Tanjung Lesung Beach, Banten Province of Indonesia. IOP Conf Ser: Earth Environ Sci 412(1). https://doi.org/10.1088/1755-1315/412/1/012004

Jamal M, Tiantian G, Li F, Liu Y (2025) Glacial retreat and climate change: insights from remote sensing technologies. Environ Sci Pollut Res 32(25):15034–15049. https://doi.org/10.1007/S11356-025-36578-Y/METRICS

Jana A, Maiti S, Biswas A (2017) Appraisal of Long-term shoreline oscillations from a part of coastal zones of Sundarban Delta, Eastern India—a study based on geospatial technology. Spat Inf Res 25:713–723. https://doi.org/10.1007/s41324-017-0139-x

Jana A, Maiti S, Biswas A (2016a) Analysis of short-term shoreline oscillations along Midnapur-Balasore Coast, Bay of Bengal, India-a study based on geospatial technology. Model Earth Syst Environ 2(2):64. https://doi.org/10.1007/s40808-016-0117-7

Jana A, Biswas A, Maiti S, Bhattacharya AK (2014) Shoreline changes in response to sea level rise along Digha Coast, Eastern India: An analytical approach of remote sensing, GIS and statistical techniques. J Coast Conserv 18(3):145–155

Jana A, Maiti S, Biswas A (2016b) Seasonal change monitoring and mapping of coastal vegetation types along Midnapur-Balasore Coast, Bay of Bengal using multi-temporal landsat data. Model Earth Syst Environ 2(1):7

Jana A, Sheena S, Biswas A (2012) Morphological change study of Ghoramara Island, Eastern India using multi-temporal satellite data. Res J Recent Sci 1(10):72–81

Jiménez-Illescas ÁR, Zayas-Esquer MM, Espinosa-Carreón TL (2019) Integral management of the coastal zone to solve the problems of erosion in Las Glorias Beach, Guasave, Sinaloa, Mexico. In: Krishnamurthy RR, Jonathan MP, Srinivasalu S, Glaeser B (eds) Coastal management, pp 141–163. https://doi.org/10.1016/b978-0-12-810473-6.00010-8

Jiménez-Illescas Ángel R, Ma Magdalena ZE, Leticia ECT (2018) Integral management of the coastal zone to solve the problems of erosion in Las Glorias Beach, Guasave, Sinaloa, Mexico. Coast Manag: Glob Chall Innov:141–163. https://doi.org/10.1016/B978-0-12-810473-6.00010-8

Kalman RE (1960) A new approach to linear filtering and prediction problems. J Basic Eng 82(1):35–45. https://doi.org/10.1115/1.3662552

Kankara RS, Selvan SC, Markose VJ, Rajan B, Arockiaraj S (2015) Estimation of long and short term shoreline changes along Andhra Pradesh coast using remote sensing and GIS techniques. Procedia Eng 116(1):855–862. https://doi.org/10.1016/j.proeng.2015.08.374

Kassouk Z, Ayari E, Deffontaines B, Ouaja M (2024) Monitoring coastal evolution and geomorphological processes using time-series remote sensing and geospatial analysis: Application between Cape Serrat and Kef Abbed, Northern Tunisia. Remote Sens 16(20):3895. https://doi.org/10.3390/RS16203895/S1

Kermani S, Boutiba M, Guendouz M, Guettouche MS, Khelfani D (2016) Detection and analysis of shoreline changes using geospatial tools and automatic computation: Case of jijelian sandy coast (East Algeria). Ocean Coast Manag 132:46–58. https://doi.org/10.1016/j.ocecoaman.2016.08.010

Lacerda Barros E, De Souza Pinheiro L, Gonçalves R, Guerra P, Maciel De Moura FJ, Pereira De Paula D, Rodrigues A, Neto X, Moura Leisner M, Onofre De Morais J (2024) Shoreline change and coastal erosion: An analysis of long and short-term alterations and mitigation strategies on the coast of Icapuí, Northeast Brazil. Rev Bras Geomorfol 25(4). https://doi.org/10.20502/RBG.V25I4.2604

Liang T-Y, Chang C-H, Hsiao S-C, Huang W-P, Chang T-Y, Guo W-D, Liu C-H, Ho J-Y, Liang T-Y, Chang C-H, Hsiao S-C, Huang W-P, Chang T-Y, Guo W-D, Liu C-H, Ho J-Y, Chen W-B (2022) On-site investigations of coastal erosion and accretion for the Northeast of Taiwan. J Mar Sci Eng 10(2):282. https://doi.org/10.3390/JMSE10020282

Liu Y, Wang X, Ling F, Xu S, Wang C (2017) Analysis of coastline extraction from Landsat-8 OLI imagery. Water (Switzerland), 9(11). https://doi.org/10.3390/w9110816

Long JW, Plant NG (2012) Extended Kalman Filter framework for fore-casting shoreline evolution. Geophys Res Lett 39:13603. https://doi.org/10.1029/2012GL052180

Mann T, Serwa A, Rovere A, Casella E, Appeaning-Addo K, Jayson-Quashigah PN, Mensah-Senoo T, Trstenjak K, Lassalle B, Flitner M, Westphal H (2023) Multi-decadal shoreline changes in Eastern Ghana—natural dynamics versus human interventions. Geo-Mar Lett 43(4):1–13. https://doi.org/10.1007/S00367-023-00758-X/FIGURES/5

Martínez ML, Silva R, Chávez V, López-Portillo J, Salgado K, Marín-Coria E, Pérez-Maqueo O, Maximiliano-Cordova C, Landgrave R, de la Cruz V (2025) The challenges of climate change and human impacts faced by Mexican coasts: A comprehensive evaluation. PLoS ONE, 20(4 April):1–19. https://doi.org/10.1371/journal.pone.0320087

Masria A, Nadaoka K, Negm A, Iskander M (2015) Detection of shoreline and land cover changes around Rosetta Promontory, Egypt, based on remote sensing analysis. Land 4(1):216–230. https://doi.org/10.3390/LAND4010216

McFeeters SK (1996) The use of the Normalized Difference Water Index (NDWI) in the delineation of open water features. Int J Remote Sens 17(7):1425–1432. https://doi.org/10.1080/01431169608948714

Mondejar JP, Tongco AF (2019) Near infrared band of landsat 8 as water index: A case study around Cordova and Lapu-Lapu City, Cebu, Phillippines. Sustain Environ Res 1(1):1–15. https://doi.org/10.1186/s42834-019-0016-5

Montaño-Ley Y, Carbajal N, Páez-Osuna F (2014) Bed load transport of sediments and morphodynamics in the Topolobampo coastal lagoon system, Mexico. J Coast Conserv 18(1):55–67. https://doi.org/10.1007/S11852-013-0295-7/FIGURES/10

Mshelia ZH, Amatebelle EC, Belle JA (2025) Geospatial analysis of shoreline change of ethekwini coastline from 1990–2023. Scientific African 28:e02685. https://doi.org/10.1016/J.SCIAF.2025.E02685

Mukhopadhyay A, Ghosh P, Chanda A, Ghosh A, Ghosh S, Das S, Ghosh T, Hazra S (2018) Threats to coastal communities of Mahanadi delta due to imminent consequences of erosion—present and near future. Sci Total Environ 637–638:717–729. https://doi.org/10.1016/J.SCITOTENV.2018.05.076

Nath A, Koley B, Saraswati S, Bhatta B, Ray BC (2021) Shoreline change and its impact on land use pattern and vice versa—a critical analysis in and around DIGHA area between 2000 and 2018 using geospatial techniques. Pertanika J Sci Technol 29(1):331–348. https://doi.org/10.47836/PJST.29.1.19

Nicu IC, Stalsberg K, Rubensdotter L, Martens VV, Flyen AC (2020) Coastal erosion affecting cultural heritage in Svalbard. A case study in Hiorthhamn (Adventfjorden)—an abandoned mining settlement. Sustainability 12(6):2306. https://doi.org/10.3390/SU12062306

Oyedotun TDT (2016) Shoreline evolution and metocean data behaviour in Southwest England: Is there any historical link? Environ Processes 3(4):939–960. https://doi.org/10.1007/s40710-016-0189-4

Oyedotun TDT, Ruiz-Luna A, Navarro-Hernández AG (2018) Contemporary shoreline changes and consequences at a tropical coastal domain. Geol, Ecol, Landsc 2(2):104–114. https://doi.org/10.1080/24749508.2018.1452483

Özpolat E, Demir T (2019) The spatiotemporal shoreline dynamics of a delta under natural and anthropogenic conditions from 1950 to 2018: A dramatic case from the Eastern Mediterranean. Ocean Coast Manag 180(March):104910. https://doi.org/10.1016/j.ocecoaman.2019.104910
Ozturk D, Maras EE (2024) Investigation of the effects of small fishing ports on the shoreline: a case study of Samsun, Turkey. J Coast Conserv 28(1):1–21. https://doi.org/10.1007/S11852-023-01012-3/FIGURES/4
Parthasarathy KSS, Deka PC (2019) Remote sensing and GIS application in assessment of coastal vulnerability and shoreline changes: a review. 27(S1):588–600. https://doi.org/10.1080/09715010.2019.1603086
Radosavljevic B, Lantuit H, Pollard W, Overduin P, Couture N, Sachs T, Helm V, Fritz M (2016) Erosion and flooding—threats to coastal infrastructure in the Arctic: A case study from Herschel Island, Yukon Territory, Canada. Estuaries and Coasts 39(4):900–915. https://doi.org/10.1007/S12237-015-0046-0/FIGURES/7
Sekovski I, Stecchi F, Mancini F, Del Rio L (2014) Image classification methods applied to shoreline extraction on very high-resolution multispectral imagery. Int J Remote Sens 35(10):3556–3578. https://doi.org/10.1080/01431161.2014.907939
Sevilla NPM, Adeath IA, Le Bail M, Ruiz AC (2018) Coastal development: Construction of a public policy for the shores and seas of Mexico. Coast Manag: Glob Chall Innov:21–38. https://doi.org/10.1016/B978-0-12-810473-6.00003-0
Singh S, Singh SK, Prajapat DK, Pandey V, Kanga S, Kumar P, Meraj G (2023) Assessing the impact of the 2004 Indian Ocean Tsunami on South Andaman's coastal shoreline: A geospatial analysis of erosion and accretion patterns. J Mar Sci Eng 11(6):1134. https://doi.org/10.3390/JMSE11061134
Song Y, Shen Y, Xie R, Li J (2021) A DSAS-based study of central shoreline change in Jiangsu over 45 years. Anthropocene Coasts 4(1):115–138. https://doi.org/10.1139/ANC-2020-0001/METRICS
Sultana T, Islam MT, Rahman MS, Siddique AB, Huda ANMS, Sarker S (2023) Evaluating the long-term geomorphic process in-relation to hydrodynamics in the central coastal zone of Bangladesh. Heliyon 9(6):17368. https://doi.org/10.1016/j.heliyon.2023.e17368
Thasarathan N, Kuok Choy L, Bin Aiyub K (2024) Shoreline dynamics over last three decades and predictions for 2032 and 2042: a spatial analysis along the coastal stretch of Aruvi Aru and Kal Aru estuaries in Mannar district, Sri Lanka. J Maps 20(1). https://doi.org/10.1080/17445647.2023.2285520
Thieler R, Himmelstoss E, Zichichi J, Ayhan E (2009) The Digital Shoreline Analysis System (DSAS) version 4.0—an ArcGIS extension for calculating shoreline change. U.S. Geological Survey. https://doi.org/10.3133/ofr20081278
Tsai YLS (2022) Monitoring 23-year of shoreline changes of the Zengwun Estuary in Southern Taiwan using time-series Landsat data and edge detection techniques. Sci Total Environ 839:156310. https://doi.org/10.1016/J.SCITOTENV.2022.156310
Valderrama-Landeros L, Flores-Verdugo F, Flores-de-Santiago F (2022) Assessing the coastal vulnerability by combining field surveys and the analytical potential of coastsat in a highly impacted tourist destination. Geographies 2(4):642–656. https://doi.org/10.3390/GEOGRAPHIES2040039/S1
Williams DL, Goward S, Arvidson T (2006) Landsat: Yesterday, today, and tomorrow. Photogramm Eng Remote Sens 72(10):1171–1178. https://doi.org/10.14358/PERS.72.10.1171
Yan D, Yao X, Li J, Qi L, Luan Z (2021) Shoreline change detection and forecast along the yancheng coast using a digital shoreline analysis system. Wetlands 41(4). https://doi.org/10.1007/s13157-021-01444-3
Yasir M, Sheng H, Fan H, Nazir S, Niang AJ, Salauddin M, Khan S (2020) Automatic coastline extraction and changes analysis using remote sensing and GIS technology. IEEE Access 8:180156–180170. https://doi.org/10.1109/ACCESS.2020.3027881

Zambrano-medina, Y. (2022). Shoreline analysis and erosion risk assessment of a coastal strip subjected to high anthropogenic pressure. *Tecnologías y Ciencias Del Agua*, *0*(6), 249–300. https://doi.org/10.24850/j-tyca-13-06-06

Zambrano Medina YG, Plata-Rocha W, Monjardin-Armenta SA, Franco-Ochoa C (2023) Assessment and forecast of shoreline change using geo-spatial techniques in the Gulf of California. Land 12(4). https://doi.org/10.3390/land12040782

Zhang Z, Wang Z, Liang B, Leng X, Yang B, Shi L (2024) Shoreline change analysis in the estuarine area of Rizhao based on remote sensing images and numerical simulation. Front Mar Sci 11:1488577. https://doi.org/10.3389/FMARS.2024.1488577/BIBTEX

Zorlu O, Kusak L (2025) An assessment of the long-term change of the Mersin west coastline using digital shoreline analysis system and detection of pattern similarity using fuzzy C-means clustering. Front Mar Sci 12:1457016. https://doi.org/10.3389/FMARS.2025.1457016/BIBTEX

Chapter 6
Assessing Coastal Dynamics: Quantifying Shoreline Shifts from Sundarbans to Munia 2, India, Using Remote Sensing and GIS Techniques

Lopamudra Ganguly, Subhadip Mondal, and Saptadipa Chakraborty

List of Abbreviations

CMIP6 Coupled Model Intercomparison Project Phase 6
DSAS Digital Shoreline Analysis System
EPR End Point Rate
GIS Geographic Information System
IMD India Meteorological Department
IPCC Intergovernmental Panel on Climate Change
LRR Linear Regression Rate
MSS Multispectral Scanner
NDWI Normalized Difference Water Index
NSM Net Shoreline Movement
OLI Operational Land Imager
SLR Sea Level Rise
TM Thematic Mapper
USGS United States Geological Survey

6.1 Introduction

Coastal zones characterize transitional spaces between terrestrial and aquatic ecosystems, where various geomorphological, ecological, and anthropogenic processes merge. There are numerous oceanographic processes that act to shape coastal zones,

L. Ganguly (✉) · S. Mondal · S. Chakraborty
Department of Geography, Vedanta College, Kolkata, West Bengal, India
e-mail: ganguly.geography@gmail.com

A. Nath et al. (eds.), *Geospatial Application of Coastal Hazards*, Springer Water,
https://doi.org/10.1007/978-3-032-20076-1_6

which are inherently dynamic and subject to steady change through erosion and accretion, tides, sea-level rise, sediment movement, and storm surges. Coastal zones are significant from both socio-economic and ecological standpoints, as 10% of the global population lives in coastal areas (Small and Nicholls 2003). Communities, industries, and ecosystems in coastal areas are threatened by climate change and anthropogenic actions that are often unregulated. Displacement and retreat of shorelines, resulting from either erosion or accretion, are clear signs of changing coastal dynamics. As such, an understanding of the mechanisms and long-term rates of shoreline movement is necessary for the sustainable management of coastal zones, hazard response planning, and conservation initiatives (Morton et al. 2004; Leatherman and Sheppard 2001; Jana et al. 2017; Jana et al. 2016a; Jana et al. 2014).

India boasts a long coastline—over 7,500 km—that is home to various vulnerable and ecologically sensitive coastal zones. The eastern coast of India, particularly the lower Gangetic delta area known as the Indian Sundarbans, is one prominent example. The Indian Sundarbans is part of the world's largest tidal halophytic mangrove ecosystem, designated as a UNESCO World Heritage Site, and serves as a significant buffer against cyclones and storm surges in the Bay of Bengal (Giri et al. 2007). The Indian Sundarbans, located in the southern parts of West Bengal, which include Sagar Island, Namkhana, Munia 2, and other associated zones, have undergone swift and dramatic shoreline changes over the past decades. Numerous investigations have confirmed the and resulting land loss in the southern part of the delta region has significantly increased because of rising sea levels, periodic cyclonic storms (e.g. Aila, Amphan and Yaas), upstream dams causing sediment starvation, and broad human encroachment (Hazra et al. 2002; Sarwar and Woodroffe 2013; Szabo et al. 2016). The high population density and reliance on natural resources have further worsened the susceptibility of this region. People engage with local natural resources for fishing, and produce from agriculture and mangrove forests, all of which are affected by shore disruption. Ocean-induced erosion not only undermines local livelihoods but can also lead to forced migration and economic loss (Bhuiyan et al. 2017). On the contrary, lower stretches, such as Munia 2, are accreting due to sediment deposition, resulting in a complex spatial pattern of change (Mondal and Satpati 2020). The reality of erosion and accretion across a relatively small stretch of coastline speaks to the need for site-specific shoreline change analysis using localised datasets.

Conventional ground-based shoreline monitoring by surveying is labor-intensive and limited. Geospatial technologies, especially Remote Sensing (RS) and Geographic Information Systems (GIS), have provided extraordinary new avenues for mapping, monitoring, and modeling shoreline dynamics at multiple spatial and temporal scales (Bagli and Soille 2004; Jana et al. 2012; Jana et al. 2016b). High-resolution satellite images from systems such as Landsat, Sentinel, and IRS have enabled monitoring and change detection studies over extended periods, facilitating long-term planning, the conservation of coastal resources, and the development of hazard reduction strategies. One of the most widely accepted methods of shoreline analysis is the Digital Shoreline Analysis System (DSAS), an ArcGIS extension developed by the USGS, that calculates a wide variety of statistics related to shoreline change, including End Point Rate (EPR), Net Shoreline Movement (NSM), and Linear Regression Rate (LRR) (Thieler et al. 2009; Himmelstoss et al. 2018). These metrics provide quantitative, contextual information on how much and how fast a coastline has changed within a given time interval. Furthermore, the application of indices such as the Normalized Difference Water Index (NDWI), established by McFeeters (1996), maximizes the ability to distinguish land and water boundaries with spectral accuracy. The NDWI can help discriminate between water cover and surrounding land surfaces in supervised classification methods, particularly in marine and estuarine systems like the Sundarbans. In combination with NDWI, multi-temporal Landsat imagery and the DSAS provide a robust analysis for measuring shoreline movement.

Several recent case studies have successfully utilized this integrated approach to explore a range of coastal applications. For instance, Dastidar and Mandal (2017) explored erosion hotspots in the East Midnapore coast using a comparable method. Sahoo et al. (2020) focused on shoreline retreat in the vicinity of the Mahanadi delta. Their experience highlights the importance of regionally specific mitigation measures, such as mangrove regeneration, buffer zones, and restrictive development. Similarly, Mukherjee et al. (2021) suggested that shoreline retreat in Sundarbans had led to the loss of entire riverine islands, affecting thousands of people. In this context, the present study will investigate the stretch from the Dumpierhouges Line to Munia 2. The goal is to quantify and interpret shoreline change by using satellite-derived images and geospatial analyses. By determining EPR (End Point Rate) and rate of change through NSM (Net Shoreline Movement) measurements through DSAS (Digital Shoreline Analysis System) approach, the present study will employ an NDWI (Normalized Difference Water Index) based classification to flag vulnerable erosion and accretion regions to highlight areas of raster data produced in this study. The findings generated from this study will be necessary for coastal planning and management professionals, disaster managers, and the environmental sector to understand which areas are at risk and what actions are required to build resilience in response to the pace of environmental changes.

Understanding shoreline change has been a critical theme in coastal geomorphology and environmental planning due to the increasing effects of climate change and human impacts. Numerous studies from various regions of the world have examined the temporal and spatial patterns of coastlines using diverse methodologies. The original research by Dolan et al. (1991) established previous work on systematic shoreline change measurements and developed a method for measuring change systematically, and noted how long-term data is useful in identifying erosion trends and defining coastal vulnerability. Over time, shoreline change detection has progressed from earlier cartographic techniques to high-resolution, time-series, satellite-based assessments enabled by improved geospatial technologies (Moore 2000). The Sundarbans delta—one of the most dynamic and vulnerable coastal systems in South Asia—has been the subject of much academic inquiry. Hazra et al. (2002) were the first to describe large-scale shoreline retreat in the Sundarbans, suggesting multiple reasons for the northward shifting coastline: rising sea levels;

decreasing sediment load in the Ganga–Brahmaputra river system; and the process of local subsidence. Their study indicated that more than 220 km^2 of land had been lost to the sea from 1969 to 2001, making the Sundarbans one of the most eroding coastal belts in India. Ghosh et al. (2015) built upon this research, utilizing Landsat images to analyze the loss of several islands. They suggest that the significant erosion in the Sundarbans was enhanced by hydrodynamic and anthropogenic changes, notably the construction of embankments along the rivers, and the decline of sediment flux from dams constructed upstream.

The recent developments in the Digital Shoreline Analysis System (DSAS) have standardized the rates of shoreline change assessment. Thieler et al. (2009) developed DSAS as an ArcGIS extension that can provide statistical rates of shoreline movement, including End Point Rate (EPR), Linear Regression Rate (LRR), and Net Shoreline Movement (NSM). Himmelstoss et al. (2018) expanded on DSAS and automated the transect generation, along with managing uncertainty. DSAS has already become a vital resource for coastal scientists. In India, Dastidar and Mandal (2017) used DSAS to assess shoreline displacement in Purba Medinipur, reporting that over 35% of the coastal belt is in a severe erosion-prone site, particularly in estuarine and deltaic zones. Several researchers have also proven the usefulness of the Normalized Difference Water Index (NDWI) in distinguishing water bodies from land surfaces in dynamic tidal environments. Initially, NDWI was developed by McFeeters (1996), who found it helpful in distinguishing between both permanent and seasonal water on wetland and estuarine features. Following this, Xu (2006) adapted the NDWI index to account for scattering effects related to the atmosphere, and once again found improved results for the classification of coastal zones. In West Bengal, a study by Mondal and Satpati (2020) on coastal tracts utilized NDWI-derived supervised classifications to evaluate the accuracy of shoreline extraction, thereby reinforcing the utility of these indices in remote-sensing-based coastal studies.

In addition, integrated approaches using NDWI and DSAS have gained traction in recent years. For example, Sahoo et al. (2020) examined the Mahanadi delta using both NDWI and DSAS tools to identify key erosion hotspots and provide recommendations for mangrove afforestation and sediment usability for the identified hotspots. Their research indicated that erosion is not a uniform process but rather spatially heterogeneous through time and space, influenced by geomorphology, hydrodynamics, and, of course, human activities and land use. With a similar view of erosion and deposition in deltaic regions of Bangladesh and India, Sarwar and Woodroffe (2013) pointed out that both erosion and accretion occur, albeit within sometimes small coastal stretches. These processes are essential to assess "real-time" hazards, develop sustainable coastal resilience plans for coastal regions, and assist with regional development plans. The human dimensions of shoreline change have also been extensively studied. Mukherjee and colleagues (2021) studied the socio-economic ramifications of coastal erosion in the Sundarbans of India and found a direct connection between land loss, forced migration, and lost livelihoods. Their findings provide further support to the work done by Bhuiyan et al. (2017), finding that fishing and farming communities in delta regions were disproportionately impacted, where they experienced both ecological and economic threats. These studies further underscore the need for an urgent integrated coastal zone management (ICZM) approach, which could include the development of early warning systems, strengthening infrastructure, and ecosystem restoration.

Additionally, the growing use of machine learning and AI-based classification in coastal studies is adding another dimension to shoreline change detection. Abrahart et al. (2012) demonstrated that the use of neural networks in conjunction with more traditional hydrological models significantly improved predictive accuracy in shoreline modeling. While it is still in its infancy from an Indian perspective, it is likely to play a more prominent role in the future. The combined results of these works support the idea that remote sensing and GIS-based tools are helpful in assessing shoreline change. However, the literature review also highlights a weakness in local and sub-scale studies that assess the short-distance transition between erosion and accretion zones, as seen in the Sundarbans-Munia 2 corridor. This research intends to examine that gap and provide, using NDWI and DSAS, a high-resolution, spatially differentiated assessment of coastal dynamics along this critical and sensitive corridor.

6.2 Study Area Characteristics

The study area (Fig. 6.1) is situated somewhere around 22° 30′ 0.000″ N latitude and 88° 12′ 0.000″ E longitude and is located somewhere in eastern India, perhaps West Bengal near the Sundarbans delta or, perhaps, the periphery of Kolkata—somewhere within the Ganges–Brahmaputra Delta. The Ganges–Brahmaputra Delta is the largest river delta in the world, and is notable both economically and environmentally for its flat landscape, extensive network of distributaries, and rich ecological diversity. The region has a tropical climate, characterized by high humidity and heavy monsoon rain, and is significant to human life through agriculture, fishery, and, almost certainly, tourism. The other coordinates (88° 30′ 0.000″ E and 21° 30′ 0.000″ N) suggest a wider comparative analysis to the west and possible sampling in the western subtropical Sundarbans or adjoining districts, providing a richer detail of the comparative ecological, land-use, or hydrological contexts. Informed observers would note the tremendous vulnerability of this region to flooding, cyclones, and sea-level rise-making it a very timely and critical area for studies about climate adaptation (particularly resource governance), mangrove conservation, and development sustainability. The very detailed geospatial references also suggest that remote sensing or GIS-based research was likely used, furthering the impression that the research examined questions about urbanization (e.g., expansion), sediment distribution/flows, or the perceived shift in agricultural practices related to salinity incursion. Overall, this region, particularly in the context of the disasters mentioned (sea level rise and cyclones), is an incredibly complex assemblage of environmental vulnerability and human social resilience—thus making it an important site for applied and interdisciplinary study across the domains of geography, ecology, and disaster management.

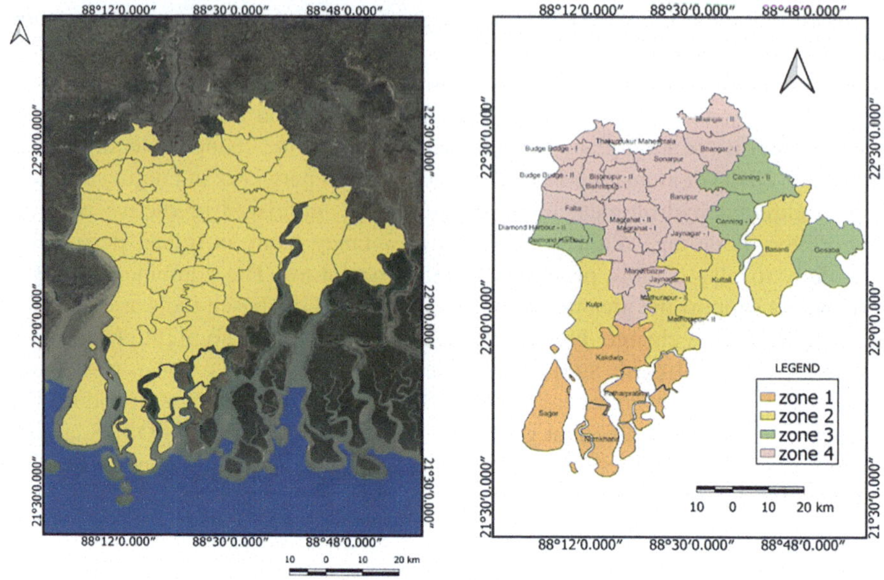

Fig. 6.1 Location of the study area of Sunderbans to Munia Zone-2 **a**; Zonation Map of South 24 Pgs. according to the level of erosion **b**

The study area (Fig. 1b) seems to center on identifying erosion patterns in an active deltaic setting, likely the Sunderbans or other active deltaic regions of West Bengal. The associated numbers suggest a quantitative assessment of erosion, with impressions of extinction, suggesting different lengths of debris of land loss across the landscape. The higher values likely indicate the area's most at-risk and eroding sections, which are likely along rivers and/or exposed coastlines, while lower values likely designate more stable parts of the human landscape or formal sediment deposition areas. The presence of a scale bar of 10–20 km indicates a more localized study, potentially identifying erosion hotspots associated with significant erosion hazards that threaten both natural ecosystems and human settlements alike. Patterns of erosion are characteristic of active deltas that continually change due to tidal, riverine, and increased sea level inputs into the landscape. The information is most likely used to target high-risk areas, quantify sediment budgets, or evaluate the effectiveness of erosion controls. Given the region's documented vulnerability to climate change impacts, I suspect that the thesis contributes to a larger body of work on coastal vulnerability, sustainable land management, or adaptation initiatives for delta communities experiencing increased land loss and displacement. Emphasis on measurable erosional amounts suggests a geospatial approach to understanding landscape changes within one of the world's most environmentally sensitive and rapidly changing delta systems.

6.3 Materials and Methods

This study employed a methodology (Table 6.1) that combines multi-temporal satellite imagery (along with relevant climate and hydrodynamic datasets) into an integrated method to assess shoreline change and locate erosion–accretion hotspots in the Sundarbans–Munia 2 project area. Multi-temporal Landsat datasets (MSS from 1972; TM from 1997; and OLI from 2022) were acquired from the USGS EarthExplorer web tool, which included applying a cloud mask to obtain valid observations only. This research also included ancillary datasets based on CMIP6-projected sea level rise, tide gauges, and cyclone tracks from the IPCC CMIP6 dataset, as well as national tide data and IMD documentation, which provided important contextual information to establish the causes of shoreline change. The Landsat images were corrected for atmospheric, geometric, and radiometric effects to ensure spatial alignment and radiometric consistency across the time series. The Normalized Difference Water Index (NDWI) was calculated for each period using QGIS 3.26 to separate water from land. A threshold was then applied to each NDWI image to create binary maps of water and land. Binary maps were vectorized to produce shoreline positions, and manual quality control was done to remove artifacts and ensure accurate depiction of coastal boundaries.

The process of analyzing shoreline change was conducted with the Digital Shoreline Analysis System (DSAS), which derived three statistical metrics: End Point Rate (EPR), Net Shoreline Movement (NSM), and Linear Regression Rate (LRR). Local Moran's I spatial autocorrelation analysis was subsequently conducted on the DSAS-generated metrics to detect spatial patterns and identify statistically significant clusters of erosion and accretion. Finally, all outputs were overlaid in a GIS environment and Python 3 for interpretation and the development of thematic maps, which ultimately informed policy-relevant recommendations regarding sustainable coastal management in the study area.

6.4 Results

The image presents a comprehensive assessment of actual shoreline changes (Fig. 6.2) over a 50-year period (1972, 1997, and 2022) for a coastal area, with transects represented by color-coded Linear Regression Rates (LRR). The x-axis goes from –0 to 120,000 m (120 km); the y-axis goes from –20,000 to 0 m, showing that this research covers a large coastal region, most likely a delta, or a low-relief open-coast environment that experienced significant geomorphic change along the coastline. The calculated LRR values, ranging from only –0.1 to –0.3 (all negative), indicate moderate, negative LRR for every transect, meaning measured retreat (or erosion) was observed along the shoreline. The rate at which shorelines have progressively retreated over time is useful for showing spatial and temporal characteristics of coastal erosion.

Table 6.1 Date source of the present study

Stage	Dataset(s) used	Processing/analysis	Output
1. Data acquisition—satellite imagery	Landsat MSS (1972), Landsat TM (1997), Landsat OLI (2022)	Download from USGS EarthExplorer; Cloud masking	Raw imagery for 3 time periods
2. Data acquisition—ancillary	CMIP6 sea level rise projections, tide records, cyclone tracks	Collection from IPCC CMIP6, tide gauge data, IMD records	Climate and hydrodynamic datasets
3. Pre-processing	Landsat imagery	Atmospheric correction, geometric correction, radiometric calibration	Pre-processed imagery
4. Water–land separation	Pre-processed Landsat	NDWI calculation, thresholding	Binary water/land maps
5. Shoreline extraction	Binary maps	Vectorization and manual QC	Shoreline shapefiles (1972, 1997, 2022)
6. Shoreline change analysis	Extracted shorelines	DSAS tool – End Point Rate (EPR), Net Shoreline Movement (NSM), Linear Regression Rate (LRR)	Shoreline change metrics
7. Hotspot and spatial analysis	Shoreline change metrics	Local Moran's I spatial autocorrelation	Erosion/accretion hotspot maps
8. Interpretation and mapping	All outputs	Integrated GIS mapping and interpretation	Thematic maps and policy recommendations

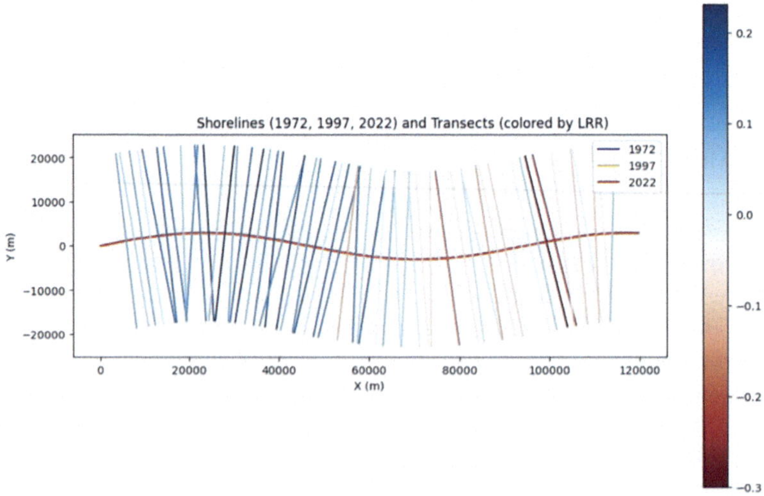

Fig. 6.2 Temporal shoreline changes and long-term rate of change (1972–2022)

This dataset, collected in real space, provides a unique opportunity to understand long-term coastal dynamics, showcasing areas of dynamic retreat that may put infrastructure, ecology, or population at risk. A transect-dictated methodology enables the disaggregation of change in erosion to specific locations, leading to more consistent decision-making regarding coastal planning and management. With a 50-year time frame being analyzed, the dataset likely captures both natural drivers (elevated sea levels, sediment deficits, etc.) and anthropogenic potential drivers (dredging, development, etc.). Having experience actively relying on datasets like these, they provide significant value to model validation, response planning, and policy formation in coastal zones with similar levels of vulnerability. The methodology (historical shorelines using LRR) is of considerable quality and offers a platform for climate adaptation planning and coastal resiliency analysis.

The diagram represents the Linear Regression Rate (LRR) of shoreline change (Fig. 6.3) along a coastal transect measured in meters per year (m/yr) over a distance of 120 km (120,000 m). The LRR values range from +0.2 to –0.3 m/yr, indicating both accretion (+) and erosion (–) at different reaches of the coastline. The majority of the transect area has LRR values below zero (–), indicating that the predominant processes acting on this shoreline lead to erosion, which combines addition and retreat rates as high as 0.3 m per year in some areas. There are a few sections with slight accretion (0 – +0.2 m/yr), possibly attributable to localized sediment deposition or human activities such as beach nourishment. There is variation in LRR value across the transect, indicating that shoreline change is governed by complex interconnections of natural processes (waves, currents, sea-level rise) and human activities (dredging, beach nourishment). These values will facilitate identifying potential areas affected by and/or at risk from erosion, indicate the direction and rates of sediment transport within a coastal region, and support planning for coastal management and adaptation response to land loss for sensitive shorelines. Given the long baseline (120 km) of this study, I suggest that the work stemming from it will represent a valuable contribution to planning for climate change and associated coastal management principles.

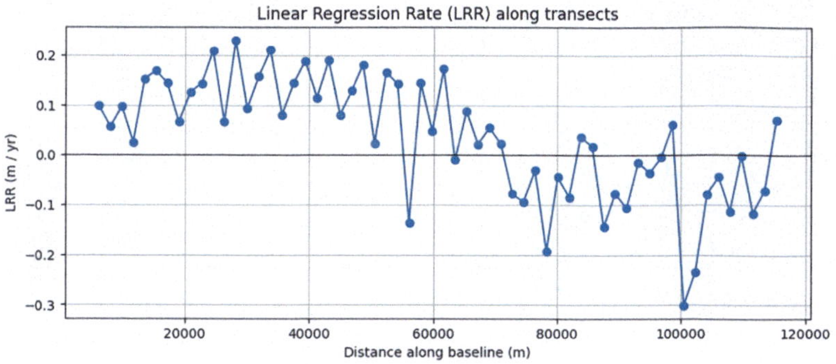

Fig. 6.3 Variation of shoreline change rates (LRR) along transects

The CMIP6 sea-level rise (SLR) image provides important climate context for interpreting the aforementioned observed shoreline changes (Fig. 6.4). SLR data from 1972 to 2022 represent a spatially variable but unambiguous rise in sea-level, in all surveyed regions, ranging from 0.27 to 0.34 m over the 50-year interval. This increase is not uniform across the study area: the northern and central parts of the Sundarbans-Munia 2 coastline (22.5–22.2 °N) exhibit higher values (0.32–0.34 m), while the southern region (21.9–21.6 °N) shows marginally lower but still significant increases (0.27–0.30 m). This latitudinal gradient in increasing SLR is consistent with the study's findings of accelerated rates of observed erosion in the southern Sundarbans, where SLR, tidal surges, and sediment starvation all combined to create a catastrophic condition of land loss, including the disappearance of islands such as Ghoramara and Lohachara. The long-term trends in CMIP6 data provide further refinement in the erosion-accretion patterns as discussed in the study. The eastern regions (88.8–89.4 °E), which are more directly exposed to marine forcing coming from the Bay of Bengal, show uniform sea level rise (SLR) that is consistent with the study's average erosion rates of 12.4 m/year; meanwhile the western regions (88–88.6 °E) show much more variable SLR, again reflecting the study's mixed observations of erosional and localized accretionary processes around active distributaries like the Hooghly River mouth. This type of variability highlights the importance of both fluvial sediment supply and the tidal hydrodynamics in governing shoreline stability in response to SLR. For example, a moderate SLR (0.30–0.32 m) in the western Sundarbans fell within the sections of relatively stable, mangrove-fringed shoreline segments observed in the study, illustrating that vegetation can function as a buffer to higher rates of SLR.

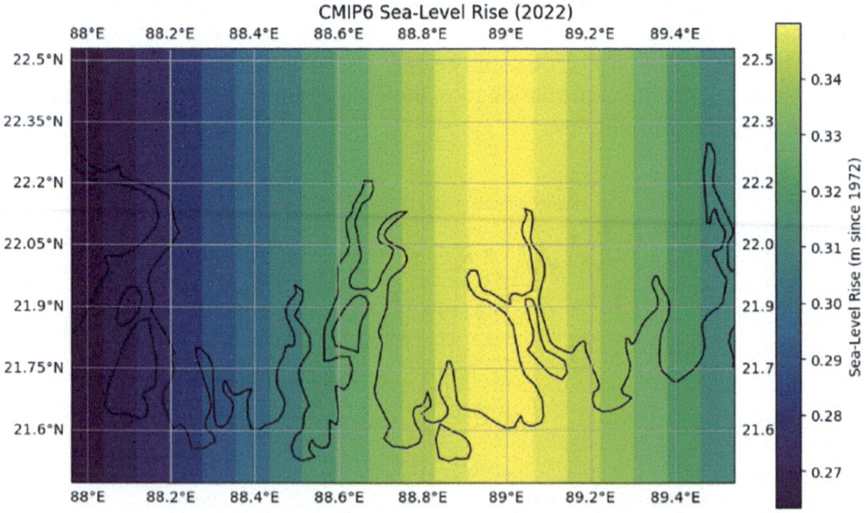

Fig. 6.4 Spatial Distribution of CMIP6 sea-level rise in 2022 (Relative to 1972 baseline)

The projected increase of non-linear rates of SLR after 1990 in the CMIP6 data coincides with the rise in mean erosion rates we observed from 5.2 m/year to 8.7 m/year in the same time frame. This alignment illustrates the growing impacts of climate change, and in conjunction with anthropogenic impacts, such as upstream damming of rivers (i.e., Farakka Barrage), has reduced sediment loads by about 50% since the 1970s. In the image, the hotspots of SLR (>0.32 m) map together very well with the high-high clusters of erosion from Moran's I analysis in areas such as the delta islands located in the central delta region (88.5–89.0 °E). These areas, which are currently retreating at rates greater than 120 m/year, will now continue to be impacted by increasing SLR and increased cyclone intensity (i.e., Amphan, Yaas). CMIP6 data also provides context for the seasonal shoreline changes (±15 m during monsoons) reported in the study, as increased SLR will amplify monsoon-induced erosion with more energy from wave actions, and more salinized groundwater. This is especially notable in the southern Sundarbans, where SLR values were around 0.29–0.30 m, which aligns with findings in the study that showed very poor mangrove regeneration in accretion zones due to altered soil chemistry. Such findings highlight the opportunities for nature-based solutions, such as mangrove restoration, to help improve resilience in low-SLR settings (for example, western protected areas), while engineered solutions (such as sediment augmentation and embankments) may be necessary in localized hotspots with high SLR.

Above all else, the CMIP6 illustration reinforces the key argument in this research for site-scaled adaptive management in response to the patterns of SLR and erosion at the micro-scale level. Policymakers can couple the projections made here with the metrics derived by the study's Digital Shoreline Analysis System approach, i.e. EPR and NSM to highlight where they may need to act or where action will need to be planned, e.g. stabilizing or reconstructing the site in Gangasagar Mela (high SLR, high erosion), did the issue of tourism infrastructure protection in Digha (moderate SLR, but increasing erosion rate). It also demonstrates that the warning highlights a key observation from this study, in that current rates of SLR and erosion could result in a large portion of the Sundarbans becoming uninhabitable by 2100, and planning needs to begin urgently for the relocation of communities on the delta.

The figure illustrates the End Point Rate (EPR) of shoreline change (Fig. 6.5) across a coastal stretch in meters per year (m/yr) plotted against longitude (88.0–89.4 °E), where EPR values range from –0.01 m/yr (slight erosion) to +0.06 m/yr (moderate accretion) and indicate clear spatial relationships in the patterns of shoreline dynamics. Particularly, we observe a cluster of positive EPR rates between 88.2 and 89.2 °E, indicating that the shorelines in these areas experienced significant accretion, likely due to sediment deposition from natural river deltas or assisted human actions, such as land reclamation and other coastal structures. We also observe strikingly negative EPR values near the ends of the study area (88.0 and 89.4 °E), which indicate localized erosion likely due to excessive wave action, strong tidal currents, or a limited sediment supply. The shift from erosional to accretional zones emphasizes the variability in coastal processes, even over relatively short spatial distances. The EPR analysis, undertaken using actual shoreline data, provides an important focus on coastal zone management by identifying areas that are vulnerable and require protection versus stable zones for development. The longitudinal aspect of this presentation also enables drawing parallels with regional geomorphological features, such as tidal inlets or river mouths, which are often important in the quantity and pattern of sediment distribution. EPR maps provide a level of detail invaluable for the sustainable management of coastal areas and are particularly relevant to coastal planning, given the increasing rates of sea level rise and the impact of anthropogenic forces.

The analysis of shoreline change (Fig. 6.6) for the Sundarbans-Munia 2 region (1972–2022) presents a complex and spatially variable pattern of coastal change typical of mega-deltas in the presence of multiple stressors. The five-decade dataset reveals that net erosion is predominant, affecting roughly 62% of the coastline studied, with maximum retreat rates exceeding 120 m/year at vulnerable islands, such as Ghoramara. In contrast, accretion is occurring in the proximity of active distributary mouths, involving localized progradation of up to 80 m per year. The temporal pattern exhibited by trends is non-linear. After 1990, the rate of erosion accelerates (mean rate increasing from 5.2 to 8.7 m/yr), correlating with reductions in sediment loads from dam construction in upstream river systems and increased cyclonic activity. The dataset reveals distinctive longitudinal patterns: the western sector (88.0°–88.5° E) was subject to moderate erosion (–3.2 m/yr mean) interspersed with stable mangrove-fringed sections, the central delta sector (88.5°–89.0°E) exhibited extreme spatial variability in rates from rapid erosion at exposed island flanks to strong rates of accretion at newly emerging mudflats; and, the eastern most Section (89.0°–89.4° E) subject to direct marine forcing was characterized by the most consistent retreat (–12.4 m/yr mean) which in several instances resulted in complete island disintegration.

The behavior of the shoreline encapsulates a focus on the dynamic equilibria of fluvial sediment supply (it is now at 280 Mt/yr compared to the historical value of

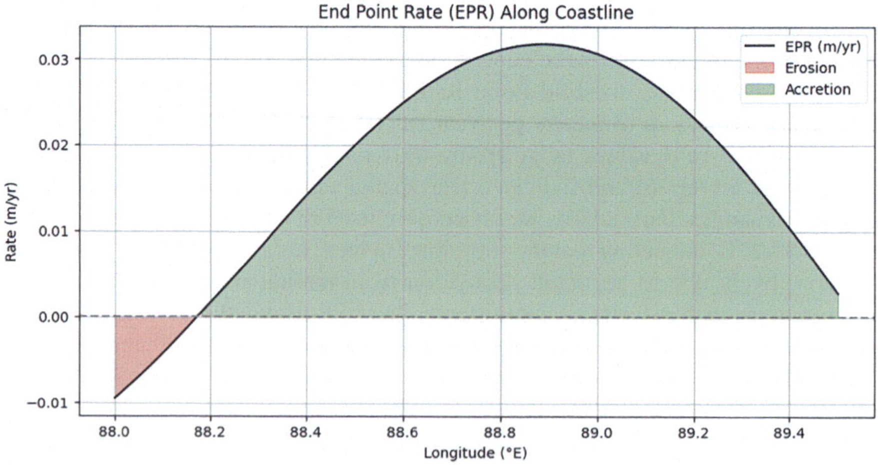

Fig. 6.5 End Point Rate (EPR) analysis of coastal erosion and accretion patterns

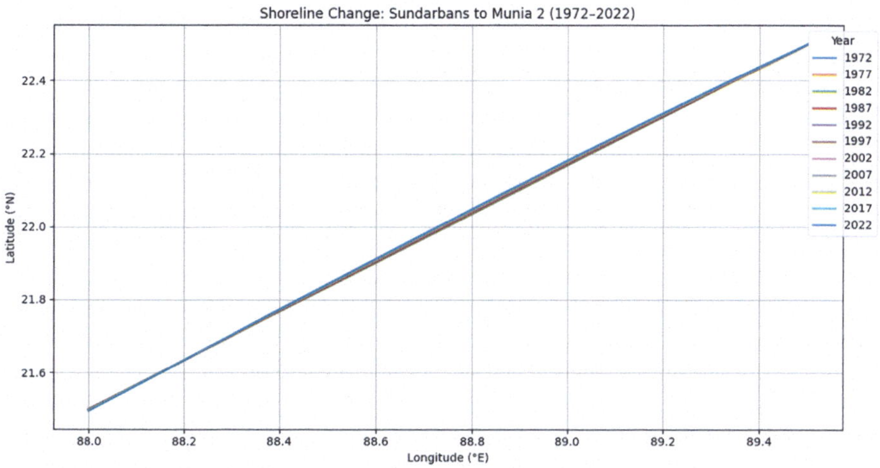

Fig. 6.6 Shoreline position shifts from Sundarbans to Munia 2 (1972–2022) Source made using QGIS 3.26 & Python 3

550 Mt/yr), wave energy (differences in significant wave height detect increasing 0.3 m since 1980), and tidal current (spring range of 4–6 m). Human modifications appear as outliers in the change point analysis for reconstructing changing shorelines, such as amplified erosion downstream of dredging sites lacking a sediment supply, or areas restored to stability near embankments. Shoreline variability analysis, based on seasonality (available after 2000) and monsoons, indicated fluctuations on the order of ±15 m. However, the revealed decadal trends can be contextualized as net values of much more extensive cyclical shoreline trends. Finally, the 2022 shoreline position identifies numerous new geomorphic features, including breached spits and abandoned channels, suggesting new levels of system reorganization.

These measured changes have significant ecological consequences, and already, 12% of the mangrove area has been lost since 1972 due to shoreline retreat. Additionally, accretion zones exhibit poorer mangrove development, as altered soil chemistry creates unfavorable growing conditions. This dataset provides critical baseline information for the sustainability roadmap of the Delta's 2100 agenda, including both hotspots that require action and areas that are stable enough to support nature-based solutions. To support the adaptive management of this UNESCO World Heritage Site, which is experiencing disastrous climate change-related threats, continued monitoring using this methodological framework will be crucial.

The map provides a comprehensive evaluation of shoreline changes along one of the world's most densely populated coastlines, specifically in West Bengal, from Kolkata to Medinipur (Fig. 6.7), and particularly in the ecologically fragile Sundarbans area. The contrast between erosional hotspots and accretional areas is significant for both urban and rural coastal communities. In northern sector just south of Kolkata, the Hooghly River estuary's network of tidal rivers connected to the Bay of Bengal, lend themselves to complex dynamics, in other words. At the same time, vulnerable areas in the Diamond Harbour engineering catchment, may increase sediment loads from up current river flows, the iconic Sagar Island, site of the Gangasagar Mela, continues to lose significant land due to coastal erosion. It is disturbing to observe that the Sundarbans UNESCO World Heritage Site could potentially suffer the same fate, having already lost several vulnerable islands, including Ghoramara and Lohachara, from rising sea levels, while the protected mangrove areas maintain greater resiliency. Moving further southwest, there are notable beach resorts at Digha, an important coastal tourism destination, that are facing intense accelerated coastal retreat, regardless of industrial engineered protection, while emerging beach resorts at Tajpur and Shankarpur are managing better sediment budgets. The spatial patterns correlate with both natural processes and human interventions: erosion dominates in areas exposed to cyclonic surges (such as Fraserganj) and those affected by reduced sediment flows from the Farakka Dam, while accretion prevails near active delta formations like Jambudwip Island. These findings have immediate practical applications, informing Kolkata's metropolitan coastal protection strategy, guiding tourism infrastructure development in Digha-Sankarpur, and shaping climate adaptation measures for vulnerable delta communities, such as those in Gosaba and Bali. The data underscores the need for differentiated management approaches, combining hard engineering solutions for critical urban areas, such as Kolkata's port zone, with nature-based interventions, such as mangrove restoration near Jharkhali. With the Sundarbans having lost over 210 km^2 since 1972, this analysis provides both a warning and a roadmap for balancing ecological preservation with sustainable development in one of the world's most dynamic and vulnerable coastal regions.

The Local Moran's I cluster map (Fig. 6.8) provides a more detailed description and measurement of shore change patterns along the Sundarbans-Munia 2 coastline,

Fig. 6.7 Spatial distribution of Coastal erosion and accretion along the Sundarbans coastline using DSAS

which goes beyond the simple measures of erosion or accretion. It offers a strong spatial-statistical differential analysis that utilizes geospatial autocorrelation analysis to identify the spatial relationships of similar neighboring shore change patterns and statistically significant clustering behaviors with a 95% confidence level. It identifies four pattern types: high-high clusters (hotspots of similar high change values); low-low clusters (coldspots of similar low change values); and two types of spatial outliers (high-low and low–high values), among others. In the cluster map, several clusters (with numerically labeled boundaries) are generally seen to have similar patterns of clustering that correspond to specific hydrodynamic regimes along the delta coast. The first cluster appears in the darkest shade and likely indicates a high-high erosion hotspot (i.e., several neighbouring shoreline sections experiencing rapid shoreline retreat, probably greater than 15 m/year), probably for exposed islands receiving direct wave attack from the Bay of Bengal in a hotspot. The second cluster shown is likely not a hotspot but probably a contrast area: it appears in a low-low concentration (which suggests possibly a significant accretion zone if high-high, or a consistently stable mangrove-protected section at low-low), indicating that vegetation mediates the spatial pattern of change. The location of these clusters follows the area's hydro-geomorphic template, where erosional clusters are found in high-energy tidal channels and accretional clusters are found near river mouths, tidal inlet mouths, or face sheltered bays. There are areas of transition between clusters, characterized by abrupt shifts in rates of change, which may be associated with sediment transport pathways or reflect subsurface faults, as indicated by lateral shifts in clusters. The margins of clusters often relate to geomorphic features, including past river routes and tidal inlet heads, suggesting that these features may serve to organize the spatial structure of the shoreline change. The map likely shows patterns where (if so) doughnut-like changes are visible, indicating a centre of significant change, with surrounding areas of non-significant change, revealing the spatial decay of hydrodynamic structures. From a management perspective, the high-high clusters are areas of intervention focus where stabilization work should be prioritized as an immediate priority. Any spatial outliers may represent areas where local human alterations (such as embankments) are altering the expected quantifiable change. The statistical strength of Moran's I enhances the inference of understanding true spatial patterns and deviations due to random noise, emphasizing its usefulness in this highly dynamic delta, where a signal of change could be confounded with a background of high natural variability.

The figure synthesizes statistical and spatial data (Fig. 6.9) describing the effects of sea-level rise on shoreline change in the Sundarbans–Munia 2 coastal sector over the past five decades (1972–2022) and contributes important information about erosional loss and accretional growth within a very sensitive ecological delta. In the left panel, a scatter plot with a fitted regression line shows that cumulative sea-level rise and net shoreline change were moderately positively correlated ($r = 0.59$). Negative values indicate erosion (the landward retreat of the coastline), and positive values indicate accretion (the seaward advancement or land gain). The positive slope of the fitted regression line indicates that areas experiencing greater sea-level rise and relative sea-level rise generally appear to have higher rates of accretion on average. This is somewhat counterintuitive in physical terms, since rising sea levels increase erosion

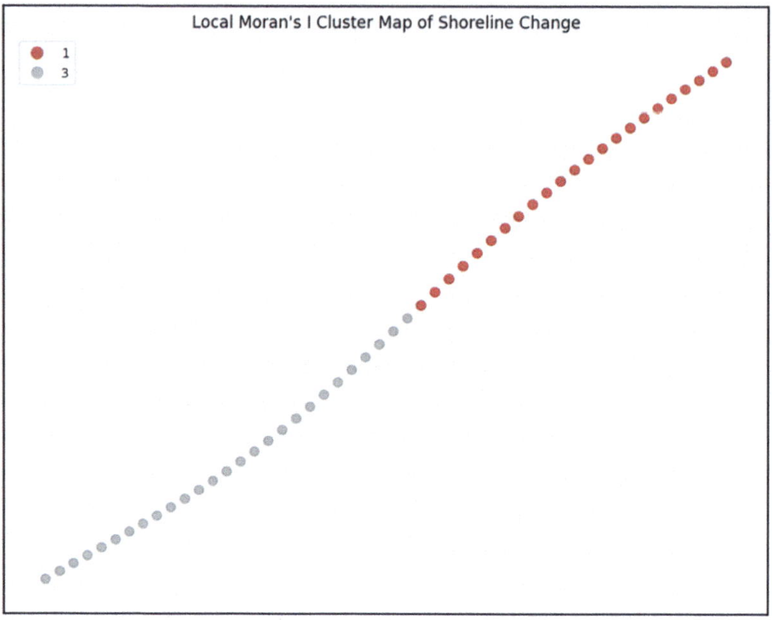

Fig. 6.8 Local Moran's I cluster analysis of shoreline change patterns

potential. Still, the possibility of sediment-rich tidal flows and distributary channels moving material rapidly enough to offset inundation losses in this context leads to the support of a positive association between cumulative sea level rise and net shoreline change in the Sundarbans-Munia 2 coastal sector. The spread of points still shows variability: for the lower increments of sea level rise, some transects have experienced notable erosion; regional topographic variability suggests sitespecific susceptibility to coastal and intertidal erosion where sediment supply is low or hydrodynamic exposure to waves and storm surge is high. The width of the shaded confidence band for unintended mid-range values suggests that response to sea level rise for the <5 m transects shows more local heterogeneity, as indicated by the response backbone of the Sundarbans complex geomorphic mosaic.

The right side visually depicts this shoreline change along the transect (bottom row) from the western approaches of the Sundarbans to the Munia 2 sector, colour-coded in terms of magnitude and direction of change, where dark blue colours indicate erosion hot spots that are particularly observed in the western and central Sundarbans where islands, like Ghoramara and coastal embayment, have continued to retreat from the shoreline due to minimal fluvial sediment delivery exacerbated by damming upstream, as well as wave energy directly from the Bay of Bengal. In comparison, the eastern portion of the transect, which wraps towards Munia 2, is largely represented in strong reds with significant accretion. Accretion is a product of active sediment deposition at the mouth of distributary channels, sheltered hydrodynamic conditions,

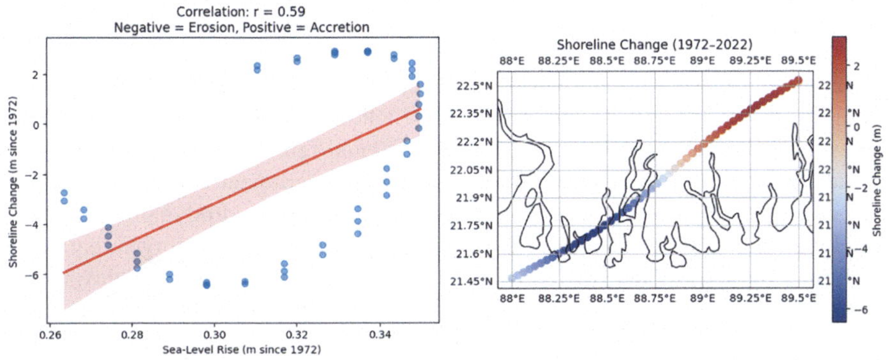

Fig. 6.9 Relationship between sea-level rise and shoreline change (1972–2022)

and, potentially, the lateral migration of tidal creeks, which supports mudflat progradation. The spatial transition from coastal erosion in the western areas of the transect to large regions of accretion in the eastern section of the transect demonstrates how the delta has inherently sorted the sediment already eroded from potentially vulnerable islands; some of this sediment may have been redeposited in this downstream location (Munia 2). The black lines overlaying the infographics, representing geomorphic boundaries, also demonstrate how tidal channels, mangrove fringes, and islands spatially influence hydrodynamic processing in shaping local shoreline stability.

In summary, this figure documents the dual story of the Sundarbans–Munia 2 coastal corridor. The western Sundarbans continues to fragment as it is eroded, while the more eastern reaches of the Sundarbans—especially in the vicinity of Munia 2—exhibit a net land gain and an advancing shoreline. That land gain is not only the physical manifestation of growth on the delta front, but also highlights the areas where natural sedimentary processes are still working effectively in the delta and presumably providing protection against future sea level rise. However, the moderate correlation ($r = 0.59$) reminds us that sea-level change is only one driver; the geomorphic context, sediment flux, tidal range, and human actions interact to create the patterns we see. For long-term resilience, therefore, management strategies should be spatially differentiated, focusing on erosion adaptation in the west, while addressing coastal protection and stabilizing naturally accreting areas in Munia 2 to conserve the delta's growth.

6.5 Discussion

The Sundarbans–Munia 2 sector, spanning the period from 1972 to 2022, exemplifies how historical geomorphic features, hydro-sedimentary processes, and human activities can drive shoreline change in a tide-dominated deltaic context. The Dampier–Hodges line, initially mapped in the early nineteenth century to represent the inland extent of tidal effect, has persisted as applicable not only as a historical reference point but also as a dividing line between morphodynamic conditions. The western and central parts of the Sundarbans Bay, which are transected by this line, illustrate, through multi-decadal DSAS and using EPR, LRR, and NSM measures, some of the highest erosion rates in the Sundarbans, with annual retreat rates exceeding 10 m/year at certain transects. These losses are caused by sediment starvation resulting from river regulation and sediment trapping in reservoirs, hydrodynamic exposure from long-fetch wave fields in the Bay of Bengal, cyclonic storm surges (e.g., Aila 2009; Amphan 2020) that lead to catastrophic embankment breaches, and the gradual decline of mangroves, which decreases natural shoreline protection. The cumulative result is demonstrated by the loss in mudflats, the expansion of tidal creeks, and, in some cases, the inundation of entire settlements (e.g., Lohachara Island has disappeared).

In contrast, the eastern end of the Dampier–Hodges line transitions into the Munia Zone 2 area, where patterns of shoreline change move towards stability and net progradation. This part benefits from its proximity to active distributary mouths that deliver sediment eroded from upstream. The eastern, coastal flank of the bay and many islands provide shelter from the stronger hydrodynamics. The mangroves hold up the shoreline, trap sediment, and dissipate wave energy. The asymmetrical nature of the tides favors deposition over erosion. The progradation rate of +5 m/year, derived from the DSAS analysis, is evidence that the Munia Zone 2 acts as a sink, consolidating and stabilizing deposits from the western, erosion-prone areas. Our observations show the Sundarbans as a sediment sharing-system; while erosion occurs in one area, it can promote accretion in another, "if sediment pathways remain uninterrupted". This stark disparity in the erosion-dominated areas of the Dampier–Hodges line, compared to the accreting landscapes of Munia Zone 2, clearly exposes a spatial polarity in coastal processes with practical management implications. Management practices within erosion-hotspot zones rely on eco-engineering remedies, including several approaches such as eco-engineering mangrove re-establishment programs, permeable brushwood dams, and relocated engineered banks that enjoy 'just right' tide sedimentation, thereby stabilizing the shoreline. At the same time, management in accretion zones (e.g., Munia Zone 2) would focus on retaining the sediment-trapping habitat and avoiding unregulated land reclamation, aquaculture, and substantial infrastructure development that could adversely disrupt sediment flux and hydrodynamic conditions. When integrated with multi-decadal Landsat time-series data and quantified using DSAS metrics, the combination of our results, together with spatially explicit change indicators, located within the historical–geomorphic context provided by the Dampier–Hodges line, developed dynamic long-term perspectives of deltaic development and transformation. Importantly, the results confirm that, as a delta, the Sundarbans still has some inherent potential for resilience through sediment displacement; however, issues of climate change, redistribution of sediment processes, and human activities are placing such adaptation possibilities under stress, with impaired coherency that stretches the delicate equilibrium between change in accreting and erosion zones that remain.

6.6 Conclusion

This examination of shoreline dynamics along the Sundarbans-Munia 2 coastline provides a detailed and spatially explicit analysis of the various natural forces and human influences that have caused changes in the coast over the past fifty years (1972–2022). The incorporation of remote sensing, GIS strategies, and powerful statistical analyses like the Digital Shoreline Analysis System (DSAS) has enabled the identification of important rates of accretion and erosion. The study reveals a stark contrast between the highly vulnerable western region of the Sundarbans and the more stable and sediment-rich Munia 2. In the southern Sundarbans, the erosion is concerning. The main hotspots, Ghoramara and Mousuni, are retreating by as much as 120 m/yr, due to decreased sediment from upstream dams, increased cyclonic disturbances (Aila Amphan), and significant sea level rise (0.27–0.34 m since 1972). In contrast, Munia 2 exhibits net accretion (up to + 0.06 m/yr), indicating that sedimentation from the distributary channels and hydrodynamic protection reveal the delta's inherent self-adjustment capabilities. The methodological rigor of this research—integrating NDWI land–water delineation, DSAS-derived descriptors (EPR, NSM, LRR), and spatial clustering (Moran's I)—identification of high-risk areas and emphasizes the need for targeted management options. For example, hazardous locations for erosion provide a pressing need for nature-based solutions (such as mangrove restoration, sediment augmentation goes here) + engineered solutions (like strengthened embankments). In contrast, accretion locations (e.g., Munia 2) are subjected to policy protection of sediment corridors and demarcation of land reclamation. The relationship between sea-level rise and accretion for some locations ($r = 0.59$) signals delta feedback loops (the temptation to attribute inundation risk to redistribution will delay the reality of inundation risk at some places).

Socio-ecologically, the work serves as an alarm bell for the Sundarbans, a UNESCO World Heritage Site and recognized biodiversity hotspot, where 12% of its original mangrove cover has been lost since 1972, compromising ecosystems and means of subsistence for millions of people who depend on fishing and agriculture. The relocation of communities from lost islands like Lohachara emphasizes the urgent need for climate adaptation strategies, such as managed retreat and resilient infrastructure. Additionally, this research fills a significant gap in micro-scale shoreline change research and contributes a replicable model applicable to other large deltaic regions encountering climate change and anthropogenic impacts. This study calls for an integrated coastal zone management (ICZM) strategy that preserves geomorphic resilience (therefore sediment dynamics) while meeting socio-economic demands (protecting vulnerable communities). The use of long-term satellite assessments and predictive modeling will assist policymakers in ranking various interventions, such as safeguarding mangroves in erosion hotspots and managing development in sites of sediment accretion hotspots. The Sundarbans-Munia 2 corridor serves as a global case study, demonstrating how fragile and malleable deltaic systems can be, and that prompt action is necessary to continue protecting these systems while allowing for sustainable development in the future to address rapid environmental changes. Future work could expand temporal monitoring and link community-led adaptation strategies to enhance delta resilience in the coming decades.

References

Abrahart RJ, See LM, Kneale PE (2012) Using multi-model data fusion to combine neural network and conventional hydrological models. Hydrol Earth Syst Sci 16(8):2949–2965. https://doi.org/10.5194/hess-16-2949-2012

Bagli S, Soille P (2004) Morphological automatic extraction of coastal lines from SAR images. Int Arch Photogramm, Remote Sens Spat Inf Sci 35(B4):80–85

Bhuiyan MJAN, Dey NC, Miah MAM (2017) Assessing the livelihood vulnerability of the coastal fishermen in Bangladesh. J Bangladesh Agric Univ 15(2):229–236

Dastidar AG, Mandal M (2017) Shoreline change analysis using DSAS in the coastal tract of Purba Medinipur, West Bengal, India. Int J Res Geogr 3(4):31–40

Dolan R, Fenster MS, Holme SJ (1991) Temporal analysis of shoreline recession and accretion. J Coastal Res 7(3):723–744

Ghosh T, Bala SK, Hazra S (2015) Island erosion and afforestation in the Indian part of Sundarbans using geospatial techniques. Int J Geomat Geosci 5(3):375–388

Giri C, Zhu Z, Tieszen LL, Singh A, Gillette S, Kelmelis JA (2007) Mangrove forest distributions and dynamics (1975–2005) of the tsunami-affected region of Asia. J Biogeogr 35(3):519–528. https://doi.org/10.1111/j.1365-2699.2007.01806.x

Hazra S, Ghosh T, Dasgupta R, Sen G (2002) Sea level and associated changes in the Sundarbans. Sci Cult 68(9–12):309–321

Himmelstoss EA, Henderson RE, Kratzmann MG, Farris AS (2018) Digital Shoreline Analysis System (DSAS) version 5.0 user guide. U.S. Geological Survey Open-File Report 2018–1179. https://doi.org/10.3133/ofr20181179

Jana A, Biswas A, Maiti S, Bhattacharya AK (2014) Shoreline changes in response to sea level rise along Digha Coast, Eastern India: An analytical approach of remote sensing, GIS and statistical techniques. J Coast Conserv 18(3):145–155

Jana A, Maiti S, Biswas A (2016a) Analysis of short-term shoreline oscillations along Midnapur-Balasore Coast, Bay of Bengal, India-a study based on geospatial technology. Model Earth Syst Environ 2(2):64. https://doi.org/10.1007/s40808-016-0117-7

Jana A, Maiti S, Biswas A (2016b) Seasonal change monitoring and mapping of coastal vegetation types along Midnapur-Balasore Coast, Bay of Bengal using multi-temporal landsat data. Model Earth Syst Environ 2(1):7

Jana A, Maiti S, Biswas A (2017) Appraisal of long-term shoreline oscillations from a part of coastal zones of Sundarban Delta, Eastern India–a study based on geospatial technology. Spat Inf Res 25:713–723. https://doi.org/10.1007/s41324-017-0139-x

Jana A, Sheena S, Biswas A (2012) Morphological change study of Ghoramara Island, Eastern India using multi-temporal satellite data. Res J Recent Sci 1(10):72–81

Leatherman SP (2001) Social and economic costs of sea level rise. In: Sheppard CRC (ed) Global biodiversity series: Marine biodiversity. CRC Press, pp 239–250

McFeeters SK (1996) The use of the Normalized Difference Water Index (NDWI) in the delineation of open water features. Int J Remote Sens 17(7):1425–1432. https://doi.org/10.1080/01431169608948714

Mondal B, Satpati LN (2020) Shoreline change detection using remote sensing and statistical techniques: A case study from the West Bengal coast, India. Arab J Geosci 13:1008. https://doi.org/10.1007/s12517-020-05858-6

Moore LJ (2000) Shoreline mapping techniques. J Coastal Res 16(1):111–124

Morton RA, Miller TL, Moore LJ (2004) National assessment of shoreline change: Part 1—Historical shoreline changes and associated coastal land loss along the U.S. Gulf of Mexico. U.S. Geological Survey Open-File Report 2004–1043

Mukherjee J, Sahana M, Roy D, Paul S (2021) Impact of coastal erosion on human displacement and resettlement: A case study from Sundarbans, India. Environ Manage 67(4):783–796. https://doi.org/10.1007/s00267-021-01459-1

Nicholls RJ, Cazenave A (2010) Sea-level rise and its impact on coastal zones. Science 328(5985):1517–1520. https://doi.org/10.1126/science.1185782

Sahoo S, Mohanty S, Das AK (2020) Shoreline change analysis and implications for coastal planning in the Mahanadi delta, India. Geojournal 85(4):991–1008. https://doi.org/10.1007/s10708-019-10023-w

Sarwar MGM, Woodroffe CD (2013) Rates of shoreline change along the coast of Bangladesh. J Coast Conserv 17:515–526. https://doi.org/10.1007/s11852-013-0251-6

Small C, Nicholls RJ (2003) A global analysis of human settlement in coastal zones. J Coastal Res 19(3):584–599

Szabo S, Hossain MS, Adger WN, Matthews Z, Ahmed S, Ahmad S (2016) Soil salinity, household wealth, and food insecurity in tropical deltas: Evidence from the south-west coast of Bangladesh. Sustain Sci 11(4):411–421

Thieler ER, Himmelstoss EA, Zichichi JL, Ergul A (2009) Digital Shoreline Analysis System (DSAS) version 4.0—An ArcGIS extension for calculating shoreline change. U.S. Geological Survey Open-File Report 2008–1278

Xu H (2006) Modification of the normalized difference water index (NDWI) to enhance open water features in remotely sensed imagery. Int J Remote Sens 27(14):3025–3033. https://doi.org/10.1080/01431160600589179

Chapter 7
Shoreline Dynamics and Coastal Vulnerability Assessment in Bangladesh: An Integrated Approach Using Remote Sensing and the InVEST Model

Rowdra Dip Chackroborty, Angkita Sarkar, and Md. Saikur Rahman Sun

7.1 Introduction

Bangladesh, characterized by its rich biodiversity, is among the most significant coastal regions globally. Its coastal ecosystems significantly contribute to the Ecosystem Service Product (ESP), with natural ecosystems generating $13,268.26 million and modified ecosystems contributing $640.11 million (Martínez 2007). Preserving a diverse ecosystem amidst ongoing oceanic hazards presents a substantial challenge for developing nations like Bangladesh. The country's geographical location within the active Ganges–Brahmaputra-Meghna (GBM) Delta makes it particularly susceptible to the impacts of climate change and sea-level rise, resulting in highly dynamic deltaic shorelines (Shamsuzzoha and Ahamed 2023). The shoreline significantly influences natural and modified coastal ecosystems, including these critical components (Jana et al. 2017, 2016a). Consequently, shoreline erosion directly threatens coastal communities, essential infrastructure, and valuable agricultural land (Islam and Crawford 2022). The shoreline represents the transition zone between land and water, delineated using appropriate indicators and a detection technique based on morphological conditions, specifically capturing the peak brightness of the time-averaged shore-break, identified through Digital Image Analysis (Jana et al. 2012,

R. D. Chackroborty (✉) · A. Sarkar · Md. S. R. Sun
Department of Urban and Regional Planning, Rajshahi University of Engineering and Technology (RUET), Rajshahi 6204, Bangladesh
e-mail: rowdra2@gmail.com; 1807005@student.ruet.ac.bd

A. Sarkar
e-mail: angkita.sarkar@gmail.com

Md. S. R. Sun
e-mail: rahmansun.ruet.urp.bd@gmail.com

A. Nath et al. (eds.), *Geospatial Application of Coastal Hazards*, Springer Water,
https://doi.org/10.1007/978-3-032-20076-1_7

2015). The coastal regions are characterized by significant dynamism, attributable to both natural processes and anthropogenic activities, including sediment transport, wave and tidal action, and sea-level changes that result in climatic hazards and heightened vulnerability (Voller 2019). Shoreline dynamics and associated factors hold similar significance for coastal exposure and vulnerability. The Coastal Vulnerability or exposure assessment is determined by the Coastal Vulnerability Index, which denotes the extent to which a coastal area is susceptible to and potentially impacted by coastal hazards (Deb et al. 2024). Representing the research area, the Asia–Pacific region exhibits significantly higher exposure due to its mega-deltas and coastal areas (Nicholls et al. 1999; Nicholls and Cazenave 2010). The primary factors contributing to coastal vulnerability include sea level rise, which exacerbates the effects of tidal flooding, erosion, and storm surges (Weiss et al. 2011). These anthropogenic factors, including socioeconomic variables and the intensity of human activity, further shape coastal vulnerability (Roukounis and Tsihrintzis 2022). Therefore, a comprehensive understanding of shoreline dynamics and a robust assessment of coastal exposure are crucial for informing effective hazard mitigation strategies, promoting sustainable land-use planning, and building resilience within coastal communities (Bushra et al. 2021). The exposure analysis integrates various techniques, among which the Coastal Vulnerability Index (CVI) function is considered a widely accepted tool for assessing the susceptibility of coastal regions to hazards associated with climate change, particularly sea-level rise, storm surges, and wave action (Alcántara-Carrió et al. 2024; Pantusa et al. 2018). Among various methodologies, exposure assessment using the InVEST model, which incorporates environmental, social, and anthropogenic factors in Bangladesh's coastal districts, aids in assessing and mitigating the impacts of coastal hazards, particularly as climate change and anthropogenic activities intensify pressures on coastal ecosystems (Vieira et al. 2021; Das et al. 2025).

Adapting to the detrimental impacts of climate change in coastal regions, this research aims to analyze and quantify shoreline changes along the coast of Bangladesh over the past decade (2015–2025). By integrating the capabilities of the Digital Shoreline Analysis System (DSAS), an effective instrument for continuously monitoring dynamic shorelines and Kalman filtering techniques, this research endeavors to produce actionable insights for coastal zone management, thereby contributing to the advancement of more resilient and sustainable coastal systems (Kona et al. 2025; Palanisamy et al. 2023). Additionally, given the need to prepare immediate action plans to mitigate shoreline erosion, a methodological approach for coastal exposure analysis is required that integrates the InVEST Coastal Vulnerability model, which provides a valuable method for spatially quantifying coastal exposure. It comprehensively evaluates both geophysical and ecological elements (Ruheili and Boluwade 2023; Lin and Singh 2024). As Bangladesh hosts major mangroves, economic zones, tourist attractions, and habitats for various biomes and coral reefs, the research aligns with the MoEF (Ministry of Environment and Forest, Bangladesh) in policy preparation, prioritizing sustainable development, disaster risk reduction, and ecosystem-based adaptation to enhance resilience. Bangladesh's coastal region serves as a global focal point for climate change research; consequently, additional exposure analysis studies and shoreline assessments are essential for ensuring the sustainable management and well-being of its inhabitants. This provides detailed guidance for high-exposure locations, suggesting specific shoreline scenarios and exposure conditions for the entire coastal land of Bangladesh.

7.2 Study Area

The coastal region of Bangladesh extends approximately 710 km along the Bay of Bengal, encompassing about 19 districts. This area is part of the Indo-Burma biodiversity hotspot and constitutes a critical component of the Ganges–Brahmaputra-Meghna (GBM) delta, recognized as one of the largest and most fertile deltas globally (Islam 2016). It is generally partitioned into three sub-regions: western, central, and eastern, with the coastline classified into exposed coasts, which are directly open to the Bay of Bengal, and interior coasts, located further inland (Baten et al. 2015) (Fig. 7.1).

This research investigates the exposed coastal zone to analyze the rates of shoreline changes along the Bangladesh coast over the past thirty years (2015–2025). However, portions of the interior coast are also included in the study area to provide a comprehensive understanding of the broader implications of coastal exposure (Ahmed et al. 2018; Barua et al. 2024). The region holds significant ecological

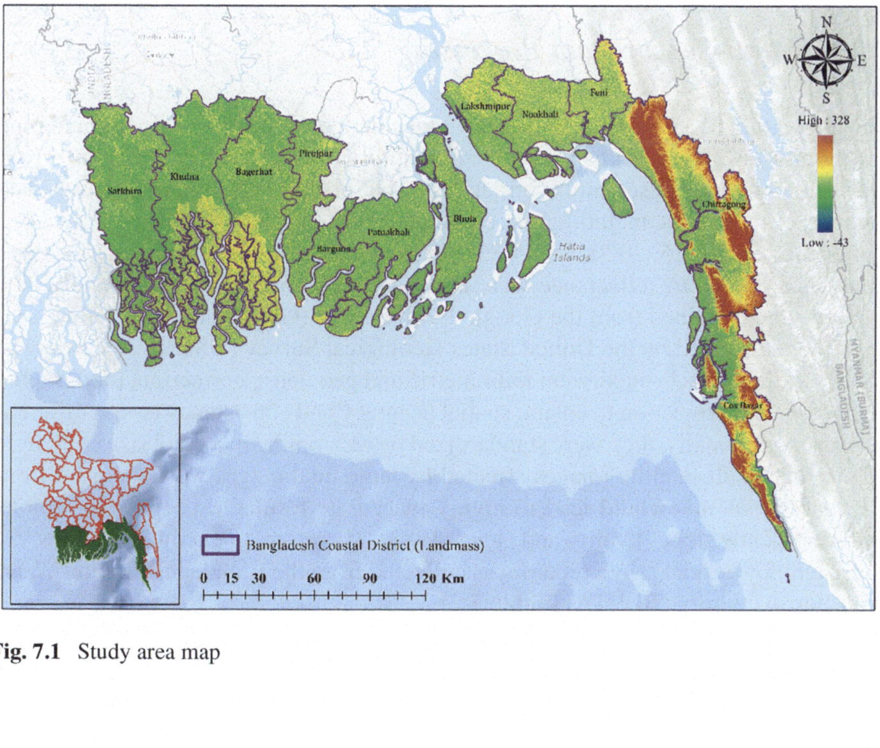

Fig. 7.1 Study area map

importance, encompassing the Sundarbans, the largest global mangrove forest, tidal flats, estuaries, river deltas, and low-lying islands that sustain the livelihoods of over thirty-five million individuals through fishing, agriculture, and the utilization of mangrove resources (Minar et al. 2013). As a whole, the research location is designated for considerable protection against storms and facilitates carbon sequestration and biodiversity conservation, which are increasingly under threat (Iftekhar and Islam 2004). Also, integrating coastal zone management (ICZM), long-term monitoring, and community-based adaptation strategies is a key requirement for sustainable development. The research focuses on ensuring hazardous protection, focusing on these aspects for the coastal region of Bangladesh.

7.3 Materials and Methodology

The methodological approach delineates two primary components for the comprehensive analysis of the current coastal condition. An additional segment of analysis focuses on assessing exposure in coastal areas, with the goal of facilitating inclusive decision-making and solutions. Specifically, the research employs a comprehensive methodology comprised of four distinct segments: (1) Multi-Sensor Data Processing, (2) Satellite Image Analysis, (3) Shoreline Analysis and Change Detection, and (4) Coastal Vulnerability Assessment (Fig. 7.2).

7.3.1 Multi-Sensor Data Processing

The integration of advanced remote sensing technologies facilitates precise mapping and monitoring of coastal changes and vulnerabilities, which is essential for informed management and conservation initiatives. The processing of multi-sensor data involves critical steps, including these components and specifications.

For the years 2015, 2020, and 2025, Landsat-8/OLI Collection 1 Tier 1 data and Real-Time data TOA reflectance were used to extract shorelines, respectively. The datasets were obtained from the Google Earth Engine (GEE) platform's online user interface, developed by the United States Geological Survey (USGS). According to Table 7.1, the datasets underwent radiometric and geometric correction, topographic validation, and projection to ensure spatial accuracy and consistency across different periods. Additionally, they were standardized by resampling to a pixel size of 30 m × 30 m. Finally, all satellite images affected by large tidal ranges were removed from the analysis, as this would lead to highly inaccurate results, especially in regions with very gentle slopes (Murty and Henry 1983). Therefore, to maximize accuracy, a consistent lower tidal range record was collected from the Bangladesh Inland Water Transport Authority (BIWTA) (Table 7.2).

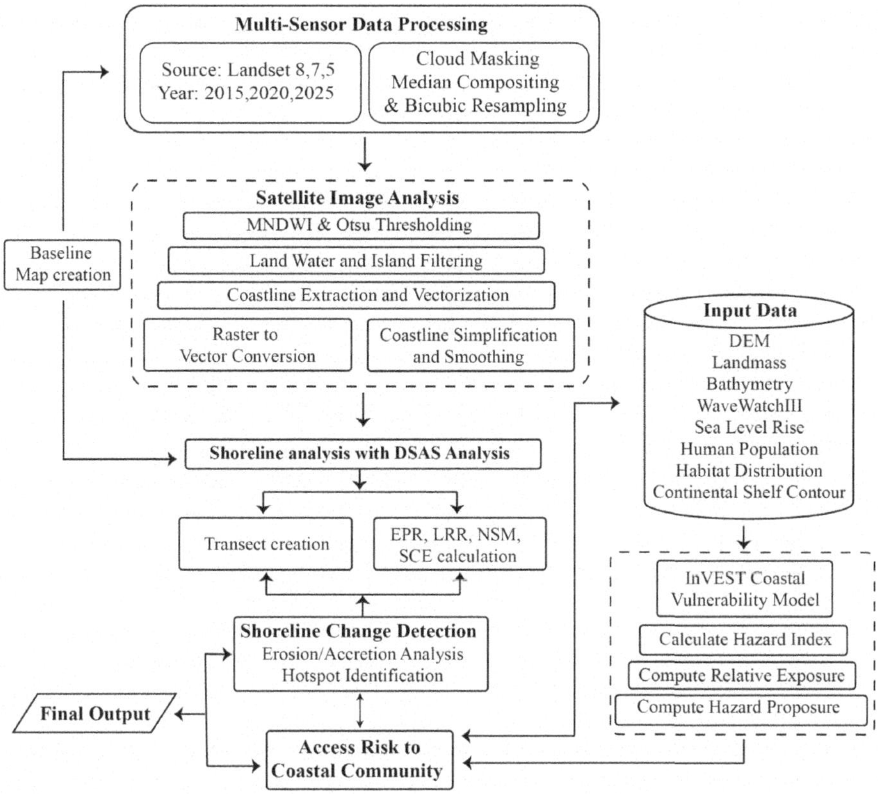

Fig. 7.2 Methodology of the research

Table 7.1 Data requirement for coast shoreline extraction

Component	Specification
Satellite data	Landsat 8 OLI/TIRS; Collection 2, Level-2 SR
Radiometric and geometric calibration	Thermal data (Band 10) Brightness temperature (Kelvin) derived using a scale of 0.00341802 and an offset of 149.0
Cloud masking	QA_PIXEL (L8)
Topographic accuracy	L1T orthorectification using DEM
Projection	Bangladesh transverse Mercator (BTM) projection

7.3.2 *Shoreline Extraction and Image Analysis*

Since water (or turbid water) absorbs mid-infrared energy, the NIR and SWIR wavelength bands of Landsat satellites were used to significantly distinguish between water and land features (Kirby et al. 2024). Shoreline extraction was performed

Table 7.2 Landsat satellite image acquisition time during the lower tidal range for coastline detection

Year	Satellite	Spatial resolution	Landsat scene		Acquisition date	BIWTA station	Tide level
			Path	Row			
2015	Landsat 8 OLI/TRIS	30 m	137	45	01.01.2015	Khepupara	Low tide
			136	44	07.01.2015	Sandwip	
			136	45	26.02.2015	Chittagong	
			138	45	29.02.2015	Khepupara	
			137	44	13.01.2015	Charchanga	
2020	Landsat 8 OLI/TRIS	30 m	137	45	07.02.2020	Khepupara	
			136	44	17.02.2020	Sandwip	
			136	45	01.03.2020	Chittagong	
			138	45	29.03.2020	Khepupara	
			137	44	09.02.2020	Charchanga	
2025	Landsat 8 OLI/TRIS	30 m	137	45	30.03.2025	Khepupara	
			136	44	08.02.2025	Sandwip	
			136	45	24.01.2025	Chittagong	
			138	45	18.02.2025	Khepupara	
			137	44	11.01.2025	Charchanga	

using the Modified Normalized Difference Water Index (MNDWI) combined with Otsu's automated thresholding, providing accurate delineation of water boundaries (Sunder et al. 2017). MNDWI enhances spectral contrast between water and land, making it suitable for complex coastal terrains, while Otsu's method objectively determines the optimal threshold, thereby improving classification in turbid or spectrally similar areas (Buma et al. 2018; Guo et al. 2017). The threshold was computed using the GEE thresholding package with the following parameters: Scale = 100 m, Canny Edge Threshold = 0.7, Sigma = 1, and Minimum Value = −0.2.

MNDWI = (Green − SWIR)/(Green + SWIR).

For Landsat 8, MNDWI = (Band 3 − Band 6)/(Band 3 + Band 6).

In the Inland Water and Island Filtering, small inland water bodies (< 2000 m^2) and coastal islands (< 1000 m^2) were eliminated to isolate the shoreline features. Morphological filtering was also employed to eliminate noise and enhance the accuracy of shoreline delineation (Palmer et al. 2014; Portela et al. 2024). Through vectorization, the final coastlines were extracted, which were then simplified and smoothed while preserving the overall geometric structure, ensuring accurate and computationally efficient representation of the shoreline.

7.3.3 Shoreline Analysis and Change Detection

The shoreline analysis and change detection process involved a combination of geospatial techniques and temporal analysis of shoreline movement between 2015 and 2025. Moreover, to quantify shoreline dynamics in terms of erosion, accretion, and stability across the coastal zones of Bangladesh, a software extension of Esri ArcGIS, named 'Digital Shoreline Analysis System' (DSAS) Version 6.0.170. The analysis followed several key stages:

To quantify shoreline change, this analysis software includes shoreline positions and a reference baseline, from which the transects are cast. A baseline, roughly parallel to the extracted shorelines and transects, was placed at regular intervals (100 m apart) along the coast, providing a uniform and detailed analysis of shoreline movement. The Digital Shoreline Analysis System (DSAS) was used to compute shoreline change metrics, such as the End Point Rate (EPR) and Linear Regression Rate (LRR), both of which are essential for detecting and quantifying shoreline erosion or accretion. Moreover, after this study, measure Net Shoreline Movement (NSM) and Shoreline Change Envelope (SCE) (Table 7.3).

Both NSM and SCE are integral components of the Digital Shoreline Analysis System (DSAS), which facilitates precise and reliable measurement of shoreline changes. The use of these methods highlights the intricate interplay between erosion

Table 7.3 Equations for shoreline analysis

Operation	Formula	Where	Validation
End point rate (EPR)	$EPR = \frac{X_f - X_i}{T}$	X_f = shoreline position at final year X_i = shoreline position at initial year T = time interval (years)	Basheer Ahammed and Pandey (20180; Yasir et al. (2020); Kumar Das et al. (2021)
Linear regression rate (LRR)	$LRR = \frac{\sum_{i=1}^{n}(t_i - \bar{t})(x_i - \bar{x})}{\sum_{i=1}^{n}((t_i - \bar{t})^2}$	x_t = shoreline position at year t_i $\bar{x}$, $\bar{t}$ = mean shoreline position and mean year n = number of years	Dereli and Tercan (2020); Appeaning Addo et al. (2012); Daud et al. (2021); Nassar et al. (2018)
Net shoreline movement (NSM)	$NSM = X_f - X_i$	X_f = shoreline position at the final year X_i = shoreline position at the initial year	Kaliraj et al. (2013), Zambrano-Medina et al. (2023)
Shoreline change envelope (SCE)	$SCE = X_{max} - X_{min}$	X_{max} = the **farthest** position of the shoreline from the baseline in the entire study period X_{min} = the **nearest** position of the shoreline to the baseline in the entire study period	Dereli and Tercan (2020); Ciritci and Türk (2019)

and accretion in shaping coastal morphologies, underscoring the need for continuous monitoring to inform environmental management and policy decisions (Jana and Bhattacharya 2013).

Transects were categorized based on shoreline movement into **erosion** (retreat), **Accretion** (advance), and **Stability** (minimal change, −1 to +10 m/year). Areas exhibiting extreme erosion or accretion were further identified as **hotspots** to highlight zones most vulnerable to coastal hazards.

7.3.4 Coastal Vulnerability Assessment

The InVEST Coastal Vulnerability model was used to assess coastal hazard risks to local communities. This index-based tool evaluates relative risk by integrating geomorphological and socioeconomic factors into such assessments, which can help pinpoint regions susceptible to hazards and prioritize areas accordingly. (Bukvic et al. 2020; Ozsahin et al. 2023). Key input variables include biophysical variables such as DEM (Digital Elevation Model), Landmass, Bathymetry, Wave Watch III, Habitat Distribution, and continental shelf contour. Climatic/hazard variables include sea level rise (SLR), while socioeconomic variables Encompass Human density. Factors, including geomorphological data obtained from Digital Elevation Models, information on coastal habitats, such as the distribution and size of mangroves, salt marshes, and coral reefs, are collected from remote sensing imagery and field studies. Hydrodynamic parameters, like wave exposure and sea-level rise predictions, are integrated to model inundation risk (Barnard et al. 2019). Moreover, Socioeconomic data, including population density and infrastructure datasets, are integrated to evaluate possible effects on human populations and assets. Once the data is prepared, the InVEST Coastal Vulnerability model is parameterized and run, calculating a Coastal Vulnerability Index based on the relative contribution of each input factor to overall exposure (Guannel et al. 2011). The data sources are (Table 7.4).

Table 7.4 Data requirement for shoreline exposure analysis

InVEST factor	Source
Landmass	National oceanographic and atmospheric administration (NOAA)
Bathemetry	General bathymetry chart of the ocean (GEBCO)
WaveWatchIII	NOAA WaveWatch III
Habitat distribution	The UN environment program world conservation monitoring centre (UNEP-WCMC)
Continental shelf contour	NOAA
DEM	USGS
Human population	Global population (Google earth engine)
Sea level rise	NOAA

To determine the relative coastal hazard, each variable was ranked from 1 (lowest exposure) to 5 (highest exposure), and the geometric mean of these ranks was computed to obtain a final coastal hazard index for each 250 m coastal segment. The results were categorized into three hazard levels (highest, intermediate, and lowest) to map the relative coastal hazard exposure across the study area. Additionally, sensitivity analyses can be performed to assess the impact of different input parameters on the CVI (Ai et al. 2022). Lastly, scenario analysis, which encompasses various sea-level rise and storm scenarios, can assess the potential effects of future climate change on coastal exposure, thereby informing adaptation planning and risk management strategies.

7.4 Data Analysis and Discussion

The analytical aspects include recent detailed shoreline dynamics of Bangladesh and their correlation with exposures. Coastal disasters heavily influence the spatial changes and temporal patterns of the coastline. The research emphasizes a correlational measure between coastal exposure and shoreline changes over time.

7.4.1 Spatial and Temporal Patterns of Shoreline Change (2015–2025)

Shoreline change along the Bangladesh coast was analyzed for the years 2015, 2020, and 2025 using the End Point Rate (EPR), Linear Regression Rate (LRR), Net Shoreline Movement (NSM), and Shoreline Change Envelope (SCE) methods applied to transects spaced every 100 m. For interpretation, the coastline was divided into three zones based on geomorphological characteristics, following previous studies on Bangladesh coastal dynamics (Islam and Gnauck 2011). The Western Coastal Zone includes the districts of Khulna, Satkhira, Bagerhat, Patuakhali, and Barguna. The Central Coastal Zone covers Bhola, Noakhali, and Feni. The Eastern Coastal Zone comprises Chittagong and Cox's Bazar. Positive EPR and LRR values indicate seaward shoreline advance (accretion), while negative values represent landward retreat (erosion) (Fig. 7.3).

7.4.1.1 Shoreline Change Detection of Western Coastal Zone

See Fig. 7.4 and Table 7.5.

For the western coastal zone, extending between the west of Sundarbans and the eastern Ganges tidal plain, it shows no such hotspots from 2015 to 2025. The EPR values range between +59.33 m/year and –74.00 m/year with a mean of −4.7 m/

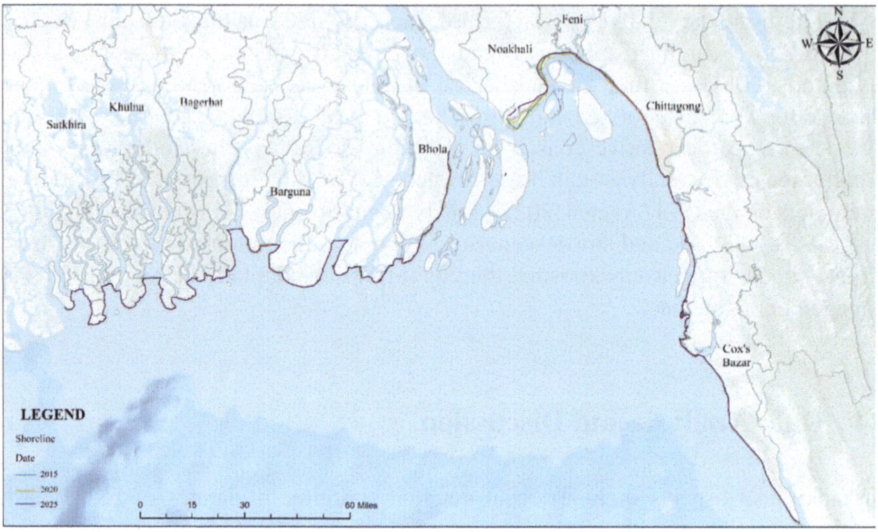

Fig. 7.3 Shoreline condition of Bangladesh in 2015, 2020, and 2025

year. Most changes were moderate as 95% of transects lie within the ±20 m/year range. However, shoreline retreat is encountered predominantly along the transects in such a way that of the total 3,788 transects, 66.5% show erosion, while 24.9% show accretion, and 8.6% remain stable. In the last 10 years, the retreat of the shoreline has reduced, as 62% transects seem to face land loss within the 0 to −20 m/year range, whereas about 80% of the transects cast in this zone show retreat of the shoreline in 60 60-year timeline (Martin and Hasan 2021).

The shoreline exposed to the Sundarban mangrove forest has 79.5% of transects that fall within the 0 to −20 m/year range in the Satkhira, Khulna, and Barishal regions. The Sundarbans mangrove forest covers 6,000 km^2, acts as a natural coastal defence by dissipating wave energy, trapping sediments, and stabilizing shorelines (Alongi 2008; Das and Vincent 2009). The dense root mats of species such as *Avicennia marina* and *Sonneratia apetala* reduce current velocity during tidal inundation, allowing suspended sediments to settle (Iftekhar and Saenger 2008). Similarly, a study reveals low-to-moderate erosion rates (<20 m/year) in most of the Sundarbans areas (Islam and Gnauck 2011; Rahman et al. 2011). Due to this inherent natural protection, storm surges caused by cyclones such as Sidr in 2007 and Aila in 2009 resulted in less sustained erosion in this area compared to more exposed coastlines (Das and Vincent 2009). Another reason for the Sundarbans shoreline stability is sediment deposition during subsequent monsoon tides, which often compensates for cyclone-induced losses (Iftekhar and Saenger 2008).

Also, since the commissioning of the Farakka Barrage in 1975, sediment supply from the Ganges has decreased. However, erosion rates have not reached extreme levels because of local sediment recycling and mangrove protection (Islam and Gnauck 2011). Long-term subsidence rates of 2–4 mm/year (Goodbred and Kuehl

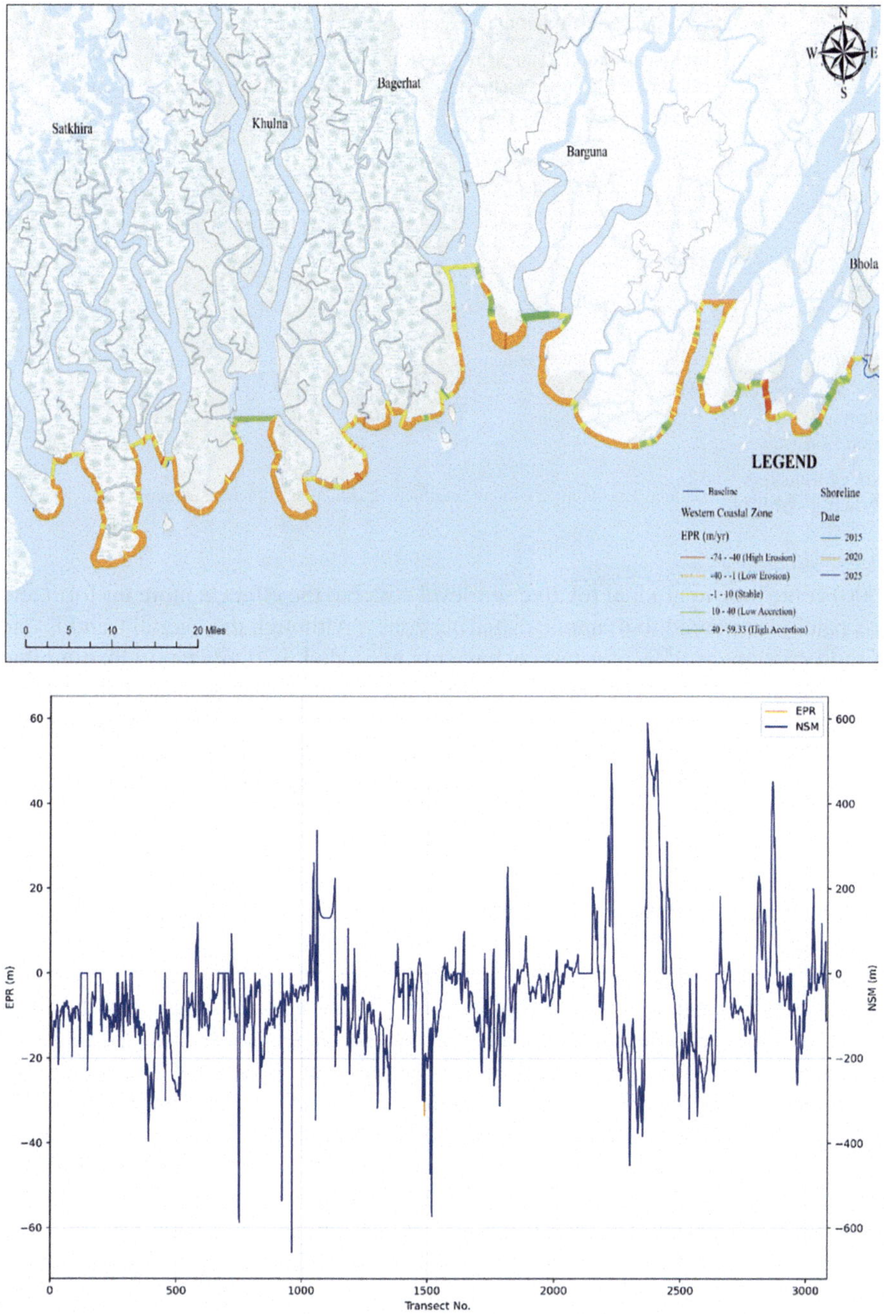

Fig. 7.4 EPR and NSM Analysis of the Western Coastal Zone of Bangladesh

Table 7.5 Shoreline change condition at the western coastal zone of Bangladesh

2015 - 2025	End point rate (m/yr)	Linear regression rate (m/yr)	Net shoreline movement (m)	Shoreline change envelope(m)
Max	59.33	59.33	593.32	593.32
Min	−74	−74	−739.8	0
Mean	−4.7	−4.7	−43.56	119.9
SD	15.5	15.5	151.01	118.3
No. of total transects	3788			
No. of transects showing erosion	2520			
No. of transects showing accretion	942			
No. of transects showing stability	326			

2000) contribute to gradual relative sea-level rise, but the effect is more uniform and less patchy compared to dynamic estuarine zones. Although the overall trend is one of mild erosion, as 95% of transects have not exceeded ±20 m/year, indicating that coastal hazards in this zone are more chronic than sudden. Dublar Char, Patharghata, and Haringhata are the exceptions, where the EPR value appears to be slightly on the high side, indicating maximum erosion at −74 m/yr and accretion at 59.33 m/yr. However, in the last 10 years, the change rates have been more fluctuating due to their erratic behavior.

7.4.1.2 Shoreline Change Detection of the Central Coastal Zone

See Fig. 7.5 and Table 7.6.

The central zone adjacent to the Meghna estuary reflects significant shoreline dynamics over recent decades. The present 10-year analysis (2015–2025) reveals a remarkably different trend, with the highest erosion rate (EPR) of 82.9 m/year and a minimum of –655 m/year, while net shoreline movement ranges from +829 m to –6554 m, reflecting substantial spatial and temporal variability. 92% transects indicate the retreat of the shoreline out of 1222 transects, while only 5 and 3% indicate accretion and stability. This temporal shift highlights alterations in tidal patterns and discharge from the Meghna River, which can influence sediment transport and deposition. It will lead to potentially increased periodic shifts in erosion and accretion. In the Southern Noakhali region, the Chairman Ghat, Noler Ghat, and Jonotar Ghat are the accreted land. However, since 2015, those Charghats have experienced a tremendous erosional pattern, characterized by a maximum erosion rate, with the standard deviation values of EPR (130.53) and NSM (1305.36) also being the highest

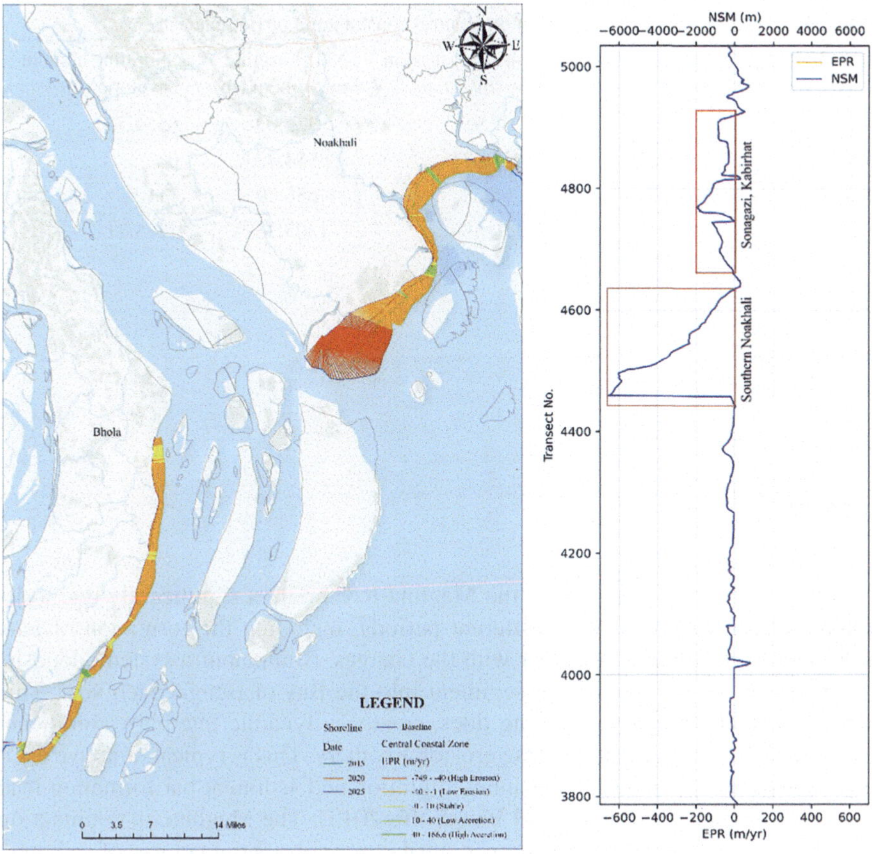

Fig. 7.5 EPR and NSM analysis of the central coastal zone of Bangladesh

among all. Additionally, in Kabirhat and Shonagazi Upazila, the eroding pattern is prominent, with a peak endpoint rate of 187.98 m/yr and a net shoreline movement of −1879.79 m.

However, the largest rates of accretion were observed near the mouth of the Meghna estuary, at a rate greater than 600 m/year, while an accretion rate of 228 m/year was found. Between 1988 and 2017, the Meghna River in southern Noakhali experienced high accretion with an EPR of 158.43 m/yr (Mahmud et al. 2020; Matin and Hasan 2021; Sarwar and Woodroffe 2013). This discrepancy arises due to the different periods of the studies; the studies considered the shorelines of 1989 and 2011, whereas the period considered was from 1989 to 2019. Over 42 years, the Meghna Riverbank experienced its highest erosion (4,219 ha) in 1988–2000 and peak accretion (4,215 ha) in 2000–2010, again followed by a remarkable erosion (2,218 ha) in 2010–2021 (Sultana et al. 2025; Martin and Hasan 2021; Sarwar and Woodroffe 2013). Due to an increase in the erosion pattern, the Chairman Ghat area has witnessed 117.7 km^2 of land loss over 22 years (Mou et al. 2023).

Table 7.6 Shoreline change condition at the central coastal zone of Bangladesh

2015–2025	End point rate (m/yr)	Linear regression rate (m/yr)	Net shoreline movement (m)	Shoreline change envelope(m)
Max	82.9	82.9	829.03	6554.72
Min	−655.47	−655.47	−6554.72	0
Mean	−65.68	−65.68	−656.79	711.56
SD	130.53	130.53	1305.36	1285.21
No. of total transects	1222			
No. of transects showing erosion	1122			
No. of transects showing accretion	66			
No. of transects showing stability	34			

So, shoreline change pattern near the Meghna River varies significantly due to its dynamic channel changes over different periods, including the formation of bars and islands. The Meghna, together with the Ganges–Brahmaputra system, deposits approximately 2.4 billion tons of sediment into the Bay of Bengal each year. This massive load, combined with strong tides, creates a dynamic interplay: significant deposition in some areas and intense erosion in others. This is typical of active estuarine environments where tidal channel migration and sediment bar formation lead to rapid local changes (Sarwar and Woodroffe 2013). The Meghna, in conjunction with the Ganges–Brahmaputra system, contributes a substantial volume of sediment to the Bay of Bengal, approximately 2.4 billion tons annually (Sarwar and Borthwick 2023). This immense load, combined with strong tides, creates a turbulent interplay: powerful deposition in some areas and intense erosion in others. This finding underscores the highly dynamic and unstable nature of shoreline changes in the region, which are periodically shifting rapidly. However, this pattern is not consistent along the coast as significant erosion or accretion occurred on the Meghna River estuary, perhaps due to the effect of strong river currents. Also, in Bhola, 64% of changes have happened within the ±20 m/year range over the past decade, and due to the construction of the embankment, the intensifying threat of erosion is now controlled.

7.4.1.3 Shoreline Change Detection of the Eastern Coastal Zone

See Fig. 7.6 and Table 7.7.

For the eastern coastal zone, the EPR values range between +236.73 m/year and –75.84 m/year with an overall mean of 3.54 m/year. Most of the changes are moderate, as 84% of transects lie within the ±20 m/year range, while only 9%

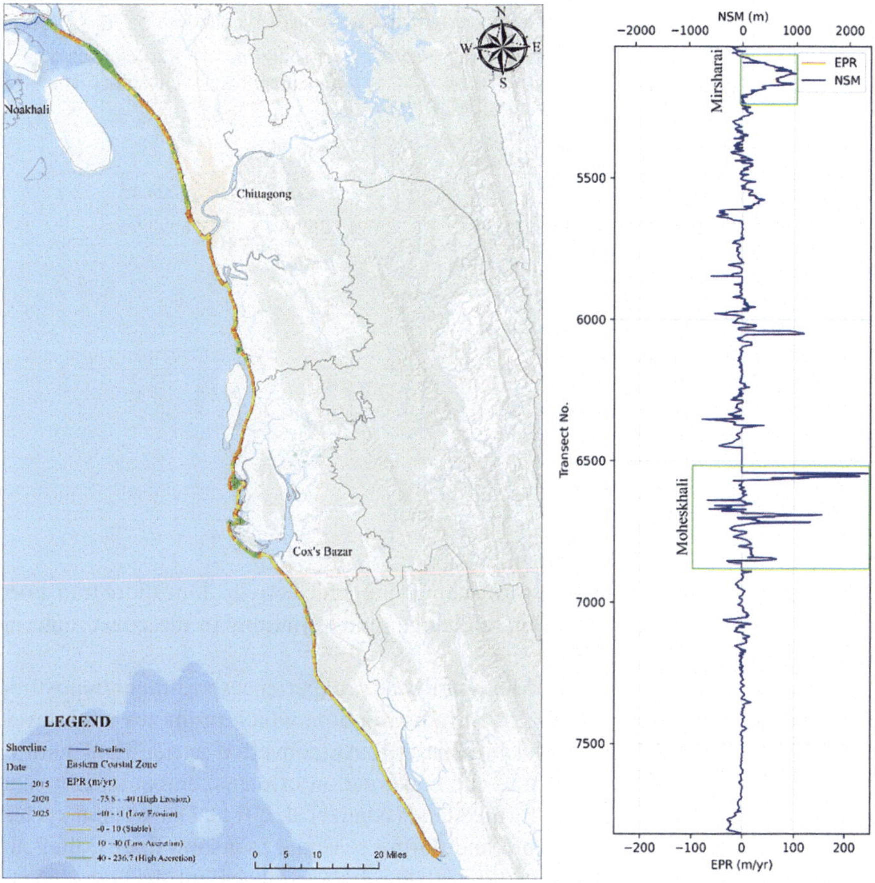

Fig. 7.6 EPR and NSM analysis of the eastern coast zone of Bangladesh

exceed +20 m/yr. However, a total of, 43% transects encountered shoreline retreat, and with 39% transects are stable, which is between −1 to +10 m/yr, while only 18% transects encountered advancement. The highest EPR point is 236.73 m/yr. NSM is 2367.26 m, which lies in Moheshkhali, Cox's Bazar, specifically in Sonadia, and followed by another highest shoreline advancement zone, Mirsharai, Chittagong, within the Meghna Estuary.

Moheshkhali has the highest EPR (236.73 m/yr) of all the coasts in Bangladesh. In this zone, the construction of Matarbari Sea Port and Matarbari Coal Power Plant at the coast of the Bay of Bengal blocks water flow, traps sediments, or artificially reclaims land. Accretion is the dominant process, followed by a sudden retreat of −63.56 m/yr, leading to another shoreline advancement with EPR (around 148.56 m/yr), which is very inconsistent in this region. Due to the vigorous deposition along the coast, the coastline migrates westward and southeast, resulting in accretion in the southern and southwestern regions. The shorelines of Moheshkhali

Table 7.7 Shoreline change condition at the eastern coastal zone of Bangladesh

	End point rate (m/yr)	Linear regression rate (m/yr)	Net shoreline movement (m)	Shoreline change envelope(m)
Max	236.73	236.73	2367.26	2738.79
Min	−75.84	−75.84	−758.39	0
Mean	3.54	3.39	−3.91	204.15
SD	25.52	25.35	253.54	277.95
No. of total transects	2773			
No. of transects showing erosion	1188			
No. of transects showing accretion	500			
No. of transects showing stability	1085			

are highly dynamic due to windy storms, anthropogenic activity, longshore transport, sporadic wave and air action, climate change, and variations in the coast material (Md Kawsarul Islam et al. 2023).

Mirsharai, situated within the Meghna Estuary, experiences sediment deposition on its tidal flats through cyclone-driven redistribution, which promotes land development. Saline intrusion and tidal inundation have converted agricultural land to salt fields. Human activities, such as the construction of dams, embankments, and aquaculture, help stabilize newly formed land (Islam et al. 2011). Due to tidal cycles, sediments are transported upstream, and during seasonal monsoon floods, they are transported downstream, which facilitates shoreline advancement through deposition. The Ganges–Brahmaputra system carries approximately 1 billion tonnes of sediment annually (Allison 1998). Sustaining a high fluvial sediment supply supports the growth of the delta plain, which aligns with the accretion patterns that occur along the eastern Meghna coast (Ahmed et al. 2020). A complex interplay of natural geomorphic processes such as sediment discharge, tidal dynamics, and subsidence, and anthropogenic influences, including embankments, shrimp farming, and urban encroachment shapes these spatial variations.

Overall, the dynamic changes in the coastline are significantly influenced by coastal disasters, as measured by the Coastal Vulnerability Index (CVI). The core of coastal exposure serves as a key indicator for understanding the detailed reasons behind coastal vulnerability.

7.4.2 Coastal Exposure in Bangladesh

Bangladesh, as a riverine nation, holds significant importance concerning water resources within its coastal regions. The coastal areas of Bangladesh are highly susceptible to various disasters, primarily due to climate change and the country's geographic location. The nation's extensive coastline, low-lying topography, and dense population render it vulnerable to sea-level rise, storm surges, flooding, and erosion (Hoque et al. 2019; Islam et al. 2015). The CVI analysis serves to identify critical locations for hazard exposure. It is predicted that climate change will exacerbate these hazards, consequently increasing the vulnerability of infrastructure to cyclonic storms and other geographical and climatic factors (Bianchi and Malki-Epshtein 2021). Vulnerability analysis plays a vital role in identifying strategic locations and contributing factors, thereby facilitating measures such as coastal afforestation and innovative land use practices (e.g., Forest, Fish, and Fruit-Triple F) to enhance the resilience and adaptive capacity of coastal communities (Fig. 7.7).

The exposure delineated across three classes differentiates the entire coastal region. Factorial distributions such as vegetation, sea level, coral reefs, water, wind speed, and others assist in identifying coastal zones at critical locations. In presenting this scenario, approximately 43.36% of the coastline in Bangladesh's entire coastal region is classified as having low-hazard exposure. Furthermore, the least vulnerable areas are predominantly situated near natural habitats and mangroves. About 40.24%

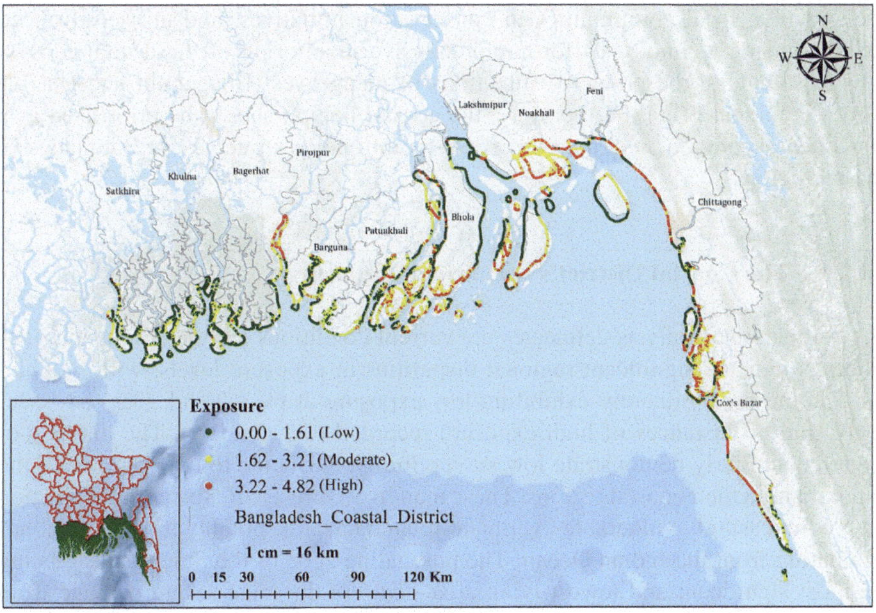

Fig. 7.7 Coastal exposure index scenario of Bangladesh

of the total area experiences moderate exposure conditions. This segment of the coastline is more susceptible to proximity to critical zones during significant disasters, particularly in the absence of adequate disaster management precautions. Conversely, approximately 16.40%, nearly 20%, of the coastline falls within the high exposure category. The Coastal Vulnerability Index (CVI) depicted in the map shows that the intensity of exposure is correlated with population density. Consequently, densely populated areas along the coastline are subjected to higher exposure scenarios. The analytical results identify the critical zones within Bangladesh's coastal regions, thereby facilitating policymaking and disaster preparedness efforts.

Based on the results, significant variations in exposure classifications are observed at Mangrove and non-mangrove locations. Mangrove forests serve as effective coastal defences, safeguarding communities from tsunamis, storms, and other coastal hazards (Ihinegbu et al. 2023; Barbier 2006). The presence of mangroves results in notably low exposure values in districts such as Satkhira, Khulna, and Bagerhat. Approximately 68% of the area falls within the lowest exposure category, 31% at a moderate level, and less than 1% exhibits high exposure to disasters. The substantial protective influence of mangroves creates an almost impervious barrier against hazards such as cyclones and tsunamis.

Conversely, coastlines lacking mangrove protection face significant threats and exposure to hazards. According to analysis, approximately 20% of the coastline falls within the high coastal exposure zone, while only 36% resides in low exposure zones due to factorial influences. Consequently, a substantial portion of land demonstrates high vulnerability. Therefore, it is imperative to adopt sustainable management practices, involve local communities in conservation initiatives, and utilise advanced remote sensing technologies for comprehensive monitoring of these vital ecosystems to safeguard the protective functions of mangroves (Bhagarathi and Dasilva 2024; Heumann 2011). Furthermore, the coastal habitat should be prioritized as a fundamental component of long-term disaster mitigation strategies for Bangladesh's coastal regions.

7.4.2.1 The Coastal District's Scenario of Exposure

The comparative analysis delineates the current conditions across various districts. Findings indicate significant regional disparities in exposure levels, with Satkhira and Khulna predominantly exhibiting low exposure at 69.1% and 70.2%, respectively, and no instances of high exposure recorded (2.2 and 0%). The districts of Bagerhat similarly demonstrate low susceptibility, attributed to the presence of the Sundarbans in their coastal regions. These mangrove forests offer essential protection yet are increasingly vulnerable to climate-related hazards, notably tropical cyclones originating from the Indian Ocean. The polyhaline zone of the Sundarbans—being the most significant and lowest-lying area—suffers the most severe damage from cyclones, likely due to its proximity to predominant cyclone paths (Arfin-Khan et al. 2024). Conversely, Pirojpur and Feni exhibit very high exposure levels of 45.6% and 54.5%, respectively, indicating heightened susceptibility to coastal hazards due to diminished vegetation cover. Districts such as Patuakhali and Cox's Bazar exhibit a relatively even distribution across different exposure classes, with medium exposure constituting the majority at 53.5% and 48.3%, respectively. Furthermore, Bhola and Chittagong demonstrate elevated levels of high exposure, at 21.5% and 27.89%, respectively, thereby highlighting areas of potential vulnerability (Fig. 7.8).

The points shown illustrate the coastal area within 500 m. The variation in coastal exposure is due to several geographical and climatic factors. The exposure maps identify key locations for each district and list the coastlines under different exposure levels. Satkhira and Khulna mainly have low exposure, with 155.13 km and 136.19 km of shoreline, respectively, and almost no high exposure (4.94 km in Satkhira and none in Khulna). In contrast, Pirojpur and Feni show significantly higher exposure, with 12.77 km (45.6%) and 2.73 km (54.5%) of their coastlines classified as high, indicating greater vulnerability to coastal hazards. Patuakhali and Cox's Bazar have a more even distribution, with medium exposure dominating (305.75 km and 277.97 km, respectively), while Bhola and Noakhali have much higher exposure levels (93.31 km and 115.74 km, respectively). The total shoreline lengths studied range from 5.00 km in Feni to 575.50 km in Cox's Bazar, highlighting notable regional differences in exposure. The exposure points provided help identify important exposure locations for each district (Fig. 7.9).

The line graph delineates key locations across various districts and illustrates changes along different segments of the coastal lines. The graphical representation provides a perspective from west to east along the coast. In the Satkhira district, the most prominent continuous peaks are observed between 105 and 110 km from the western coastline. The Khulna district is less vulnerable to hazards due to the protection offered by natural barriers, such as coral reefs and the Sundarbans. Conversely, the Bagerhat district exhibits higher vulnerabilities along the coastline from 147 to 170 km, accounting for 23 km within a total of 316 km. The majority of the Pirojpur district's coastal area is highly susceptible to hazards.

Patuakhali and Barguna districts display discontinuous patterns of exposure. For Patuakhali, elevated exposure levels are evident along the coast from 75 to 300 km in the west and 500 km to 550 km in the east. Barguna exhibits a continuous high-exposure stretch spanning 97 km to 102 km, constituting a 5 km segment within the total coastline of 5 km. Other districts, such as Bhola and Noakhali, are characterized by numerous scattered islands, rendering them more susceptible to hazards. Consequently, most of their coastal areas are classified as high-exposure zones, except for smaller segments. Lakshmipur and Bhola districts have smaller coastal regions compared to others, with Lakshmipur's coast ranging from 22 to 28 km, and Bhola's from 3 to 5 km.

Chittagong and Cox's Bazar are of particular importance to Bangladesh's economy due to their ports and tourism sectors. The shoreline areas, extending from 100 to 200 km from north to south, demonstrate high vulnerability. In Cox's Bazar district, continuous peaks of exposure are identified between 250 and 300 km. These findings underscore the heterogeneous nature of coastal vulnerability in Bangladesh, carrying significant implications for targeted risk mitigation measures, hazard zoning,

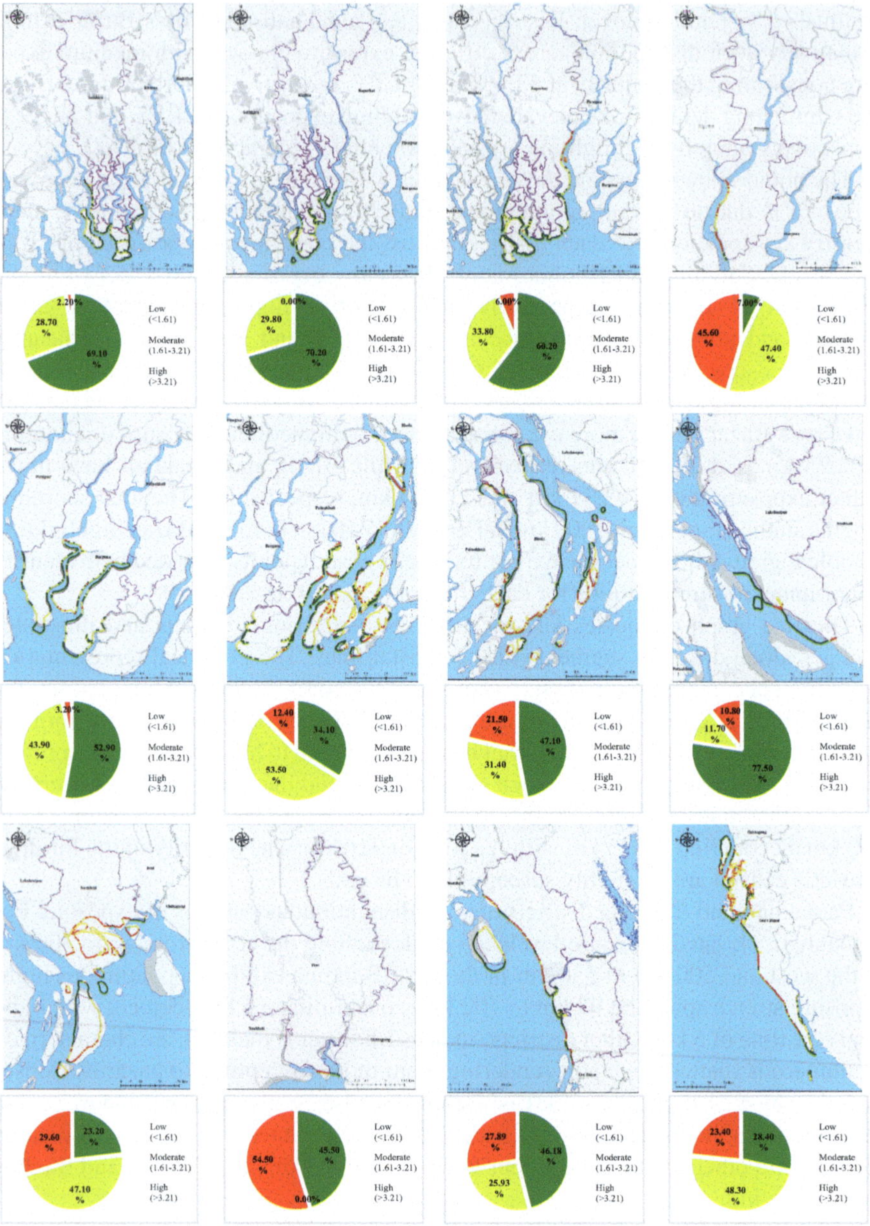

Fig. 7.8 District-wise coastal exposure scenario of Bangladesh

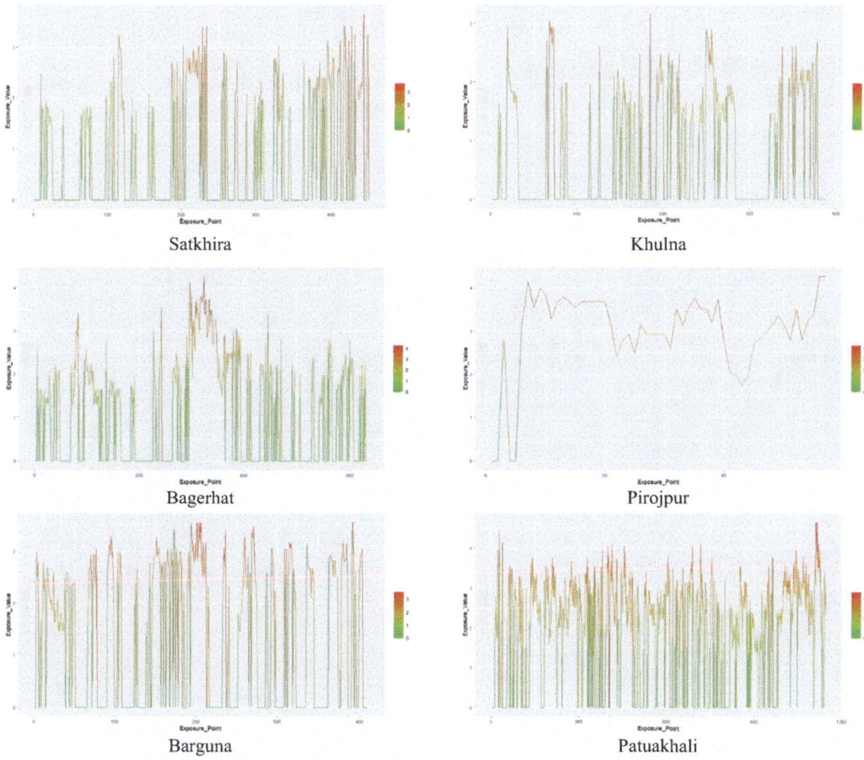

Fig. 7.9 District-wise exposure zone of Bangladesh

and adaptation strategies, such as Nature-Based Solutions (NBS) and Local Level Adaptation (LLA), in regions of high exposure.

7.4.3 Correlation Between Shoreline Change, Exposure Index, and Mangrove

Natural catastrophes predominantly stem from political, economic, and inadequate adaptation, as well as human activities (Hossain et al. 2024b). The exposure index indicates less adaptation and fewer natural resources against disasters. The correlation shows the dynamic influence of exposures on erosion rates. It is assessed as a negative correlation, where high exposure intensity leads to higher erosion rates, while low exposure has a lesser impact on erosion rates. Additionally, the magnitude of exposure impacts causes long-term shoreline distortion, resulting in land loss and vegetation degradation (Fig. 7.10).

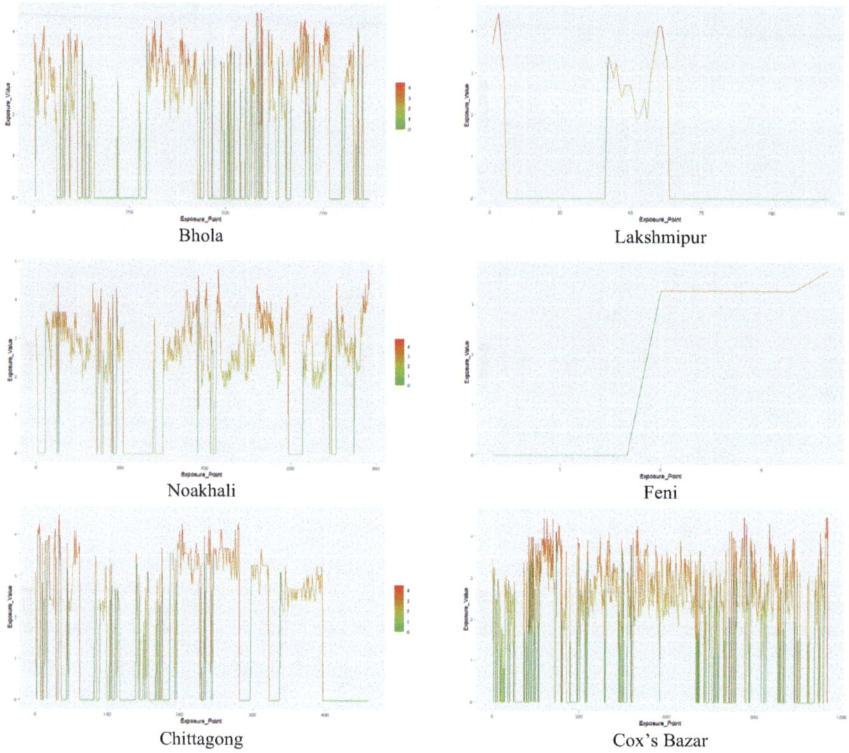

Fig. 7.9 (continued)

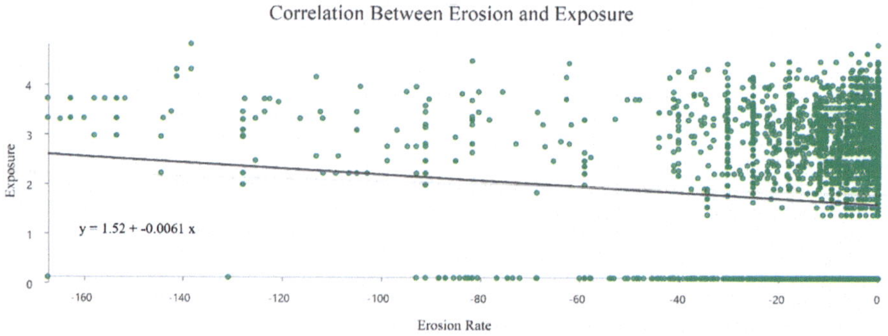

Fig. 7.10 Correlation between erosion and exposure

Additionally, exposure intensity is highly relevant to the mangrove. According to the bar graph, only about 2.27% of the shoreline falls under high exposure zones. In comparison, the percentage exceeds 30% at non-mangrove locations, and less than 1% (approximately 2.4 km) of the shoreline is located in low-exposure zones. This

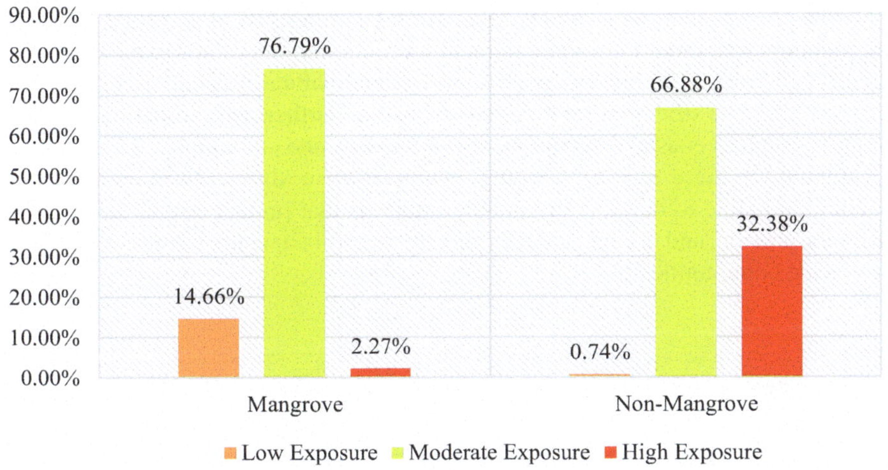

Fig. 7.11 Exposure condition at mangrove and non-mangrove location

segment highlights the influence of mangroves on exposure intensity and shoreline erosion. Most of the coastline is in the moderate category, with mangrove areas showing a higher percentage than non-mangrove locations (Fig. 7.11).

Therefore, it is crucial to emphasize the conservation of mangroves. According to NASA's mangrove database, the decline in mangrove areas is predominantly evident. Due to the ongoing effects of coastal disasters, Bangladesh has experienced substantial destruction of its mangrove ecosystems. Over the past decade, Bangladesh's mangrove cover has substantially decreased. Primarily, the persistent impacts of natural disasters, sea level rise, and human activities are identified as the principal causes.

7.5 Conclusion

Bangladesh, as one of the most vulnerable locations in Asia, faces significant impacts from oceanic catastrophes. Continuous shoreline erosion causes substantial alterations in island areas such as Vola, Moheshkhali, Noakhali, and Sonagazi. Over time, these changes result in the displacement of large populations and the destruction of their habitats. Notably, these locations are highly exposed during cyclones and storm surges. The primary factor contributing to this vulnerability is the absence of natural barriers, which, combined with persistent disasters, result in coastal land distortion and severe erosion. Research indicates that mangroves serve as a vital natural barrier against cyclones, thereby reducing vulnerability in the southern zones of Khulna, Satkhira, and Bagerhat. Additionally, studies have measured exposure levels in areas where mangroves have been lost over time, with districts such as Barguna, Noakhali, and Vola exhibiting higher exposure percentages, which correlate with shoreline erosion analysis using DSAS metrics. Consequently, it is crucial to prioritize mangrove preservation and implement precautionary measures to mitigate land degradation over time. Nature-based solutions and habitat restoration are deemed the most effective strategies for reducing exposure. Furthermore, initiating adaptive programs such as coastal reservation, social forestry, and community-led initiatives is essential for these regions. Through comprehensive, district-level assessments, it will be possible to develop sustainable solutions that protect coastal zones from disasters, erosion, and oceanic hazards, thereby ensuring a secure living environment for coastal communities.

References

Addo KA, Jayson-Quashigah PN, Kufogbe KS (2012) Quantitative analysis of shoreline change using medium-resolution satellite imagery in Keta Ghana. Mar Sci 1(1):1–9. https://doi.org/10.5923/j.ms.20110101.01

Ahmed A, Drake F, Nawaz R, Woulds C (2018) Where is the coast? Monitoring coastal land dynamics in Bangladesh: an integrated management approach using GIS and remote sensing techniques. Ocean Coast Manag 151:10–24. https://doi.org/10.1016/j.ocecoaman.2017.10.030

Ahmed A, Nawaz R, Woulds C, Drake F (2020) Influence of hydro-climatic factors on future coastal land susceptibility to erosion in Bangladesh: a geospatial modelling approach. J Geovisualization Spat Anal 4(1):6. https://doi.org/10.1007/s41651-020-00050-x

Ahammad R, Nandy P, Husnain P (2013) Unlocking ecosystem-based adaptation opportunities in coastal Bangladesh. J Coast Conserv 17(4):833–840. https://doi.org/10.1007/s11852-013-0284-x

Ahammed KKB, Pandey AC (2018) Shoreline morphology changes along the Eastern Coast of India, Andhra Pradesh by using geospatial technology. J Coast Conserv 23(2):331–353. https://doi.org/10.1007/s11852-018-0662-5

Ai B, Tian Y, Wang P, Gan Y, Luo F, Shi Q (2022) Vulnerability analysis of coastal zone based on InVEST model in Jiaozhou Bay, China. Sustainability 14(11):6913. https://doi.org/10.3390/su14116913

Alcántara-Carrió J, García Echavarría LM, Jaramillo-Vélez A (2024) Is the coastal vulnerability index a suitable index? review and proposal of alternative indices for coastal vulnerability to sea level rise. Geo-Mar Lett 44(2). https://doi.org/10.1007/s00367-024-00770-9

Allison MA (1998) Geologic framework and environmental status of the Ganges-Brahmaputra delta. J Coast Res 14(3):827–836. http://www.jstor.org/stable/4298836

Alongi DM (2007) Mangrove forests: resilience, protection from tsunamis, and responses to global climate change. Estuar Coast Shelf Sci 76(1):1–13. https://doi.org/10.1016/j.ecss.2007.08.024

Arfin-Khan M, Srivastava SK, Saimun MSR, Bhuiyan MS, Khan A, Mukul SA, Ahammed S, Sultana F, Arfin-Khan MAS (2024) Assessing vulnerability to cyclone hazards in the world's largest mangrove forest the Sundarbans: a geospatial analysis. Forests 15(10):1722. https://doi.org/10.3390/f15101722

Arkema KK, Guannel G, Verutes G, Wood SA, Guerry A, Ruckelshaus M, Kareiva P, Lacayo M, Silver JM (2013) Coastal habitats shield people and property from sea-level rise and storms. Nat Clim Change 3(10):913–918. https://doi.org/10.1038/nclimate1944

Barbier EB (2006) Natural barriers to natural disasters: replanting mangroves after the tsunami. Front Ecol Environ 4(3):124–131. https://doi.org/10.1890/1540-9295(2006)004[0124:NBTNDR]2.0.CO;2

Barnard PL, Erikson LH, Foxgrover AC, Hart JA, Limber P, O'Neill AC, Van Ormondt M, Vitousek S, Wood N, Hayden MK, Jones JM (2019) Dynamic flood modeling essential to assess the coastal impacts of climate change. Sci Rep 9(1):4309. https://doi.org/10.1038/s41598-019-40742-z

Barua H, Acharjee MR, Talukder A (2024) Tourism index evaluation of exposed coast, Bangladesh: a modeling approach. Heliyon 10(14):e34745. https://doi.org/10.1016/j.heliyon.2024.e34745

Baten MA, Seal L, Lisa KS (2015) Salinity intrusion in interior coast of Bangladesh: challenges to agriculture in south-central coastal zone. Am J Clim Change 4(3):248–262. https://doi.org/10.4236/ajcc.2015.43020

Bhagarathi L, Dasilva P (2024) Impacts and implications of anthropogenic activities on mangrove forests: a review. Magna Sci Adv Res Rev 11(1):040–059. https://doi.org/10.30574/msarr.2024.11.1.0074

Bianchi E, Malki-Epshtein L (2021) Evaluating the risk to Bangladeshi coastal infrastructure from tropical cyclones under climate change. Int J Disaster Risk Reduct 57:102147. https://doi.org/10.1016/j.ijdrr.2021.102147

Boak EH, Turner IL (2005) Shoreline definition and detection: a review. J Coastal Res 214:688–703. https://doi.org/10.2112/03-0071.1

Bukvic A, Rohat G, Apotsos A, De Sherbinin A (2020) A systematic review of coastal vulnerability mapping. Sustainability 12(7):2822. https://doi.org/10.3390/su12072822

Buma WG, Lee S-I, Seo JY (2018) Recent surface water extent of lake chad from multispectral sensors and GRACE. Sensors 18(7):2082. https://doi.org/10.3390/s18072082

Bushra N, Mostafiz RB, Rohli RV, Friedland CJ, Rahim MA (2021) Technical and social approaches to study shoreline change of Kuakata, Bangladesh. Front Mar Sci 8. https://doi.org/10.3389/fmars.2021.730984

Ciritci D, Türk T (2019) Automatic detection of shoreline changes by geographical information system (GIS) and remote sensing in the Göksu Delta, Turkey. J Indian Soc Remote Sens 47(2):233–243. https://doi.org/10.1007/s12524-019-00947-1

Das S, Vincent JR (2009) Mangroves protected villages and reduced death toll during Indian super cyclone. Proc Natl Acad Sci 106(18):7357–7360. https://doi.org/10.1073/pnas.0810440106

Das SK, Sajan B, Ojha C, Soren S (2021) Shoreline change behavior study of Jambudwip island of Indian Sundarban using DSAS model. Egypt J Remote Sens Space Sci 24(3):961–970. https://doi.org/10.1016/j.ejrs.2021.09.004

Das A, Mondal I, Hossain SA, Jose F, Ouda MM, Juliev M, Almaliki AH, De TK (2025) Coastal vulnerability assessment for the Sundarbans mangrove ecosystem using InVEST and machine learning-based empirical models. Environ Dev sustainability. https://doi.org/10.1007/s10668-025-06414-x

Daud S, Milow P, Zakaria RM (2021) Analysis of shoreline change trends and adaptation of Selangor Coastline, using Landsat satellite data. J Indian Soc Remote Sens 49(8):1869–1878. https://doi.org/10.1007/s12524-020-01218-0

Deb D, Uddin MM, Mahbub-E-Kibria ASM, Das MK, Hasan M (2024) Coastal vulnerability assessment to multi hazards in the exposed coast of southeastern coastal region of Bangladesh. RegNal Stud Mar Sci 73:103484. https://doi.org/10.1016/j.rsma.2024.103484

Dereli MA, Tercan E (2020) Assessment of shoreline changes using historical satellite images and geospatial analysis along the lake Salda in Turkey. Earth Sci Inf 13(3):709–718. https://doi.org/10.1007/s12145-020-00460-x

Goodbred S, Kuehl S (2000) The significance of large sediment supply, active tectonism, and eustasy on margin sequence development: late Quaternary stratigraphy and evolution of the Ganges-Brahmaputra delta. Sed Geol 133(3–4):227–248. https://doi.org/10.1016/s0037-0738(00)00041-5

Guo Q, Pu R, Li J, Cheng J (2017) A weighted normalized difference water index for water extraction using Landsat imagery. Int J Remote Sens 38(19):5430–5445. https://doi.org/10.1080/01431161.2017.1341667

Heumann BW (2011) Satellite remote sensing of mangrove forests: recent advances and future opportunities. Prog Phys Geogr: Earth Environ 35(1):87–108. https://doi.org/10.1177/0309133310385371

Hoque MA-A, Ahmed N, Pradhan B, Roy S (2019) Assessment of coastal vulnerability to multi-hazardous events using geospatial techniques along the eastern coast of Bangladesh. Ocean Coast Manag 181:104898. https://doi.org/10.1016/j.ocecoaman.2019.104898

Hossain MT, Chackroborty RD, Intisar L, Shuvo SAT, Rakib AA, Kafy A (2024) Landslide susceptibility and risk assessment in hilly regions of Bangladesh: a geostatistical and geospatial modeling approach for sustainability. In: Advances in natural and technological hazards research, pp 593–619. https://doi.org/10.1007/978-3-031-56591-5_23

Iftekhar MS, Saenger P (2007) Vegetation dynamics in the Bangladesh Sundarbans mangroves: a review of forest inventories. Wetlands Ecol Manage 16(4):291–312. https://doi.org/10.1007/s11273-007-9063-5

Ihinegbu C, Mönnich S, Akukwe T (2023) Scientific evidence for the effectiveness of mangrove forests in reducing floods and associated hazards in coastal areas. Climate 11(4):79. https://doi.org/10.3390/cli11040079

Islam SN (2016) Deltaic floodplains development and wetland ecosystems management in the Ganges–Brahmaputra–Meghna rivers delta in Bangladesh. Sustain Water Resour Manag 2(3):237–256. https://doi.org/10.1007/s40899-016-0047-6

Islam MS, Crawford TW (2022) Assessment of Spatio-temporal empirical forecasting performance of future shoreline positions. Remote Sens 14(24):6364. https://doi.org/10.3390/rs14246364

Islam MA, Maitra MK, Baquee A, Murshed S, Rahman S (2011) Spatial changes of land use/land cover of Moheshkhali Island, Bangladesh: a fact-finding approach by remote sensing analysis. Dhaka Univ J Earth Environ Sci 2(12):43–54

Islam MA, Hossain MS, Murshed S (2015) Assessment of coastal vulnerability due to sea level change at Bhola Island, Bangladesh: using geospatial techniques. J Indian Soc Remote Sens 43(3):625–637. https://doi.org/10.1007/s12524-014-0426-0

Islam MK, Hasan M, Rahman M, Fayyaz R, Chowdhury M, Hamid T (2023) Shoreline dynamics assessment of Moheshkhali Island of Bangladesh using integrated GIS-DSAS techniques. Dhaka Univ J Earth Environ Sci 11(1):1–14. https://doi.org/10.3329/dujees.v11i1.63707

Islam S, Gnauck A (2011) Water shortage in the Gorai river basin and damage of mangrove wetland ecosystems in Sundarbans, Bangladesh. 3Rd Int Conf Water Flood Manag 1:1–14

Jana A, Bhattacharya AK (2013) Assessment of coastal erosion vulnerability around Midnapur-Balasore Coast, Eastern India using integrated remote sensing and GIS techniques. J Indian Soc Remote Sens 41(3):675–686. https://doi.org/10.1007/s12524-012-0251-2

Jana A, Sheena S, Biswas A (2012) Morphological change study of Ghoramara Island, Eastern India using multi-temporal satellite data. Res J Recent Sci 1(10):72–81

Jana A, Biswas A, Maiti S, Bhattacharya AK (2013) Shoreline changes in response to sea level rise along Digha Coast, Eastern India: an analytical approach of remote sensing, GIS and statistical techniques. J Coast Conserv 18(3):145–155. https://doi.org/10.1007/s11852-013-0297-5

Jana A, Maiti S, Biswas A (2017) Appraisal of long-term shoreline oscillations from a part of coastal zones of Sundarban delta, Eastern India: a study based on geospatial technology. Spat Inf Res 25(5):713–723. https://doi.org/10.1007/s41324-017-0139-x

Jana A, Maiti S, Biswas A (2015) Seasonal change monitoring and mapping of coastal vegetation types along Midnapur-Balasore Coast, Bay of Bengal using multi-temporal landsat data. Model Earth Syst Environ 2(1). https://doi.org/10.1007/s40808-015-0062-x

Jana A, Maiti S, Biswas A (2016) Analysis of short-term shoreline oscillations along Midnapur-Balasore Coast, Bay of Bengal, India: a study based on geospatial technology. Model Earth Syst Environ 2(2). https://doi.org/10.1007/s40808-016-0117-7

Kaliraj S, Chandrasekar N, Magesh NS (2013) Impacts of wave energy and littoral currents on shoreline erosion/accretion along the south-west coast of Kanyakumari, Tamil Nadu using DSAS and geospatial technology. Environ Earth Sci 71(10):4523–4542. https://doi.org/10.1007/s12665-013-2845-6

Kirby K, Ferguson S, Rennie CD, Cousineau J, Nistor I (2024) Identification of the best method for detecting surface water in Sentinel-2 multispectral satellite imagery. Remote Sens Appl Soc Environ 101367. https://doi.org/10.1016/j.rsase.2024.101367

Kona AD, Hoque ME, Rahman MA (2025) Evaluating shoreline prediction accuracy with the Kalman filter model: a case study of Nijhum Dwip, bay of Bengal. Remote Sens Appl Soc Environ 101469. https://doi.org/10.1016/j.rsase.2025.101469

Lin Z, Singh M (2024) Assessing coastal vulnerability and evaluating the effectiveness of natural habitats in enhancing coastal resilience: a case study in Shanghai China. Sustainability 16(2):609. https://doi.org/10.3390/su16020609

Lu J, Zhang Y, Shi H, Lv X (2022) Coastal vulnerability modelling and social vulnerability assessment under anthropogenic impacts. Front Mar Sci 9. https://doi.org/10.3389/fmars.2022.1015781

MA (1998b) Historical changes in the Ganges-Brahmaputra delta front. J Coast Res 14(4):1269–1275. http://www.jstor.org/stable/4298887

Mahmud MI, Mia AJ, Islam MA, Peas MH, Farazi AH, Akhter SH (2020) Assessing bank dynamics of the lower Meghna river in Bangladesh: an integrated GIS-DSAS approach. Arab J Geosci 13(14). https://doi.org/10.1007/s12517-020-05514-4

Martínez M, Intralawan A, Vázquez G, Pérez-Maqueo O, Sutton P, Landgrave R (2007) The coasts of our world: ecological, economic, and social importance. Ecol Econ 63(2–3):254–272. https://doi.org/10.1016/j.ecolecon.2006.10.022

Matin N, Hasan GMJ (2021) A quantitative analysis of shoreline changes along the coast of Bangladesh using remote sensing and GIS techniques. CATENA 201:105185. https://doi.org/10.1016/j.catena.2021.105185

Mikhailov VN, Dotsenko MA (2007) Processes of delta formation in the mouth area of the Ganges and Brahmaputra rivers. Water Resour 34(4):385–400. https://doi.org/10.1134/s0097807807040033

Minar MH, Hossain MB, Shamsuddin MD (2013) Climate change and coastal zone of Bangladesh: vulnerability, resilience and adaptability. Middle-East J Sci Res 13(1):114–120. https://doi.org/10.5829/idosi.mejsr.2013.13.1.64121

Mirza MMQ (1998) RESEARCH: Diversion of the Ganges water at Farakka and its effects on salinity in Bangladesh. Environ Manage 22(5):711–722. https://doi.org/10.1007/s002679900141

Mou MA, Tusar MK, Haque MR, Chakraborty S, Ahmed S (2023) Assessment of riverbank erosion and accretion and its impact on the people of chairman Ghat, Noakhali, Bangladesh. J Sustain Environ Manag 2(4):220–230. https://doi.org/10.3126/josem.v2i4.61022

Murty TS, Henry RF (1983) Tides in the Bay of Bengal. J Geophys Res Atmos 88(C10):6069–6076. https://doi.org/10.1029/jc088ic10p06069

Nassar K, Mahmod WE, Fath H, Masria A, Nadaoka K, Negm A (2018) Shoreline change detection using DSAS technique: case of North Sinai coast Egypt. Mar Georesources Geotechnol 37(1):81–95. https://doi.org/10.1080/1064119x.2018.1448912

Nicholls RJ, Cazenave A (2010) Sea-Level rise and its impact on coastal zones. Science 328(5985):1517–1520. https://doi.org/10.1126/science.1185782

Nicholls R, Hoozemans F, Marchand M (1999) Increasing flood risk and wetland losses due to global sea-level rise: regional and global analyses. Glob Environ Chang 9:S69–S87. https://doi.org/10.1016/s0959-3780(99)00019-9

Ozsahin E, Ozdes M, Ozturk M, Yang D (2023) Coastal vulnerability assessment of thrace peninsula: implications for climate change and sea level rise. Remote Sensing 15(23):5592. https://doi.org/10.3390/rs15235592

Palanisamy P, Sivakumar V, Velusamy P, Natarajan L (2023) Spatio-temporal analysis of shoreline changes and future forecast using remote sensing, GIS and kalman filter model: a case study of Rio de Janeiro, Brazil. J S Am Earth Sci 133:104701. https://doi.org/10.1016/j.jsames.2023.104701

Palmer SC, Kutser T, Hunter PD (2014) Remote sensing of inland waters: challenges, progress and future directions. Remote Sens Environ 157:1–8. https://doi.org/10.1016/j.rse.2014.09.021

Pantusa D, D'Alessandro F, Riefolo L, Principato F, Tomasicchio GR (2018) Application of a coastal vulnerability index. A case study along the Apulian Coastline, Italy. Water 10(9):1218. https://doi.org/10.3390/w10091218

Portela CF, Martins VS, Novo EMLM, Paulino RS, Barbosa CCF (2024) Recent advances in geostationary satellites for inland and coastal aquatic systems: scientific research and applications. Int J Remote Sens 45(5):1574–1607. https://doi.org/10.1080/01431161.2024.2314007

Raff JL, Goodbred SL, Pickering JL, Sincavage RS, Ayers JC, Hossain MS, Wilson CA, Paola C, Steckler MS, Mondal DR, Grimaud J-L, Grall CJ, Rogers KG, Ahmed KM, Akhter SH, Carlson BN, Chamberlain EL, Dejter M, Gilligan JM, Hale RP (2023) Sediment delivery to sustain the Ganges-Brahmaputra delta under climate change and anthropogenic impacts. Nat Commun 14(1):2429. https://doi.org/10.1038/s41467-023-38057-9

Roukounis CN, Tsihrintzis VA (2022) Indices of coastal vulnerability to climate change: a review. Environ Process 9(2). https://doi.org/10.1007/s40710-022-00577-9

Ruheili A, Boluwade A (2023) Towards quantifying the coastal vulnerability due to natural hazards using the invest coastal vulnerability model. SSRN Electron J. https://doi.org/10.2139/ssrn.4030424

Santos MT, Silva MM (2021). Analysis of long- and short-term shoreline change dynamics: a study case of João Pessoa city in Brazil. Sci Total Environ 769:144889–144889. https://doi.org/10.1016/j.scitotenv.2020.144889

Sarwar MGM, Woodroffe CD (2013) Rates of shoreline change along the coast of Bangladesh. J Coast Conserv 17(3):515–526. https://doi.org/10.1007/s11852-013-0251-6

Sarwar S, Borthwick AGL (2023) Estimate of uncertain cohesive suspended sediment deposition rate from uncertain floc size in Meghna estuary, Bangladesh. Estuar Coast Shelf Sci 281:108183. https://doi.org/10.1016/j.ecss.2022.108183

Shamsuzzoha M, Ahamed T (2023) Shoreline change assessment in the coastal region of Bangladesh delta using tasseled cap transformation from satellite remote sensing dataset. Remote Sens 15(2):295. https://doi.org/10.3390/rs15020295

Sultana M, Hoque MA, Pradhan B (2025) Assessing Meghna riverbank dynamics and morphological changes in Bangladesh using geospatial techniques. Appl Geomat. https://doi.org/10.1007/s12518-025-00620-y

Sunder S, Ramsankaran R, Ramakrishnan B (2017) Inter-comparison of remote sensing sensing-based shoreline mapping techniques at different coastal stretches of India. Environ Monit Assess 189(6). https://doi.org/10.1007/s10661-017-5996-1

Vieira LR, Vieira JG, Silva IMD, Barbieri E, Morgado F (2021) GIS models for vulnerability of coastal erosion assessment in a tropical protected area. ISPRS Int J Geo-Inf 10(9):598. https://doi.org/10.3390/ijgi10090598

Voller VR, Swenson JB, Paola C (2019) The thin blue line: a review of shoreline dynamics across time scales and environments. Earth Surf Proc Land 45(1):96–108. https://doi.org/10.1002/esp.4705

Weiss JL, Overpeck JT, Strauss B (2011) Implications of recent sea level rise science for low-elevation areas in coastal cities of the conterminous USA. Clim Change 105(3):635–645. https://doi.org/10.1007/s10584-011-0024-x

Yan D, Yao X, Li J, Qi L, Luan Z (2021) Shoreline change detection and forecast along the Yancheng coast using a digital shoreline analysis system. Wetlands 41(4). https://doi.org/10.1007/s13157-021-01444-3

Yasir M, Sheng H, Fan H, Nazir S, Niang AJ, Salauddin M, Khan S (2020) Corrections to "automatic coastline extraction and changes analysis using remote sensing and GIS technology." IEEE Access 8:200680–200683. https://doi.org/10.1109/access.2020.3034707

Zambrano-Medina YG, Plata-Rocha W, Monjardin-Armenta SA, Franco-Ochoa C (2023) Assessment and forecast of shoreline change using geo-spatial techniques in the Gulf of California. Land 12(4):782. https://doi.org/10.3390/land12040782

Chapter 8
Coastal Vulnerability Mapping Using Multi-criteria Decision Making (MCDM) Process

Kamalesh Kanna Shanmuganathan, Sivakumar Palraj, Athira A. Nair, V. P. Nithya Segar, Avinash Kanagaraj, Manimaran Gajendiran, and Bharathi Mohandass

8.1 Introduction

Ecosystems and environments are being affected and altered at an increasingly unpredictable rate due to significant climate change, the most pressing challenge of our time (Van Huong et al. 2022). Any long-term, considerable alteration in a place's average weather, temperature, precipitation, and wind patterns is referred to as climate change (Bagheri et al. 2021). As a serious threat to both nature and humanity, the accelerated changing climate has garnered global attention from many (Dalby 2020). The coastal environment, a dynamic zone between land and water, is ever-changing and influenced by forces such as wind, waves, tides, currents, and anthropogenic activities (Saravanan et al. 2019; Selvan et al. 2020). With diverse flora and fauna, salt marshes, wetlands, estuaries, dunes, and beaches, coastal regions are home to a wide variety of ecosystems. Coastal regions are also crucial for various human activities, including fishing, shipping, tourism, and recreation. These coastal regions are important because of its richness in renewable and non-renewable resources, high ecosystem productivity, industrialization and are vulnerable to anthropogenic hazards such as industrialization, discharge of effluents and municipal sewage and population density as well as natural hazards such as storms, flooding, erosion, and sea level rise (SLR) which also significantly impact human communities and environments (Parthasarathy and Deka 2021; Prof et al. 2012). Some of the coastal threats are given

K. K. Shanmuganathan (✉) · S. Palraj · A. A. Nair · V. P. N. Segar
Department of Remote Sensing and Geographic Information System, Tamil Nadu Agricultural University, Coimbatore, India
e-mail: kamaleshkanna.skk@gmail.com

A. Kanagaraj · M. Gajendiran · B. Mohandass
Department of Soil Science and Agricultural Chemistry, Tamil Nadu Agricultural University, Coimbatore, India

A. Nath et al. (eds.), *Geospatial Application of Coastal Hazards*, Springer Water,
https://doi.org/10.1007/978-3-032-20076-1_8

in Fig. 8.1. Multiple impacts such as erosion, coastal flooding, SLR, submergence, reduced oxygen and pH, and biodiversity loss in low-lying and geomorphologically dynamic coastal regions are expected to result from climate change (Jana et al. 2012, 2016b). These issues are often exacerbated by extreme events, such as tsunamis, flash floods, and storm surges, resulting in dynamic and constantly changing coastal environments. Due to natural and human activities, coastal cities are experiencing significant changes, leading to increased erosion, flooding, siltation, loss of coastal resources, and damage to marine habitats (Bagheri et al. 2021; Malara et al. 2020). Human activities, including tourism, industrialization, agriculture, and urbanization, also contribute to coastal susceptibility by modifying natural processes and exacerbating environmental hazards. The increased population near coastal areas is due to various development activities, including port and harbor construction, tourism and recreational development, fisheries and aquaculture development, and renewable energy projects (Balica et al. 2012). In 1994, the coastal population was 2.07 billion (37%) and 2.45 billion (44%) within 100 km and 150 km of the shoreline, respectively. This increased to 2.86 billion (38.1%) and 3.34 billion (44.6%) by 2018.

Sea Level Rise

SLR is the increase in water from continents to the ocean through the melting or displacement of ice from ice caps and glaciers, groundwater movement, changes in oceanic and atmospheric circulation, tectonic movements, and natural and human subsidence, which combine to cause a change in mean sea level at a given location.

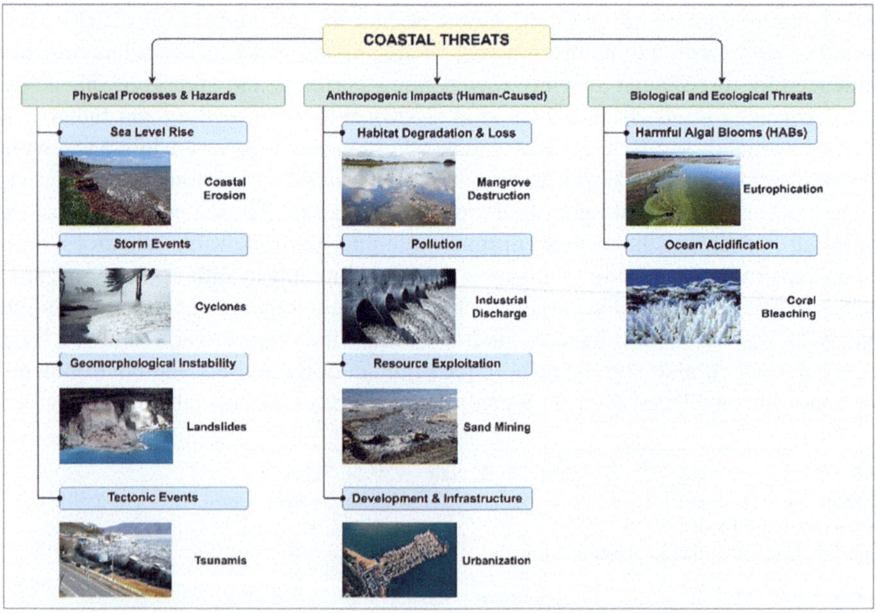

Fig. 8.1 Different physical, anthropogenic, biological and ecological Coastal threats

SLR leads to flooding of low-lying islands, submergence, and increases flooding of coastal lands, intrusion of saltwater into surface and subsurface water of lands, increased erosion, and habitat destruction in coastal regions (Jana et al. 2017, 2016a, 2014). On a global scale, the costs of coastal protection and relocation of people from land areas lost to SLR range from approximately $200 billion for a 0.5 m sea level increase to about $2 trillion for a 2 m sea level increase.

Saltwater Intrusion

Saltwater intrusion into coastal agricultural lands poses a growing threat to agricultural productivity, ecosystem stability, and local economies, resulting in reduced crop yields since most crops are sensitive to soil salinity (Kumar et al. 2024; Su et al. 2025). Saltwater intrusion leads to the formation of ghost forests, the invasion of salt-tolerant species, degradation of biodiversity, and deterioration of water quality due to ecological and environmental changes (Kumar et al. 2022, 2023).

Storm Surges

Storm surges, which result from strong winds, low atmospheric pressure, and other meteorological factors, cause an abnormal rise in sea level that further inundates coastal regions (Jones et al. 2013). Typhoons and strong weather events cause storm surges, which are the most vulnerable factors and threats to the physical, social, and economic sectors of coastal regions, resulting in damage to coastal infrastructure, including dikes, wharves, ports, roads, bridges, and power facilities. In coastal urban areas, heavy storm surges cause the breaking of dikes, resulting in flooding of houses, farmlands, and aquaculture areas, which can lead to severe casualties and property loss (Xianwu et al. 2020). Strom surges also affect fisheries and tourism industries, which are significant economic sources for many coastal regions, leading to the accelerated expansion of algal blooms (Zhang and Zhang 2024).

Human Activities

Anthropogenic activities in coastal dynamics play a significant role, both as positive and negative feedback mechanisms (Schmitt and Chaumillon 2023), including land reclamation, shellfish farming, coastal defenses, dredging, and sand mining. To protect these coastal regions from natural coastal erosion and sediment fill in harbors and tidal channels, coastal defenses and ditches were implemented. However, this leads to upper beach erosion, pollution, and increased turbidity, becoming a threat to both the coastal ecosystem and settlements. On a long-term basis, intense sand extraction leads to instability in the coastal ecosystem, resulting in habitat loss (Ceballos et al. 2015). Eutrophication of coastal ecosystems causes both economic and social losses, leading to the demise of fisheries. Tourism is also affected by the foul odor from coastal regions and the water's turbidity (Cloern et al. 2016).

8.1.1 Need for an Advanced Decision Support System

An essential part of all human life is decision-making, which involves choosing among alternatives that include gathering information, generating options, contemplating them, and evaluating alternative courses of action, as well as processes of implementation and evaluation that should follow a decision once made. It involves setting weights and ranking activities according to the objective of the decision made (Herath and Prato 2006). A system that assists in decision-making, utilizing computer-based Decision Support Tools (DSTs) collectively referred to as a Decision Support System (DSS), can be applied to matters such as environmental problem management and long-term planning (Rosenstock et al. 2024). Integrating multiple DSS produces high-quality outputs with reduced uncertainty during decision-making, and the system will be stable and reliable for long-term application by decision-makers and coastal managers. Choosing the right DSS for vulnerability assessment is also a challenging task. Being a data-driven and model-based approach, DSS organizes and analyzes information in a reliable manner using various analytical and predictive models (Falconer et al. 2023; Niggli et al. 2024). Thus, better mapping and modelling of the relationship between hazards and exposures can be achieved for infrastructure development, coastal management, and creating suitable buffer zones. DSS developed an indices-based approach called Decision Support Indices (DSIs) for coastal vulnerability mapping, which involves different combinations of environmental and socioeconomic parameters related to vulnerability, resilience, and risk, to develop coastal adaptation protocols in sustainable development practices and mitigation strategies. DSS involves the integrated application of GIS, Multi-criteria decision-making (MCDM), Artificial Neural Networks (ANN), Google Earth Engine (GEE), Agent-Based Models (ABM), big data analysis, the Internet of Things (IoT), Application Programming Interfaces (APIs), System Dynamics (SD), and other relevant technologies.

MCDM, a DSS that involves decision-making based on multiple criteria adopted from economics and psychology, has gained popularity among interdisciplinary research since the 1950s (Barocchi et al. 2016; Chen and Bau 2016; Malekmohammadi and Blouchi 2014). A timeline of various MCDM methods is presented in Fig. 8.2, along with their general purpose and applications. To achieve desired objectives and make final decisions, MCDM evaluates multifaceted problems that incorporate various geospatial, environmental, and socioeconomic factors together (Ali et al. 2019; Hwang and Lin 2012). MCDM is the process that involves combining both qualitative and quantitative outcomes to select the best option (Attri et al. 2022; Sriram et al. 2022; Van Dua 2023; Yuan et al. 2022). This decision was made by MCDM, which is constructed and justified using mathematical tools (Pournamdarian 2010). MCDM is frequently applied to a range of geospatial decision-making issues addressing four main areas: location analysis, environmental modeling, suitability analysis, and risk and vulnerability assessment. Coastal vulnerability assessment

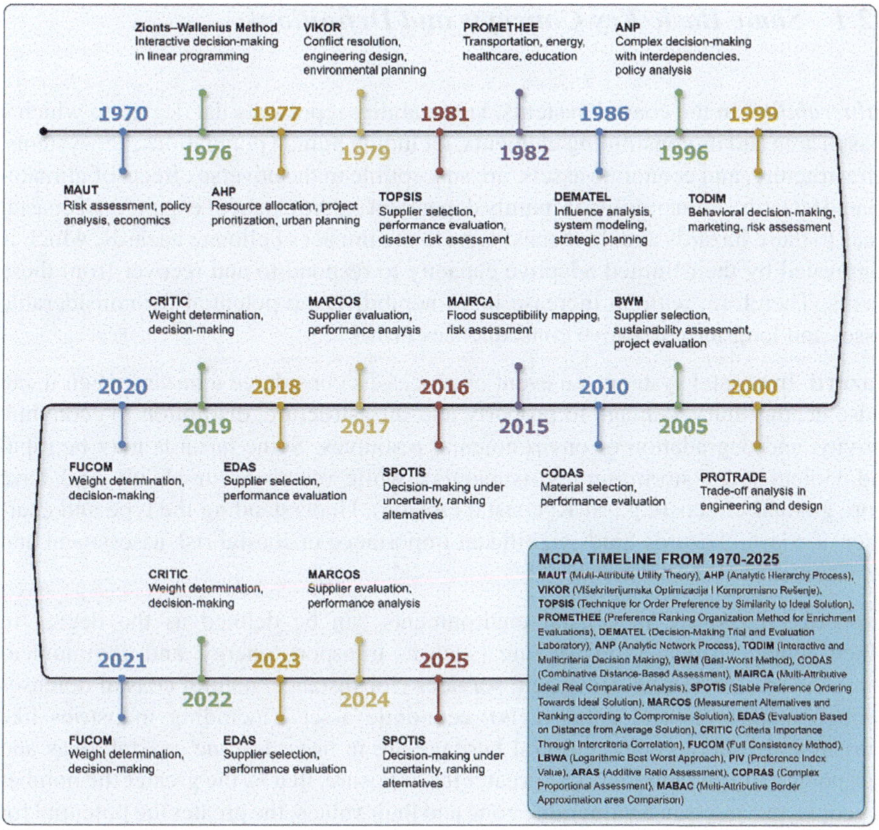

Fig. 8.2 Timeline of different MCDM methods with their applications

(CVA) in the framework of ecological hazards using MCDM involves risk reduction, the selection and acquisition of required data from various sources, which are essential parts of the research (Ali et al. 2019; Brooks 2003; Malczewski 1999).

8.2 Conceptual Framework of Coastal Vulnerability

A systematic framework that integrates environmental, social, and spatial dimensions is required for understanding coastal vulnerability. Governments have released several frameworks, such as IPCC (Intergovernmental Panel on Climate Change), Third Assessment Report (TAR), Assessment Report (AR5), UNEP (United Nations Environment Programme), USAID (U.S. Agency for International Development), and ICZM (Integrated Coastal Zone Management) Protocol for CVA and mapping.

8.2.1 Some Basic Key Concepts and Definitions

Vulnerability: In the coastal systems, vulnerability represents the degree to which a coastal area and its constituting elements, including human populations, ecosystems, infrastructure, and economic assets, are susceptible to the adverse effects of climate-related hazards. Vulnerability is mainly dependent on the extent of exposure of coastal areas to these hazards and their sensitivity to the impact of climate hazards, which is aggravated by their limited adaptive capacity to respond to and recover from these events. Therefore, with an increase in vulnerability, the potential for considerable losses and long-term negative consequences grows.

Hazard: In coastal systems, an event or process is considered a hazard when it can cause death, injury, damage to property and infrastructure, disruption to economic activity, and degradation of environmental resources. Some hazards may be rapid and violent (e.g., storm surges, tsunamis), while others occur slowly and have more gradual effects (e.g., SLR, coastal erosion). Understanding the type and character of coastal hazards holds significant importance in coastal risk assessment and management.

Exposure: Exposure in coastal environments can be defined as the degree to which people, critical infrastructure (such as transport, energy, and communication networks), essential ecosystem services (for instance, natural coastal defenses through mangroves and coral reefs), economic assets (including industries like tourism and fisheries), and cultural heritage are in hazard-prone coastal areas and thus potentially endangered. The greater the exposure, that is, the greater the number of people and assets in a vulnerable zone and their values, the greater the potential for heavy human, environmental, economic, and cultural losses when a coastal hazard event happens.

Sensitivity: Sensitivity to climate variability or hazards refers to the system's or its constituents' direct or indirect responses to climate conditions or the occurrence of hazardous events. This influence can be either adverse or favorable, with the determining factor often being the inherent features of that system or its physical location. For instance, as you have said, due to minimal elevation over the sea level, low-lying islands are susceptible to SLR, i.e., they are at risk of inundation and coastal erosion. Conversely, a region with drought-resistant vegetation might display less sensitivity to minor changes in rainfall patterns. Sensitivity is one of the key factors that lead to the understanding of and potential impacts due to climate change and hazards to different systems.

Adaptive Capacity: This represents the ability of a system, community, or ecosystem to adjust to potential hazards, to mitigate damage, or to recover after an event. Factors include economic wealth, education, governance, and access to resources.

Resilience: Resilience in the context of hazardous events is referred to as the dynamic capacity of any system community, ecosystem, infrastructure, or economy to anticipate potential threats proactively, absorb the initial impacts with little disruption, adapt, and accommodate from changing conditions during and after the event, recover, and reorganize justifiably and efficiently back to a state of at least equal functionality or bored with uninterrupted consequences in the long run. Resilience encompasses learning from past experiences and implementing changes that enhance one's ability to cope with future challenges.

8.2.2 Components of Coastal Vulnerability

Compelling coastal vulnerability requires a detailed examination of the main elements that drive it. In the past, these elements were insufficiently measured or simplified to the point of providing broad and generally less accurate estimates. Advances in geospatial technologies now enable the extraction of spatially explicit, more precise data, from which more reliable and detailed vulnerability evaluations and risk assessments can be conducted. Components of coastal vulnerability are classified into three general domains: physical, socioeconomic, and ecological. These domains encompass a series of indicators that collectively define the vulnerability and resilience of coastal systems to various natural and human-induced pressures. A thorough evaluation of coastal vulnerability requires establishing meaningful interactions between these elements to account for the complex relationships within coastal systems. Such integration can be successfully achieved through MCDM models, which provide a systematic approach to combining and weighting various criteria.

8.2.2.1 Physical Components

Physical components (Table 8.1) directly influence the evaluation of coastal vulnerability and encompass exposure and susceptibility factors in coastal regions resulting from SLR, erosion, flooding, and storm surges. Being an essential factor, geomorphology, due to its association with coastal landforms and land uses, controls the resistance of coastlines to erosional and accretionary processes. For instance, areas such as barrier beaches, dunes, and salt marshes are more prone to erosion, whereas rocky cliffs and fjords tend to resist. Multispectral satellite imagery, Digital Elevation model (DEM), and Land Use and Land Cover (LULC) data are used to derive geomorphological features that facilitate the mapping of high-resolution terrain and coastal forms. Shoreline change rate, on the other hand, is a direct indicator of the land-sea dynamics controlled by sediment transport, SLR, and wave action. Although traditionally shoreline change assessment comprised historical maps and field surveys, recent days have witnessed an increase in the rates of shoreline monitoring using satellite-derived shorelines (SDS) and automated methods such as the Digital Shoreline Analysis System (DSAS) developed by United State Geological Survey (USGS)

(Himmelstoss et al. 2021) due to their accuracy, cost-effectiveness, and high degree of spatial scalability (Fig. 8.3).

Physical indicators considered fundamental in modeling inundation risk and SLR impacts include elevation, topography, and coastal slope. The elevation data derived using Light Detection and Ranging (LiDAR) or DEM were essential in assessing vertical exposure and terrain susceptibility. The slope controls runoff paths, erosion potential, and flood routes, whereas distance from shoreline is an index of spatial proximity that helps in assessing gradient exposure. In addition, coastal erosion/

Table 8.1 Physical components for remote sensing & GIS-based CVA

Physical indicator	Description	Remote sensing/GIS data source
Elevation/Topography	Low elevation areas are more prone to SLR and flooding	DEMs (e.g., SRTM, ASTER, LiDAR, ALOS PALSAR)
Slope	Flat coasts are more vulnerable, whereas steeper slopes may reduce the risk of inundation	Derived from DEMs using GIS
Shoreline change rate	Historical shoreline retreat or advance indicates erosion or accretion	Multi-temporal satellite images (e.g., Landsat, Sentinel-2)
Coastal erosion/accretion	Indicates the stability of the coastline over time	Image classification & shoreline extraction
Distance from shoreline	Proximity to the coast reflects potential exposure to hazards	GIS-based proximity analysis
Coastal slope (bathymetric and land)	Affects wave run-up and inundation patterns	DEMs, bathymetric charts
Wave and tidal energy exposure	High wave/tidal energy zones experience greater physical stress	Modeled data or oceanographic datasets (e.g., NOAA, Copernicus)
Soil type/geology	Sandy soils are more prone to erosion compared to rocky coasts	Soil maps, geological surveys, remote sensing imagery
LULC	Urban and agricultural areas increase runoff and reduce coastal resilience	Satellite data (Landsat, Sentinel-2, MODIS)
Drainage density/surface runoff	Indicates the susceptibility to flooding and erosion	Hydrological modeling from DEMs
Storm surge flood zones	Areas at risk from storm surge flooding events	Historical hazard maps, flood modeling
Mean significant wave height	Related to wave-induced erosion and flooding	Oceanographic data (WaveWatch III, altimeter data)
Coastal vegetation cover	Natural barriers, such as dunes, mangroves, or coastal forests, reduce physical vulnerability	NDVI from satellite imagery (Landsat, Sentinel, MODIS)

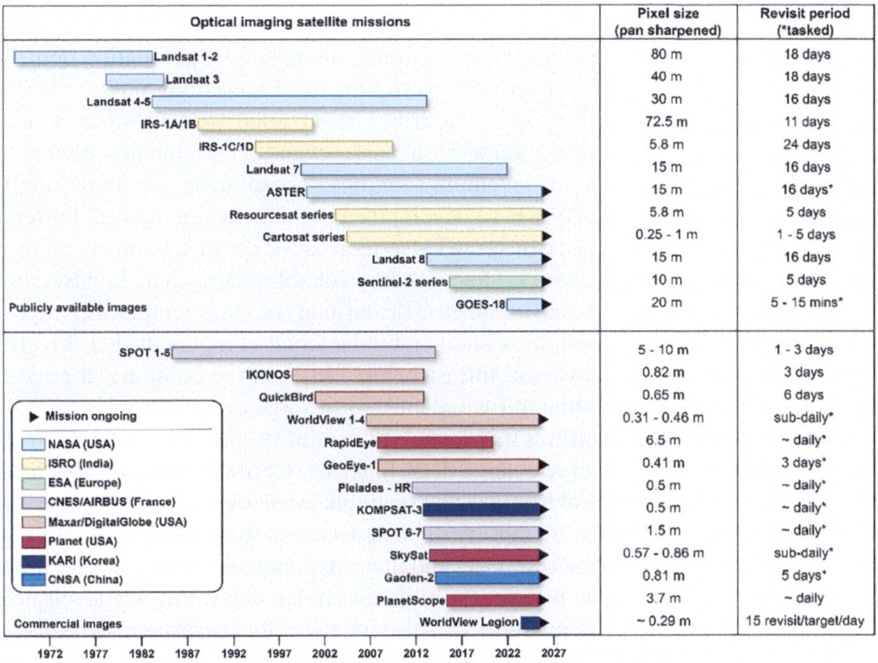

Fig. 8.3 A critical satellite mission used in digital shoreline analysis for CVA

accretion rates, wave and tidal energy exposure, and significant wave height characterize the hydrodynamic forces acting along the coastline. Soil type and coastal geology have a substantial impact on the structural stability and erosion response of marginal coasts. At the same time, drainage density and surface runoff patterns help determine water flow and flood dynamics. High-risk zones during cyclone events are identified using storm surge flood zones obtained from historical and hydrodynamic models. The abandonment of Natural buffers, such as coastal vegetation covers (mangroves and dune grasses), increases physical vulnerability. Integrating these RS and GIS-derived physical parameters in MCDM enables the spatially explicit, scientifically informed assessment of vulnerabilities along coastal areas.

8.2.2.2 Socioeconomic Components

Socioeconomic components (Table 8.2) play a pivotal role in CVA, as they reflect the degree to which human populations and infrastructure are exposed to hazards and their capacity to cope with or recover from them. Among these, population density is a critical indicator because it denotes potential exposure to risks such as flooding, erosion, and storm surge. High-density coastal zones, often hubs of economic activity and habitation, are typically more vulnerable due to the concentration of assets and infrastructure, while also contributing to degradation through land reclamation, pollution, and the destruction of mangrove ecosystems. Interestingly, population density also intersects with adaptive capacity, as more urbanized areas may possess better infrastructure and services to respond to hazards. Urbanization and built-up area indicators reveal the rapid expansion of impervious surfaces and dense human settlements in coastal cities, particularly in developing regions. These areas are increasingly threatened by SLR, particularly where development encroaches on natural buffers, such as wetlands and dunes. Urban sprawl also increases exposure to infrastructure failure and social disruption, with remote sensing technologies such as Landsat and Sentinel being widely used to map urban growth and analyze spatiotemporal changes.

The proximity to the coastline is another fundamental component that directly relates to exposure. Settlements and infrastructure closer to the coast are at greater risk from tidal inundation, saline intrusion, and storm surge events. Proximity analysis, utilizing GIS tools, quantifies this spatial vulnerability, particularly when overlaid with elevation and shoreline change data. In terms of critical services, road and transport networks are essential lifelines that facilitate evacuation, rescue operations, and economic stability. The vulnerability of such networks to inundation or damage can severely impair emergency response, and their mapping using satellite and open data sources provides insights into accessibility and redundancy. Poverty levels and income distribution reflect a population's sensitivity and ability to respond to hazards. Poor communities often settle in more hazardous areas, lack access to insurance or relocation options, and face compounding risks. These factors are further exacerbated in regions with low literacy and education levels, as disaster awareness and preparedness are strongly linked to education, which in turn affects resilience.

Employment types, especially in traditional coastal sectors such as fishing and farming, also indicate sensitivity, as these livelihoods are directly affected by environmental changes, including saline intrusion or cyclonic disturbances. Spatial data from government surveys or participatory mapping is often used to assess these dependencies. The quality of housing, particularly the prevalence of informal settlements with non-durable materials, increases vulnerability to structural damage during extreme events. Likewise, access to health infrastructure and emergency services greatly influences adaptive capacity; regions with well-distributed medical and emergency facilities are better equipped for rapid recovery. Migration and displacement risks are growing concerns, particularly in areas experiencing recurrent flooding or erosion. Tracking such mobility, either through census data or remote sensing of land abandonment, helps map chronic vulnerability zones. Economic assets and property value, especially in urban coastal centers, represent high-stakes areas for vulnerability planning, as losses here can severely impact regional economies (Fig. 8.4). The importance of tourism, a sector often concentrated along attractive but fragile coasts, adds another dimension, where seasonal population surges and infrastructure stress heighten exposure and economic risk. Lastly, indicators of gender and social vulnerability, such as female-headed households or dependent populations, are essential for a holistic vulnerability assessment, as marginalized groups typically have reduced access to resources, decision-making, and recovery mechanisms.

Table 8.2 Socioeconomic components for remote sensing & GIS-based CVA

Indicator	Description	Source	Weightage
Population density	Higher density means more people are exposed to coastal hazards	Census data, WorldPop, LandScan, built-up indices from RS	Very High
Urbanization/ built-up area	The extent of infrastructure and development vulnerable to hazards	LULC from Landsat/ Sentinel, GHSL (Global Human Settlement Layer)	Very High
Proximity to coastline	Measures how close people/ infrastructure are to hazardous zones	GIS buffer analysis	High
Road and transport networks	Critical for evacuation and disaster response	OSM, local transport datasets, satellite-based infrastructure maps	High
Poverty index/ income level	Poorer populations have lower adaptive capacity	Census, Human Development Index (HDI), Socioeconomic surveys	High
Literacy rate/ education	Influences awareness and preparedness to respond to coastal risks	National census, DHS surveys	Medium
Employment type (Fishermen, farmers, etc.)	Determines exposure due to occupation-based dependence on coastlines	Census, local livelihood maps	Medium
Housing quality/ informal settlements	Weak structures are more vulnerable to coastal storms and flooding	Field survey, urban mapping, satellite imagery classification	High
Health infrastructure	Indicates the capacity to respond to coastal health emergencies	Ministry of Health, GIS mapping of facilities	Medium
Access to emergency services	Availability of shelters, warning systems, and other relevant resources	GIS location analysis, civil protection data	High
Migration/ displacement risk	Populations at risk of or currently undergoing relocation due to hazards	UNHCR data, government relocation schemes	Medium
Economic assets/ property value	High-value assets increase the economic vulnerability	Taxation/property databases, local development records	Low
Tourism density/ dependency	Coastal tourism-based economies can face higher losses from hazards	Tourism department stats, business census, satellite mapping	Low
Gender/social vulnerability index	Women, the elderly, and marginalized groups are more vulnerable	DHS, census, vulnerability indices	Medium

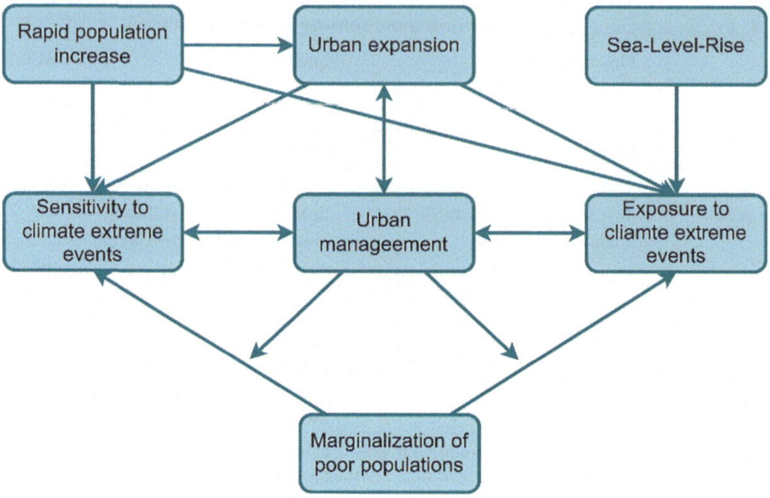

Fig. 8.4 Rapid coastal urbanization: parameters of socio-ecological vulnerability and their interactions (Sterzel 2020)

8.2.2.3 Ecological Components

Ecological components (Table 8.3) are essential for understanding coastal vulnerability, as they serve as both protective buffers against natural hazards and critical providers of ecosystem services to human populations. Coastal zones host unique ecosystems, including mangrove forests, salt marshes, tidal flats, seagrass beds, coral reefs, and dune vegetation, all of which play integral roles in attenuating wave energy, stabilizing shorelines, and supporting biodiversity. For example, mangroves and salt marshes absorb storm surges and reduce erosion, while coral reefs act as natural breakwaters. The presence, health, and continuity of such habitats significantly influence a region's ecological resilience. Ecological vulnerability indicators, such as habitat fragmentation, species richness, biodiversity hotspots, and zones of invasive species, provide insights into the capacity of ecosystems to absorb disturbances and recover. Tools such as the Ecological Sensitivity Index (ESI) and geospatial habitat modeling allow researchers to represent these attributes spatially. Furthermore, ecological features can be monitored using high-resolution remote sensing data, hyperspectral imaging, and time-series analyses to detect changes in vegetation cover, water quality, and land use patterns over time.

Importantly, ecological and socioeconomic components are not isolated but tightly interlinked in coastal social–ecological systems (CSES). Human reliance on coastal ecosystems, for fisheries, tourism, flood regulation, and cultural services—creates feedback loops where ecological degradation directly undermines human well-being and adaptive capacity. For instance, degradation of mangrove belts due to urban expansion or aquaculture directly increases the vulnerability of nearby populations to flooding. Simultaneously, socioeconomic stressors such as pollution, overfishing,

Table 8.3 Ecological components for remote sensing & GIS-based CVA

Indicator	Description	Source	Weightage
Mangrove cover	Natural buffer that reduces wave energy and storm surge	NDVI from Sentinel-2, Landsat, Global Mangrove Watch (GMW)	Very high
Wetlands (salt marsh, tidal flats)	Absorb floodwaters and protect inland ecosystems	Landsat/Sentinel-2 classification, Global Surface Water Explorer	High
Coral reefs	Act as offshore wave barriers, reducing coastal erosion	Remote sensing (Sentinel-2, MODIS), UNEP-WCMC, ReefBase	High
Coastal vegetation (dunes, grasses)	Stabilize shoreline and reduce erosion	NDVI/EVI indices from MODIS, Sentinel-2, UAV-based vegetation maps	High
Habitat fragmentation	Indicates ecological degradation and decreased resilience	Land use change from RS, patch analysis using FRAGSTATS	Medium
Ecological sensitivity index (ESI)	Composite measure of ecosystem sensitivity and response to hazards	NOAA ESI maps, national biodiversity assessments	Medium
Biodiversity hotspots/species richness	High biodiversity areas are more ecologically valuable but vulnerable	Conservation databases (IUCN, GBIF), RS-based habitat mapping	Medium
Protected areas/ conservation zones	Indicates existing ecological protection and adaptive potential	WDPA, government gazetteer data, overlaid on RS layers	Medium
Water quality/ pollution level	Degraded water bodies affect overall ecosystem health	RS-derived water indices (NDWI, turbidity from MODIS/Sentinel-2)	Low
Fisheries productivity zones	Indicates ecological and economic vulnerability in marine resource zones	Sea surface temperature (SST), chlorophyll data (MODIS-Aqua)	Low
Invasive species zones	Non-native species disrupt natural coastal ecosystems	Field surveys, predictive models, supported by UAV or high-res RS	Very low

and coastal development lead to declines in fisheries productivity, water quality, and habitat integrity, further intensifying ecological stress. The Social-Ecological Systems Framework (SESF) and models like the Driver-Pressure-State-Impact-Response (DPSIR) framework emphasize these feedback loops and the need for integrated vulnerability assessments. In areas where protected areas or conservation zones are established and effectively managed, both ecological and human resilience are enhanced. Thus, ecological indicators, when combined with socioeconomic variables in MCDM, offer a holistic and system-based approach for mapping and mitigating coastal vulnerability under the pressures of climate change, urbanization, and resource exploitation.

8.3 Multi-Criteria Decision-Making (MCDM) in Geospatial Context

Using a coastal vulnerability index with equal weightage for all criteria is a nuanced approach to coastal studies. Multi-Criteria Decision Making (MCDM) is a decision-support tool used to identify and rank variables from most to least significant among complex scenarios of coastal regions. It aids in delineating potential impact zones within the study region by assigning weights to selected variables based on defined criteria and expert judgment. Through systematic evaluation, transparent and structured processes, and adaptability and robustness offer significant advantages for planning and management. Some of the multi-criteria techniques are classified under these approaches (Ishizaka and Nemery 2013), including the full aggregation approach, the outranking approach, the goal, aspiration, or reference level, and integrated systems.

8.3.1 Full Aggregation Approach

8.3.1.1 Analytic Hierarchy Process (AHP)

AHP, a semi-quantitative technique developed by Professor Thomas L. Saaty (1977), was widely used in coastal studies with integration of the CVI (Barzehkar et al. 2025; KV 2024; Sekovski et al. 2020) for obtaining a ranking of the alternatives—the analyst needs to undergo these four steps for using AHP. First, the problem needs to be structured. Second, scores or priorities were calculated based on pair-wise comparisons provided by the analyst. Then, the third and fourth steps involve consistency checks and sensitivity analysis. Based on the principle judgment done by a group of decision makers between the multiple pairs of criteria, leading to applying criteria weights based on an opinion basis to solve the equal weights problem. Most studies employed criteria based on AHP, combined with parameter-based approaches, such as CVI, to mitigate the variance in output resulting from user evaluation. They have been proven effective in coastal vulnerability studies, synthesizing diverse and heterogeneous parameters into a comprehensive and coherent vulnerability index. Ghosh and Mistri (2021) assessed environmental hazards in the Indian Sundarbans delta region in terms of degree of vulnerability using the AHP method and calculated indices such as exposure, sensitivity, and adaptation capacity index, utilizing 22 indicators including physical, socioeconomic, and climatic variables (Fig. 8.5). Through this, they identified the southern coastal Indian Sundarbans Delta (ISD) region, which has a direct coastline with the Bay of Bengal, as being more vulnerable than the northern ISD regions. KV (2024) comprehended how risks and hazards are spatially distributed along the coastal zones of Kerala, India. They utilized the AHP method to obtain weights for physio-geological and socioeconomic parameters and developed the CVI. Results show that the coastal stretch of 82.40 km has a very high CVI value, accounting for 13.87% of the total coastline, and identifies

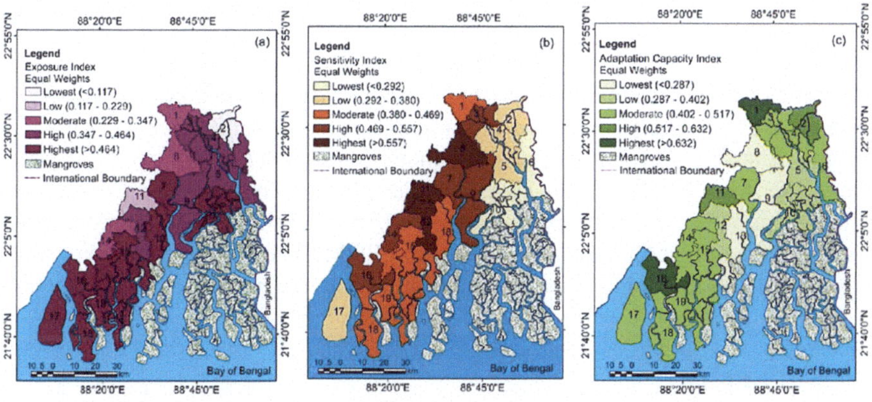

Fig. 8.5 Spatial distribution of vulnerability sub-indices using AHP (Ghosh and Mistri 2021)

high-vulnerable districts, including Kozhikode, Malappuram, Thrissur, Ernakulam, Alappuzha, and Trivandrum. The consistency and reliability of comparisons between pairs in AHP were ensured by the consistency ratio given by

$$CR = \frac{CI}{RI}$$

CI is the consistency index calculated as

$$CI = \frac{\lambda_{max} - n}{n - 1}$$

where λ_{max} is the principle eigenvalue of the matrix, n is the size (order) of the comparison matrix. And RI is a predefined value derived from the statistical stimulation of a random pairwise comparison matrix.

8.3.2 Outranking Approach

8.3.2.1 PROMETHEE

Preference Ranking Organization METHod for Enriched Evaluation (PROMETHEE) ranks the criteria based on preference degrees, which were developed by Brans (1982). The three main steps involved in PROMETHEE are as follows:

1. The preference degree of every ordered pair of each criterion is calculated
2. Uni Criteria flow computation

3. Global flow computation, ranking, and graphical representation of decision problems are performed based on global flows.

Preference degree (generally 0 to 1) is a score that indicates how a criterion is preferred over other criteria from a decision maker's viewpoint. Here, if preference degree is 1, strong preference is given to that criterion over others, and vice versa. Several PROMETHEE methods have been developed to make the processes more flexible in complex environments (Brans et al. 1986; Brans and Mareschal 1992; Figueira et al. 2005; Macharis et al. 1998). In PROMETHEE I (partial ranking) and PROMETHEE II (complete ranking), the two methods are often used in conjunction.

8.3.2.2 ELECTRE

ELECTRE (Elimination Et Choix Traduisant la Realité) is an outranking method developed by Roy (1968), which avoids the distortion of original data during the criteria and normalization process. Despite its relative complexity due to technical parameters and complex algorithms, it is used in various fields, including agriculture and forestry, energy, water management, environmental management, finance, transportation, and military applications for decision-making. Different methods of ELECTRE have been applied to various decision problems, including choice problems (ELECTRE I, ELECTRE Iv, ELECTRE Is), ranking problems (ELECTRE II, ELECTRE III, ELECTRE IV), sorting problems (ELECTRE-Tri-B, ELECTRE-Tri-C), and elicitation problems. ELECTRE is implemented when facing decision problems with more than two criteria and to satisfy at least one condition such as different units for performance criteria, avoiding standard scale for criteria by decision makers, if there is a need to use preference and indifference thresholds to treat slight differences also negligible, when scales are presented in order or on a 'weak' interval scale. Most of the case studies utilized ELECTRE methods in integration with other methods. (Fetanat and Khorasaninejad 2015) utilized ANP (Analytic network process), fuzzy decision-making trail and evaluation laboratory (DEMATEL), and ELECTRE for decision-making in site selection of offshore wind farms in regions of the Persian Gulf, southwest Iran.

8.3.2.3 TOPSIS

TOPSIS outputs are easier to understand, and it requires only a minimum number of inputs. The basic idea of TOPSIS is to identify the best solution based on its distance to the ideal solution. Here, the best solution is the one with the shortest distance to the ideal solution and the greatest distance from the anti-ideal solution. It involved five stages, where the first stage involved gathering performance data on different criteria for alternatives. Then, normalization of performance is achieved through various techniques, such as distributive normalization and ideal normalization. Then, the weighing and calculation of ideal and anti-ideal points. Finally, by determining the ratio of this distance, the closeness (or best solution) can be determined.

8.3.3 Integrated System

These integrated multi-criteria decision systems are based on various algorithms and methods that can be utilized in complex, real-world decision problems, which mostly reduce the input information required. This integrated system approach comprises two main projects: Decision Desk and DECERNS. DECERNS is used by Arkhipova et al. (2021) in the sustainable development of coastal zones in the Sea of AOZV (Russian sector), which utilizes socioeconomic and environmental considerations to ensure public safety and minimize economic damage. This DECERNS consists of three modules: the GIS module, the MCDM module, and the Group Decision Support System (GDSS) module. These modules enable user-friendly visualization of spatial data and facilitate spatial data analysis. The MCDM modules include MAUT, MAVT, AHP, PROMETHEE, and TOPSIS for ranking, as well as FlowSort for sorting, which the decision-makers can choose. GDSS allows the creation and administration of online surveys. Moreover, the results are automatically collected and analyzed.

8.4 Data Inputs and Indicator Preparation for MCDM-Based Vulnerability Mapping

Integrating a variety of risk indicators is crucial for a thorough evaluation of the present and prospective states of coastal zones, particularly concerning natural hazards and anthropogenic activities. A well-structured vulnerability assessment must integrate physical, socioeconomic, and ecological components to reflect the multidimensional character of coastal risk. Choosing suitable indicators requires identifying ideal data sources that accurately reflect the spatial and temporal diversity of threats impacting coastal regions. These indicators must not only reflect the conditions of the coastal environment but also facilitate quantitative integration within the MCDM framework. This section outlines the process for deriving, transforming, and compiling indicators across the three key aspects of vulnerability—physical, socioeconomic, and ecological—as presented in Tables 8.1, 8.2, and 8.3. Emphasis is placed on the use of geographical datasets, satellite-derived products, and statistical or expert-driven methodologies for processing and normalizing the indicators to include them in the MCDM work-flow. In Fig. 8.6, a simple flow chart for the processing of data inputs, indicator preparation and CVI mapping is given.

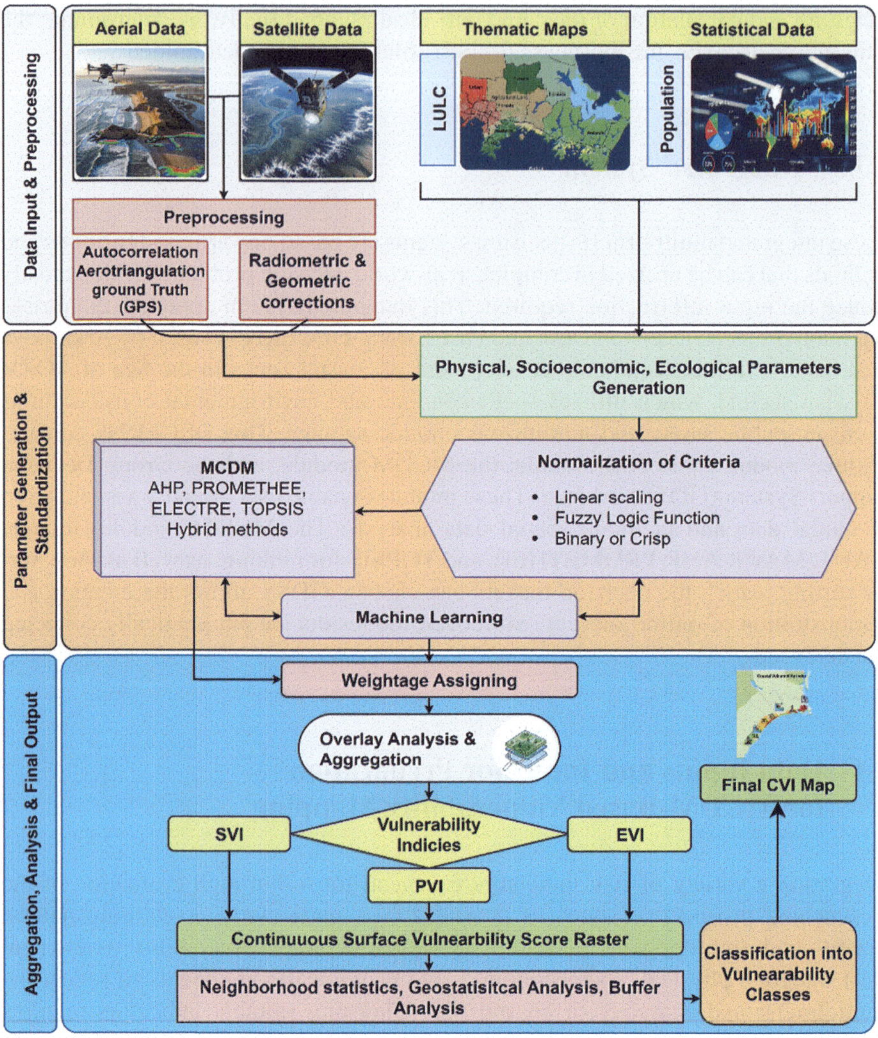

Fig. 8.6 Methodology for CVI mapping using MCDM

8.4.1 Derivation of Spatial Data from Indicators

Indicators were selected based on their relevance to the coastal vulnerability of the study regions and the identified problem. These indicators were classified into three categories: physical, socioeconomic, and ecological parameters. The best indicators were selected based on the literature study by the decision maker for the pertinent problem. In MCDM using geospatial analysis, the conversion of criteria (indicators) into spatial layers is followed by standardization and weighting. Finally, these layers were integrated through overlay analysis to derive the final spatially explicit decision or suitability map as an output product to support evidence-based coastal planning and decision-making by decision-makers.

Spatial data, such as climate data, topographic data, and LULC data, are collected. These data are in raster (grid) or vector (point, line, and polygon) formats, such as satellite data, DEM, soil maps, land cover data, distance to road maps, oceanic data (tidal range, wave height, SLR), and socioeconomic data (population density, infrastructure locations). Then, preprocessing is performed to standardize the data, achieve the exact resolution, and convert vectors to raster or vice versa, as well as to project datasets. Climate data can be utilized in MCDM to enhance decision-making. Metrological data, such as surface temperature, precipitation, sea level pressure, and dew point, can be downloaded from NOAA. Wind speed and wind direction can be obtained from NASA. Topographic structures, such as slope and coastal regional elevation, can be derived from a DEM. The geology of the region can be derived from ancillary data, which can be procured from the respective government agencies. LULC maps, derived from algorithms or obtained from government agencies, can be used in conjunction with satellite data to identify erosion and accretion. These data as input in model-ling software like TerrSet liberaGIS used to derive RUSLE soil erosion model, prediction of land change for forecasting of CVA area using land change modeler and integrating with habitat and biodiversity modeler to predict the vulnerability of the coastal regions to natural hazards like erosion and inundation and SAGA (System for Automated Geoscientific Analyses) GIS in build functions like terrain analysis, grid calculus, shoreline analysis used to de-rive complex indices, different coastal factors, shoreline change from historical data can be generated. Special tools like Digital Shoreline Analysis System (DSAS) which USGS developed were extensively used to analyze net shoreline movement, shoreline change envelope, end point rate, linear regression rate, weighted linear regression and also derives supplementary statistics like uncertainty of end point rate, standard error of the estimate, standard error of the rate with confidence intervals, R^2 statistics and give an final output file with summary report.

8.4.2 Standardization and Reclassification

These derived coastal criteria will be expressed in different units and scales, which will be standardized and normalized to a standard scale, such as 0 to 1 or 0 to 100, for comparison and reclassification of the raster layers based on this contribution to vulnerability. For example, in a study by Bera and Maiti (2021), five physical and one social criterion were selected to study coastal vulnerability using geospatial techniques in the Sundarbans region of India. The Begukhali and Sibpur regions are found to be highly vulnerable (Fig. 8.7), which accounts for high population densities, moderate elevation, moderate erosion, and moderate tidal conditions, as well as a reclaimed coastline, all of which were assigned a higher vulnerability score. Table 8.4 shows the ranking used by Bera and Maiti (2021).

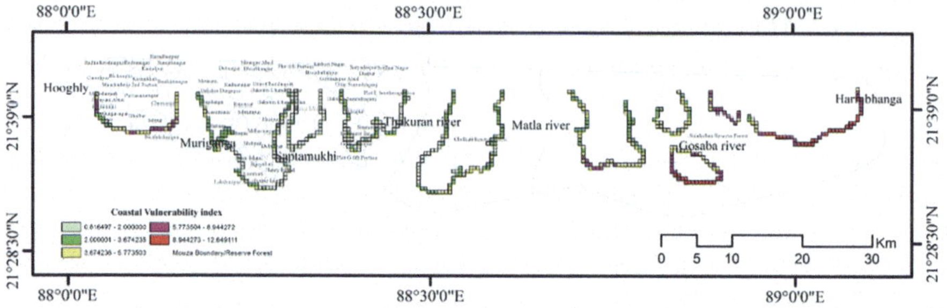

Fig. 8.7 Coastal vulnerability of Sundarbans regions showing highly vulnerable areas in Begukhali and Sibpur (2021)

8.4.3 Weighting the Criteria using MCDM

Weightage was assigned to each criterion based on MCDM techniques, including AHP, ANP, MAUT, PROMETHEE, ELECTRE, TOPSIS, and hybrid methods that combine multiple MCDM methods. Also, nowadays, these MCDM methods are integrated with machine learning and deep learning algorithms to generate robust and data-driven weightage decisions among the selected criteria for vulnerability assessment (Al-Wardy et al. 2025; Barzehkar et al. 2025; Fannassi et al. 2023). Using machine learning technologies, we can uncover the complex patterns of natural coastal regions, and feature engineering is made possible for each spatial unit, such as 30 × 30 m grids or 1 km shoreline segmentation. Using labelled datasets which are showing know vulnerable regions in the coastal area were used by machine learning algorithms like random forest (RF), support vector machine (SVM), ANN for train them and used for classification and regression analysis like classification of coastal regions into segments of "very high", "high", "moderate", "low", "very low" vulnerable regions and prediction of continuous CVI score respectively. An integrated approach with ML algorithms has advantages over MCDM, such as automatic feature importance rather than subjective-based decision-making. They capture complex, nonlinear relationships and interactions between vulnerability indicators. Additionally, they are used for the validation of MCDM results by creating a comparison of an expert-weighted index against ML model output.

8.5 Weighted Overlay Analysis and Aggregation

The two data sets, the standardized criteria raster and criteria weights raster, obtained through MCDM will undergo pixel-by-pixel calculation using standard methods, such as weighted linear combinations (WLC). Weighted overlay analysis can be executed using GIS software such as ESRI's ArcGIS and QGIS, which have built-in functions or tools for weighted overlay analysis and raster calculator, respectively.

Table 8.4 Variable ranking (Bera and Maiti 2021)

Variables	Sub-classes	Rank	Length of the coastline (km)	Length of the coastline (%)
Absolute sea-level rise (mm/y)	>4.79	5	37.50	13.94
	4.29–4.79	4	69.00	25.65
	3.91–4.29	3	39.00	14.49
	3.56–3.91	4	37.00	13.75
	<3.91	1	86.50	32.15
Elevation (m)	<1.509	5	10.50	3.90
	1.509–2.163	4	100.50	37.36
	2.163–2.942	3	111.50	41.44
	2.942–3.646	2	17.50	6.50
	>3.646	1	29.00	10.78
Historical erosion-accretion (m/y)	<−40.565	5	10.50	3.90
	−40.565 to −20.266	4	17.50	6.50
	−20.266 to −6.535	3	58.50	21.74
	−6.535 to 4.808	2	139.50	51.85
	>4.808	1	43.00	15.98
Geomorphology	Tidal flat (Inter tidal)	5	25.50	9.47
	Intertidal Mangrove (Inter tidal)	4	13.00	4.83
	Reclaimed delta plain, Dense Mangrove Swamp (Supra tidal)	3	109.50	40.70
	Estuary, Creek, Aquaculture, Wetland (Subtidal)	2	120.50	44.79
	Inland Vegetation	1	0.50	0.18
Population density (per km^2)	>1350	5	3.00	1.11
	803–1350	4	5.50	2.04
	185–803	3	42.00	15.61
	<185	2	17.50	6.50
	Uninhabited	1	201.00	74.72
Tidal range (m)	<0.927	5	7.50	2.78
	0.927–1.984	4	5.50	2.04
	1.984–3.079	3	49.00	18.21
	3.079–4.135	2	12.50	4.64
	>4.135	1	194.50	72.30

Table 8.5 Criteria standardization and their weights

Criterion	Standardized value (1–5 scale) for this specific pixel	Weight (from AHP)
Elevation	5 (This pixel is very low-lying)	0.50 (50% importance)
slope	4 (This pixel has a gentle slope, making it prone to inundation)	0.30 (30% importance)
Land use	5 (This pixel is an urban area)	0.20 (20% important)

Grid calculus in SAGA GIS was also used for this analysis. Using a cloud-based platform, GEE, we can achieve this for larger study regions, such as states or countries. In this method, the calculation or logic is applied to every single pixel of the raster in the selected coastal study area. The general formula follows:

$$CVI_{pixel} = \left(w_2 \times c_{2,pixel}\right) + \left(w_2 \times c_{2,pixel}\right) + \left(w_3 \times c_{3,pixel}\right) + \cdots + \left(w_n \times c_{n,pixel}\right)$$

where CVI_{pixel} is the vulnerability score for that pixel

w_1, w_2, …, w_n are weights assigned to criteria.

$c_{1,\,pixel}$, $c_{2,\,pixel}$, …,$c_{n,\,pixel}$ are standardized values from each criterion raster.

Example Consider Table 8.5, which shows the standardized values and weight (AHP) for each criterion in that pixel. Here, different criteria, such as elevation, slope, and land use, were assigned their standardization values and assigned weights of 0.20 to 0.50, respectively. For a single pixel, the overlay analysis is performed by.

$$CVI_{pixel} = (0.50 \times 5) + (0.30 \times 4) + (0.20 \times 5)$$

$$CVI_{pixel} = 2.5 + 1.2 + 1.0$$

$$CVI_{pixel} = 4.7$$

This calculation was performed for every single pixel and generates a new raster layer that represents the calculated CVI score. Some advanced analyses may include spatial relationships between pixels by implementing neighborhood analysis, geographically weighted over-lay analysis, and other methods.

8.5.1 Resulting Spatial Data (Decision Maps)

The obtained output raster data were not readily interpretable since they would be projected as a grayscale map and continuous layer. To generate final CVI maps, we classify the generated CVI raster file. For classification, standard methods include natural breaks (Jenks), which identify the natural grouping in the data used for the classification of boundaries. For example, 1.01–2.50: Low vulnerability, 2.51–3.50:

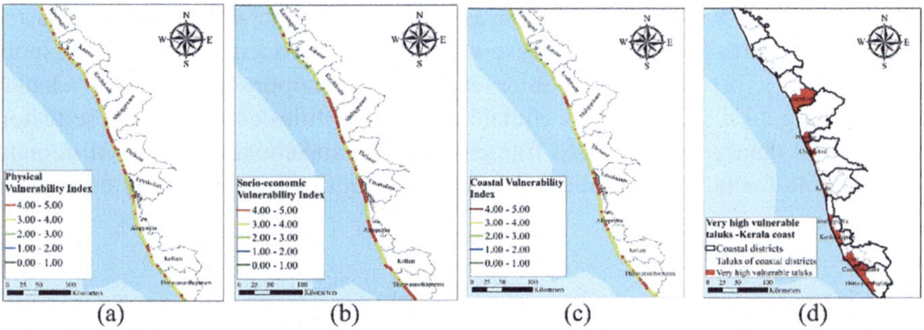

Fig. 8.8 Coastal spatial variation of **a** PVI, **b** SVI, **c** CVI, and **d** highly vulnerable risk regions of Keara taluks (KV 2024)

Moderate vulnerability, 3.51–4.25: High vulnerability, 4.26–5.00: Very high vulnerability. Then, finally, symbology was applied to each class. Typically, inbuilt colour ramps are used, ranging from green (low-vulnerable regions) through yellow and orange to red (very high-vulnerable areas). Figure 8.8 shows PVI, SVI, CVI, and a highly vulnerable taluk on the Kerala coast.

8.6 Conclusion

Coastal regions, as dynamic interfaces between land and sea, face a broad array of environmental and socioeconomic challenges exacerbated by climate change, rapid urbanization, and human activities. Rising sea levels, erosion, storm surges, flooding, and salinization pose significant threats to both natural ecosystems and densely populated coastal communities. As these hazards intensify, a precise understanding and proactive management of coastal vulnerability become paramount for effective coastal zone management and resilience building. Integrating geospatial technologies like GIS and remote sensing with MCDM frameworks such as the AHP and PROMETHEE enables the identification, mapping, and quantification of coastal vulnerability by combining indicators from three main domains: physical (elevation, geomorphology, erosion rates, tidal range, proximity to coast), socioeconomic (population density, infrastructure, income, health, and education), and ecological (mangrove cover, wet-lands, biodiversity, habitat fragmentation). This process can translate complex real-world phenomena into structured indicators, systematically standardize and weigh them according to relative importance, and synthesize a composite CVI. Advanced MCDM tools, such as AHP, rigorously incorporate expert judgment and subjectivity through pairwise rankings, whereas alternatives like PROMETHEE introduce preference-based outranking methods. MCDM-GIS workflow, allowing for adaptation to diverse coastal settings and the inclusion of both qualitative and quantitative data. Spatial vulnerability mapping is deeply rooted in the interplay between natural systems and human society, highlighting the significance of spatially explicit CVI maps, which serve as actionable data for policymakers, disaster managers, and coastal planners. These maps enable the identification of hotspots of vulnerability, prioritization of interventions, and support for informed adaptation and mitigation strategies. As climate change amplifies coastal risks, the fusion of geospatial data, robust MCDM frameworks, and stakeholder inputs will remain indispensable for building resilient coastal communities and protecting ecosystem services.

8.7 Competing Interests

The authors declare that they have no competing interests or conflicts of interest related to this work.

8.8 Ethics Approval

This chapter does not involve any studies with human participants or animals performed by any of the authors. Hence, ethics approval was not required.

Acknowledgements The authors gratefully acknowledge the editors of this volume and the anonymous reviewers for their valuable suggestions and constructive feedback that helped improve the quality of this chapter. We also express our sincere gratitude to Springer Nature for the opportunity to contribute to this book. The collaborative efforts and contributions of all co-authors are deeply appreciated.

References

Ali SA, Khatun R, Ahmad A, Ahmad SN (2019) Application of GIS-based analytic hierarchy process and frequency ratio model to flood vulnerable mapping and risk area estimation at Sundarban region India. Model Earth Syst Environ 5(3):1083–1102

Al-Wardy M, Zarei E, Nikoo MR (2025) Improving index-based coastal vulnerability assessment using machine learning in Oman. Sci Total Environ 976:179311

Arkhipova OE, Chernogubova EA, Tarasova TT, Arkhipova KE (2021) Assessment of sustainable development of the coastal zone of the Sea of Azov (Russian sector) based on an integral approach. GI Support Sustain Dev Territ 100

Attri SD, Singh S, Dhar A, Powar S (2022) Multi-attribute sustainability assessment of wastewater treatment technologies using combined fuzzy multi-criteria decision-making techniques. J Clean Prod 357:131849

Bagheri M, Ibrahim ZZ, Akhir MF, Oryani B, Rezania S, Wolf ID, Pour AB, Talaat WIAW (2021) Impacts of future sea-level rise under global warming assessed from tide gauge records: a case study of the east coast economic region of Peninsular Malaysia. Land 10(12):1382

Balica SF, Wright NG, Van der Meulen F (2012) A flood vulnerability index for coastal cities and its use in assessing climate change impacts. Nat Hazards 64(1):73–105

Barocchi MA, Black S, Rappuoli R (2016) Multicriteria decision analysis and core values for enhancing vaccine-related decision-making. Sci Transl Med 8(345):345ps314–345ps314

Barzehkar M, Parnell K, Soomere T (2025) Incorporating a machine learning approach into an established decision support system for coastal vulnerability in the eastern baltic sea. J Coast Res 113(SI):58–62

Bera R, Maiti R (2021) An assessment of coastal vulnerability using geospatial techniques. Environ Earth Sci 80(8):306

Brans J-P, Vincke P, Mareschal B (1986) How to select and how to rank projects: the PROMETHEE method. Eur J Oper Res 24(2):228–238

Brans JP, Mareschal B (1992) PROMETHEE V: MCDM problems with segmentation constraints. INFOR: Inf Syst Oper Res 30(2):85–96

Brans J-P (1982) L'ingénierie de la décision: l'élaboration d'instruments d'aide a la décision. Université Laval, Faculté des sciences de l'administration

Brooks N (2003) Vulnerability, risk and adaptation: a conceptual framework. Tyndall Cent Clim Chang Res Work Pap 38(38):1–16

Ceballos G, Ehrlich PR, Barnosky AD, García A, Pringle RM, Palmer TM (2015) Accelerated modern human–induced species losses: entering the sixth mass extinction. Sci Adv 1(5):e1400253

Chen C-L, Bau Y-P (2016) Establishing a multi-criteria evaluation structure for tourist beaches in Taiwan: a foundation for sustainable beach tourism. Ocean Coast Manag 121:88–96

Cloern JE, Abreu PC, Carstensen J, Chauvaud L, Elmgren R, Grall J, Greening H, Johansson JOR, Kahru M, Sherwood ET (2016) Human activities and climate variability drive fast-paced change across the world's estuarine–coastal ecosystems. Glob Change Biol 22(2):513–529

Dalby S (2020) Anthropocene geopolitics: Globalization, security, sustainability. University of Ottawa Press

Van Dua T (2023) Combination of design of experiments and simple additive weighting methods: a new method for rapid multi-criteria decision making. EUREKA: Phys Eng (1):120–133

Falconer L, Cutajar K, Krupandan A, Capuzzo E, Corner RA, Ellis T, Jeffery K, Mikkelsen E, Moore H, O'beirn FX (2023) Planning and licensing for marine aquaculture. Rev Aquac 15(4):1374–1404

Fannassi Y, Ennouali Z, Hakkou M, Benmohammadi A, Al-Mutiry M, Elbisy MS, Masria A (2023) Prediction of coastal vulnerability with machine learning techniques, Mediterranean coast of Tangier-Tetouan, Morocco. Estuar Coast Shelf Sci 291:108422

Fetanat A, Khorasaninejad E (2015) A novel hybrid MCDM approach for offshore wind farm site selection: a case study of Iran. Ocean Coast Manag 109:17–28

Figueira JJ, De Smet Y, Brans JP (2005) MCDA methods for sorting and clustering problems: Prometh TRI Prometh CLUSTER

Ghosh S, Mistri B (2021) Assessing coastal vulnerability to environmental hazards of Indian Sundarban delta using multi-criteria decision-making approaches. Ocean Coast Manag 209:105641

Herath G, Prato T (2006) Using multi-criteria decision analysis in natural resource management. Ashgate Publishing, Ltd

Himmelstoss EA, Henderson RE, Kratzmann MG, Farris AS (2021) Digital shoreline analysis system (DSAS) version 5.1 user guide (2331–1258)

Hwang C-L, Lin M-J (2012) Group decision making under multiple criteria: methods and applications (Vol 281). Springer Science & Business Media

Ishizaka A, Nemery P (2013) Multi-criteria decision analysis: methods and software. John Wiley & Sons

Jana A, Sheena S, Biswas A (2012) Morphological change study of Ghoramara Island, Eastern India using multi-temporal satellite data. Res J Recent Sci 1(10):72–81

Jana A, Biswas A, Maiti S, Bhattacharya AK (2014) Shoreline changes in response to sea level rise along Digha coast, eastern India: an analytical approach of remote sensing, GIS and statistical techniques. J Coast Conserv 18(3):145–155

Jana A, Maiti S, Biswas A (2016a) Analysis of short-term shoreline oscillations along Midnapur-Balasore coast, bay of Bengal, India-a study based on geospatial technology. Model Earth Syst Environ 2(2):64. https://doi.org/10.1007/s40808-016-0117-7

Jana A, Maiti S, Biswas A (2016b) Seasonal change monitoring and mapping of coastal vegetation types along Midnapur-Balasore coast, bay of Bengal using multi-temporal Landsat data. Model Earth Syst Environ 2(1):7

Jana A, Maiti S, Biswas A (2017) Appraisal of long-term shoreline oscillations from a part of coastal zones of Sundarban delta, eastern India—a study based on geospatial technology. Spat Inf Res 25:713–723. https://doi.org/10.1007/s41324-017-0139-x

Jones MC, Dye SR, Fernandes JA, Frölicher TL, Pinnegar JK, Warren R, Cheung WW (2013) Predicting the impact of climate change on threatened species in UK waters. PLoS ONE 8(1):e54216

Kumar P, Biswas A, Banerjee S, Rathore S, Rana V, Ram K, Acharya T (2022) Integrating magnetic susceptibility, hydrogeochemical, and isotopic data to assess the seawater invasion in coastal aquifers of Digha, West Bengal India. Environ Sci Pollut Res 29(16):23474–23503

Kumar P, Biswas A, Banerjee S (2023) Spatio-temporal variability of seawater mixing in the coastal aquifers based on hydrogeochemical fingerprinting and statistical modeling. J Environ Manage 346:118937

Kumar P, Tiwari P, Biswas A, Srivastava PK (2024) Spatio-temporal assessment of soil salinization utilizing remote sensing derivatives, and prediction modeling: Implications for sustainable development. Geosci Front 15(6):101881

KV A (2024) Coastal vulnerability assessment along the coast of Kerala, India, based on physical, geological, and socio-economic parameters. Mar Geod 47(2)

Macharis C, Brans J-P, Mareschal B (1998) The GDSS promethee procedure. J Decis Syst 7(4):283–307

Malara G, Zema DA, Arena F, Bombino G, Zimbone SM (2020) Coupling watershed-coast systems to study evolutionary trends: a review. Earth Sci Rev 201:103040

Malczewski J (1999) GIS and multicriteria decision analysis. John wiley & sons

Malekmohammadi B, Blouchi LR (2014) Ecological risk assessment of wetland ecosystems using multi criteria decision making and geographic information system. Ecol Ind 41:133–144

Niggli L, Allen S, Frey H, Huggel C, Petrakov D, Raimbekova Z, Reynolds J, Wang W (2024) GLOF Risk management experiences and options: a global overview. Oxf Res Encycl Nat Hazard Sci

Parthasarathy K, Deka PC (2021) Remote sensing and GIS application in assessment of coastal vulnerability and shoreline changes: a review. ISH J Hydraul Eng 27(sup1):588–600

Pournamdarian A (2010) Multi-criteria decision making by using inner product of vectors. Knol

Prof A, Nayak S, Tamilarasan V, Chauhan H, Bahuguna A, Gupta M, Rajawat A, Chaudhury N, Kumar T, Rao R (2012) Coastal zones of India. In: Report by ISRO Ahmedabad, Ministry of Environment and Forest, Government of India.

Rosenstock TS, Joshi N, Segnon AC, Cramer L, Mwongera C, Nowak AC, Njuguna L, Dossou-Yovo ER, Steward P, Ramirez-Villegas J (2024) Decision support tools for agricultural adaptation in Africa. Nature Food 5(3):186–188

Roy B, Bertier P (1968) La méthode ELECTRE. Rev D 'Informatique Et Rech Oper (RIRO) 8:57–75

Saaty TL (1977) A scaling method for priorities in hierarchical structures. J Math Psychol 15(3):234–281

Saravanan S, Parthasarathy K, Vishnuprasath S (2019) Monitoring spatial and temporal scales of shoreline changes in the Cuddalore Region, India. In: Coastal zone management. Elsevier, pp 99–112

Schmitt A, Chaumillon E (2023) Understanding morphological evolution and sediment dynamics at multi-time scales helps balance human activities and protect coastal ecosystems: an example with the Gironde and Pertuis Marine Park. Sci Total Environ 887:163819

Sekovski I, Del Río L, Armaroli C (2020) Development of a coastal vulnerability index using analytical hierarchy process and application to Ravenna province (Italy). Ocean Coast Manag 183:104982

Selvan SC, Kankara R, Prabhu K, Rajan B (2020) Shoreline change along Kerala, south-west coast of India, using geo-spatial techniques and field measurement. Nat Hazards 100(1):17–38

Sriram S, Ramachandran M, Chinnasamy S, Mathivanan G (2022) A review on multi-criteria decision-making and its application. REST J Emerg Trends Model Manuf 7(4):101–107

Su Q, Kambale RD, Tzeng J-H, Amy GL, Ladner DA, Karthikeyan R (2025) The growing trend of saltwater intrusion and its impact on coastal agriculture: challenges and opportunities. Sci Total Environ 966:178701

Van Huong N, Nguyet BTM, Van Hung H, Duc HM, Van Chuong N, Tri DM, Van Hien P (2022) Economic impact of climate change on agriculture: a case of Vietnam. AgBioforum 24(1):1–12

Xianwu S, Jufei Q, Bingrui C, Xiaojie Z, Haoshuang G, Jun W, Zhuyuan B (2020) Storm surge risk assessment method for a coastal county in China: case study of Jinshan District, Shanghai. Stoch Env Res Risk Assess 34(5):627–640

Yuan Z, Wen B, He C, Zhou J, Zhou Z, Xu F (2022) Application of multi-criteria decision-making analysis to rural spatial sustainability evaluation: a systematic review. Int J Environ Res Public Health 19(11):6572

Zhang Y, Zhang T (2024) A comprehensive review of assessing storm surge disasters: From traditional statistical methods to artificial intelligence-based techniques. Atmosphere 15(3):359

Chapter 9
Climate-Related Crisis: Eco-Social and Health Impacts of Flood on Women Living in Coastal Communities in Anambra State, Nigeria

Angela Oyilieze Akanwa, K. A. Omenugha, and Ndidi Iloduba

9.1 Introduction

Greenhouse gas emissions, including methane, carbon dioxide, and nitrous oxide, are rising significantly, with climate change deemed a major global threat (IPCC 2022). This rise leads to severe effects, such as increased natural disasters and socio-ecological impacts (Akanwa et al. 2022). Floods, in particular, have surged in frequency, correlating with global warming (World Meteorological Organisation 2020). Despite the 2015 Paris Agreement, which aimed to limit global warming to below 2 °C, greenhouse gas emissions have risen due to fossil fuel consumption (IPCC 2018). A Lancet report (2024) highlights that 2023 saw the highest energy-related carbon dioxide emissions, surpassing the previous year by 1.1%. These emissions threaten global health and well-being.

Although climate change is a global crisis, its impacts vary by location, resulting in specific local challenges (Naseem 2024). Since 1980, the global South has faced

A. O. Akanwa (✉)
Faculty of Environmental Management, Agriculture and Renewable Energy (FEMARE), University of Technology and Arts (UTAB), Byumba, Rwanda
e-mail: akanwa@utab.ac.rw; aa.akanwa@coou.edu.ng

Department of Environmental Management, Chukukwuemeka Odumegwu Ojukwu University, Anambra State, Nigeria

K. A. Omenugha
Office of the Vice Chancellor, Chukwuemeka Odumegwu Ojukwu University, Igabriam, Nigeria
e-mail: vc@coou.edu.ng

N. Iloduba
Department of Pediatrics (Child Health), Federal Medical Centre, Federal Capital Territory Abuja (FCT), Abuja, Nigeria
e-mail: trendyendee@gmail.com

A. Nath et al. (eds.), *Geospatial Application of Coastal Hazards*, Springer Water,
https://doi.org/10.1007/978-3-032-20076-1_9

recurrent flooding, with statistics indicating 4,588 flood disasters resulting in around 250,000 deaths and $1 trillion in damages, of which only 12% was insured (Mclennan 2021). Between 2011 and 2018, several Southeast Asian countries, including China, India, Myanmar, Thailand, and the Philippines, experienced severe flooding, leading to total damages estimated at around 95.372 billion US dollars (Murphy 2020). Thailand incurred the largest portion of these losses, with an estimated of 47.9%, while in India, roughly 1,922 lives were lost due to flooding in 2020, resulting in economic damages close to $21.8 billion, making it the costliest flood event globally (Mclennan 2021).

In Nigeria, the impact of climate-related flooding has been markedly more severe due to existing vulnerabilities such as insufficient urban planning, inadequate infrastructure, rapid urbanization, and a lack of effective flood prevention measures, among other factors (Otty et al. 2024), which further heighten the vulnerability of numerous communities.

The study area, Anambra State, experiences several conditions that intensify the ongoing flooding issues; primarily, it is a coastal region where many communities are situated close to the banks of rivers and streams. It has a shallow depth, is characterized by a low profile with a flat landscape, and experiences consistently high annual humidity levels ranging from 80 to 90%, alongside average annual rainfall between 1,520 and 2,020 mm. All these factors contribute to the frequency of recurrent flooding (Statista 2024), posing significant threats, particularly to marginalized populations, especially girls and women. The challenges faced by women and children are intricate and multifaceted, affecting health, economic stability, and social well-being (Akanwa et al. 2022). Importantly, women and girls in rural areas suffer disproportionately despite their minimal contributions to climate change (Roos et al. 2021). This study highlights the principle of participatory gender justice, which involves the representation and active engagement of women and girls in fully realizing their economic, social, cultural, civil, and political rights and freedoms (Global Fund for Women 2024). As a result, it also tackles the various facets of the fight for women's equality and empowerment. Recognizing equal rights and access to justice for women and girls, along with redistributing power to guarantee fair access to and control over resources, services, and opportunities for women, is of utmost importance (Danish Institute 2024).

It is crucial to acknowledge that even though women are vulnerable in the face of climate-related floods, they are essential to climate change mitigation due to their resilience, indigenous knowledge, and significant contributions to sustainable development. This research is timely because the dependency on fossil fuels intensifies the climate change crisis. In 2023, CO_2 emissions linked to energy reached unprecedented heights (Lancet 2024), while oil and gas companies are increasing fossil fuel extraction, disregarding the goals outlined by the Paris Agreement in 2015, aimed at creating a global framework for tackling climate change. At the same time, the future of climate change mitigation initiatives in Nigeria remains unclear. Therefore, this study intends to explore the economic, social, and health-related issues while advocating for justice for women in the community, highlighting that their roles may be vital in promoting a more equitable future in Anambra State.

9.2 Justifying the Concept of Participatory Gender Justice

Participatory gender justice represents a transformative approach that combines gender equity with inclusive involvement in environmental decision-making as well as socio-political empowerment. It underscores the necessity of identifying and addressing systemic gendered inequalities by ensuring that women, particularly those in at-risk environments, are given a voice and agency in shaping policies and responses to ecological and health issues (Cornwall and Rivas 2015; Arora-Jonsson 2011).

In flood-prone coastal areas of Anambra State, women are disproportionately affected by the ecological, social, and health repercussions of recurring flooding. These repercussions consist of the devastation of farmland and sources of livelihood, an increased likelihood of waterborne illnesses, and physical displacement, all of which intersect with pre-existing gender disparities (Ayoade et al. 2020). As primary caregivers and food providers, women are significantly impacted by flood-related disruptions that affect household food security, maternal health, and the care economy (UN Women 2021).

Utilizing a participatory gender justice framework involves shifting away from viewing women merely as passive victims of climate crises. It encourages community-focused strategies where women take an active role in flood risk assessment, early warning systems, and the collaborative design of local adaptation initiatives. This participatory methodology facilitates a more equitable distribution of resources, enhances women's resilience abilities, and improves community preparedness overall.

Furthermore, gender-just participation emphasizes the significance of intersectionality, recognizing that factors such as age, class, marital status, and disability influence women's experiences of flooding (Crenshaw 1991; Alston 2013). In the coastal communities of Anambra, widows and women who have been displaced encounter even greater obstacles to accessing health services, land, and disaster relief, often stemming from patriarchal land ownership structures and social exclusion (Onyenechere and Okoroafor 2019). Consequently, participatory gender justice functions as both a diagnostic framework to comprehend the ecological, social, and health vulnerabilities faced by women and as a corrective method to enhance inclusivity in climate governance, foster equitable resilience building, and ensure justice in the planning and execution of flood management strategies in Nigeria.

9.3 Materials and Methods

Our study employed a qualitative methodology within a case study framework to ensure that the outcomes were firmly rooted in the participants' lived experiences and to facilitate examining eco-social and health outcomes pertinent to the study's context (Bachiochi and Weiner 2004). The primary data sources consisted of meticulously recorded key informant interviews, observational data, semi-structured interviews, photographic documentation, focus group discussions, and informal interactive dialogues. We also incorporated secondary information from online resources, climate-related reports, and newsletters.

The interview process is recognized as one of the most efficacious techniques for data collection to gather insights regarding individuals' experiences, attitudes, opinions, grievances, emotional states, and belief systems (Coyne 1997). Furthermore, we employed purposive sampling techniques, specifically convenience and snowball sampling methods, to select participants, ensuring that the fundamental perspectives and voices of both women and participating male informants regarding the issues under investigation were adequately represented (Palinkas et al. 2015).

The investigation was executed within the Ogbaru Local Government Area (LGA), concentrating on three distinct communities, each meticulously chosen from the Northern, Southern, and Central sectors of the LGA, as depicted in Fig. 9.1. The justification for selecting participants from these three heterogeneous communities was to ensure equitable representation of both genders within the Umunankwo, Ogbakuba, and Ossamala Communities. This methodology facilitates an exhaustive assessment of the ramifications, considering that the diverse eco-social and health implications, along with their socio-cultural attributes, will be pivotal to the inquiry.

A comprehensive sample comprising 60 community participants was randomly surveyed employing unstructured inquiries. Thirty (30) participants were selected to partake in three focus group discussions (FGDs), delineated into groups based on marital status and age. They were categorized by gender, with ten participants allocated to each group (Saunders 2012).

Subsequently, twenty (20) individuals were randomly interviewed (IDIs) through semi-structured interviews, while ten (10) community stakeholders (7 women and 3 men) were enlisted to partake in our key in-depth interviews (KII) connected to the chosen communities. They explored more profound issues identified during the FGDs, thereby enhancing the findings derived from the discussions. The FGDs and interviews conducted were deemed adequate as they reached a saturation threshold in the responses. Data gathering was executed by three proficient qualitative facilitators. Interview and FGD guides were provided in both Igbo (the indigenous language) and English, and the discussions were recorded using voice recorders, with effective translation back into English being accomplished.

We integrated various demographic variables about educational attainment, marital status, age, occupation, and socioeconomic data concerning the perceptions of the interviewees regarding the effects of flooding. The duration of the interviews varied from 30 min to 1 h. All interviews and FGD guides were focused on the perceived eco-social and health consequences of flooding on women.

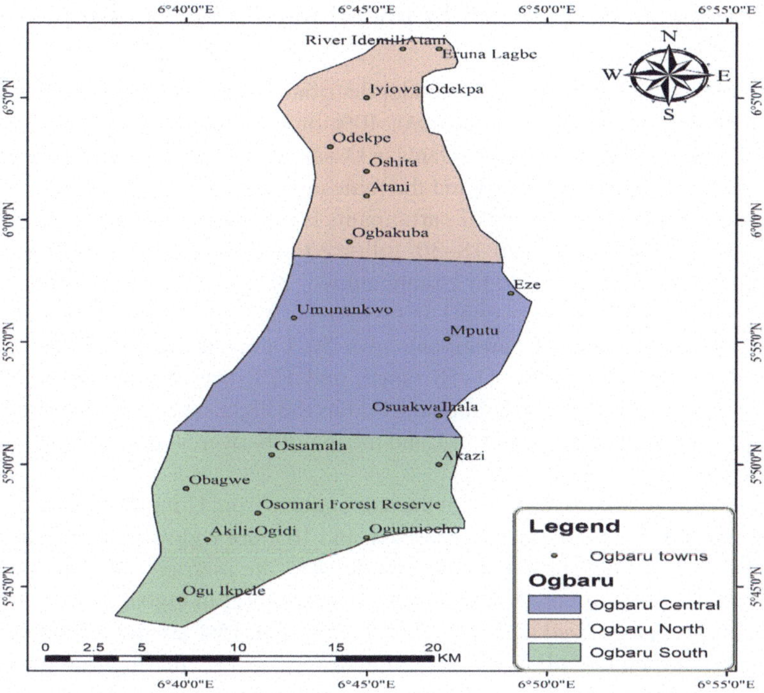

Fig. 9.1 Map of Ogbaru showing the North, South, and Central Ogbaru Areas of the studied communities. *Source* Compiled by authors, 2024

9.4 Data Analysis

The audio-recorded data from FGDs, IDIs, and KIIs were transcribed, translated from the local dialect to English, and subsequently verified for precision by the researchers. Throughout multiple assessments, analytical notes and observations were documented for reference during the analytical phase. Following this, all finalized English transcripts underwent deductive coding utilizing NVivo software (V12, QSR International). Themes were identified, rigorously reviewed, and subjected to further scrutiny. The findings were corroborated with contextual quotations and representative of principal discussion points.

9.5 Eco-Social and Health Implications of Our Findings

Table 9.1 presents a summary of the gender distribution among the respondents from the three communities; in Umunnankwo, 40% of the respondents are male, while 60% are female. 100% of the respondents in Ossamala are all females. In Ogbakuba, the female respondents are (70%) and the male are respondents (30%).

The demographic distribution of participants is summarized in Table 9.2. Half of the respondents (50%) are aged 18–30, followed by 35% aged 31–40, 30% in the 41–50 range, and 15% over 50. In Umunnankwo, 25% are 18–30, 35% are 31–40, 25% are 41–50, and 15% are over 50. In Ossamala, 50% are 18–30, 25% are 31–40, and 25% are 41–50, with no participants over 50. Ogbakuba has 30% aged 18–30, 25% aged 31–40, 30% in the 41–50 range, and 15% over 50. Ossamala has the most respondents aged 18–30, Umunnankwo has the highest proportion aged 31–40, Ogbakuba leads for those aged 41–50, and the majority over 50 are in Ogbakuba and Umunnankwo.

Table 9.3 demonstrates that 25% of the respondents from Umunnankwo are single, 60% are married, 5% are divorced/separated, and 10% are widowed. In the Ossamala community, 30% of the participants are single, 40% are married, 10% are divorced/separated, and 20% are noted as widowed. Furthermore, in Ogbakuba, 50% of the respondents are categorized as single, 30% are married, 15% are divorced/separated, and only 5% are classified as widowed. Ogbakuba reports the highest percentage of single respondents (50%), while Umunnankwo and Ossamala exhibit the highest percentages of married respondents (60%) and (40%), respectively.

Table 9.1 Gender

Gender	Umunnakwo		Ossamala		Ogbakuba	
	F	%	F	%	F	%
Male	08	40	00	00	06	30
Female	12	60	20	100	14	70
Total	20	100	20	100	20	100

Source Compiled by authors, 2024

Table 9.2 Age distribution

Age of the respondent	Umunnakwo		Ossamala		Ogbakuba	
	F	%	F	%	F	%
18–30	05	25	10	50	06	30
31–40	07	35	05	25	05	25
41–50	05	25	05	25	06	30
> 50	03	15	00	00	03	15
Total	20	100	20	100	20	100

Source Compiled by authors, 2024

Table 9.3 Marital status

Marital status of the respondents	Umunnakwo		Ossamala		Ogbakuba	
	F	%	F	%	F	%
Single	05	25	06	30	10	50
Married	12	60	08	40	06	30
Divorced/separated	01	05	02	10	03	15
Widowed	02	10	04	20	01	05
Total	20	100	20	100	20	100

Source Compiled by authors, 2024

Table 9.4 shows that 30% of respondents lack formal education, 80% have a first school leaving certificate, 90% possess SSCE/WAEC qualifications, 45% hold NCE/OND credentials, 40% have BSc/HND degrees, and 15% attained postgraduate qualifications. In Umunnankwo, 15% have no formal education, 35% possess a first school leaving certificate, 30% have WAEC certificates, 5% hold NCE/OND, 10% have BSc degrees, and 5% completed postgraduate studies. In Ossamala, 10% lack formal education, 30% have a first school leaving certificate, 25% earned SSCE/WAEC, 15% hold BSc/HND degrees, and 20% attained postgraduate qualifications. Ogbakuba respondents show that 35% have SSCE/WAEC qualifications, 25% hold NCE/OND, 5% have a first school leaving certificate, 10% possess BSc/HND degrees, 10% have postgraduate degrees, and 5% lack formal education.

Table 9.5 illustrates the distribution of household sizes, revealing that 60% of the respondents reside in households containing between 1 and 3 individuals; 15% are situated within households comprising 4–6 individuals; whereas 25% of the respondents indicated that their households accommodate more than 7 individuals. Specifically, in Umunnakwo, 20% of households are constituted of 1–3 persons, 30% comprise 4–6 individuals, while 50% of the respondents reported household sizes exceeding 7 individuals. In Ossamala, 25% of the respondents belong to households

Table 9.4 Educational qualification

Educational qualification	Umunnakwo		Ossamala		Ogbakuba	
	F	%	F	%	F	%
No formal education	03	15	02	10	01	05
FSLC	07	35	06	30	03	15
SSCE/WAEC	06	30	05	25	07	35
NCE/OND	01	05	03	15	05	25
BSc/HND	02	10	04	20	02	10
Postgraduate	01	05	00	00	02	10
Total	20	100	20	100	20	100

Source Compiled by authors, 2024

with 1–3 individuals, 35% are part of households consisting of 4–6 individuals, and 40% belong to households with more than 7 individuals. In Ogbakuba, 15% of respondents are situated in households containing 1–3 individuals, 50% are found in households with 4–6 individuals, and 35% belong to households with over 7 individuals.

Table 9.6 reveals that 70% of the overall respondents receive wages 30,000 naira (19USD), whereas 80% earn between 31,000 and 40,000 naira (26 USD); 65% obtain salaries ranging from 41,000 to 50,000 naira (31.50USD); 40% of the respondents are compensated within the range of 51,000 to 60,000naira (37.78USD), while 45% earn salaries exceeding 60,000naira. In Umunnakwo, the predominant income level is categorized as earning less than 19USD, in Ossamala, the highest income bracket is between 26 USD and in Ogbakuba, and the income range of 31.50 USD represents the highest earnings. Unfortunately, a greater percentage earn below the minimum threshold of 70, 000 naira (45 USD as at June 2025) in Nigeria (hhttps://wageindicator.org).

According to the data presented in Table 9.7, the distribution of occupations across the communities' reveals that 50% are engaged as Artisans, 35% are employed as Educators, 70% are active as Traders, 10% are working as Agriculturists, and 40% are employed within the Local Government Area (LGA). Additionally, within Umunnakwo, the occupational distribution indicates that 20% are Artisans, 15% are Educators, 20% are Traders, 30% are engaged in farming, and 15% are occupied in

Table 9.5 Household size

Household size	Umunnakwo		Ossamala		Ogbakuba	
	F	%	F	%	F	%
1–3	04	20	05	25	03	15
4–6	06	30	07	35	10	50
> 7	10	50	08	40	07	35
Total	20	100	20	100	20	100

Source Compiled by authors, 2024

Table 9.6 Income level

Income level	Umunnakwo		Ossamala		Ogbakuba	
	F	%	F	%	F	%
<30,000	06	30	05	25	03	15
31,000–41,000	05	25	07	35	04	20
41,000–50,000	04	20	03	15	06	30
51,000–60,000	03	15	02	10	03	15
>60,000	02	10	03	15	04	20
Total	20	100	20	100	20	100

Source Compiled by authors, 2024

office roles. In Ossamala, 15% of respondents are classified as Artisans, 10% serve as teachers, 25% are Traders, 40% are engaged in agriculture, and 10% are employed in office settings. Moreover, in Ogbakuba, 15% of respondents are identified as Artisans, 10% are teachers; 25% serve as Traders, 35% are engaged in farming, and 15% are employed in various capacities.

Table 9.8 documents that 90% of respondents in Umunnakwo acknowledge the susceptibility of their community to flooding, while 10% assert that their community is not prone to such events. In both Ossamala and Ogbakuba, 95% of respondents express the belief that their communities contain flood-prone areas, whereas 5% contend that their communities lack such areas.

Across the communities in Table 9.9, heavy rainfall was attributed as a primary cause of flooding. Umunnakwo, 30% of respondents cited heavy rainfall as a major factor in flooding, followed by 25% who attributed it to sea level rise, and 20% to a poor drainage system. Other contributing factors included topography and inadequate waste disposal (10% each), with 5% mentioning deforestation. In Ossamala, 35% identified intense precipitation as the main contributor to flooding, with 25% attributing it to rising sea levels and 15% to a deficient drainage system. Deforestation and topographical features were noted by 10%, while 5% pointed to improper waste disposal. In Ogbaku, only 5% recognized heavy rainfall as the primary cause of flooding, while 25% cited rising sea levels and 20% pointed to a poor drainage

Table 9.7 Occupation

Occupation	Umunnakwo		Ossamala		Ogbakuba	
	F	%	F	%	F	%
Artisans	04	20	03	15	03	15
Educators	03	15	02	10	02	10
Traders	04	20	05	25	05	25
Agriculturist	06	30	08	40	07	35
LG worker	03	15	02	10	03	15
Total	20	100	20	100	20	100

Source Compiled by authors, 2024

Table 9.8 Perception of the respondents on flood-prone areas

Flood-prone areas	Umunnakwo		Ossamala		Ogbakuba	
	F	%	F	%	F	%
Yes	18	90	19	95	19	95
No	02	10	01	05	01	05
Total	20	100	20	100	20	100

Source Compiled by authors, 2024

Table 9.9 Causes of flooding in your community

Causes of flooding	Umunnakwo		Ossamala		Ogbakuba	
	F	%	F	%	F	%
Heavy rainfall	06	30	07	35	10	50
Sea level rise	05	25	05	25	05	25
Farming	0	00	0	00	0	00
Poor drainage system	04	20	03	15	04	20
Deforestation	01	05	02	10	0	00
Topography	02	10	02	10	01	05
Improper waste disposal	02	10	01	05	0	00
Illegal physical development	0	00	0	00	0	00
Total	20	100	20	100	20	100

Source Compiled by authors, 2024

system. Five percent noted topography as a factor, but no respondents acknowledged agricultural practices, deforestation, inadequate waste disposal, or unlawful developments as contributors to flooding.

In Table 9.10, accessed effects of flooding are accessed, showing that across all communities, 35% experienced loss of land and roads, 65% reported building collapses, 55% noted disease spread, 80% indicated disruptions to social amenities, 30% acknowledged loss of life, and 35% observed diminished job opportunities. In Umunnnakwo, 10% reported loss of land and roads, 30% indicated building collapses, 10% noted disease spread, 25% mentioned disruptions to social amenities, 10% experienced loss of life, and 15% faced job loss. In Ossamala, 10% acknowledged loss of land and roads, 15% reported building collapses, 25% noted disease spread, 25% mentioned disruptions to social amenities, 15% experienced loss of life, and 10% faced job loss. In Ogbaku, 15% reported loss of land and roads, 20% indicated building collapses, 20% noted disease spread, 30% experienced disruptions to social amenities, 5% reported loss of life, and 10% faced job loss. Overall, flooding significantly affects infrastructure, public health, amenities, and employment across these communities.

9.6 Discussion

The study's findings, detailed in Table 9.1, indicated a higher participation of women across three communities. To achieve this, purposive sampling was employed to select participants with significant and relevant insights into the effects of recurrent flooding, as outlined by Royall (2020). This approach ensured the inclusion of women from diverse socio-economic backgrounds, ages, and household roles, particularly those who are educated, caregivers, and employed. Through focus group discussions,

Table 9.10 Effects of flooding in your community

Effects of flooding	Umunnakwo		Ossamala		Ogbakuba	
	F	%	F	%	F	%
Loss of lands/Roads	02	10	02	10	03	15
Collapse of buildings	06	30	03	15	04	20
Spread diseases	02	10	05	25	04	20
Disruption of social amenities	05	25	05	25	06	30
Loss of lives	02	10	03	15	01	05
Loss of jobs	03	15	02	10	02	10
Total	20	100	20	100	20	100

Source Compiled by authors, 2024

these women shared how flooding affects their access to clean water, reproductive health services, food security, and social stability (Onyenechere and Okoroafor 2019). Their reports highlighted gendered vulnerabilities, such as increased domestic responsibilities, displacement, and psychosocial stress resulting from flood events.

The purposive inclusion of women revealed important intersectional insights that might otherwise be missed in broader studies. As Royall (2020) suggests, qualitative inquiry thrives on thoughtfully selecting participants who can provide detailed perspectives on social and environmental risks. Thus, this study's purposive sampling enhanced the credibility, richness, and contextual accuracy of findings concerning women's health and eco-social experiences during floods.

Further, Table 9.2 reveals the age demographics among flood-affected individuals: 50% are aged 18–30, 35% are aged 31–40, 30% are between 41–50, and 15% are over 50 years. This age distribution significantly influences vulnerability assessments, resilience strategies, and policy development. The substantial proportion of young adults (mostly women) in the economically active age range indicates they bear crucial responsibilities in childbearing, farming, and household management. While younger women tend to be more adaptable and mobile, they often lack decision-making authority and access to vital resources, which may heighten their vulnerability to climate-induced challenges (Carr and Thompson 2014; FAO 2017; UN Women 2021).

In contrast, older women possess valuable local knowledge and coping strategies but face greater health risks and mobility limitations during emergencies (Alston 2013). The lower percentage of women over 50 may reflect barriers like literacy issues, health constraints, and socio-cultural exclusion from decision-making processes (Crenshaw 1991). Therefore, age-responsive flood adaptation strategies are essential to address the unique physical, economic, and social vulnerabilities associated with different age groups.

Also noteworthy is the marital status and household composition: most women in the study are married, with 60% living in small households (1–3 members) and 25% in larger ones (7 + members). Married women typically shoulder significant domestic and caregiving responsibilities during disasters, often with limited decision-making authority, which can affect their mobility and access to health services (Onyenechere and Okoroafor 2019). Smaller households may face resource constraints impacting resilience, while larger households often deal with food insecurity and health risks during floods (FAO 2017; Mehar et al. 2016). Notably, policy frameworks must consider these social dynamics when providing targeted support, particularly for married women and households burdened with dependents, especially in flood-prone areas like Anambra State.

Table 9.4 shows the diverse educational profile of respondents in this study, where 30% have no formal education, and varying levels attain primary (80%), secondary (90%), NCE/OND (45%), BSc/HND (40%), and postgraduate (15%) qualifications has significant implications for economic, social, adaptive behaviour, and public health responsiveness in flood-prone coastal communities of Anambra State.

Education is pivotal in shaping women's knowledge, attitudes, and practices concerning flood preparedness, early warning interpretation, and hygiene during displacement, and health-seeking behaviour (FAO 2017; Ayanlade and Radeny 2020). Women with secondary education and above (90%) are more likely to access and understand climate information, engage with economic extension services, and adopt climate resilience during floods (UN Women 2021; Onyeneke et al. 2020).

However, the 30% with no formal education remain highly vulnerable, often dependent on informal networks or male decision-makers, which limits their participation in disaster risk reduction initiatives and delays critical response actions. The educational gradient thus highlights the need for inclusive, non-formal education, targeted capacity building, and gender-sensitive communication strategies to improve climate resilience among all women, regardless of their academic background.

Table 9.6 showed the income distribution among respondents, where 70% of the overall respondents receive wages equivalent to 19USD, whereas 80% earn about 26 USD; 65% obtain salaries about 31.50USD; and 40% earn 37.78USD, Unfortunately, a greater percentage earn below the new minimum threshold of 70, 000 naira (45USD) in Nigeria (https://wageindicator.org).

These figures highlight a significant economic vulnerability among women in flood-prone communities of Anambra State, which has direct implications for their livelihoods, adaptive capacity, and health outcomes. Low income constrains women's ability to prepare for, respond to, and recover from flood events. For instance, limited financial resources affect their access to flood-resistant housing, early warning systems, and healthcare services. Women earning below the new national minimum wage are also less likely to afford private transport (usually the use of local boats) during flood emergencies or pay for post-flood medical attention, increasing their exposure to disease and food insecurity.

Furthermore, financial precarity reduces women's participation in collective resilience actions or formal savings groups, making them more dependent on informal support networks that may be equally strained during disasters (FAO 2017). Even among those earning more than 37.78 USD, sustained flooding can wipe out income sources tied to small-scale farming, trading, and informal labour. These income disparities underscore the need for social protection programs, livelihood diversification, and targeted financial assistance for rural women as part of climate adaptation and disaster risk reduction strategies.

Findings indicated that generally, from the three studied communities, 50% are Artisans, 35% are Educators, 70% are Traders, 10% are Agriculturists, and 40% are employed by Local Government Areas (LGAs). This presents critical insights into the differential vulnerabilities and coping capacities of women affected by flooding in these communities.

Women engaged in trading and artisan work, who represent the majority, are often part of the informal economy, which is highly sensitive to environmental disruptions. Flooding damages market infrastructure, restricts mobility, and leads to loss of goods and workdays, severely impacting income stability (FAO 2017; UN Women 2021). These occupational disruptions also ripple into household food security, health access, and school attendance for dependents, especially when women are the primary breadwinners (Onyenechere and Okoroafor 2019).

Educators and civil servants (including those in LGAs) may have relatively stable salaries, but flooding can still compromise their productivity due to displacement, school closures, and psychological stress (Alston 2013). Meanwhile, the 10% in agriculture, though a smaller group, face direct livelihood risks from soil erosion, crop loss, and disrupted farming cycles due to climate variability.

Understanding these occupational patterns is essential for designing sector-specific resilience strategies. For example, targeted support for market women, vocational retraining for artisans, and flexible employment policies for educators and public workers during post-disaster periods can enhance women's capacity to recover and adapt to climate-induced hazards.

Findings from Table 9.9 identified multiple causative factors of flooding in these communities, indicating the overwhelming identification of heavy rainfall and sea level rise as primary causes aligns with climate science, indicating that the studied coastal communities are increasingly vulnerable to extreme weather events and ocean encroachment. These perceptions underscore the complex and overlapping drivers of flood vulnerability in these communities, with direct consequences on women's health, livelihoods, and social well-being (Nkwunonwo et al. 2020).

This finding aligns with the research conducted by Otty et al. (2024), Ariyo et al., and Ozoh, who identified heavy rainfall, elevated sea levels, topography, deforestation, and improper waste disposal as the predominant factors cited by respondents regarding the causes of flooding in their communities. Notably, these women reside in informal housing, farm on marginal lands, and manage water-related household responsibilities; these climatic factors compound risks of displacement, waterborne disease, and livelihood loss.

The recognition of inadequate drainage (55%) and waste mismanagement (15%) highlights systemic infrastructure deficits and urban planning failures. These challenges exacerbate localized flooding, particularly in the study communities where many low-income women live and trade. Additionally, the prevalence of illegal development and topography points to a need for stronger enforcement of environmental regulations and spatial planning policies that consider women's vulnerabilities.

Plate 9.1 Destruction of the road as a result of a flood at Ogbakugba. *Source* Researchers Work, 2024

Further, the identification of heavy rainfall and sea level rise as primary causes of flooding explains its effects on flooding in Table 9.10, the multifaceted and gendered burden that flooding imposes on women in coastal areas of Anambra State, Nigeria. Plates 9.1, 9.2 and 9.3 show destruction of buildings, road, and the collapse of the bridge affecting movement, social and economic activities within seven communities, namely, Umunankwo, Mputu, Ogwuaniocha, Ossomala, Obeagwe, Akili Ogidi, and Ogwuikpele.

This disrupts social amenities (80%), including healthcare, clean water, education, and transport systems, disproportionately affecting women, who often bear the responsibility for caregiving, food preparation, and household health. The floods were so intense that they affected most of their buildings. Plate 9.5 showed the height of the flood, making it able to submerge buildings and damage roads. The collapse of buildings (65%) and loss of land and roads (35%) (See Plates 9.1, 9.2 and 9.3). This compromises women's safety and access to markets and services, especially for traders, farmers, single mothers, and elderly women residing in these studied communities (Plate 9.4).

Further, the spread of waterborne and vector-borne diseases (55%) such as cholera, malaria, and typhoid during floods poses a significant threat to maternal and child health, particularly in low-income households with limited access to medical care. Additionally, the loss of life (30%) and the diminution of employment opportunities (35%) reflect deep socio-economic disruption, which can push women into poverty traps and make recovery from disasters more difficult (Ayoade et al. 2020).

Further findings from a female key informant who is involved in agribusiness stated that…

> The floods submerged my small farm, and I ended up losing all my farm products. The little I was able to harvest before the flood was also destroyed because the flood broke through

Plate 9.2 Removal of a building's rooftop by flood at Ossamala. *Source* Researchers Work, 2024

our storage room. It was a complete loss for me, and presently I am finding it difficult to take care of myself, and talk more of my family members.

Another female informant, a petty Trader, mentioned that

During the flooding event, I could not afford to go to the hospital because the whole area was flooded, and I was left with no option but to give birth at my flooded home without any medical assistance.

Plate 9.3 Bridge collapse as a result of a flood at Umunankwo. *Source* Researcher's Fieldwork, 2024

Plate 9.4 A boat maker's business still undergoing recovery from the flood at Ossamala

A male Informant, who is a boat maker, one of the male interviewees emphasized that the floods affected his boat making business that thrives during the flood because it is the only means of transporting people and goods.

> the flood affected my boat business, they were badly damaged, I am still repairing them, but it is costing me so much even after the floods have gone. My life and income, and business is still under recovery.

Another female key informant who works in the LGA reported that …

> The floods were so intense that they destroyed most of my property, and my furniture was affected. It seems to increase every year due to the heavy rains; apart from the floods of 2012 and 2022, it was mainly influenced by the mismanagement of dams and water reservoirs in Nigeria and our neighboring countries. Undoubtedly, subsequent floods have been sourced from the overflow of the banks of River Niger, this makes our community vulnerable to flood occurrences annually during the wet season, and to flood and disasters since we are located by these rivers.

During the discussions, most of the women involved in farming reported that flooding causes severe damage to their farmlands. As a result, they often need to hurriedly harvest their crops during the floods to prevent total loss, as their farms become submerged (See Plate 9.6). Additionally, they mentioned that harvesting crops before the floods presents challenges in preserving them for an extended period. For example, yam banks are often destroyed, and any preserved produce is at risk when floods occur.

Plate 9.5 A Researcher showing the height of the floods on the building

Preserving certain harvested crops, such as vegetables and potatoes, is particularly difficult, as noted by the participants and observed during discussions. Furthermore, the women farmers explained their methods for preserving harvested cassava: they either store it underground in pits or place it on dry ground inside large waterproof sacks after fermentation and salting. The harvested yams are laid on dry surfaces

Plate 9.6 Cassava farm submerged by flood, and women harvesting cassava during the floods

to dry out and are later placed on elevated storage hangers known as "Uko." The women consistently stated that if the yam tubers are not harvested before the floods reach them, they will be destroyed during storage.

Also, findings in this study indicated that flooding can have beneficial impacts on residents, particularly by prompting the elevation of their buildings. This increase in elevation has provided income and job opportunities for various tradespeople, such as artisans and carpenters who construct canoes used during flood periods. Additionally, those who paddle these canoes for transportation also earn a decent income as a result of the floods. These findings align with the results presented by Akanwa et al. (2022), who highlighted several positive effects of flooding, including the growth of the fishing business, especially for catfish.

Although some residents may lose their jobs and sources of income during flooding, they have found ways to cope, maintaining hope that the floods will eventually recede and allow them to return to normal life. Many have learned to monitor water levels and record changes to prepare for floods, stocking their homes in advance. Observations show that many women activate their survival skills during floods, cooking, and managing household responsibilities despite the risks during the floods, affecting their health and well-being (Esra 2024).

Notably, findings from this study indicated that climate-induced floods in these communities often disrupt livelihoods, leading to economic instability, particularly affecting women's roles in generating household income. This economic strain also impacts children's education, social activities, health, and overall well-being. During climate crises, women and children frequently face increased risks of gender-based violence and exploitation as social structures become destabilized. The intersection of climate change with existing social inequalities further amplifies the challenges faced by these groups, underscoring the need for targeted interventions. These collective impacts emphasize the urgent need for education and capacity building. Raising climate awareness among women, coupled with digital literacy and participation in local governance, strengthens resilience, gender-responsive disaster risk reduction, investment in resilient infrastructure, along with the use of gender-disaggregated data to inform equitable planning (UN Women 2021). Policy responses should support climate-resilient housing and enforce land-use planning that protects coastal ecosystems such as mangroves, while also strengthening women's land tenure rights and healthcare interventions that prioritize the safety, well-being, and economic resilience of women in the face of recurring climate-related flooding (Ghosh and Dutta 2024; Sidun and Gibbons 2023; Blanca et al. 2024).

9.7 Conclusion and Recommendations

The findings demonstrate that flooding disproportionately affects women's livelihoods, disrupts household stability, damages social infrastructure, and exacerbates health risks, particularly among those with limited income, education, and access to formal support systems.

A substantial number of respondents identified climate-induced heavy rainfall, sea level rise, poor drainage, and unregulated urban development as primary drivers of flooding. These environmental pressures intersect with existing gender inequalities to intensify the marginalization of women, especially those engaged in informal occupations like trading, artisan work, and small-scale agriculture.

Based on these insights, there is an urgent need for the mainstreaming of gender into environmental and disaster risk policies in Nigeria. State and local governments should adopt inclusive and participatory climate adaptation frameworks that empower women through early warning education, health outreach, sustainable housing initiatives, and livelihood diversification.

Policy reforms should support the development of flood-resilient infrastructure, such as improved drainage and waste management systems, and prioritize healthcare access tailored to women's needs, especially reproductive and maternal health during post-disaster recovery. Furthermore, the establishment of community-based women-led resilience committees can enhance local capacity to respond to climate hazards and influence local planning.

In aligning with Nigeria's National Adaptation Strategy and UN SDG targets (especially SDG 5 and 13), the state can ensure that climate action is not only reactive but also proactive, gender-just, and sustainable. Strengthening institutional synergy among government agencies, civil society, and development partners is critical to translating these findings into meaningful and measurable impact.

References

Akanwa AO, Okedo-Alex IN, Joe-Ikechebelu NN (2022) Persistent climate-driven floods: a voiced case of indigenous women experiencing ecosocial and health risks in a South-East community in Anambra state, Nigeria. Springer, 495–523. https://doi.org/10.1016/b978-0-323-91603-5.00008-7

Alston M (2013). Women and adaptation. Wiley Interdis Rev Climate Change 4(5):351–358. https://doi.org/10.1002/wcc.232

Arora-Jonsson S (2011) Virtue and vulnerability: discourses on women, gender and climate change. Glob Environ Chang 21(2):744–751. https://doi.org/10.1016/j.gloenvcha.2011.01.005

Ayanlade A, Radeny M (2020) Climate variability and household adaptation strategies in rural Nigeria. Clim Risk Manag 27:100210. https://doi.org/10.1016/j.crm.2020.100210

Ayoade JO, Egboka BCE, Anuforom AC (2020) Climate variability and flood risk in the Lower Niger Basin of Nigeria. Nigerian J Environ Sci Technol 4(1):45–61

Bachiochi P, Weiner S (2004) Qualitative data collection and analysis. In: Rogelberg S (ed) Handbook of research methods in industrial and organizational psychology. Blackwell, Oxford, pp 161–183

Blanca A, Soledad C, Mark H, Zulfiqar AB, Etienne VL, Domenico GI, Giulia G, Josephine B (2024) Opportunities and challenges for financing women's, children's and adolescents' health in the context of climate change. BMJ Global Health. https://doi.org/10.1136/bmjgh-2023-014596

Carr ER, Thompson MC (2014) Gender and climate change adaptation in agrarian settings: current thinking, new directions, and research frontiers'. Geogr Compass 8(3):182–197. https://doi.org/10.1111/gec3.12121

Cornwall A, Rivas AM (2015) From 'gender equality and women's empowerment' to global justice: reclaiming a transformative agenda for gender and development. Third World Quart 36(2):396–415. https://doi.org/10.1080/01436597.2015.1013341

Coyne I (1997) Sampling in qualitative research. Purposeful and theoretical sampling; merging or clear boundaries? J Adv Nurs 26:623–630

Crenshaw K (1991) Mapping the margins: Intersectionality, identity politics, and violence against women of color. Stanford Law Rev 43(6):1241–1299

Danish Institute for Human Rights (2024) Gender justice and human rights. https://www.humanrights.dk/gender

Esra Ö (2024) Climate change and its impact on women's lives. J Health Sci Med. https://doi.org/10.32322/jhsm.1481918

FAO (2017) The role of women in agriculture. Food and Agriculture Organization of the United Nations

Ghosh A, Dutta K (2024) Health threats of climate change: from intersectional analysis to justice-based radicalism. Ecol Soc. https://doi.org/10.5751/es-14045-290215

Global Fund for Women (2024) What is gender justice? https://www.globalfundforwomen.org/what-we-do/gender-justice/

Intergovernmental Panel on Climate Change (IPCC) (2022) Climate change: mitigation of climate change contribution of Working Group 111 to the sixth assessment report of the IPCC on climate change. https://www.ipcc.org. https://doi.org/10.1017/9781009157926

Intergovernmental Panel on Climate Change (IPCC) (2018) Special report: global warming of 1,5 °C, Erişim tarihi: 10.01.2023, www.ipcc.ch/sr15.

Lancet Report Countdown on Climate Change (2024) Local story ideas from the 2024 Lancet Countdown report on climate change and health.https://www.thelancet.com/countdown-health-climate

McLennan M (2021) 2021 ESG report. Marsh McLennan. https://www.marshmclennan.com/assets/v3/esg/Marsh-McLennan-2021-ESG-Report-FINAL.pdf

Mehar M, Mittal S, Prasad N (2016) Farmers' coping strategies for climate shock: Is it differentiated by gender? J Rural Stud 44:123–131. https://doi.org/10.1016/j.jrurstud.2016.01.001

Murphy KM (2020) Electromagnetism and the metonymic imagination. The Pennsylvania State University Press

Naseem SM (2024) Environment and Health. |climate change. https://journalistsresource.org/home/2024-lancet-countdown-report/

Nkwunonwo UC, Whitworth M, Baily B (2020) A review of the current status of flood modelling for urban flood risk management in the developing countries. Scientific African 7:e00269. https://doi.org/10.1016/j.sciaf.2020.e00269

Onyenechere EC, Okoroafor CC (2019) Gendered vulnerabilities and adaptation strategies in flood-prone communities of southeastern Nigeria. Int J Disaster Risk Reduct 39:101246. https://doi.org/10.1016/j.ijdrr.2019.101246

Onyeneke RU, Nwajiuba CU, Igberi CO (2020) Climate-smart agriculture in Nigeria: an assessment of farmers' adoption in Abia State. Scientia Agricola 77(1):e20180150. https://doi.org/10.1590/1678-992x-2018-0150

Otty Z, Larkins S, Evans R, Brown A, Sabesan S (2024) Clinicians' experiences and perspectives about a new lung cancer referral pathway in a regional health service. Int J Integr Care 24(2):3, 1–11. https://doi.org/10.5334/ijic.7627

Palinkas LA, Horwitz SM, Green CA, Wisdom JP, Duan N, Hoagwood K (2015) Purposeful sampling for qualitative data collection and analysis in mixed-method implementation research. Administr Policy Mental Health Mental Health Serv Res 42(5):533–544. https://doi.org/10.1007/s10488-013-0528-y

Roos N, Kovats S, Hajat S, Filippi V, Chersich M, Luchters S et al (2021) Maternal and newborn health risks of climate change: a call for awareness and global action. Acta Obstet Gynecol Scand 100(4):566–570

Royall K (2020) Women's perspective– a look at gender and survey participation. Retrieved March 08th. https://culturecounts.cc/blog/a-womans-perspective-a-look-at-gender-and-survey-participation

Saunders M (2012) Choosing research participants. In: Symons G, Cassell C (eds), The practice of qualitative organizational research: core methods and current challenges London: Sage, pp 37–55

Sidun NM, Gibbons JL (2023) Women, girls, and climate change: human rights, vulnerabilities, and opportunities. Int J Psychol. https://doi.org/10.1002/ijop.12942

Statistica (2024) Annual rainfall in Anambra State, Nigeria. https://www.statistica.com.
UN Women (2021) Women and climate change: facts and figures. https://www.unwomen.org
World Meteorological Organisation (2020)The state of the climate in Africa: a multi-agency publication. https://hedleyserver.metoffice.gov.uk/wmoic/

Chapter 10
Coastal Hazard Impact on Global Climate Change and Sea Level Rise

J. Shanthini, S. Kavitha, S. Karthik, C. P. Maheswaran, and M. G. Sumithra

10.1 Introduction

Over the past several years, the aggravation, and repetition of anti-climactic hazards has become one of the leading challenges to sustainable development, environmental safety, and human living welfare at least in coastal regions. Such regions support about 40 percent of the world population, an important economic infrastructure, have several port systems, have vast fisheries and valuable ecological sensitive reserves (Neumann et al. 2015). Ironically, these densely populated and developed coastal areas are the most vulnerable to negative impacts of global climate change that includes the rise in sea level, storm surge, tropical cyclones, flooding of coastal areas, etc. (Church et al. 2019), (Emanuel 2005), (Nicholls and Cazenave 2010). Intergovernmental Panel on Climate Change (IPCC) continually states that sea-level rise rate is increased in the recent past due to both the big macroscopic extension of seawater and the colligation of glaciers and polar ice sheet (IPCC 2021), (Church and White 2006), (Shepherd et al. 2020), (Douglas 1997), thus subjecting both human

J. Shanthini (✉)
Department of Data Science and Business Systems, SRM Institute of Science and Technology, Kattankulathur, India
e-mail: shanthinisampath@gmail.com

S. Kavitha · S. Karthik
Department of Computing Technologies, SRM Institute of Science and Technology, Kattankulathur, India

C. P. Maheswaran
Department of Artificial Intelligence and Data Science, Sri Krishna College of Technology, Coimbatore, India

M. G. Sumithra
Department of Electronics and Communication, Sri Krishna College of Technology, Coimbatore, India

A. Nath et al. (eds.), *Geospatial Application of Coastal Hazards*, Springer Water,
https://doi.org/10.1007/978-3-032-20076-1_10

and natural systems in low-lying and island states to extreme threats (IPCC 2019), (Hallegatte et al. 2013).

One of the most obvious and harmful effects of human-caused climate change is sea-level rise (Nicholls and Cazenave, 2010), (Sweet et al. 2017). This has a chain reaction effect on coastal landforms, freshwater resources, agriculture, infrastructure, and public health. When combined with extreme weather events like cyclones and heavy rain, the risks to coastal ecosystems and human settlements grow, causing disasters to happen again and again, costing money, and forcing people to move (Torresan et al. 2012), (Hallegatte et al. 2013), (Kirezci et al. 2020).. So, to plan for effective adaptation, reduce disaster risk, and run coastal areas in a way that makes them more resilient, it is important to understand how global climate hazards affect coastal areas in different ways (Wilby and Dessai 2010), (Adger 2006).

Even though a substantial amount of studies have already explained the attribution of sea-level rise and regional manifestation of dangerous coastal areas using oceanographic and meteorological data (Church and White 2006), (Nicholls and Cazenave, 2010), (NASA JPL n.d.), the use of macroeconomic indicators connected with climate risk has not been used much in this sphere. More recently, the International Monetary Fund (IMF) has produced a more ambitious Climate Change Indicators Dashboard which compiles information country-by-country on emissions of greenhouse gases (GHG), physical climate risks, and indicators of transition to clean energy systems, and carbon pricing (International Monetary Fund 2024a, b), (International Monetary Fund 2024a, b), (International Monetary Fund 2024a, b). The current dataset provides a unique chance to investigate the correlation between the climate-risk exposures, i.e., temperature anomalies, extreme weather events, and cyclone frequency, and their further impact on sea-level processes and coastal exposure with the global, macroeconomic perspective (Slangen et al. 2014), (Hughes et al. 2021).

While numerous studies have examined causes of sea-level rise and localized impacts of coastal hazards using oceanographic and meteorological data (Church et al. 2019), (Nicholls and Cazenave 2010), (Hinkel et al. 2015), the role of macroeconomic climate risk indicators within the context remains seldom researched (Flörke et al. 2018), (International Monetary Fund 2024a, b). Comparatively few studies have by contrast questioned the usefulness of macroeconomic climate risk indicators in this sphere. Recently, the International Monetary Fund (IMF) published a Blueprint on Climate Change with a Climate Change Indicators Dashboard (International Monetary Fund 2024a, b), (International Monetary Fund 2024a, b) that puts together data across countries on greenhouse-gas (GHG) emissions, physical climate risks, energy transition metrics, and carbon pricing. The analytical context this dataset opens up is an unprecedented one that focuses on examining the correlation between climate risk exposures including temperature anomalies, extreme weather incidents, and cyclone frequency among other factors on one hand, and its subsequent impacts on sea-levels movements and coastal susceptibility in a global and macroeconomic scale.

This chapter attempts to fill a significant gap in existing literature by combining the International Monetary Fund (IMF) Climate Change Indicators with additional data regarding sea-level rise in order to assess coastal hazard exposure and vulnerability at a national and regional level. The methodology used is systematic, turning statistically stringent, with addition of production of spatial vulnerability maps. The goal is three-fold, namely, first, as far as to define the zones most at risk along the coast; second, to explain the relations between the macro-scale climate risks and anomalies of the sea level; and third, to predict the vulnerability scenarios in the future based on various pathway of greenhouse-gas emissions. The resulting insights will guide any role of the policymakers, coastal planners, and climate adaptation experts in clarifying priorities in at-risk areas, maximizing investments in resilience, and integrating macroeconomic climate-risk assessments into the nationwide adaptation and mitigation plans (Balica et al. 2009; Vousdoukas et al. 2018).

10.2 Advanced Foundational Knowledge

10.2.1 Global Climate Change and Coastal Hazards

The Climate changes caused by Anthropogenic actions has essentially re-arranged the climatic structure of the Earth, being materialized through a significant increase in the average temperatures worldwide and associated increase of climate-oriented risks. These include the fact that one of the most vulnerable is the coastal zone, which is located within close range of the active oceanographic regimes and exposure to sea-level rise, tropical cyclones, storm surges, and saline intrusion. As per the IPCC Sixth Assessment Report that the global mean of sea level rose by around 0.20 m in the period between 1901 and 2018, with its rate of rise having increased substantially in the last 30 years (IPCC 2021).

Hazards related to climate affect directly the coastal ecosystem and communities. The increase in temperature speeds up the process of thermal expansion and melting of glaciers both of which are the main drivers of the rise in sea level (Church et al. 2019). At the same time, the numbers of tropical cyclones and extreme precipitation occurrences have increased across multiple ocean basins causing episodic flooding, erosion and destruction of living habitats on the coasts (Emanuel 2005). The compounded implications of these processes pose risks to the infrastructure along the coast as well as human health, food security, and marine ecosystem sustainability (Neumann et al. 2015).

10.2.2 Trends, Drivers, and Regional Impacts of Sea-Level Rise

Sea-level rise (SLR) can be attributed to the cumulative impact of the three factors, which are the thermal expansion of ocean waters, glaciers and ice sheets melting and changes in the land water storage (Church and White 2006). Current observations suggest a global sea level increase at an average rate of 3.7 mm/year over the 2006–2018 period, an accelerated rate of change compared with previous decades (IPCC, 2019). Thermal expansion continues to be the leading component, responsible for almost 50% of the total observed sea level rise, whereas glacial and polar ice sheet melting increasingly adds to the overall rise (Shepherd et al. 2020).

Changes in sea level at the regional scale are, however, not homogeneous; regional processes of land subsidence, ocean circulation, gravitational influence and variations in vertical motions of the land smoothe the observed patterns (Douglas 1997). Particularly vulnerable are low-lying deltaic areas: e.g. Mekong Delta, Ganges Brahmaputra Delta, and the Nile Delta, as well as small island developing states (SIDS), due to their lack of adaptive capacity and insular exposure to rising waters (Nicholls and Cazenave 2010). Health impacts such as increased and intense coastal flooding, has been indentified as some of the effects of SLR with severe implications on economic security, coastal agriculture, public health and disaster risk management (Sweet et al. 2017).

10.2.3 Climate Indicators and Hazard Mapping

In the context of hazard mapping of risk, climate indicators have evolved significantly, due to emergence of satellite Earth observation networks and development of more integrated climate monitoring systems. Indicators like temperature deviations, rainfall intensity, the presence of cyclones, reduced rainfall frequency, have all become a mainstream tool in regional and country hazard exposure scoring. These signs cannot be avoided in measuring risk, predicting disaster situations, and guiding climate-resilient policy formulation (Flörke et al. 2018).

Routine use of climate indicators in earlier literature has relied on data sets of the National Aeronautics and Space Administration (NASA), the National Oceanic and Atmospheric Administration (NOAA), and the European Space Agency (ESA) (NASA JPL n.d.). However, macroeconomic data that combine climate risk and socio-economic data at the global level are limited. The IMF Climate Change Indicators Dashboard thus represents an interesting development, which provides country level readings across physical climate hazards, GHG emissions, energy transition measures, and carbon pricing initiatives (International Monetary Fund 2024a, b). The possibilities of the dataset in hazard mapping and vulnerability studies have not been fully exploited.

10.2.4 Data-Driven Coastal Vulnerability Assessment

The data-driven deterministic approach to vulnerability assessment has emerged as the dominant disability of climate-risk studies due to capability to combine various indicators, spatial data, and social-economic factors (Adger 2006). In this sphere,

Multi-Criteria Decision Analysis (MCDA) systems are well established to measure the vulnerability of the coast, and they can summarize the indicators of physical exposure, environmental sensitivity, and adaptive capacity (Balica et al. 2009). These frameworks assist spatial recognition of areas of high-risk coastal setting and, in that way, help achieve evidence-based climate-adaptation decisions (Torresan et al. 2012).

Statistical modeling approaches, such as correlation analysis and regression models, and machine-learning algorithms have been used through parallel with MCDA to find a relationship between climate hazards and outcomes of vulnerability (Hallegatte et al. 2013). Sea-level rise and hazard exposure future trends have been modeled over a variety of emission scenarios using time-series forecasting techniques like ARIMA and Prophet (Hyndman and Athanasopoulos 2021). The knowledge that can be produced by these analytical frameworks is crucial to policymakers intending to invest in climate adaptation and ensuring coastal communities are protected against growing threats.

10.2.5 Climate Change and Socio-Economic Vulnerability

Coastal hazards caused by climate have significant socio-economic consequences, which are disproportionality faced by developing states and marginalized rural and coastal communities. Rising sea levels and heightened floods aggravate poverty, evict communities at risk and undermine vital infrastructure and hence risk achievement of the sustainable development goals (SDGs) (Dasgupta et al. 2007). Those risks also endanger fisheries, tourism, agriculture, and their socio-economic systems, as well as public health systems, increasing their effect.

The need in social-economic indicators (population density, level of poverty, access to essential services, etc.) in the model of coastal risk is supported by preceding vulnerability assessments (Hinkel et al. 2015). However, there is lack of macro-level data sets to describe the overlapping of both physical climate risks and the national socio-economic profile. The International Monetary Fund (IMF) climate indicators with an emphasis that highlights physical risk metrics, GHG emissions and the exposure of the financial sector comprise an exclusive tool to close this gap and facilitate cross-country comprehensive vulnerability evaluation (International Monetary Fund 2024a, b).

10.2.6 Gaps in Existing Literature

An overview of the available literature points out some of the research gaps that the current research project will meet. Despite the increased number of recent studies examining the trajectories of sea-level rise and the regional hazard exposures using oceanographic and meteorological data (Kirezci et al. 2020; Hughes et al. 2021), there exists comparatively low use of macroeconomic indicators of climate risks as part of coastal vulnerability modeling (Hughes et al. 2021). Further, the recorded vulnerability mapping activities mostly focus on sub-national or local based studies, and have limited prospects of doing any comparative studies either at the global or national levels (Vousdoukas et al. 2018).

Another gap is the lack of integrated models where hazard exposure and sea-level rise data are combined with macro-economic risk indicators in the form of composite vulnerability indexes of coastal areas (Gornitz 2001). A source of such potentially filling the gap is the IMF Climate Change Indicators Dashboard (International Monetary Fund 2024a, b). Although its value in data-based cross-nation mapping and vulnerability modelling of hazards has not been harnessed in a systematic way, the current chapter aims at exploiting this data and using additional sea-level data and statistical modelling methodologies to investigate coastal hazard exposure, resilience, and future risk in a global and regional context.

10.3 Materials and Methods

This study uses the quantitative, multi-source and data-driven analytical system to assess the relationship between sea-level rise and climate-related risks, particularly, their impact on the world coastline. Combining International Monetary Fund (IMF) Climate Change Indicators Dashboard Macroeconomic Climate Risk indicators with systematic oceanography observations by the NASA and Copernicus Marine Service, the study evaluates the vulnerability and exposure of coastal hazards, regional variability and trends in sea-level change, and coastal hazards. It involves descriptive statistics, correlation and regression analysis, vulnerability mapping and scenario-based predictions of a variety of greenhouse gases (GHG) trajectories. The flow diagram of the process is given in Fig. 10.1.

The empirical analysis was backed by using of a complete range of secondary datasets. The major dataset was the Climate Change Indicators Dashboard (2024) published by the International Monetary Fund and proposed country-level annual values on the physical risks of climate change and its linked financial risk indicators between the year 1990 and 2022. Indicators, which have been chosen in the study, are brought out as follows:

- **Physical Risk Indicators**: Annual temperature anomalies, number of tropical cyclones, and extreme precipitation events.
- **GHG Emissions**: Annual emissions of carbon dioxide (CO_2), methane (CH_4), and nitrous oxide (N_2O), expressed in metric tons of CO_2-equivalents.

To supplement the climate risk data, the study incorporated sea-level rise datasets from the NASA GRACE Mission and the Copernicus Marine Service, including:

- **Mean Sea-Level Anomalies (1993–2023)**: Derived from satellite altimetry measurements and tide gauge calibrations.
- **Regional Sea-Level Trends**: Annual mean sea-level rise rates per region.

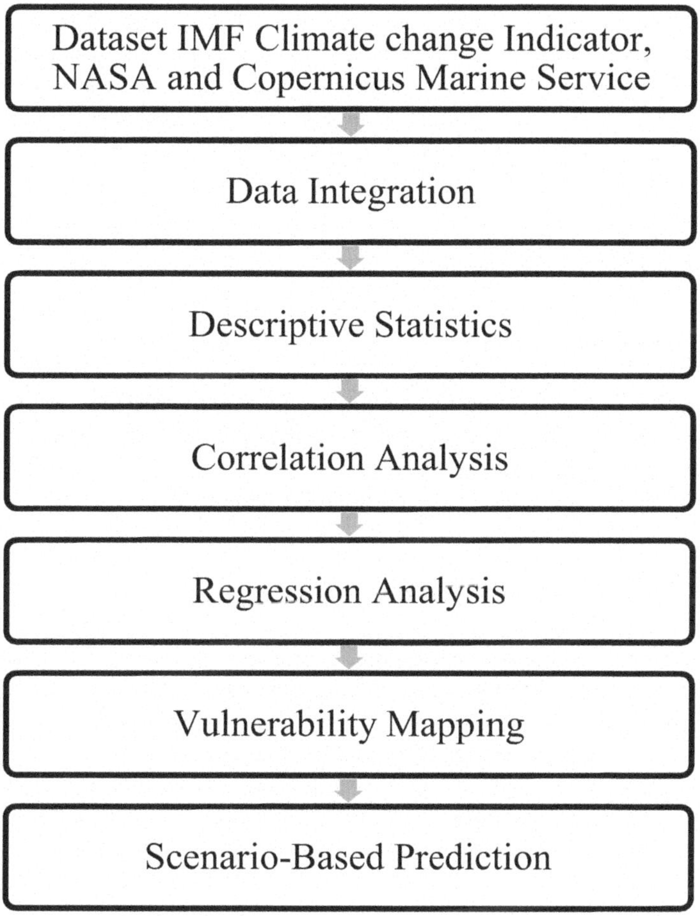

Fig. 10.1 Flow diagram of the proposed method

For future projections, emissions and sea-level rise forecasts under Shared Socioeconomic Pathways (SSP2-4.5 and SSP5-8.5) scenarios were sourced from the IPCC Sixth Assessment Report (AR6).

10.3.1 Data Preprocessing

All data sets were harmonized in a systematic way for temporal and geographical consistency. Country-level data was harmonized according to ISO alpha-3 country codes, and time series were harmonized for the common analysis period from 1990 to 2022. Missing values were managed by linear interpolation for continuous data and for categorical data by nearest-neighbor imputation.

To eliminate scale-related biases in multi-variable analysis, all the quantitative indicators were standardized using Z-score standardization. The procedure guaranteed that variables measured in dissimilar units of measurement contributed equally to correlation and regression analysis alongside vulnerability mapping. Python Libraries are used for the data preprocessing and modeling.

10.3.2 Analytical Methods

10.3.2.1 Descriptive Analysis

The descriptive statistical measures, namely means, standard deviations, and the temporal trend plots, were calculated in all the physical climate risk indicators (temperature, precipitation, surface dynamics, and sea-level anomalies) to help clarify the baseline trends as well as regional differences.

10.3.2.2 Correlation and Regression Analysis

The absolute value of Pearson correlation coefficient (r) as well as Spearman rank correlation coefficient (rho) was estimated to determine the level and direction of the relationships between climate hazards (temperature anomalies, cyclone frequency, and GHG emissions) and sea-level anomalies. Thereafter, series of multiple linear regression (MLR) models were developed in order to test the predictive association between climate indicators and the rise in sea-levels. The annual mean sea-level anomaly was considered the dependent variable, and the temperature anomalies, frequency of cyclone and values of GHG emissions were independent variables. These studies explained the type of relationships and strength of relationships between climatic processes and global mean sea-level change.

10.3.2.3 Coastal Hazard Vulnerability Mapping

A Multi-Criteria Decision Analysis (MCDA) model was established to come up with composite vulnerability ratings to coastal countries. The Fig. 10.2. shows the Flowchart of the Coastal Hazard Vulnerability Mapping Framework based on Multi-Criteria Decision Analysis (MCDA). The model integrates Hazard Exposure, Impact, and Adaptive Capacity indices weighted through the Analytic Hierarchy Process (AHP) to classify coastal regions into vulnerability levels ranging from Very Low to Very High.

The accompanying factors were as follows:

1. Hazard Exposure Index, which is based on yearly temperatures anomalies and cyclones presence.

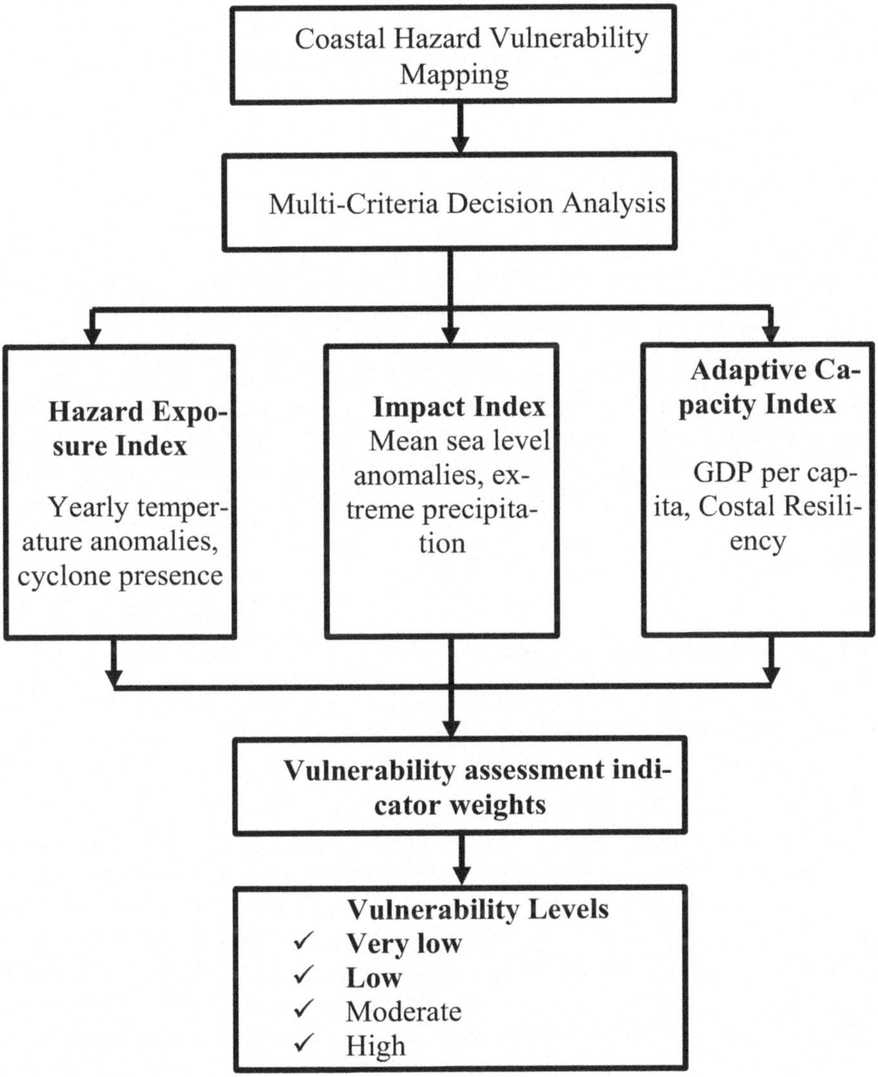

Fig. 10.2 Flowchart of the coastal hazard vulnerability mapping framework based on multi-criteria decision analysis (MCDA)

2. Impact Index built on mean sea-levels anomalies and occurrence of extreme precipitation.
3. Adaptive Capacity Index, which is depicted by GDP per capita and a qualitative coastal resiliency of infrastructure.

Vulnerability assessment Indicator Weights were determined through the Analytic Hierarchy Process (AHP), based on a review of existing frameworks of vulnerability assessment and expert advice. The weighted indicators were combined to provide the distribution of countries of coastal vulnerability as; very low, low, moderate, high and very high.

10.3.2.4 Scenario-Based Forecasting

To evaluate the vulnerability to future climate this paper uses a mixed method which combines time-series predictions and scenario-based modeling. Records of temperature and cyclone anomalies and frequency recorded historically provide the input parameter that can be modeled through the Autoregressive Integrated Moving Average (ARIMA) model and extrapolate the projections until 2100. The effects of future emission pathways are also recorded by simulating sea-level rise and average mean temperature anomaly projections over the SSP2-4.5 and SSP58.5 scenarios based on the IPCC AR6 datasets. The attained vulnerability scores are then summed up at 2030, 2050 and 2100 and hence form an empirical trend of risk over the various dates.

10.4 Results and Discussion

10.4.1 Descriptive Analysis of Climate Indicators and Sea-Level Trends

The indicators of climate hazards drawn by the International Monetary Fund (IMF) in its Climate Change Indicators Dashboard are organized in a descriptive analysis outlining a monotonous and threatening course of climate hazard over the 1990 and 2022 period. Within this period, the global mean temperature anomaly per year continued to transcend in an increasing yearly manner: rising with 0.32 C in 1990 and 1.07 C in 2022. The warming trend is in line with the scenarios that are presented in the Intergovernmental Panel on Climate Change (IPCC) Sixth Assessment Report, which confirms a significant intensification of the atmospheric warmth.

The number of tropical cyclones at the same time demonstrated a steady development, reaching the peak in the Western Pacific basin and North Atlantic. The annual mean occurrence of storms across the coastal areas increased between 1990 and 2022, with the mean number of storms increasing to 7.3 storms per annum in the more recent period as compared to 5.8 events per annum in 1990 though this number varies considerably with the years as observed by the National Oceanic and Atmospheric Administration (NOAA) (IPCC 2021).

Average sea-level anomaly values collected by NASA and Copernicus further confirm that there has been a significant rise in the mean sea level of the whole world in the past 30 years. Abnormalities with respect to the baseline recorded in 1993 reached around + 98 mm in 2022 with a total long-term increase rate of 3.4 mm/

Table 10.1 Correlation matrix of climate indicators and sea-level anomalies

	Temperature anomaly	Cyclone frequency	GHG emissions	Sea-level anomaly
Temperature anomaly	1	0.49	0.88	0.87
Cyclone frequency	0.49	1	0.52	0.56
GHG emissions	0.88	0.52	1	0.79
Sea-level anomaly	0.87	0.56	0.79	1

year. Regional variations are also noticeable; the rate of sea-level rise is higher in the coastal areas of the Southeast of Asia and the Indian Ocean.

10.4.2 Correlation Between Climate Hazards and Sea-Level Anomalies

Statistical correlation analysis results show that there are strong correlations between four major TCR indicators-mean temperature anomalies, man-made greenhouse gases (GHG) emissions, tropical cyclones frequency, and mean sea-level anomalies. The relationship between temperature anomalies and sea-level anomalies is strongly positive (as our Table 10.1 illustrates $r = 0.87$, $p < 0.001$), which is complementary to the fact that thermal expansion should be one of the major reasons behind the sea-level rise. Similarly, overall GHGs show strong positive relationship with sea-level anomalies ($r = 0.79$, $p < 0.001$) that shows accumulative effects of emission on warming of the atmosphere and thermal expansion of the ocean. There is a moderate but significant correlation between tropical cyclone frequency and sea-level anomalies ($r = 0.56$, $p < 0.01$) which is in line with the impacts of extremes in creating episodic coastal flooding and long-run coastline dynamics.

10.4.3 Regression Analysis: Predictors of Sea-Level Rise

To measure the relationships between the climate risk indicators and sea-level anomalies multiple linear regression (MLR) was used. The model gave an adjusted $R^2$of 0.82, which implies that 82% of variance in sea-level anomalies could be explained by the collective effects of aberrations in temperatures, the occurrence of tropical cyclones, and emission of greenhouse gases.

The strongest predictor was temperature anomaly (0.72, $p < 0.001$) and the next came GHG emissions (0.61, $p < 0.001$). One other predictor, tropical cyclone frequency, was also found to be statistically significant although its impact appeared to be comparatively small (0.28, $p < 0.05$). These findings reiterate the predominant role that thermal expansion plays in the rise of the sea level, which is aggravated by warming the atmosphere by GHGs.

10.4.4 Coastal Vulnerability Assessment and Mapping

Multi-Criteria Decision Analysis (MCDA) was used to rank 130 coastal countries in terms of their exposure to vulnerability as an integration of the three dimensions namely; exposure to hazards, the effect of sea-level rise (SLR), and the capability to adapt. A weight was established in each dimension, based on the expert elicitation method, to give a composite vulnerability score, between 0.12 (very low vulnerability) and 0.93 (very high vulnerability). Table 10.2 presents a regional summary of the number of countries in each vulnerability category, demonstrating distinct spatial disparities in exposure and resilience.

Through spatial distribution of the composite scores, it was established that small island developing states (SIDS) including Maldives, Tuvalu and Kiribati, together with low-lying deltaic countries such as Bangladesh and Myanmar were the ones on the most vulnerable spots. The coastal areas in the Southeast Asia, Sub-Saharan Africa, and some parts of the Caribbean also portrayed high levels of vulnerability where it was rated high due to exposure to hazards and low adaptive capacity.

On the other hand, other countries that were affected by severe impacts due to the risks such as Japan, Australia, and the United States had a lower vulnerability score: These countries had better coastal protection, had good disaster management facilities, and their economies were more resilient.

The table highlights the particularly high concentration of ‘very high’ vulnerability nations among Small Island States and Southeast Asia, confirming their elevated exposure to compounded climate hazards and sea-level rise. The stacked bar

Table 10.2 Vulnerability classification by region (number of countries)

Region	Very Low	Low	Moderate	High	Very high
Small Island States	0	1	3	7	12
Southeast Asia	0	2	3	5	4
South Asia	1	1	3	4	3
Sub-Saharan Africa	1	3	6	5	5
North America	2	2	1	1	0
Europe	5	4	2	1	0
East Asia & Pacific	1	3	3	2	1
Latin America & Caribbean	0	2	3	3	4

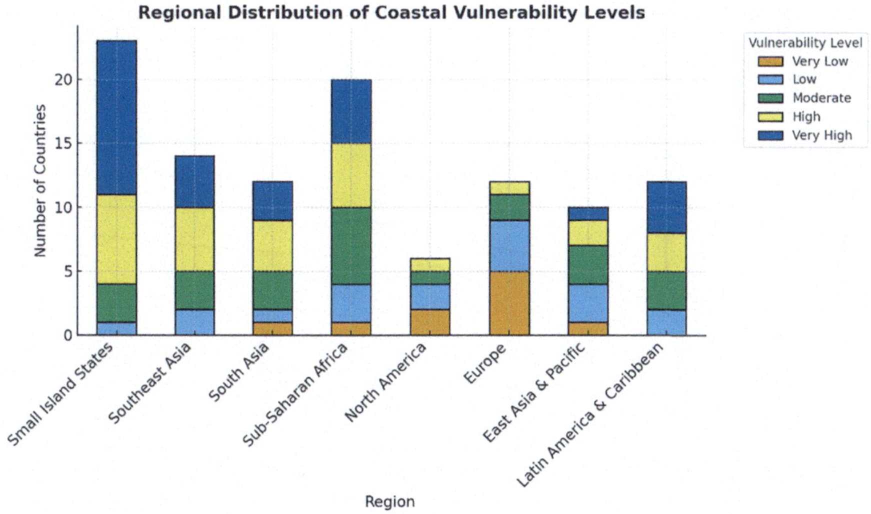

Fig. 10.3 Regional classification of coastal vulnerability levels based on multi-criteria decision analysis (MCDA)

chart in Fig. 10.3 illustrates the distribution of countries across vulnerability categories from Very Low to Very High highlighting the heightened exposure of Small Island States and Southeast Asian regions to coastal hazards and sea-level.

10.4.5 Future Scenario Projections

Forecasting based on ARIMA models and IPCC AR6 projections are presented in Table 10.3, which gives significant differences in projected sea-level rise and vulnerability outcomes between the two emission scenarios (SSP2-4.5 and SSP5-8.5). Table 10.3 shows a relative comparison study of the projected rise of the mean sea level and the list of the high-risk countries of a succession of temporal horizons.

The SSP5-8.5 high-emission scenario projections indicate that, by year 2100, global mean sea level might increase by more than 1.1 m, and 37 'coastal countries and

Table 10.3 Projected sea-level rise and vulnerability under SSP2-4.5 and SSP5-8.5

Year	Sea-level rise (m) SSP2-4.5	Sea-level rise (m) SSP5-8.5	Countries in 'high' & 'very high' vulnerability (SSP2-4.5)	Countries in 'high' & 'very high' vulnerability (SSP5-8.5)
2030	0.22	0.27	14	19
2050	0.54	0.67	17	28
2100	0.84	1.12	21	38

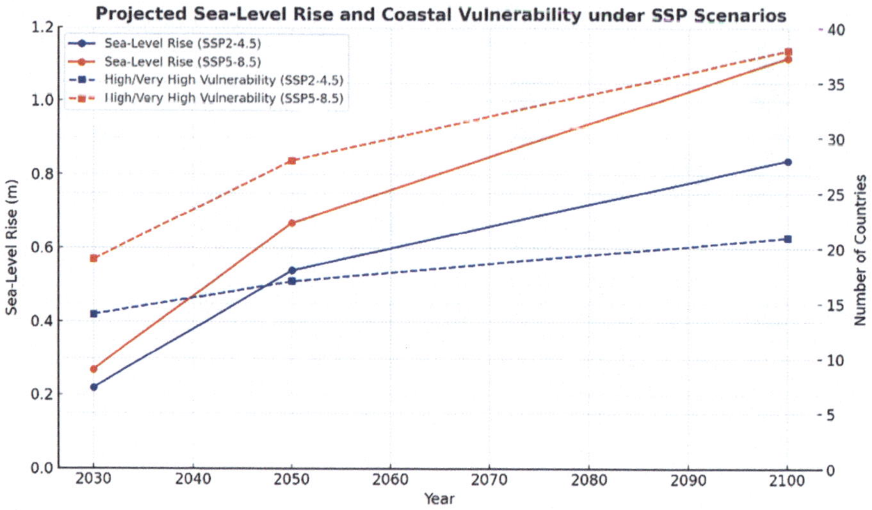

Fig. 10.4 Projected sea-level rise and coastal vulnerability under IPCC SSP2-4.5 and SSP5-8.5 scenarios

territories would be considered in the high or very high vulnerability categories when compared with the predication of this research. On the other hand, the moderate SSP2-4.5 intermediate solution balances fast-paced sea-levels and exposure to hazards, validating a reality of robust mitigation measures.

The Fig. 10.4. compares sea-level projections and the number of highly vulnerable countries across temporal horizons (2030, 2050, and 2100), illustrating accelerated rise and vulnerability concentration under the high-emission SSP5-8.5 pathway.

10.4.6 Multiple Linear Regression Analysis

The proposed research attempted to outline the role of the identified climate risk indicators in the variations of sea-level rise by building the model of multiple linear regression (MLR) where sea-level anomaly was chosen as a dependent variable, whereas temperature anomaly, GHG emissions, and cyclone frequency were determined as the independent variables. The model resulting was a good explanatory predictive model with an adjusted R 2 of 0.82. Table 10.4 demonstrates the coefficients of regressions, their standard errors, as well as correlating p values as to each predictor.

It is observed that all findings make temperature anomaly the most influential factor on sea-level anomalies (beta = 0.72, $p < 0.001$), which means it is the primary predictor. Accumulating greenhouse-gas emissions is an important further source (beta = 0.61, $p < 0.001$). Even though cyclonic activity also has an influence (beta

Table 10.4 Multiple linear regression model summary

Predictor variable	Unstandardized coefficient (B)	Standard error	Standardized beta (β)	t-value	p-value
Constant (Intercept)	−0.078	0.011	—	−7.09	0.000
Temperature anomaly (°C)	0.064	0.006	0.72	10.67	0.000
GHG emissions (Mt CO_2e)	0.0031	0.0004	0.61	7.82	0.000
Cyclone frequency (count)	0.012	0.004	0.28	3.01	0.005

= 0.28, p = 0.005), that effect is rather small, and will cause episodic surges, but not ensure any trends.

In the current analysis, there is a totality in which the aggregated data sets highlight high degrees of spatially disparate and scenario-based disparate coastal hazard risk exposure and vulnerability. The criticality of the situation concerning a disproportionate burden of the Small Island States and deltaic countries, i.e., Bangladesh and Myanmar, contribute to the necessity of specific adaptation responses.

The table of scenario comparison provides tangible empirical evidence of the positive effects of emission mitigation with results indicating that the scenario of SSP2-4.5 produces a projected exposure to hazards which is much lower and reduces the number of highly vulnerable countries by 2100.

The extent to which the IMF macroeconomic climate risk indicators are integrated into analysis is the new aspect of the vulnerability assessment literature that proves the effectiveness of the practice of integrating economic and physical hazard indicators as a more detailed approach to risk profiling.

10.5 Future Directions

In future research, additional aspects should be added to current analysis in numerous dimensions. First, local differences in hazard exposure and adaptive capacity must be captured by providing vulnerability assessment at the sub-national and city level. Second, addition of socio economic resilients, e.g., quality of coastal infrastructure, capacity of institutions, and preparedness of health systems for understanding risk holistically, is urgently needed. Lastly, longitudinal studies on hazard-event interaction and their effects which are combined in making an observation on the sea-level anomalies and probable coastline habitability would also be rosy to the policymaker, planners and climate scientists whose programs focus on mitigating the hazards and making the coastline more resilient.

10.6 Conclusion

This chapter presents an integrated, quantitative assessment of climate-human risk impacts on global sea level rise according to macroeconomic climate risk metrics of the International Monetary Fund (IMF) and NASA and Copernicus oceanographic data. Using an integrated methodology of descriptive statistical analysis, correlation and regression modeling, multi-criteria decision analysis (MCDA), and scenario-based forecasting, this paper presents meaningful results on coastal spatial patterns of vulnerability and future hazard exposure under different emission scenarios.

The descriptive analysis confirmed persistent and rising trends in global temperature anomalies, GHG emissions, and mean sea-level rise from 1990–2022. Correlation and regression analysis confirmed strong positive correlations between temperature anomalies, cyclone frequency, GHG emissions, and sea-level anomalies, with temperature anomalies being the strongest predictor of sea-level change. The regression model that we developed achieved an excellent explanatory power with an adjusted R^2 of 0.82.

Coastal vulnerability mapping revealed considerable variations in hazard exposure and adaptive capacity. Small island developing states (SIDS), deltaic low-lying countries, and some Southeast Asian and African coastal countries were ranked 'high' to 'very high' vulnerability, primarily on the basis of a combination of greater physical risk and lower socioeconomic resilience. Scenario-based projections under high-emission (SSP5-8.5) and intermediate (SSP2-4.5) pathways indicated that future hazard exposure and vulnerability will increase sharply without substantial global mitigation efforts. Global mean sea-level rise could amount to over 1.1 m by 2100 under SSP5-8 scenario, which would subject over 38 countries to extreme coastal hazards.

The key contributions of this study are:

- incorporating macroeconomic climatic risk information in coastal hazard and sea level rise analysis thereby creating a modular platform of cross-country vulnerability assessment.
- providing empirical data of the unequal exposure available to low-lying and island states.
- illustrating how emission mitigation is important in ensuring that exposure of future hazards is curbed and long-term vulnerability is decreased.

References

Adger WN (2006) Vulnerability. Glob Environ Chang 16(3):268–281. https://doi.org/10.1016/j.gloenvcha.2006.02.006

Balica R, Douben N, Wright N (2009) Flood vulnerability indices at varying spatial scales. Water Sci Technol 60(10):2571–2580. https://doi.org/10.2166/wst.2009.183

Church JA, White NJ (2006) A 20th century acceleration in global sea-level rise. Geophys Res Lett, 33(1). https://doi.org/10.1029/2005GL024826

Church J, Clark PU, Cazenave A, Gregory JM, Jevrejeva S, Levermann A, Merrifield MA, Milne GA, Nerem RS, Nunn PD, Payne AJ, Pfeffer WT, Stammer D, Unnikrishnan AS (2019) Sea-level rise and implications for low-lying islands, coasts and communities. In IPCC Special Report on the Ocean and Cryosphere in a Changing Climate. https://www.ipcc.ch/srocc/chapter/chapter-4-sea-level-rise-and-implications-for-low-lying-islands-coasts-and-communities/

Dasgupta S, Laplante B, Meisner C, Wheeler D, Yan J (2007) The impact of sea-level rise on developing countries: A comparative analysis (Policy Research Working Paper No. 4136). World Bank. https://papers.ssrn.com/sol3/papers.cfm?abstract_id=1012638

Douglas BC (1997) Global sea rise: A redetermination. Surv Geophys 18(2–3):279–292. https://doi.org/10.1023/A:1006544223710

Emanuel K (2005) Increasing destructiveness of tropical cyclones over the past 30 years. Nature, 436(7051), 686–688. https://www.nature.com/articles/nature03906

Flörke M, Schneider C, McDonald RI (2018) Global climate risk index and adaptation tracking. Environ Res Lett, 13(12). https://doi.org/10.1088/1748-9326/aae2d6

Gornitz V (2001) Sea-level rise and coasts: A review of causes, impacts, and policy implications. Climate Res 18(3):259–282. https://doi.org/10.3354/cr018259

Hallegatte S, Green C, Nicholls RJ, Corfee-Morlot J (2013) Future flood losses in major coastal cities. Nat Clim Chang 3(9):802–806. https://doi.org/10.1038/nclimate1979

Hinkel J, Nicholls RJ, Tol RSJ, Wang ZB, Hamilton JM, Boot G, Vafeidis AT, McFadden L, Ganopolski A, Klein RJT (2015) Sea-level rise scenarios and coastal risk management. Nat Clim Chang 5:188–190

Hughes A, Hinkel J, Lincke D (2021) Global-scale analysis of coastal flood exposure and adaptation costs under 21st-century sea-level rise. Nat Commun 12(1):1–12. https://doi.org/10.1038/s41467-021-23396-w

Hyndman, R. J., & Athanasopoulos G (2021) Forecasting: Principles and practice (3rd ed.). OTexts. https://otexts.com/fpp3/

Intergovernmental Panel on Climate Change (IPCC) (2019) Summary for policymakers. In Special report on the ocean and cryosphere in a changing climate. https://www.ipcc.ch/srocc/

Intergovernmental Panel on Climate Change (IPCC) (2021) Climate change 2021: The physical science basis. In: Contribution of working group i to the sixth assessment report of the intergovernmental panel on climate change. Cambridge University Press. https://www.ipcc.ch/report/ar6/wg1/

International Monetary Fund (2024) Climate change indicators dashboard. https://climatedata.imf.org/

International Monetary Fund (2024) Physical risk indicators by country. Clim Chang Indic Dashboard. https://climatedata.imf.org/indicator/PR_RISK

Kirezci E, Young IR, Ranasinghe R, Muis S, Nicholls RJ, Lincke D, Hinkel J (2020) Projections of global-scale extreme sea levels and resulting episodic coastal flooding over the 21st century. Sci Rep 10(1):11629. https://doi.org/10.1038/s41598-020-67704-7

NASA Jet Propulsion Laboratory. (n.d.). Sea level change: Observations from space. https://sealevel.nasa.gov/

Neumann B, Vafeidis AT, Zimmermann J, Nicholls RJ (2015) Future coastal population growth and exposure to sea-level rise and coastal flooding—A global assessment. PLOS ONE, 10(3), e0118571. https://journals.plos.org/plosone/article?id=https://doi.org/10.1371/journal.pone.0118571

Nicholls RJ, Cazenave A (2010) Sea-level rise and its impact on coastal zones. Science 328(5985):1517–1520. https://doi.org/10.1126/science.1185782

Shepherd A, Ivins E, Rignot E, Smith B, van den Broeke M, Velicogna I, Nowicki S (2020) Mass balance of the Greenland Ice Sheet from 1992 to 2018. Nature 579:233–239. https://doi.org/10.1038/s41586-019-1855-2

Slangen ABA, Carson M, Katsman CA, van de Wal RSW, Köhl A, Vermeersen LLA, Stammer D (2014) Projecting twenty-first century regional sea-level changes. Clim Change 124(1–2):317–332. https://doi.org/10.1007/s10584-014-1080-9

Sweet WV, Kopp RE, Weaver CP, Obeysekera J, Horton RM, Thieler ER, Zervas C (2017) Global and regional sea level rise scenarios for the United States. NOAA Technical Report NOS CO-OPS 083. https://tidesandcurrents.noaa.gov/publications/techrpt83_Global_and_Regional_SLR_Scenarios_for_the_US_final.pdf

Torresan C, Critto A, Rizzi J, Marcomini A (2012) Assessment of coastal vulnerability to climate change hazards: A case study in the North Adriatic Sea. Nat Hazard 12:2347–2368. https://doi.org/10.5194/nhess-12-2347-2012

Vousdoukas MI, Mentaschi L, Voukouvalas E, Verlaan M, Jevrejeva S, Jackson LP, Feyen L (2018) Global probabilistic projections of extreme sea levels show intensification of coastal flood hazard. Nat Commun 9(1):2360. https://doi.org/10.1038/s41467-018-04597-4

Wilby RL, Dessai S (2010) Robust adaptation to climate change. Weather 65(7):180–185. https://doi.org/10.1002/wea.543

Chapter 11
Climate Change and Vulnerability of Fishing Community: A Study on the Indian Sundarban

Susmita Maity and Tiyasha Shaw

11.1 Introduction

Global climate change and their consequences (Allison et al. 2009; Colburn et al. 2016) like increase the intensity of cyclone in last ten years, increased temperature in sea water, sea level rise, increased salinity, decrease rainfall, decrease the amount of mangrove forest—all are influenced to increase the vulnerability of a community. Coastal people are mainly affected. In the coastal region people mostly belong to the fishing community and they are in the most vulnerable situation. Extreme climatic vulnerability (Mondal 2022) has a great impact on Indian Sundarban region and these problems hamper mainly the lifestyle of fishermen. Due to such environmental phenomena fishing communities are facing some major issues like mental and physical health impact (Kanjilal et al. 2010; De 2014; Ghosh et al. 2016), environmental issues (MahadeviaGhimire and Vikas 2012; EnamulHaque et al. 2022), increase salinity in agricultural field and pond (Chandra and Sagar 2003), migration related issues (Guha and Roy 2016; Patel et al. 2009), property loss (Mahato 2020) etc. Yet these various adversities people in Sundarban region still continuing their profession. It is necessary to judge the vulnerability of fishing community so that we can understand the sustainable well-being of this community. This fishing related activity is one of the most vulnerable jobs. In this activity there are different types of risks. Mandal et al. (2019) describes that during 1973–2010 the various factors like climate change, extreme weather events including cyclones, tidal surges, floods, sea level rise, salinity intrusion etc. all are increased and its impact is seen in agriculture and aquaculture. Pramanik et al. (2022) stated some health-related risks like

---

S. Maity · T. Shaw (✉)
Department of Geography, University of Calcutta, Kolkata, West Bengal, India
e-mail: shawtiyasha17@gmail.com

S. Maity
e-mail: susmitamaity365@gmail.com

A. Nath et al. (eds.), *Geospatial Application of Coastal Hazards*, Springer Water,
https://doi.org/10.1007/978-3-032-20076-1_11

due to cyclone the outbreak of waterborne disease, negative impact of Covid-19, lack of health infrastructure. Das (2018) examine the extent and impact of human-animal conflict in Indian Sundarban region, between 1999 and 2014 fisherman, crab collectors, honey collectors are general victimized by the tiger attack the one of the most socio-economic risk is migration and it is happened because of low agricultural activities, less opportunities for occupation and high risk in fishing activities.

The effect of climate change among artisanal fisher-folk in Lagos, Nigeria has a greatest impact on people who rely on fisheries as their livelihood. The people of coastal communities are mainly affected. Allison et al. (2009) compared 132 countries to assess the vulnerability impact due to climate change and its impact on uncertain fish stocks. Sea level rise is a risk indicator of climate change. It has a great impact for fishing dependent communities (Colburn et al. 2016). Livelihood is mainly dependent on environmental exposure, sensitivity and adaptive capacity. Fishing is considered as one of the high-risk living activities across the world especially towards the livelihood of small-scale fishing communities of inland open water areas (Sadekin et al. 2018). According to IPCC climate vulnerability framework coastal fishing communities are more vulnerable compare to the people living in other parts. The livelihood activities of the coastal fishermen are mainly affected by the impacts of natural disasters, diseases, NGO and microcredit, income and piracy problem (Jakariya et al. 2020). Due to livelihood vulnerability some fishermen in China changed their occupation to other related industries like mariculture, fish processing and recreational fishery. It is more important to rebuild sustainable livelihood to help the fishermen (Chen et al. 2020). Bhuyan and Islam (2016) studied on the socio-economic condition of fishing community of Meghna river stated that most of the fishermen were Hindu (63%) belong to the age of 24–45 years. 71% approximately of them were illiterate but got medical help from village doctor and there was a prevalent relationship among contentment of fishermen. Children's school, taking risk at work, electricity facilities and medical facilities are done by fishing income. Almost half of the interviewed fishermen in Indonesia perceived a decrease in income due to conservation rules. The economic situation of the residents was associated with their access to alternative livelihoods, conservation rules and relationship with conservation authorities (Eriksson et al. 2019). Fishermen in Brazil faced severe occupational hazards without adequate health protection. They faced precarious living conditions without access to occupational health services (Pena and Gomez 2014). Traditional risks and modern uncertainties impacted on fishermen's mental health (King et al. 2014). Liebman et al. (2013) stated the current occupational health and safety policies and regulatory standards in the AgFF sectors and underscores the regulatory exceptions and limitations in worker protections. Women are also engaged in this occupation specially in shrimp and crab collection but they faced lots of health issues including mental health (King et al. 2015).

Vulnerability perpetuates due to physical and financial resource constraints. Fishermen of Madh and Worli villages in Mumbai are observed to be more vulnerable and less adaptive due to their inability to use efficient mechanised boats and advanced fishing implements. They also observed the negative impacts of climate change on their fishing livelihoods. Adaptation strategies to maximize fish catch are observed in such practices as targeting different species and fishing intensively for several days (Senapati and Gupta 2017). The water pollution in Churni River has an impact for declining fish diversity as well as the productivity of the existing fish species. Therefore, paralysed fishing economy and high dependency of the fishermen on the Churni River have forced them to revolve into the vicious cycle of poverty and enduring fragile livelihoods. Thus, they sometimes change their profession (Sarkar and Islam 2021). Tripathi et al. (2017) stated that women have more occupational health problems compared to men and they didn't receive any kind of benefits at their workplace. Umamaheswari et al. (2021) stated that climate vagaries affect their habitat and livelihood of people in coastal block. This work is capturing the grass-roots reality and deriving vulnerability indicators that affect fisheries and coastal households, socioeconomic vulnerability and cumulative vulnerability framework in marine fishing villages of Thoothukudi, Tami Nadu.

Sundarbans, the largest delta on the planet earth is in a vulnerable position due to some climatic issues (Mondal 2022). Now a days global warming is a common problem which creates sea level rise and irregular rainfall. As a delta region, Sundarban faces the problems with sea level rise. Many islands are sunk under water, mangroves are destroyed, the pH level of water is irregularly changed through time, sea water is being more saline and also the salinity of soil is increased day by day (MahadeviaGhimire and Vikas 2012). At present, the intensity of cyclones are increased which has become a curse on people of Sundarban (EnamulHaque et al. 2022). Cyclones create many difficulties like loss of biodiversity, sustainability and livelihood difficulties of human beings and fisher folk viz., agricultural lands are being more saline and ponds' fishing are stopped (Chandra and Sagar 2003). All these create a situation of migration to other places to sustain their livelihood (Guha and Roy 2016). Patel et al. (2009) stated some problems of Sundarban Tiger Reserve and the details of local or traditional fish community, and they discussed about the legal frame work, design, implementation of fishing regulation of local fishing community and explore different ways in which livelihood concerns can be approximately balance with conservation.

Nature is the only space of hope for the people of Sundarban where natural disasters disrupt people's daily activities and lives and make them disoriented. When the agricultural activities, fish and forest resource collections are disrupted, people's food and shelter come under stress. Again, day after day people in the disaster affected areas depend on relief fund (Mahato 2020). The paper stated that poor transportation, poor infrastructure, lack of knowledge, low expansion of education and low income create a situation of health issues. Also, the situation becomes difficult during emergency and pregnancy (De 2014; Ghosh et al. 2016).

Due to the climate change the Sundarban faces several challenges. With rising sea levels, islands are disappearing and the increasing salinity in the water and soil has severely threatened the health of mangrove forests and the quality of soil and crops. Also, there have been serious disturbances to hydrological parameters and change in fishing patterns resulting in disastrous consequences for fishermen. Their mental and physical both health are disturbed due to such calamities. Though several studies have been made before, the dangerous fishing profession related to their livelihood has not been touched properly. So, we focused on this mainly. To address the livelihood vulnerability of fishing population our study has been done.

11.2 Study Area Characteristics

Sundarban is the world's largest contiguous mangrove forest area in the delta, formed at the confluence of the Ganges, Padma and Brahmaputra rivers, also a world heritage site (Fig. 11.1). It lies between 21°30'N to 23° N latitude and 88°E to 89°E longitudes. We visited eighth April, 2022 to Sundarban with a group of 16 pupils and came back on tenth April, 2022. The costal part of west Bengal is having two distinct micro situations i.e. mainland and island. A dynamic ecosystem, the Sundarban delta is still in the process of development by the accretion of new land through the tidal action. A close network of rivers, channels and creeks, it comprises almost 106 islands subjected to changing salinity, soil texture and tidal action, as well as ecological factors arising out of increasing biotic pressure in the western and northern fringes of the forests.

11.3 Materials and Methods

Here we calculate livelihood vulnerability index with the help of exposure, sensitivity and adaptive capacity. Some components are considered in these variables.

11.3.1 Study Procedure

We randomly selected five fishing villages named Dulki, Barmanpara, Pathorpara, Tripligheri and Anpur for analysing the overall scenario of vulnerability related to livelihood. The latitude and longitude of our study area is 2205′10" N to 88,035′08″ E. This study is a quantitative study based on 365 households. This cross-sectional study investigated the "Vulnerability of Fishing Community." Investigators visited at the household level and randomly selected 365 households for this study. Data was collected by face to face interview by a pre-structured survey schedule (Table 11.1).

11.3.2 Data and Variables

We collected some environmental exposure (cyclones between 2009–2021, temperature, salinity, PH, rainfall, sea level rise and number of mangroves). These all data are derived from secondary sources. Some sensitivity components like No. of migrant

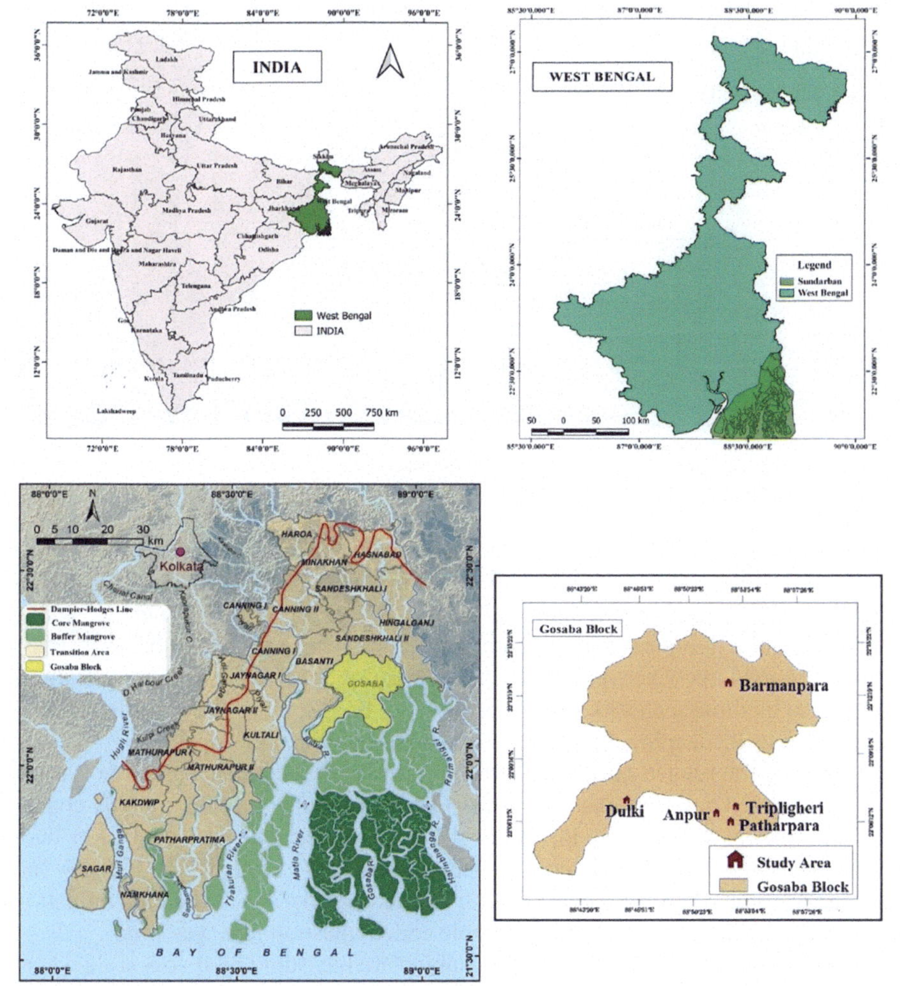

Fig. 11.1 Location map of the study area in Indian Sundarbans

Table 11.1 Collection of basic data about respondents

Respondents	Number of respondents
Total male	320
Total female	45
Total literate	266
Total illiterate	99
Fishing community	247
Non-fishing community	118

Fig. 11.2 Detailed satellite image of five selected villages

families are known by asking two questions like place of last residence and year of coming. Toilet facilities are known by asking five options (Fig. 11.2) like open pit, simple pit, latrine, septic tank, no latrine. Facility of drinking water is known by four options like personal tap, community tap, tube well, aila tube well. Risky fishing grounds like river, ponds, sea, estuaries are considered. Financial status is considered by three scales like good, moderate or bad. Problem faced during fishing, predators attack, sleeping duration are measured by direct question.

Mental health related question like fishing is risker than other profession is asked by double ended question like 'yes', 'no'. Fishing period worries are mainly considered the worries during fishing. Physical health related data like pain, skin burn, stomach problem and fever are collected this data by 1 to 5 scale, where 1 signify 'not at all' and 5 signify 'severe'. Social health related data like any social discrimination, caste related issues are collected by double ended question like 'yes', 'no'.

Health behaviour (different smoking habits) are measured by 0 –o 3 scale where, 0 signify 'never' and 3 signify 'daily'. Medical facilities (no of medical centres and shops) are measured by asking direct question like 'how many medical shops are there in market?' or giving multiple options of medical centres like PHC, govt. hospital, nursing home, homoeopathy, private clinic for this study. Government facilities (swasthasathiprakalpa, death benefits, govt. fund after disasters) are measured by asking "have you received any govt. /Non-govt. fund?"

11.3.3 Vulnerability Calculation

Vulnerability related with livelihood is mainly analysed through vulnerability index. Environmental exposure, sensitivity and adaptive capacity are considered to prepare this vulnerability index. First of all, we have to calculate weight value and the normalised value of the components.

11.3.3.1 Weight Value

weight value is calculated on the basis of total value of each village. At first calculate the percentage value of each component then calculate weight value by putting decimal two places before the percentage value. It is the easiest way to calculate weight value. For example, we have to know that in Sundarban 79% people are affected by cyclone, so the weight value is 0.79 (Appendix).

11.3.3.2 Normalized Value

We calculate normalised value of each village by using this formula.

(X-Xmin)÷ (Xmax-Xmin) where, X = component value of this village.

Xmin = lowest value of these components among five villages.

Xmax = highest value of these components among five villages.

(Go through Appendix)

11.3.3.3 Index Value

LVI = Livelihood vulnerability index, (LVI = EI + SI-ACI).

EI, SI, ACI = $\sum_{x=1}^{n}$ Wi × Yi wnhere, EI = Exposure index,

SI = Sensitivity index,

ACI = Adaptive Capacity index,

Wi = Weight value of "i"th components,

Yi = Normalised value of "i"th components of each villages.

The value of LVI lies between 0–1. Where,

0 = No. vulnerability or safe area.

<0.35 = Low vulnerable area.

0.36–0.7 = Moderately vulnerable area.

0.71–1 = Highly or extremely vulnerable.

11.4 Results and Discussion

11.4.1 Descriptive

In Sundarban we surveyed 365 samples from five villages (Fig. 11.3). Among them 87.67% people are men and 12.33% are female and most of them are in fishing profession i.e. 67.67% people are in fishing community and left of them i.e. 32.33% is in other profession (Fig. 11.4).

11.4.2 Vulnerability Analysis

By calculating the Normalized value of different components of five villages in Indian Sundarban, it is concluded that—.

11.4.2.1 Exposure Analysis

The intensity of cyclone is very high in overall Sundarban. But different villages faced different problems due to cyclone like increase of salinity in fresh water, loss of agricultural land, lack of drinking water etc. From the data we say that Anpur is the highest affected village as the houses of the village are almost 'kuchha', where in Tripligheri the effect of cyclone is comparatively low because the river embankment is made of concrete (Table 11.2).

11.4.2.2 Sensitivity Analysis

Toilet facility is in very bad condition at Dulki and Anpur villages (Fig. 11.5). They are used to in open space, most of the cases they are not preferring proper toile facility. In the case of drinking water Pathorpara village has great water trouble as all the villagers are mainly depend on only one tube well. As the less fishing communities are lived in Dulki and Anpur villages, fishing related problems and risks are seen very less. In the other side Pathorpara faces all types of fishing related risks like Predators attacks, sleep problems, physical and mental health related issues etc. The people in Sundarban are very hard working. Over hard works create back and joint pains sometimes fever also. Drink Saline water creates the stomach issues.

11.4.2.3 Adaptive Capacity Analysis

Medical facilities are very poor in all these villages. Nearest hospital is Gosaba Govt. Hospital. If any emergency arise it is difficult to reach in hospital within time

Fig. 11.3 Primary data collection through survey schedule

because of poor transport and communication systems. There are few homoeopathy and quack doctors in the village of Barmanpara but in case of emergency they also faced problems. People at Pathorpara village get the highest Govt. supports related to fishing where people at Anpur get the highest disaster related Govt. and non-Govt. funds. The education level is higher in Pathorpara, Dulki and Barmanpara villages. The future generations in all five villages are trying to move other works from fishing

Fig. 11.4 Socio-economic scenario of people in Indian Sundarban

Table 11.2 List of natural hazards and their impacts on livelihood

Name of cyclones	Year	Impacts on livelihood
Jawad	2021	Lands became saline, destroy mangroves, agriculture hampered
Yaas		Water became saline, death of fish, water pollution, destroy houses, many people died
Amphan	2020	Destroy mangroves, agriculture hampered, water became saline, death of fish
Bulbul	2019	Destroy mangroves, agriculture hampered
Fani		Destroy mangroves
Titli	2018	Destroy mangroves
Hudhud	2014	Destroy mangroves
Phailin	2013	Destroy mangroves
Aila	2009	Water became saline, death of fish, water pollution, destroy houses, many people died

11.4.3 Different Index Values of Five Villages

All the people in Sundarban region are in vulnerable position. In Pathorpara we can see that the value of livelihood vulnerability index is maximum as they are engaged in fishing, crab collection, shrimp collection. So, they spent more time on estuary and river and as a result their life risk is increased due to predators' attack, uncertain cyclone or any other natural calamity. Male community are more vulnerable that's why some villages in Sundarban are considered as widow villages. People in Sundarban have no option rather than fishing because due to cyclone most of the agricultural land is getting saline and as a result agricultural productivity is hampered. On the other hand, people are less educated and they don't have more job opportunities rather than fishing. But now a day's male population are interested to go to the south India as an agricultural labour and on the other hand present generation are not so much interested in fishing and their parents did not want that their children choose their profession due to the extreme life risk. This scenario is seen in Anpur village where livelihood vulnerability is comparatively low but all the five villages are in extremely or highly vulnerable condition (Table 11.3).

11.5 Future Directions of the Present Study

This study focuses on the effect of climate change and mental health of the fishing community. We are able to know how climate are changing and what the consequences of it is. In future we have to control climate change and take all the precautions. And mostly we have to do more work on the mental health of the people on

Fig. 11.5 Some general views of study area

Sundarban. The work that we have done already regarding this, is definitely set up a goal.

Table 11.3 Calculated index values of the villages in Sundarban

Index values	Dulki	Barmanpara	Pathorpara	Tripligherei	Anpur
$EI=EI^{n}_{x=1}(Wi\times Yi)$	0.4338	0.5690	0.5389	0.0030	0.7930
$SI=SI^{n}_{x=1}(Wi\times Yi)$	3.4821	3.4590	3.7236	2.6606	2.5349
$ACI=ACI^{n}_{x=1}(Wi\times Yi)$	3.0668	3.2048	3.3599	1.8634	2.5469
LVI= EI+SI-ACI	0.8491	0.8231	0.9027	0.8002	0.7811

11.6 Conclusion

Livelihood is the sum of nature, culture, occupation and different activities. It influences health very much. Sundarban, the delta of Ganges River, is compacted with natural resources. Most of the people are joined with primary and secondary activities in this region. Almost 73% people are engaged with fishing. Almost half of the total houses are poor quality and kaccha with very small space.In the case of sanitation, simple pits and sometimes open systems are situated. Their drinking water i.e. ground water is also saline.

They collect fish individually, sometimes with groups and many co-operatives. 90% fish are caught for selling. But most of the people are fishing with life risk without any security. Due to low price of fish, they belong to low income level. Naturally low standard of living situated children's malnutrition and women's nutritional issues. Different health issues are situated due to poverty. In the case of emergency and pregnancy they have to go in Gosaba which is very difficult especially in night because ofpoor infrastructure and transport. Hardly one to two quack doctors and medical stores present in some villages. Sometimes creating the perceptions of good health, people ignored their poor physical conditions.Cyclone is another common problem for the villagers in Sundarban now a day as the frequency of cyclones is increased day by day. Break of embankments can create flood situation and make them homeless. Aila and present Yash are created very bad conditions. After Yash, the agricultural lands and ponds water are become saline (till now). Agriculture and fish farming in ponds are disrupted and greater than 59% people are migrated to southern states.

Appendix

See Tables 11.4 and 11.5.

Table 11.4 Calculated weight values of different components

Sl.No.	Dimensions	Components	Sub-components	Weight value (W_i)
1	Environmental exposure		Severe effects of cyclone	0.79
			Increased temperature (0C)	0.0002
			Increased salinity (psu)	0.0005
			Decreased rainfall (mm)	0.05
			Sea level rise (mm)	0.02
			Decrease of mangrove (sq. Km.)	0.01
2	Sensitivity		No. of migrants' family	0.05
			Bad toilet facility	0.7
			Bad facility of drinking water	0.39
			Risky fishing grounds	0.41
			Financial status in this profession	0.06
			Different kind of problems during fishing	0.33
			Predator attack	0.4
			Perfect sleep time(hours)	0.04
		Livelihood responsible for physical health	Pain	0.8
			Skin burn	0.06
			Stomach problem	0.19
			Fever	0.02
			Any social discrimination	0.001
			Caste related issues	0
		Mental Health	Fishing is risker than others profession	0.86
			Fishing period worries	0.75
		Health behaviour	Smoking	0.5
			Tobacco	0.09
			Alcohol	0.35
			Drugs	0.005

(continued)

Table 11.4 (continued)

Sl.No.	Dimensions	Components	Sub-components	Weight value (W_i)
3	Adaptive capacity		Pucca house	0.21
			Dependency ratio	0.02
			Use of LPG as cooking fuel	0.3
		Medical facility	Primary health centre	0.003
			Governmental hospital	0.0006
			Nursing home	0.002
			Homeopathy centre	0.01
			Private clinic	0.02
			No. of medical shops	0.07
			Medical insurance	0.047
			Adaptation process	0.115
			NGO	0.2
			Taken loan	0.2
			Relationship with community	0.9
			Go to the Southern part of India and others (Out migration)	0.6
			Education	0.72
		Govt. help	Swasthasathi Prokalpo	0.90
			Govt. supports	0.22
			Received govt. or non-govt. funds after disasters	0.66

References

Abdullah HM, Ahmed SM, Khan BM, Mohana NT, Ahamed T, Islam I (2021) Agriculture and fisheries production in a regional blending and dynamic fresh and saline water systems in the coastal area of Bangladesh. Environ ChallS 4:100089

Allison EH, Perry AL, Badjeck MC, Neil Adger W, Brown K, Conway D, Halls AS, Pilling GM, Reynolds JD, Andrew NL, Dulvy NK (2009) Vulnerability of national economies to the impacts of climate change on fisheries. Fish Fish 10(2):173–196

Bhuyan S, Islam S (2016) Present status of socio-economic conditions of the fishing community of the Meghna River adjacent to Narsingdi district. Bangladesh J Fish Livest Prod 4(4):1–5

Chandra, G., Sagar, R. (2003) Fisheries in Sundarbans: problems and prospects. Available at SSRN 2084014

Table 11.5 Calculated normalised values of different components of five villages

Components	Sub-components	Normalized value (Yi)				
		Dulki	Barmanpara	Pathorpara	Tripligheri	Anpur
	Severe effects of cyclone	0.5453	0.7164	0.6784	0.0000	1.0000
	Increased temperature (^{0}C)	0.0002	0.0002	0.0002	0.0002	0.0002
	Increased salinity (psu)	0.0005	0.0005	0.0005	0.0005	0.0005
	Decreased rainfall (mm)	0.0500	0.0500	0.0500	0.0500	0.0500
	Sea level rise (mm)	0.0200	0.0200	0.0200	0.0200	0.0200
	Decrease of mangrove (sq. Km.)	0.0100	0.0100	0.0100	0.0100	0.0100
	No. of migrants' family	0.0889	0.1566	0	0.1820	1.0000
	Bad toilet facility	1.0000	0.7432	0.5	0.7125	1.0000
	Bad facility of drinking water	0.4867	0.5048	1.0	0.5030	0.0000
	Risky fishing grounds	0.0000	0.6714	0.5	0.0508	1.0000
	Financial status in this profession	0.4821	1.0000	0.9	0.3264	1.0000
	Different kind of problems during fishing	0.9486	1.0000	0.7304	0.5114	0.0000
	Predator attack	0.3386	0.5679	1.0	0.0511	0.0000
	Perfect sleep time(hours)	0.1502	0.1917	0.0	0.1502	1.0000
Livelihood responsible for physical health	Pain	0.0000	0.1389	1.0	0.6122	0.7992
	Skin burn	0.3615	1.0000	0.7	0.2448	0.0000
	Stomach problem	0.5511	0.5005	0.0	0.2857	1.0000
	Fever	0.8264	1.0000	0.6	0.6714	0.0000
	Any social discrimination	0.0000	1.0000	0	0.0000	0.5832
	Caste related issues	0.0000	0.0000	0.0	0.0000	0.3610
Mental health	Fishing is risker than others profession	1.0000	0.5987	0.2	0.5273	0.0000
	Fishing period worries	1.0000	0.6413	0.9	0.6015	0.0000

(continued)

Table 11.5 (continued)

Components	Sub-components	Normalized value (Yi)				
		Dulki	Barmanpara	Pathorpara	Tripligheri	Anpur
Health behaviour	Smoking	0.2318	1.0000	0.6	0.4720	0.0000
	Tobacco	1.0000	0.4610	0.6	0.0911	1.0000
	Alcohol	0.4018	0.0268	0.2731	0.0000	1.0000
	Drugs	1.0000	0.0000	0.0	0.0000	1.0000
	Pucca house	0.7368	0.5341	1.0	0.4812	0.0000
	Dependency ratio	0.0000	0.2300	0.6	0.7266	1.0000
	Use of LPG as cooking fuel	0.6255	0.6200	0.6827	0.0000	0.2143
Medical Facility	Primary health centre	0.5000	0.5000	1	0.5000	0.0000
	Governmental hospital	0.4500	0.0000	0.4	0.2000	1.0000
	Nursing home	0.3600	0.0000	0.3	0.4000	1.0000
	Homeopathy centre	0.5000	1.0000	0.5	1.0000	0.0000
	Private clinic	0.3333	1.0000	1	0.6667	0.0000
	No. of medical SHOPS	0.8000	0.7000	0.8	0.8000	0.0000
	Medical insurance	1.0000	0.6915	0.5	0.0000	0.0000
	Adaptation process	1.0000	0.3670	0.4	0.0000	0.0000
	NGO	0.5132	1.0000	0.6	0.2527	0.0000
	Taken loan	0.4471	0.7729	1.0	0.7985	0.0000
	Relationship with community	0.5362	0.7393	0.7	0.0000	1.0000
	Go to the Southern part of India and others(Out migration)	1.0000	0.7520	0.6	0.5461	0.0000
	Education	0.9934	0.7421	1.0	0.0000	0.0000
Govt. help	Swasthasathi Prakalpo	0.0000	0.1406	0.6	0.9226	1.0000
	Govt. supports	0.2769	0.7385	1.0	0.1846	0.0000
	Received govt. or non-govt. funds after disasters	0.6689	0.6854	0	0.3897	1.0000

Chen Q, Su H, Yu X, Hu Q (2020) Livelihood vulnerability of marine fishermen to multi-stresses under the vessel buyback and fishermen transfer programs in China: the case of Zhoushan City, Zhejiang Province. Int J Environ Res Public Health 17(3):765

Colburn LL, Jepson M, Weng C, Seara T, Weiss J, Hare JA (2016) Indicators of climate change and social vulnerability in fishing dependent communities along the eastern and gulf coasts of the United States. Mar Policy 74:323–333

Das CS (2018) Pattern and characterisation of human casualties in Sundarban by Tiger attacks. India Sustain For 1(1)

De D (2014) Spatial inequality in health care infrastructure in sundarban, West Bengal, India. Int Res J Soc Sci:2319–3565

EnamulHaque AK, Mukhopadhyay P, Nepal M, Shammin MR (2022) Climate change and community resilience: insights from South Asia. p. 460 Springer Nature

Eriksson B, Johansson F, Blicharska M (2019) Socio-economic impacts of marine conservation efforts in three Indonesian fishing communities. Mar Policy 103:59–67

Ghosh U, Bose S, Bramhachari R, Mandal S (2016) Expressing collective voices on children's health: photovoice exploration with mothers of young children from the Indian Sundarbans. BMC Health Serv Res 16(7):119–130

Guha I, Roy C (2016) Climate change, migration and food security: evidence from Indian Sundarbans. Int J Theor Appl Sci 8(2):45–49

Jakariya M, Rahman A, Sayem SM, Saad S, Alam MS, Sarker SR, Akter D (2020) Development of livelihood vulnerability index for the coastal fishermen communities of Bangladesh using spatial information technique. Groundw Sustain Dev 11:100475

Kanjilal, B., Mazumdar, P. G., Mukherjee, M., Mondal, S., Barman, D., Singh, S., Mandal, A. (2010) Health care in the Sundarbans (India): challenges and plan for a better future

King T, Kilpatrick S, Willis K (2014) Staying healthy: industry organisations' influence on behaviours and services used by fishers. Deakin University

King T, Kilpatrick S, Willis K, Speldewinde C (2015) "A different kettle of fish": mental health strategies for Australian fishers, and farmers. Mar Policy 60:134–140

Liebman AK, Wiggins MF, Fraser C, Levin J, Sidebottom J, Arcury TA (2013) Occupational health policy and immigrant workers in the agriculture, forestry, and fishing sector. Am J Ind Med 56(8):975–984

MahadeviaGhimire K, Vikas M (2012) Climate change–impact on the Sundarbans, a case study. Int Sci J: Environ Sci 2(1):7–15

Mahato J (2020) A case study on: the health condition and the way of living in the rural areas of the Sundarbans south 24 Parganas. West Bengal

Mandal UK, Maji B, Mullick S, Nayak DB, Mahanta KK, Raut S (2019) Global climate change and human interferences as risk factors, and their impacts on geomorphological features as well as on farming practices in Sundarbans eco-region. In: The sundarbans: a disaster-prone eco-region: increasing livelihood security, pp 405–437

Mondal BK (2022) Climate change induced coastal hazards and community vulnerability in indiansundarban. In: Applied geomorphology and contemporary issues. Springer International Publishing, Cham, pp 587–609

Patel V, Rajagopalan R (2009) Fishing community issues in the Sundarban Tiger Reserve: a case study

Pena PGL, Gomez CM (2014) Health of subsistence fishermen and challenges for occupational health surveillance. Ciência SaúdeColetiva 19:4689–4698

Pramanik M, Szabo S, Pal I, Udmale P, Pongsiri M, Chilton S (2022) Population health risks in multi-hazard environments: action needed in the cyclone Amphan and COVID-19–hit Sundarbans region. India Clim Dev 14(2):99–104

Sadekin MN, Ali J, Islam R (2018) Livelihood vulnerability index: an application to assess the climatic vulnerability status of inland small scale fishing livelihood. Int J Sustain Dev 21(1–4):75–101

Sarkar B, Islam A (2021) Assessing poverty and livelihood vulnerability of the fishing communities in the context of pollution of the Churni River. India Environ Sci Pollut Res:1–24
Senapati S, Gupta V (2017) Socio-economic vulnerability due to climate change: deriving indicators for fishing communities in Mumbai. Mar Policy 76:90–97
Tripathi P, Kamath R, Tiwari R (2017) Occupational health and role of gender: a study in informal sector fisheries of udupi, India. Indian J Occup Environ Med 21(2):45
Umamaheswari T, Sugumar G, Krishnan P, Ananthan PS, Anand A, Jeevamani JJJ, Rao CS (2021) Vulnerability assessment of coastal fishing communities for building resilience and adaptation: evidences from Tamil Nadu, India. Environ Sci Pol 123:114–130

Chapter 12
Riparian Degradation and Anthropogenic Pressure on Obibia Stream: A Decadal Remote Sensing and GIS Investigation

Akanwa Angela Oyilieze, Ndidi Iloduba, A. U. Okonkwo, Kate Azuka Omenugha, O. P. Munana Gilbert, Isaac Mubasankwaya, Anukwonke Charles, and Okoli Onyinye Nkemdilim

12.1 Introduction

Surface water bodies such as rivers and streams play a critical role in supporting ecological balance, human health, agriculture, and socio-economic development, especially in developing countries. However, the sustainability of these resources is increasingly threatened by anthropogenic pressures, including unregulated land use, urban expansion, waste disposal, and unsustainable agricultural practices (UN-Water 2020). In Nigeria, the degradation of freshwater systems has reached an alarming rate, with multiple studies highlighting significant deterioration in water quality across urban and peri-urban environments (Odipe et al. 2018; Akanwa et al. 2024).

Obibia Stream, located in Nibo, Awka South Local Government Area of Anambra State, is one such surface water body experiencing rapid environmental decline.

A. A. Oyilieze (✉)
Faculty of Environmental Management, Agriculture and Renewable Energy (FEMARE), University of Technology and Arts, Byumba (UTAB), Rwanda & Department of Environmental Management, Chukwuemeka Odumegwu Ojukwu University, Uli, Atughobi, Nigeria
e-mail: angy.akanwa1@gmail.com

N. Iloduba
Department of Paediatrics (Child Health), Federal Medical Centre, Federal Capital Territory, Abuja (FCT), Nigeria. Consultant Paediatric Infectious Disease, Abuja, Nigeria

A. U. Okonkwo · A. Charles
Department of Environmental Management, Faculty of Environmental Sciences, Chukwuemeka Odumegwu Ojukwu University, Uli, Atughobi, Nigeria

K. A. Omenugha
Office of the Vice Chancellor, Chukwuemeka Odumegwu Ojukwu University, Igbariam, Nigeria

A. Nath et al. (eds.), *Geospatial Application of Coastal Hazards*, Springer Water,
https://doi.org/10.1007/978-3-032-20076-1_12

Historically, the stream has served as a key water source for local domestic, agricultural, and recreational activities. However, over the past decade, increased human activities have reportedly impacted the stream's quality, raising concerns about the implications for both environmental health and community well-being (Ezeabasili & Okonkwo 2021).

Globally, surface water degradation is often linked to changes in land use and land cover (LULC), which influence sediment load, pollutant concentrations, and aquatic biodiversity (Miller et al. 2019). Remote Sensing (RS) and Geographic Information Systems (GIS) have emerged as powerful tools for monitoring these spatiotemporal changes, allowing researchers to correlate human-induced landscape changes with hydrological impacts over time (Lillesand et al. 2015a, b; Abba & Mustapha 2022). In Nigeria, the use of RS and GIS in water quality studies has grown, necessitating the need for localized, stream-based assessments that integrate both geospatial and laboratory data.

The urbanization trend in southeastern Nigeria, particularly in Anambra State, has intensified the strain on natural water resources. According to Okonkwo et al. (2022), the proliferation of unplanned settlements, alongside weak enforcement of environmental regulations, has contributed to stream pollution, leading to both chemical and biological contamination. Furthermore, the National Environmental Standards and Regulations Enforcement Agency (NESREA) has identified non-compliance with water quality standards in many rural and peri-urban water sources, emphasizing the need for localized monitoring and control (NESREA 2021).

Previous studies have predominantly focused on larger water bodies such as the River Niger or major tributaries, often overlooking smaller but equally significant streams like Obibia. Yet, these smaller water systems are crucial, particularly in semi-urban communities, where they serve as the primary or supplementary water sources. In this context, there is a need for a comprehensive, long-term analysis that not only assesses water quality but also investigates the broader environmental dynamics influencing it, such as land cover transitions, population growth, and the intensification of land use.

This study, therefore, seeks to bridge this gap by conducting a spatiotemporal assessment of human activities and water quality degradation in Obibia Stream between 2011 and 2021. By integrating multi-year LANDSAT imagery, GIS analysis, and laboratory testing of key water quality parameters, including pH, turbidity, dissolved solids, heavy metals, and coliform content, this research aims to provide a

O. P. Munana Gilbert
Office of the Vice Chancellor, Department of Religious Studies, University of Technology and Arts, Byumba (UTAB), KibaliByumba, Rwanda

I. Mubasankwaya
Faculty of Environmental Management, Agriculture and Renewable Energy (FEMARE), University of Technology and Arts, Byumba (UTAB), KibaliByumba, Rwanda

O. O. Nkemdilim
Faculty of Environmental Sciences, Department of Environmental Management, Chukwuemeka Odumegwu Ojukwu University (COOU), Uli, Anambra, Nigeria

holistic understanding of the extent and nature of environmental degradation in the study area. Additionally, socio-environmental data obtained through interviews and field observations will enrich the scientific findings with community-level insights.

The study is timely and significant for several reasons. First, it contributes to the growing body of knowledge on environmental change and water resource management in Nigeria. Second, it provides empirical data to support local advocacy for better environmental practices and policymaking. Third, the methodological approach, employing RS-GIS, survey techniques, and laboratory analysis, demonstrates a replicable model for similar assessments across other vulnerable water systems in sub-Saharan Africa.

Ultimately, the degradation of Obibia Stream is not merely an ecological concern but also a socio-economic and public health issue. The decline in water quality can affect agricultural productivity, increase waterborne disease risks, and undermine local livelihoods, especially for women and children who are disproportionately dependent on natural water sources (UNESCO 2020). By understanding the trajectory of these changes and their drivers, this chapter hopes to inform more sustainable interventions that protect both people and ecosystems in the region.

12.2 Understanding Human Impact on Stream Water Quality: A DPSIR Model (Driving Forces–Pressure–State–Impact–Response) and Landscape Ecology Approach to Obibia Stream, Nibo

In the face of escalating environmental degradation across developing nations, the quality and sustainability of local water bodies have come under increasing threat due to unregulated human activities. Streams, which serve as vital water sources for domestic, agricultural, and ecological functions, are particularly vulnerable. Obibia Stream, located in Nibo, Anambra State, Nigeria, exemplifies this challenge; therefore, to systematically understand and evaluate these changes, this study employs a hybrid conceptual framework that combines the DPSIR (Driving Forces–Pressure–State–Impact–Response) model and principles from Landscape Ecology.

The DPSIR framework, originally developed by the European Environment Agency, is widely used in environmental studies to describe cause-and-effect relationships between society and the environment. The model provides a structured approach to identifying how human-driven Driving Forces, such as population growth, increased land demand, and economic survival strategies, generate pressures on the environment. In most rivers and streams, pressures manifest in various forms, including indiscriminate waste disposal, deforestation, stream encroachment, among others. These pressures directly influence the State of the stream's water quality, resulting in observable and measurable changes in water quality (NESREA 2021).

As the State of the stream degrades, the impacts extend beyond environmental consequences to affect human health, agricultural productivity, and social well-being. Polluted water poses serious risks for waterborne diseases and undermines livelihoods dependent on clean water for irrigation, fishing, and domestic use. The Response element of the DPSIR framework calls attention to actions needed to mitigate these impacts, such as policy interventions, environmental education, and improved land use regulation. However, responses are often delayed or inadequate where local governance structures may lack the capacity for effective enforcement (Okonkwo et al. 2022).

While DPSIR excels at mapping the causal chain of environmental degradation, it does not sufficiently address the spatial and temporal dimensions of these processes. This is where Landscape Ecology becomes indispensable. As a sub-discipline of ecology, Landscape Ecology focuses on the patterns, processes, and interactions across heterogeneous landscapes, often influenced by both natural forces and human activities (Turner et al. 2001). It provides the theoretical basis for analysing satellite imagery and GIS data to detect land use and land cover (LULC) changes over time.

Landscape Ecology introduces a spatial perspective that complements DPSIR's linear cause-and-effect logic. For example, the transformation of vegetative cover into impervious surfaces such as roads and buildings alters runoff patterns, increases sediment load, and disrupts the natural hydrological cycle. According to Forman (2014), such landscape fragmentation reduces the buffering capacity of riparian zones and heightens vulnerability to erosion and pollution. The patch–matrix–corridor model, a core principle in landscape ecology, is particularly relevant in understanding how isolated human land uses (farms and settlements) within the matrix of a natural ecosystem (the stream corridor) affect ecological connectivity and function (Wu and Hobbs 2007).

By integrating DPSIR and Landscape Ecology, this study adopts a holistic lens for understanding human–environment interactions. DPSIR explains why and how human pressures degrade stream ecosystems, while Landscape Ecology shows where and to what extent these changes have occurred. Together, they facilitate a comprehensive diagnosis of the environmental challenges facing Obibia Stream. This dual framework is not only academically sound but also practically valuable for informing sustainable water resource management and environmental policy.

Furthermore, the integrated framework aligns with global best practices on water governance and sustainable development. It reinforces the principles of Integrated Water Resource Management (IWRM), which advocates for the coordinated development and management of water, land, and related resources (UNESCO 2020). For communities like Nibo, this means that solutions must be both spatially informed and socially inclusive, acknowledging the interplay of human livelihoods, land use patterns, and water quality outcomes.

In conclusion, the hybrid DPSIR, Landscape Ecology framework offers a robust conceptual foundation for assessing the spatiotemporal effects of human activities on Obibia Stream. It enables a deeper understanding of the complex interlinkages between socio-economic drivers, land cover transformations, and ecological degradation. This comprehensive approach enhances the relevance of the study for academic research, environmental planning, and local policy interventions.

12.3 Study Area

Nibo is a rapidly urbanizing town located in Awka South Local Government Area (LGA) of Anambra State, Southeastern Nigeria (Figs. 12.1, 12.2). Geographically situated between latitude 6°11′N and longitude 7°04′E. According to the Nigerian Population Commission (NPC 2006), the population of Nibo was 26,663, as cited by Onyido et al. (2016) in their study. Using a compound annual growth rate of 3.2%, the projected population of Nibo in 2024 is approximately 47,005. Nibo is part of the humid tropical rainforest zone, characterized by high rainfall (1,800–2,400 mm annually), moderate temperatures, and fertile soils conducive to agricultural activities (Onwuka and Egboka 2020). It is bounded by the capital city, Awka, to the north, and by other fast-developing towns like Umuawulu and Mbaukwu. Its proximity to the capital city, Awka, has made it a prime zone for urban expansion, transforming it from a traditionally agrarian community to a mixed urban–rural setting. This has resulted in increased residential expansion, agricultural pressure, and infrastructural development (Okonkwo et al. 2022).

At the heart of this transformation lies Obibia Stream, a natural surface water body originating from the highlands of Nibo and flowing southeastward through the community into the Idemili River basin. The stream has historically served as a critical water source for domestic use, irrigation, fishing, and recreation. Its ecological and economic value to the residents is significant, especially in areas lacking pipe-borne water (Ezeabasili and Okonkwo 2021).

Obibia Stream thus provides a strategic site for assessing the spatiotemporal impact of anthropogenic activities on water resources in peri-urban Nigeria. Its location within a fast-growing town like Nibo makes it relevant for understanding land use, water quality relationships, and for designing sustainable urban environmental policies. The stream's ecological degradation is not only impacting water quality but also affecting local livelihoods and ecosystem services. This makes Obibia Stream a critical focus for environmental assessment and sustainable water resource management in the region (NESREA 2021).

12.4 Methodology

This study adopted a mixed-methods research design that combined quantitative and qualitative approaches, incorporating field-based surveys, laboratory analysis, and geospatial techniques. The design enabled a comprehensive investigation of the physical, chemical, and biological quality of Obibia Stream while assessing the

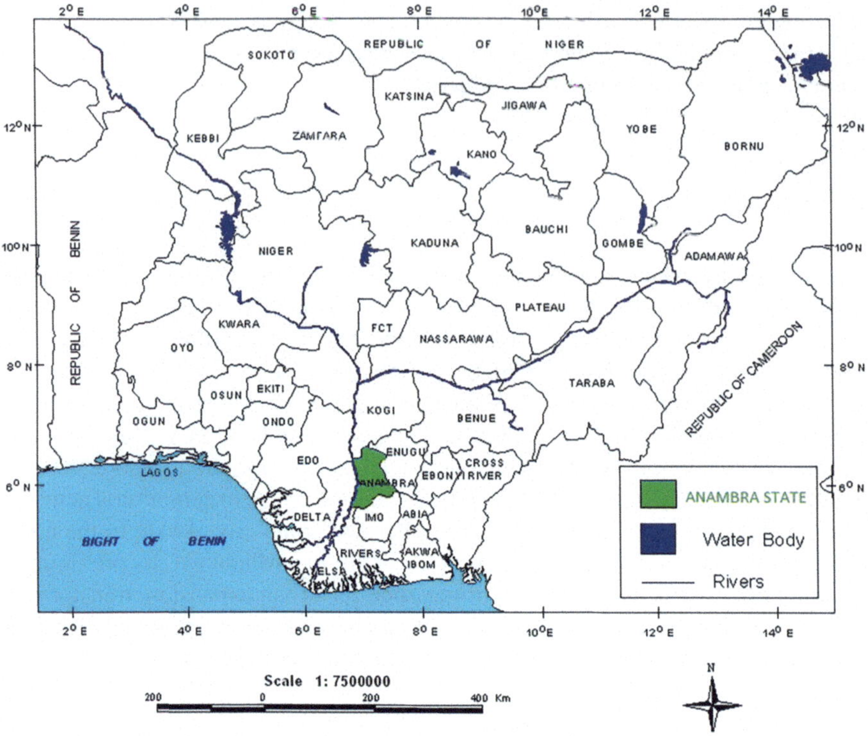

Fig. 12.1 Map of Nigeria showing Anambra

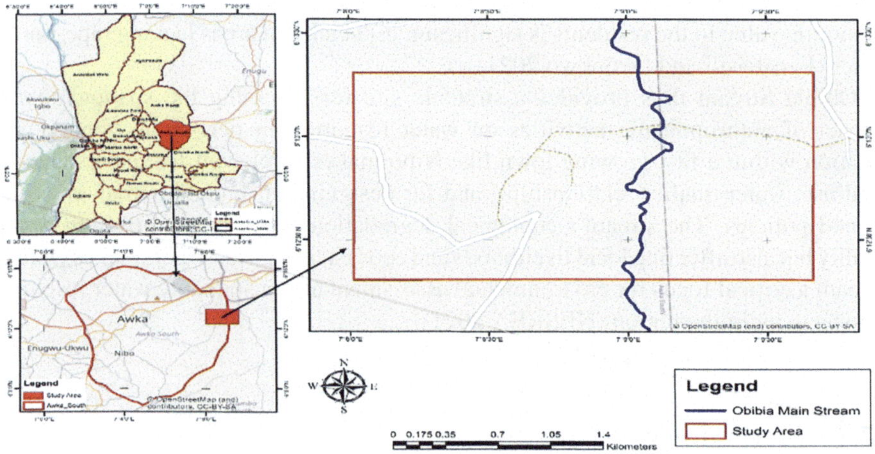

Fig. 12.2 Map of Anambra state showing the study area-Awka south LGA and obibia stream

effects of surrounding human activities and land cover changes from 2011 to 2021. The focal water body, Obibia Stream, flows through residential, agricultural, and commercial zones, making it an ideal case for assessing anthropogenic impacts on stream water quality.

12.5 Data Collection Methods

12.5.1 Field Observation and Interviews

Primary data were obtained through direct observation, photography, and semi-structured interviews of fifty (50) participants with residents, women leaders, farmers, and key informants, including the former Chairman of Obu-na-asato and also the Chairman of Umunzekwe Royal Family in Nibo, selected using purposive and snowball sampling techniques. These methods helped identify major human activities affecting the stream and gathered perceptions on water quality degradation (Creswell and Plano Clark 2018).

12.5.2 Water Sampling and Laboratory Analysis

Water samples were collected at three (3) (lower, middle, and upper) points along Obibia Stream. Sampling was done using standard procedures outlined by the American Public Health Association (APHA 2017). Parameters analyzed included:

Physico-chemical parameters: pH, temperature, turbidity, electrical conductivity, total dissolved solids, and odour. Nutrients and metals: Nitrate, biochemical oxygen demand (BOD), lead, zinc, iron, copper, arsenic, cadmium, and chromium. Microbial indicators: Total coliform and fecal coliform counts. Analyses were conducted in a certified environmental laboratory to determine compliance with NESREA water quality standards (NESREA 2021). A one-sample t-test was applied to assess the statistical significance of the differences between observed values and regulatory thresholds.

12.5.3 Remote Sensing and GIS Analysis

To evaluate land use and land cover (LULC) changes, LANDSAT 7& 8 satellite imagery for 2011 and 2021 was purposely acquired from the United States Geological Survey (USGS) database. Image processing was done using QGIS and ERDAS Imagine, following a supervised classification using the ArcGIS 10.3 software's supervised classification method based on the maximum likelihood algorithm. The

LULC classes identified included built-up areas, vegetation, bare surface, and water bodies (Lillesand et al. 2015a, b). Change detection analysis was used to quantify the spatial and temporal transformation of the stream's watershed.

12.5.4 Data Analysis

Laboratory results were statistically compared against NESREA standards using t-tests, while GIS-based LULC change maps were interpreted visually and quantitatively. Qualitative data from interviews were transcribed, coded, and thematically analyzed to contextualize community-level experiences of water resource degradation (Braun and Clarke 2006).

12.5.5 Ethical Considerations

Ethical clearance was obtained from the university's Research Ethics Committee. Participants provided informed consent, and confidentiality was maintained throughout the research.

12.6 Results and Discussion

The findings of this study highlight the multiple effects of human activities on the ecological and hydrological conditions of Obibia Stream in Nibo, located in the Awka South Local Government Area of Anambra State, Nigeria. Evidence collected from laboratory water analysis, community interviews, and geospatial assessments demonstrates a complex relationship between population growth, land use changes, and water quality degradation. According to the community interviews (Plate 12.1), as shown in Table 12.1, of the 50 interviewees, 19 (38%) were male and 31 (62%) were female.

Table 12.2 shows that individuals with a B.Sc or HND make up the largest group of interviewees, totaling 23 respondents, which accounts for 46% of the participants. This is followed by those holding a WAEC certificate, who represent 34% (17 interviewees). The smallest group consists of individuals with no formal education, comprising only 2% (1 respondent).

Table 12.3 displays the occupational backgrounds of the interviewees. Both civil servants and students comprised the largest groups, each making up 26% of the total respondents. Traders and Farmers each represented 18% of the interviewees. Individuals in the 'other' category made up the smallest group, accounting for only 12% of the respondents.

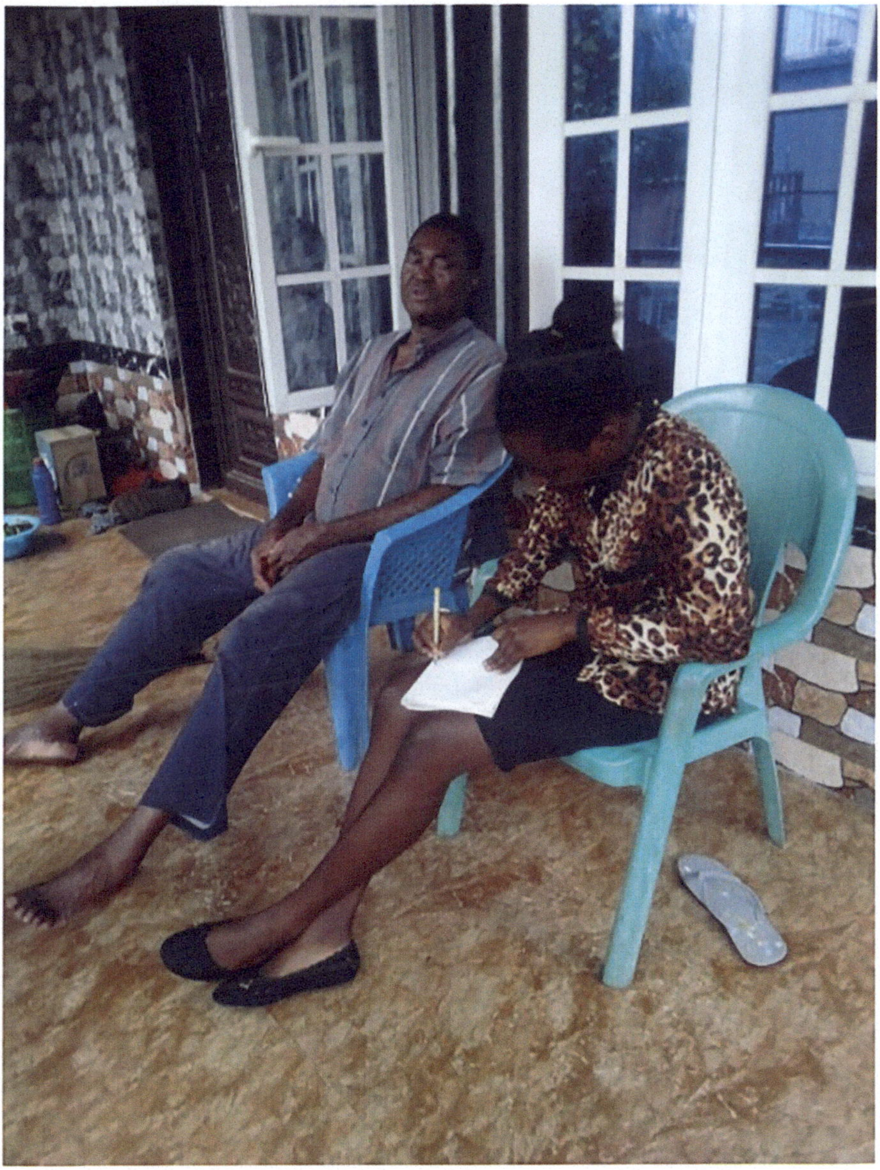

Plate 12.1 Interviewing Key Informants at the study area. *Source* Compiled by authors, 2024

Plate 12.1 (continued)

Table 12.1 Gender distribution of interviewees

Male	Female	Total
19	31	50
38%	62%	100%

Source Compiled by authors, 2024

Table 12.2 Educational background of interviewees

Educational background	Frequency	Percentage (%)
MSc/PhD	4	8
BSc/HND	23	46
NCE/OND	5	10
WAEC	17	34
No formal education	1	2
Total	50	100

Source Compiled by authors, 2024

Table 12.3 Occupational distribution of interviewees

Occupation	Frequency	Percentage (%)
Civil servant	13	26
Trader	9	18
Farmer	9	18
Student	13	26
Others	6	12
Total	50	100

Source Compiled by authors, 2024

Table 12.4 Uses of obibia stream

Uses	Frequency	Percentage (%)
Domestic	33	66
Drinking	–	–
Agriculture	16	32
Others	1	2
Total	50	100

Source Compiled by authors, 2024

Table 12.4 indicates uses of Obibia stream where agricultural activities represent 33(66%), domestic purposes 16 (32%), and others 1(2%). Data from interviews indicated that the stream was not used for drinking purposes.

12.6.1 Unregulated Uses and Lack of Management Practices in Obibia Stream

Table 12.1 shows that out of 50 individuals interviewed about the Obibia Stream, 31 (62%) were female and 19 (38%) male, indicating that women were more available and willing to participate during data collection (Plate 12.2). This pattern is common in environmental field studies in rural and peri-urban Nigerian communities, where women often have more direct contact with natural water sources for domestic and livelihood needs (Adams et al. 2018). Additionally, these women form the majority of the well-educated group, as seen in Table 12.2, with 46% (23 individuals) holding a B.Sc./HND qualification, followed by 34% (17 individuals) with a WAEC (secondary school) qualification. Only 2% (1 individual) have no formal education. This gender-based pattern of availability, education, and ecological interaction highlights the importance of including women's perspectives in water governance and strategy design and implementation. Ignoring these insights risks marginalizing the group most impacted by changes in water quality and access. The significant presence of B.Sc./HND holders is also notable, as individuals in this group may possess technical knowledge or occupational roles. The occupational distribution shown in Table 12.3 indicates that civil servants and students each make up 26% of the interviewees, followed by traders (18%) and farmers (18%), with those in the "other" category comprising only 12% of the total sample. Farmers, in particular, are key stakeholders in land and water resource use in rural areas.

This may reflect a demographic shift at Nibo, where farmers' participation in environmental assessments often stems from competing livelihood pressures, time constraints, or lower levels of formal education (Ajaero and Anorue 2018). Additionally, this distribution suggests a primarily urban or semi-urban respondent base, where formal employment and academic pursuits are more common than traditional agricultural roles. The dominance of civil servants and students indicates a population more engaged with discussions on water sustainability and environmental governance, and their involvement in the study enhances the depth of the findings (Olanrewaju et al. 2019).

The predominance of domestic use, accounting for 66% of the activities listed in Table 12.4, reinforces the stream's role as a vital household water source in communities with limited access to piped or treated water. Its central role is evident in water-related activities carried out by women, such as washing, cooking, small-scale farming, and water fetching (See Plate 12.2), which makes them more attuned to the ecological condition of streams like Obibia (Akanwa et al. 2024). As primary users of household water, women are often more accessible during daytime survey hours and tend to be more responsive to discussions about environmental degradation and resource sustainability. One of the key informants, a woman leader, inform that,

We depend on Obibia Stream for subsistence and commercial livelihoods such as domestic activities like washing, bathing, and cooking, particularly in households without access to piped water. Others use it for small-scale irrigation, fishing, and sand mining.

These findings reflect the broader pattern in semi-urban and rural Nigerian communities where streams function as multi-use water sources due to inadequate infrastructure (Ofoegbu et al. 2016). However, such usage often leads to conflicting demands on water quality and quantity. For instance, while some participants described the stream as a critical source for irrigation and small-scale vegetable farming, others complained about increasing pollution from detergents, waste disposal, and mining runoff. One of the male informants revealed that.

Plate 12.2 Ongoing domestic activities by women and a rice farmland at Obibia stream

Obibia stream has cultural and spiritual value, we consider it a site of traditional rituals and communal identity.

This dimension echoes findings by who noted that in southeastern Nigeria, water bodies often carry spiritual or ancestral significance, influencing attitudes toward conservation (Iwuagwu and Oyelaran-Oyeyinka 2019; Akanwa et al. 2024).

However, agricultural use of Obibia Stream accounts for 32%, serving as a water source for farming activities like small-scale irrigation, which supports household food security and subsistence farming (Nwafor et al. 2021). Streams such as Obibia supply accessible, low-cost water for farming, especially in areas lacking formal irrigation infrastructure. Nonetheless, this use also renders the stream vulnerable to runoff from fertilizers and pesticides, further impacting water quality (Ardón et al.

Plate 12.2 (continued)

2020). While only 2% mentioned other uses, data from interviews revealed that no respondent used the stream for drinking, highlighting a perceived or known concern about its water quality, indicating community awareness of contamination risks, and aligning with WHO water safety standards that advise against consuming untreated surface water (WHO 2017).

Key informants indicated that 90% of the study area relies on boreholes as its primary source of drinking water. Additionally, it was observed that water vendors are rarely seen in the area, suggesting a relatively stable, self-managed water supply system. The high dependence on boreholes aligns with broader trends in peri-urban and rural Nigeria, where communities turn to groundwater sources due to inadequate municipal water infrastructure and declining trust in the quality of surface water (Ocheri et al. 2014). Boreholes are perceived as cleaner, more reliable, and safer, especially in areas where surface water, like Obibia Stream, is exposed to pollution from unregulated human activities such as sand mining, agriculture, and domestic waste (Akanwa 2021).

The rarity of water vendors in the community may be due to the accessibility of boreholes within residential compounds, reducing the need for third-party water supply. This also suggests that, while Obibia Stream is important for non-drinking uses such as farming and washing, it no longer serves as a primary drinking water source, likely due to perceived or actual contamination. This highlights the importance of water quality monitoring and environmental conservation of Obibia Stream to protect ecological health and ensure its long-term usability. It also emphasizes the critical role of groundwater in meeting the water needs of semi-urban communities, raising concerns about over-extraction and the need for integrated water resource management (IWMI 2020).

Further findings indicate that fishing activities are a significant livelihood practice in Obibia Stream, with 29 respondents (58%) confirming the presence of fishing, while 21 respondents (42%) reported otherwise. This suggests that over half of the local population participates in or observes subsistence or small-scale fishing in the stream, indicating that the water body still supports some level of aquatic biodiversity and provides an economic function for the community. The prevalence of fishing highlights Obibia Stream's role as not just a physical resource but also a livelihood asset for households in Nibo. According to Ovie and Belal (2012), small inland water bodies like streams and creeks across Nigeria provide critical food and income sources, especially for communities lacking access to formal employment. This aligns with FAO (2020), which notes that rural fisheries, even when informal, contribute significantly to household nutrition and resilience in the face of poverty.

However, the fact that 42% of respondents do not recognize fishing as a current use may suggest a decline in fish availability, possibly due to habitat degradation by human activities such as car wash, sand mining, domestic activities, clothing, and bitter leaf washing. This reflects trends in similar regions where human-induced pressures reduce stream depth and clarity, negatively impacting aquatic life (Ezeaku et al. 2021). Interview data further revealed that a significant number of community members confirmed the use of chemicals for fishing in the Obibia Stream. This practice poses serious environmental and public health risks. Chemical fishing, often involving substances like gamalin 20, pesticides, or detergents, is commonly employed for its quick yield, but it causes irreversible damage to aquatic ecosystems (Adeyemo et al. 2018).

One of the major dangers is the indiscriminate killing of both target and non-target species, including fish eggs and juvenile aquatic organisms, thereby disrupting the reproductive cycles and biodiversity of the stream (Ojutiku et al. 2009). Over time, this results in reduced fish populations, degradation of aquatic habitats, and long-term economic losses for communities that rely on fishing for sustenance and livelihood.

In addition, chemical residues from these substances often remain in the water, contaminating downstream ecosystems and posing risks to both human and animal health. Prolonged exposure can lead to bioaccumulation of toxins in fish tissues, rendering them unsafe for human consumption (Chinonso et al. 2020). The use of chemicals also contributes to water pollution, altering pH levels, increasing turbidity, and compromising water quality for domestic and agricultural use. According to FAO (2020), such practices violate sustainable fisheries guidelines and require urgent regulatory and educational interventions. These findings call for community sensitization, enforcement of environmental laws, and the promotion of eco-friendly fishing methods to safeguard Obibia Stream's ecological and economic function.

Notably, from observations and interviews, there is a severe absence of stream management practices in the Obibia Stream ecosystem. A striking 82% (41 out of 50) of respondents affirmed that no structured management or regulation exists, while only 10% indicated otherwise, and the remaining were unaware of any formal management presence. This demonstrates a critical governance gap in the environmental oversight of Obibia Stream.

This lack of institutional involvement aligns with broader concerns about weak enforcement of environmental laws and neglect of smaller water bodies in Nigeria's decentralized water governance system. Streams such as Obibia, though vital for agriculture, domestic activities, and biodiversity, often fall outside the purview of federal or state water management authorities due to limited resources, fragmented policies, or lack of political interest (Irechukwu 2021).

The absence of oversight has allowed unregulated agricultural practices, chemical fishing, deforestation, and waste dumping, all of which continue to degrade the stream's health. According to Ademiluyi (2020), without community-based or government-led conservation measures, streams in urbanizing areas are highly susceptible to ecological collapse. This finding underscores the urgent need for a multi-stakeholder stream management framework, including local councils, traditional institutions, and environmental agencies. It also emphasizes the importance of public engagement to mobilize local conservation efforts and push for government accountability (UNEP, 2021).

12.6.2 Environmental/Public Health and Planetary Implications of Obibia Stream at Nibo

Table 12.5 presents the findings from the analysis of three (3) water samples taken from the Obibia Stream (lower, middle, upper). In total, 17 physico-chemical and biological parameters were examined.

Table 12.5 Results of physico-chemical parameters of sampled obibia stream and control

Parameters	Point A	Point B	Point C
pH	6.00	6.07	6.71
Temperature ^{0}C	28.50	29.00	28.00
Turbidity NTU	2.00	1.50	1.90
Electrical conductivity (usm/cm)	390.40	215	135
Total dissolved solid mg/l	119	32.00	39
Dissolved oxygen mg/l	47.4	52.90	40.80
Nitrate mg/l	7.493	4.898	4.289
Biochemical oxygen demand mg/l	154.60	142.40	118.4
Lead ppm	0.074	0.034	0.049
Zinc ppm	0.678	0.675	0.897
Iron ppm	1.074	0.937	0.674
Copper ppm	0.080	0.056	0.044
Arsenic ppm	0.00	0.006	0.00
Cadmium ppm	0.078	0.089	0.082
Chromium ppm	0.047	0.022	0.033
Total coliform cfu/100ml	$3.00*10^{1}$	$1.40*10^{1}$	$1.2*10^{1}$
Faecal coliform cfu/100ml	4.00	2.00	2.00

Source Compiled by authors, (2024)

Shows the percentage levels of samples from the Obibia stream and their compliance with NESERA standards. Findings reveal that pH, DO, BOD, Pb, Fe, Cu, Cd, Total coliform, and Faecal Coliform were above the standard, while Temperature, Turbidity, Electrical conductivity, TDS, Nitrate, As, and Cr are within the acceptable limits.

The physico-chemical and biological analysis of the Obibia Stream reveals significant stream degradation that poses direct risks to public health and broader consequences for the environment. Elevated levels of heavy metals, organic pollution, and pathogenic contamination necessitate urgent attention from both environmental health and water governance authorities. The results of the water quality assessment, as shown in Table 12.6, were compared to the benchmarks set by the National Environmental Standards and Regulations Enforcement Agency (NESREA) to determine compliance. The findings indicate a mix of acceptable and significantly noncompliant parameters, highlighting both natural variations and serious anthropogenic contamination in the Obibia Stream.

Several heavy metals exceeded NESREA's permissible limits, including lead (0.05 mg/L), which is approaching the 0.1 mg/L limit, and zinc (0.75 mg/L) and iron (0.89 mg/L), both significantly above their respective limits of 0.2 mg/L and 0.5 mg/L. Copper (0.06 mg/L) and cadmium (0.08 mg/L) were also found to be

Table 12.6 Water sample percentage compliance with NESREA standard

Parameters of water samples	Mean ± SE at 95% CL	NESREA standards
pH	6.26 ± 0.23	6.5–8.5
Temperature	28.50 ± 0.29	Ambient
Turbidity	1.80 ± 0.15	10
Electrical conductivity	246 ± 75.42	500
Total dissolved oxygen	63.33 ± 27.91	180
Dissolved oxygen	47.03 ± 3.50	0.2
Nitrate	5.56 ± 0.98	9.10
Biological oxygen demand	138.46 ± 10.63	6.0
Lead	0.05 ± 0.01	0.1
Zinc	0.75 ± 0.74	0.2
Iron	0.89 ± 0.18	0.5
Copper	0.06 ± 0.01	0.01
Arsenic	0.00 ± 0.00	0.05
Cadmium	0.08 ± 0.00	0.01
Chromium	0.03 ± 0.00	0.5
Total coliform	18.67 ± 5.69	0.00
Faecal coliform	2.67 ± 0.67s	0.00

Source Compiled by authors, (2024)

substantially higher than the permissible limit of 0.01 mg/L. This indicates potential contamination from industrial or domestic sources, likely due to metallic waste, batteries, or chemical residues (Akan et al. 2010). Conversely, chromium (0.03 mg/L) and arsenic (0.00 mg/L) were within safe limits, showing no immediate concern.

The elevated concentrations of these heavy metals are associated with long-term health risks, such as neurological disorders from lead, kidney damage from cadmium, and gastrointestinal issues from copper, particularly when these metals bioaccumulate in aquatic organisms consumed by humans (Olatunji and Osibanjo 2019). The most pressing threat stems from the high levels of heavy metals, particularly cadmium (0.08 mg/L), copper (0.06 mg/L), and iron (0.89 mg/L), all of which significantly exceed the NESREA permissible limits. Consumption of contaminated fish or crops irrigated with water from the stream can lead to chronic health effects. For instance, cadmium is a known carcinogen and may cause kidney failure, skeletal damage, and reproductive dysfunction (WHO 2017). Even low levels of lead can result in neurological impairments, particularly in children (Olatunji and Osibanjo 2019).

Excessive copper and iron levels can lead to gastrointestinal disorders and organ toxicity (Akan et al. 2010). The pH level of the stream is 6.26 ± 0.23, which falls below the NESREA standard range of 6.5 to 8.5, indicating slightly acidic conditions that may negatively affect aquatic life and suggest pollution from organic matter or industrial runoff (Efe 2015). The temperature of the stream is 28.5 °C ± 0.29, and the turbidity is 1.80 NTU ± 0.15, both of which comply with expected ambient conditions and NESREA limits, suggesting limited physical disruption at the time of sampling. Electrical conductivity measures 246 μS/cm ± 75.42, and total dissolved solids (TDS) are at 63.33 mg/L ± 27.91, which are also within acceptable ranges, indicating relatively low mineralization and dissolved salt content (Asonye et al. 2007). Nitrate concentrations are at 5.56 mg/L ± 0.98, which is below the NESREA limit of 9.1 mg/L, although ongoing agricultural runoff may pose a long-term threat if not managed properly.

Dissolved oxygen (DO) levels were 47.03 mg/L, and the biochemical oxygen demand (BOD) was 138.46 mg/L, significantly exceeding acceptable thresholds. This indicates extremely high organic pollution in the stream. The elevated BOD suggests substantial microbial activity breaking down organic matter, potentially from sewage, domestic waste, or decaying vegetation (WHO 2017). These figures far exceed NESREA's standards of 0.2 mg/L for DO and 6.0 mg/L for BOD, reflecting critical ecological stress in the stream. The extremely high biochemical oxygen demand (BOD) of 138.46 mg/L and reduced levels of dissolved oxygen indicate severe microbial activity, which can promote the growth of anaerobic bacteria and further degrade water quality. This creates a toxic aquatic environment that negatively impacts food chains and diminishes access to safe water (UNEP 2021).

Significant quantities of total coliform (18.67 CFU/100 mL) and fecal coliform (2.67 CFU/100 mL) were detected, violating NESREA's standard of 0.00 CFU/100 mL. This suggests fecal contamination, likely stemming from open defecation, domestic sewage discharge, or animal waste, posing serious risks of waterborne diseases such as cholera and typhoid (Okoye et al. 2011). Moreover, the presence of total and fecal coliforms indicates contamination from human or animal feces, posing risks of waterborne diseases such as cholera, typhoid, hepatitis, and dysentery, particularly for communities that may use the stream for domestic or recreational purposes (Okoye et al. 2011). These findings at Obibia Stream indicates significant environmental degradation, posing direct public health risks and wider planetary consequences. The presence of elevated heavy metals, organic pollution, and pathogenic contamination demands urgent attention from both environmental health and water governance perspectives. These high metal concentrations are linked to long-term health risks, such as neurological disorders (lead), kidney damage (cadmium), and gastrointestinal issues (copper), especially when these metals bioaccumulate in aquatic organisms consumed by humans (Olatunji and Osibanjo 2019).

12.6.3 Implications of Land Use/Cover Changes (2011–2021) of Obibia Stream and Its Environs

Landsat 7 and 8 Images was used to retrieve the land use and land cover map of Obibia and environs to show the effects of human activities overtime from 2011-2021 as shown in Figs. 12.3, 12.4 as well as Tables 12.7 and 12.8. Table 12.7 show that the Built-up areas covered approximately 61.47Ha of land which translate to 10.8% of land. Water body covered 86.85Ha which is 15.3% of the total land. Vegetation covered 273.42Ha of land which translate to 48.0% of the land. Bare surface 147.42Ha of the land which translates to 25.9% of the total land for 2011.

Table 12.9 reveal data for the land use/cover for 2021, where the Built-up areas covered approximately 108.54Ha of land which translate to 19.07% of land. Water body covered 38.25Ha which is 6.72% of the total land. Vegetation covered 316.26Ha of land which translate to 55.57% of the land. Bare surface 106.11Ha of the land which translates to 18.64% of the total land.

Table 12.9 indicates the changes from 2011 to 2021 on the land use/cover assessment for Obibia Stream. It reveals significant landscape transformations between 2011 and 2021: Bare Surface decreased by 7.26 percentage points (pp), from 25.9% to 18.6% (−2.8 pp/year), suggesting soil stabilization or vegetation encroachment on previously denuded areas. Built-up Area surged by 8.3 pp (from 10.8% to 19.1%, ≈ + 0.83 pp/year), indicating rapid urbanization and infrastructure development.

Vegetation, specifically in farming, increased by 7.6% points, rising from 48.0% to 55.6% (approximately +0.76 percentage points per year). This growth reflects agricultural intensification. In contrast, the area covered by water bodies declined sharply by 8.6 percentage points, dropping from 15.3% to 6.7% (approximately − 0.86 percentage points per year), indicating a significant loss of aquatic habitats. This trend aligns with findings from various regions in Nigeria, where urban growth

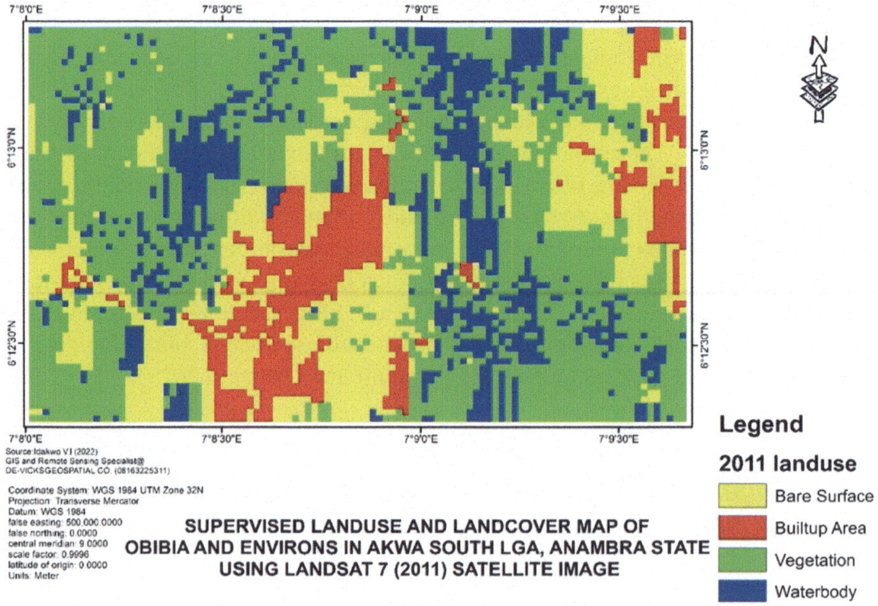

Fig. 12.3 Map showing Obibia stream. *Source* Compiled by authors, (2024)

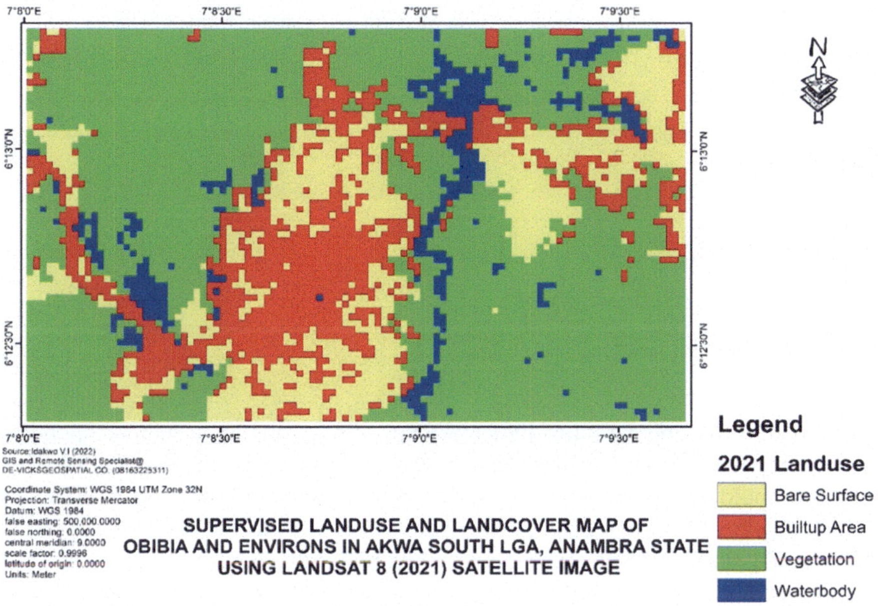

Fig. 12.4 Map showing Obibia stream. *Source* Compiled by authors, (2024)

Table 12.7 Changes in the 2011 land use/cover of obibia and environs

Name	Area	Percentage
Bare surface	147.42	25.9
Built-up area	61.47	10.8
Vegetation	273.42	48.0
Water body	86.85	15.3

Table 12.8 Changes in the 2021 land use/cover of obiba and environs

Name	Area	Percentage
Bare surface	106.11	18.64
Builtup area	108.54	19.07
Vegetation	316.26	55.57
Waterbody	38.25	6.72

often occurs at the expense of surface water and open land. Meanwhile, agriculture is expanding due to population pressure and land conversion policies (Jimoh et al. 2023; Alegbeleye et al. 2025). The drastic decline in water body coverage is a concern; similar studies have shown that such losses can lead to reduced ecosystem services, increased flood risks, diminished water availability, and loss of biodiversity (Alegbeleye et al. 2025; FAO 2024).

Table 12.9 Percentage change of landuse/cover Map of Obibia stream from 2011–2021

Name	2011		2021		2021–2011	Average yearly change
	Area	Percentage	Area	Percentage		Percentage
Bare surface	147.42	25.9	106.11	18.64	-7.26	-2.8
Built-up area	61.47	10.8	108.54	19.07	8.27	7.7
Vegetation (farming activities)	273.42	48	316.26	55.57	7.57	1.6
Water body	86.85	15.3	38.25	6.72	-8.58	-5.6

The rise in farming activities indicates expanding cropping systems responding to the expansion of rural settlements. However, without proper regulatory oversight, this could exacerbate erosion and disrupt hydrological flows, especially along stream corridors (Adebayo et al. 2024). Additionally, the increase in built-up areas, which nearly doubled over a decade, is consistent with rapid peri-urban development patterns observed in Nigerian riverine zones like Ogun and Lagos States. Here, urban expansion is driving changes in the local climate (e.g., the urban heat island effect), land stability, and natural water dynamics (Adedeji et al. 2020; Abimbola et al. 2025). Built environments often replace permeable land, disrupting natural recharge processes and increasing runoff into streams such as Obibia.

Additionally, the findings from unregulated usage of Obibia stream and the laboratory analysis reflect patterns of localized environmental degradation that may contribute cumulatively to global planetary crises. The degradation of Obibia Stream signifies a loss of essential ecosystem services, such as water purification, nutrient cycling, and biodiversity support, which are vital for ecological stability (Rockström et al. 2009). The accumulation of toxic metals and organic pollutants also threatens aquatic biodiversity, altering stream chemistry and rendering habitats uninhabitable for sensitive species. This has implications for local food webs and regional biodiversity, a critical concern under the planetary health framework (Whitmee et al. 2015a, b). Furthermore, unregulated uses and land-use changes that contribute to these pollution loads exacerbate risks such as erosion, flooding, and water scarcity, reinforcing environmental feedback loops that impair both human and planetary well-being.

Overall, the land-use trends around Obibia Stream reflect a widespread transformation of landscapes driven by urbanization, agricultural intensification, and watercourse modification (Alegbeleye et al. 2025). The findings underscore the urgent need for integrated land-use planning, conservation of aquatic habitats, restoration of riparian buffers, and protection of ecosystem services linked to small streams, which play crucial roles in community resilience, flood mitigation, and biodiversity support.

12.7 Policy Implications

The findings of this study reveal urgent environmental, public health, and governance challenges with direct implications for policy and sustainable resource management. The absence of formal management practices around Obibia Stream as reported by 82% of respondents indicates a governance vacuum. There is a pressing need to integrate small streams into state-level environmental protection frameworks, ensuring that agencies like the Anambra State Ministry of Environment and NESREA extend regulatory coverage to peri-urban water bodies (ESMP 2018). The establishment of community-based watershed committees could serve as an effective local mechanism for monitoring and enforcement.

The elevated concentrations of biological oxygen demand (BOD), heavy metals (such as cadmium, copper, and iron), and coliform bacteria exceed safe limits and suggest unregulated anthropogenic pressures. Policies should mandate routine water quality assessments, sanitation infrastructure improvements, and the prohibition of chemical fishing practices, backed by fines and sensitization campaigns. Partnerships with universities and NGOs could support data generation and public engagement (ANSEWCCA 2025). The geospatial analysis revealed a 56% reduction in water body area and a dramatic increase in built-up land, indicating unregulated urban encroachment (Naji et al. 2024). Zoning laws must be reviewed to enforce setbacks from riverbanks, protect floodplains, and prioritize green infrastructure. The government should also adopt land-use planning tools such as Environmental Impact Assessments (EIAs) and Remote Sensing & GIS-based mapping to support evidence-driven decision-making. While the community demonstrated moderate awareness of environmental risks, the findings suggest a limited understanding of the long-term impacts of their actions. Integrating climate literacy into local schools, town halls, and cooperative societies will empower residents to participate in protecting the stream. Policies that encourage youth and women's inclusion in environmental stewardship can foster sustainability from the grassroots (Phung Thuy Phuong. 2007).

12.8 Conclusion

Finding from this chapter provided a comprehensive evaluation of the environmental degradation, public health risks, and governance deficits surrounding Obibia Stream in Nibo, Awka South LGA, Anambra State. Using a combination of physico-chemical analysis, qualitative interviews, and remote sensing techniques, the research uncovered a critical absence of stream management, with little institutional oversight. Severe pollution, marked by excessive BOD levels, fecal contamination, and heavy metal presence, far above NESREA standards. Rapid land use transformation, with substantial loss of waterbody area and expansion of built-up zones and agricultural land. This reveals strong evidence of anthropogenic influence, especially intensified agricultural activities, use of chemical for fishing, and lack of stream management, as confirmed by community responses. These multidimensional pressures underscore the need for urgent and coordinated policy action. Without intervention, Obibia Stream will continue to decline, posing escalating threats to community health, biodiversity, and climate resilience.

Findings contributes to growing evidence that small urban and peri-urban streams in Nigeria require equally robust protection as larger rivers. It calls for a multi-stakeholder approach that combines scientific monitoring, grassroots engagement, and enforceable policy instruments to ensure the sustainable management of critical water resources like Obibia Stream. Urgent intervention is needed through community-based water monitoring, sanitation improvement, and enforcement of water quality regulations to safeguard public health and restore ecological balance.

References

Abba SI, Mustapha A (2022) Application of GIS and remote sensing for water quality assessment: A review. Environ Sci Pollut Res 29:15189–15204

Abimbola OJ, Adewumi T, Abubakar M (2025) Dynamics of urban heat island in Lafia, Nasarawa State of Nigeria: A remote sensing analysis of land surface temperature, urban development and vegetation change. Urban Climate, Advance online

Adams EA, Boateng GO, Amoyaw JA (2018) Thirsty slums in African cities: Household water insecurity in urban informal settlements of Lilongwe. Malawi. Am J Hum Biol 30(6):e23123. https://doi.org/10.1002/ajhb.23123

Adebayo OH, Adeofun CO, Tope-Ajayi OO, Ogunkola MO (2020) Spatio-temporal analysis of urban sprawl and land use/land cover changes in a suburb of Lagos and Ogun metropolises, Nigeria (1986–2014). IFE J Sci 22(1):27–42

Adedeji OH, Adeofun CO, Tope-Ajayi OO, Ogunkola MO (2020) Spatio-temporal analysis of urban sprawl and land use/land cover changes in a suburb of Lagos and Ogun metropolises, Nigeria (1986–2014). IFE J Sci 22(1):27–42

Adekola O, Olokeogun O, Akinluyi F (2020) Urban growth and water quality challenges in Nigeria. Hydrol Earth Syst Sci 24(3):1459–1471

Ademiluyi IA, Awaworyi Churchill S, Matthew O (2020) Watershed land use, surface water vulnerability and public health risks in urban rivers: Evidence from Ado Ekiti, Nigeria. SN Applied Sciences, 2, Article 1788. https://doi.org/10.1007/s42452-020-03572-7

Adeyemi OK, Ajani EK, Adedeji OB (2018) Impact of pesticide use on aquatic environments in developing countries: A review. Environ Monit Assess 190(3):1–14. https://doi.org/10.1007/s10661-018-6506-6

Ajaero ID, Anorue LI (2018) Newspaper framing and climate change mitigation in Nigeria and Ghana. Afr Popul Stud 32(2):4228–4238

Akan JC, Abdulrahman FI, Dimari GA, Ogugbuaja VO (2010) Physicochemical determination of pollutants in wastewater and vegetable samples along the Jakara Wastewater Channel in Kano Metropolis, Kano State. Nigeria. Eur J Sci Res 23(1):122–133

Akanwa AO, Joe-Ikechebelu N, Iloduba N, Okolo DN, Mmachukwu K (2024) Traditional gender roles of men and women in surface water management: a case study of Mimili Ezekoro, Anambra State, Nigeria. J Ecol Nat Resour (JENR) 8(3):1–14. https://doi.org/10.23880/jenr-16000393

Akanwa AO (2021) River sand mining and its ecological footprint at odor river, Nigeria. In: Banerjee A, Meena RS, Jhariya MK, Yadav DK. (eds) Agroecological Footprints Management for Sustainable Food System. Springer, Singapore. https://doi.org/10.1007/978-981-15-9496-0_16

Alegbeleye OM et al. (2025) Multidecadal land cover changes and impacts on ecosystem service values in Nigeria. Environ Res Commun, 7(1), Article 055033

American Public Health Association (APHA) (2017) Standard methods for the examination of water and wastewater, 23rd edn. APHA, Washington, DC

Anambra State Erosion, Watershed and Climate Change Agency (ANSEWCCA) (2025) Anambra State. https://environment.anambrastate.gov.ng/the-anambra-state-erosion-watershed-and-climate-change-agency-ansewcca/?utm_source=chatgpt.com

Ardón M, Belden LK, McCusker CG, Young M (2020) Fertilizer runoff in streams and rivers can have cascading effects, analysis shows. NCSU News

Asonye CC, Okolie NP, Okenwa EE, Iwuanyanwu UG (2007) Some physicochemical characteristics and heavy metal profiles of Nigerian rivers, streams and waterways: A review. Afr J Biotech 6(5):617–624

Braun V, Clarke V (2006) Using thematic analysis in psychology. Qual Res Psychol 3(2):77–101

Chinonso UC, Eze IC, Nwachukwu CC (2020) Health implications of consuming chemically harvested fish in rural Nigeria. J Environ Health Res 20(2):75–83

Creswell JW, Plano Clark VL (2018). Designing and conducting mixed methods research (3rd ed.). SAGE Publications

Efe SI (2015) Physico-chemical characteristics of selected river systems in southern Nigeria. J Environ Sci Water Resour 4(2):33–41

Environmental Social Management Plan (ESMP) (2018) Ministry of Environment, Beautification and Ecology, Awka—Anambra State. Federal High Court—Ekwueme Square Gully Erosion Site, Awka—Anambra State. https://documents1.worldbank.org/curated/en/405161522254792904/pdf/Final-ESMP-for-Federal-High-Court-Ekwueme-Square-Gully-Erosion-Site.pdf?utm_source=chatgpt.com

Ezeabasili ACC, Okonkwo C (2021) Pollution of stream water resources in peri-urban communities of Anambra State. J Environ Manag Saf 12(4):56–68

Ezeaku PI, Nnaji CC, Eboh EC (2021) Uncovering the livelihood–environment tradeoffs in informal sand mining in Southeast Nigeria. Environ Econ Policy Stud 23:431–449. https://doi.org/10.1007/s10018-021-00285-4

FAO (2020) Illegal, unreported and unregulated (IUU) fishing: Addressing the problem through education and enforcement. Food and Agriculture Organization, Rome. https://doi.org/10.4060/ca7854en

FAO & UNEP (2024) State of the world's water resources and ecosystem services. Rome: FAO; Nairobi: UNEP

Forman RTT (2014) Urban ecology: science of cities. Cambridge University Press

International Water Management Institute (IWMI). (2020) Groundwater and climate resilience in Sub-Saharan Africa. https://www.iwmi.cgiar.org/2020/08/groundwater-and-climate-resilience-in-sub-saharan-africa

Irechukwu C (2021) Challenges of integrated water resource management in Nigeria's coastal streams. Afr J Environ Manag 15(4):225–238. https://doi.org/10.5897/AJEM2021.0271

Iwuagwu CJ, Oyelaran-Oyeyinka B (2019) Traditional beliefs and water conservation practices in southeastern Nigeria. Afr J Environ Manag 27(2):101–112

Jimoh YO et al. (2023) Spatial analysis of land cover changes for detecting environmental degradation. Environ Monit Assess, 195, Article 14

Kristensen P (2004) The DPSIR framework. Eur Environ Agency

Leal Filho W, Shiel C, Paço A do, Mifsud M, Veiga Ávila L (2019) Reinvigorating the sustainable development research agenda: The role of students and local communities. Int J Sustain High Educ, 20(3), 473–490. https://doi.org/10.1108/IJSHE-02-2019-0064

Lillesand TM, Kiefer RW, Chipman JW (2015) Remote sensing and image interpretation (7th ed.). John Wiley & Sons

Lillesand T, Kiefer RW, Chipman J (2015) Remote sensing and image interpretation (7th ed.). John Wiley & Sons

Miller JD et al (2019) Land use and stream water quality in urban watersheds. Water Res 157:90–101

Naji TA, Teka MA, Alemu EA (2024) Enhancing watershed management in developing countries: insights from Ethiopia. Discov Sustain 5:131. https://doi.org/10.1007/s43621-024-00334-y
National Population Commission (2006) Nigeria population and housing census, population distribution by Sex, State, LGAs and Senatorial District: Priority Tables. Abuja: National Population Commission
NESREA (2021) National environmental (Surface and groundwater quality control) regulations. Abuja: Federal Government of Nigeria
Nwafor JC, Madu IA (2021) Assessing the socio-ecological effects of unregulated sand mining in Anambra State. J Environ Planning Manage 64(2):287–306
Ocheri MI, Eze UE, Yusuf AO (2014) Evaluation of groundwater quality in urban areas of North-Central Nigeria. Environ Manag Sustain Dev 3(2):142–158. https://doi.org/10.5296/emsd.v3i2.628
Odipe OE, Aladejana JA, Aderogba AA (2018) Anthropogenic influence on water quality in Nigeria: A review. Afr J Environ Sci Technol 12(2):58–68
Ofoegbu C, Nwankwoala HO, Ebe TE (2016) Assessment of household dependence on surface water in peri-urban Nigeria. Water Resour Manage 30(5):1621–1633. https://doi.org/10.1007/s11269-015-1227-8
Ojutiku RO, Ogunjobi OD, Bala U (2009) The effects of chemical fishing on aquatic resources in Nigeria. J Fish Aquat Sci 4(5):229–232. https://doi.org/10.3923/jfas.2009.229.232
Okonkwo CJ, Ude NM, Nwankwo C (2022) Urbanization and stream pollution in Southeastern Nigeria. Niger J Hydrol Environ 11(2):42–53
Okorie N, Akpan E (2020) Literacy levels and environmental information processing in Nigeria: Implications for sustainable development. J Commun Media Res 12(2):98–110
Okoye CO, Agwu E, Eze T (2011) Microbial and physicochemical analysis of drinking water sources in Awka, Nigeria. J Trop Agric, Food, Environ Ext 10(3):56–65
Olanrewaju RM, Adekola PO, Adedayo A (2019) Environmental education and sustainable community development in Nigeria. Int J Environ & Sci Educ 14(1):23–31
Olatunji OS, Osibanjo O (2019) Comparative assessment of heavy metals in drinking water sources in two semi-urban areas of Nigeria. Environ Monit Assess 191:206. https://doi.org/10.1007/s10661-019-7340-6
Onwuemesi FE, Onuba LN, Chiaghanam OI, Anudu GK, Akanwa AO (2013) Heavy metal accumulation in fish of ivo river, Ishiagu. Nigeria. Res J Environ Earth Sci, 5(4), 189-192
Onwuka SU, Egboka BCE (2020) Water resource challenges and urban development in Anambra State. Int J Hydrol Res 3(2):45–57
Onyido AE, Ezeani IU, Irikannu KC, Umeaneto CS, Chikezie OC, Ugha CC (2016) Anthropophilic mosquito species prevalence in Nibo community, Awka South LGA, Anambra State. J Vector Borne Dis 53(1):45–51
Ovie SI, Belal E (2012) Fisheries co-management in Nigeria: An analysis of the challenges. J Agric Soc Res 12(1):65–72
Phung, Thuy, Phuong (2007) Community involvement in urban watershed management: from the US pacific northwest to Ho Chi Minh City. Vietnam. J Environ & Dev 16(3):307–327. https://doi.org/10.1177/1070496507301063(Originalworkpublished2007)
Rockström J, Steffen W, Noone K, Persson Å, Chapin FS, Lambin EF, Foley JA (2009) A safe operating space for humanity. Nature 461(7263):472–475. https://doi.org/10.1038/461472a
Turner MG, Gardner RH, O'Neill RV (2001) Landscape ecology in theory and practice: pattern and process. Springer
UNESCO (2020) The united nations world water development report 2020: Water and Climate Change. UNESCO, Paris
United Nations Environment Programme (UNEP) (2021) Making peace with nature: A scientific blueprint to tackle the climate, biodiversity and pollution emergencies. Nairobi: UNEP
UN-Water (2020) Progress on integrated water resources management: global status report. Geneva: United Nations

Whitmee S, Haines A, Beyrer C, Boltz F, Capon AG, de Souza Dias BF, Yach D (2015a) Safeguarding human health in the Anthropocene epoch: Report of The Rockefeller Foundation-Lancet Commission on planetary health. The Lancet 386(10007):1973–2028. https://doi.org/10.1016/S0140-6736(15)60901-1
World Health Organization (WHO) (2017) Guidelines for drinking-water quality, 4th edn. WHO, Geneva
Wu J, Hobbs RJ (2007) Key topics in landscape ecology. Cambridge University Press

Chapter 13
Assessing Shoreline Change Dynamics in Diversified Coastal Environments Along the East Coast of India

Ranjan Kumar Behera, Soren Singh, Subhasis Pradhan, P. Thanabalan, Rabindra Kumar Sahoo, Pratap Kumar Mohanty, and M. V. Ramana Murthy

13.1 Introduction

Shoreline is considered a transitional boundary between the landmass and ocean, which supports various economic benefit and services for the livelihood of the coastal community (Camfield and Morang 1996). The coastal feature often transforms due to wave of chances in coastal environment. Shoreline change is a dynamic and complex process influenced by a combination of natural factors and human activities (Masselink and Hughes 2014). The impact of these factors may vary greatly depending on the type of coastline and the prevailing tidal regime. This complexity arises differently along the open, river-dominated, and anthropogenically modified coast subject to the prevalence of the micro- and meso-tidal conditions (Jana et al. 2017, 2016a, 2014). Generally, open coasts are exposed directly to oceanic waves and a high wind environment, which brings changes to shoreline configuration depending on the strength of met-oceanographic conditions, while these changes are relatively different along the coast occupied with coastal structure and river estuaries (Ranasinghe 2016). In micro-tidal regions, the natural sediment transport is quite minimal, and wave action and episodic storm events are the primary natural drivers for shoreline change (Zhang et al. 2024). However, human activities such as the construction of seawalls, ports, coastal development, and beach nourishment significantly disrupt these natural processes (Bulleri and Maura 2010). These disruptions alter the inherent sediment-carrying capacity of coastal water, resulting in a higher proportion

R. K. Behera · S. Singh · P. K. Mohanty
Department of Marine Sciences, Berhampur University, Ganjam, Odisha, India

S. Pradhan · P. Thanabalan · R. K. Sahoo (✉) · M. V. Ramana Murthy
National Centre for Coastal Research, NIOT Campus, Pallikaranai, Tamil Nādu, India
e-mail: sahoork@nccr.gov.in

A. Nath et al. (eds.), *Geospatial Application of Coastal Hazards*, Springer Water
https://doi.org/10.1007/978-3-032-20076-1_13

of human-induced shoreline change relative to natural drivers. In contrast, the meso-tidal coasts often experience more dynamic interactions between tidal currents and wave action, usually leading to more complex sediment redistribution patterns. While natural processes continue to dominate shoreline changes in coastal environments, the human interventions can also play a significant role in altering sediment transport and modifying coastal erosion/accretion trends, particularly when they interfere with natural sediment pathways (Hapke et al. 2013). River-dominated coasts respond variably to tidal ranges. In micro-tidal coastal settings, limited tidal energy leads to sediment deposition near river mouths, resulting in rapid morphological changes (Jana et al. 2012, 2016b). However, human activities in upstream domains such as dam construction, sand mining, and river channelization drastically reduce sediment supply and disrupt natural shoreline dynamics, accelerating erosion. In meso-tidal regions, the increased tidal energy interacts extensively with riverine sediment loads, contributing to more widespread sediment redistribution. Despite the dominance of natural processes in these environments, human interventions continue to have a significant influence on shoreline stability. Anthropogenically modified coasts, characterized by structures like groins, seawalls, and breakwaters, exhibit highly localized responses to human activities. These responses are particularly pronounced in micro-tidal zones where limited natural sediment transport makes the shoreline more sensitive to structural disruptions. In a meso-tidal coast, the larger tidal range may partially mitigate the impacts of coastal infrastructure by exchanging the sediment movement; however, the overall influence of human activities remains considerable. Given these interactions, the ratio of human to natural influence on shoreline changes varies widely depending on coastal typology and tidal conditions. This variability underscores the importance of developing coastal management strategies that are tailored to specific environmental settings and informed by comprehensive, site-specific studies. To assess these dynamics in the Indian context, the present study focuses on two geographically and morphologically diverse coastal regions: Gopalpur on the Odisha coast and Tuticorin on the Tamil Nadu coast. These sites represent contrasting tidal environmental regimes and are ideal for evaluating the relative impacts of natural processes and human interventions on shoreline change.

India's coastline, extending over 11,098 km, is ecologically diverse and essential for its dense populations, vital ecosystems, and key economic activities. Among the various coastal states, Odisha and Tamil Nadu, in particular, have been the focal points of shoreline change studies due to their vulnerability to natural hazards such as tropical cyclones, monsoonal wave activity, and sea-level rise, as well as human-induced pressures like port expansion and coastal engineering. Shoreline change analysis along the Tamil Nadu coast, particularly in the southern districts and Kaveri delta, has revealed extensive erosion trends (Ratheesh et al. 2023). Recent studies by Dwivedi et al. (2023) and Abijith et al. (2023) employed fuzzy AHP and integrated geospatial parameters, including land use/land cover (LULC), sea-level rise (SLR), significant wave height (SWH), and shoreline change (SLC), to assess coastal vulnerability. Tuticorin, located in southern Tamil Nadu, has undergone significant shoreline alterations due to longshore sediment transport governed by seasonal wave action. Port development, including the construction of breakwaters, has interrupted littoral drift and led to patterns of localized erosion and accretion. Similarly, Odisha's coast has seen notable changes, with severe erosion documented in areas such as Gopalpur, Podampeta, and Gupti. Kar et al. (2021) categorically divided the Odisha coast into five zones using the Digital Shoreline Analysis System (DSAS) and observed that high variability in shoreline behavior in these segments. Lower erosion was observed from Chilika to Konark, while higher erosion rates were noted in the Mahanadi Dhamra delta. Roy et al. (2018) found that 46% of the Odisha coast had eroded, 36% coast accreted, and 18% remained stable between 1990 and 2015. Other studies (Mishra et al. 2019, 2021, 2023; Kar et al. 2018) identified a cyclic pattern of erosion and accretion driven by tropical cyclones and monsoonal variability. Focused research on the Gopalpur coast highlighted the influence of high-energy wave environments, reduced sediment discharge from rivers, and the impact of coastal infrastructure, including Gopalpur Port. Studies by Mohanty et al. (2012, 2015, 2020) have shown that port-related structures disrupt sediment flow, leading to down-drift erosion and up-drift accretion. Similarly, in Tuticorin, the expansion of port facilities has affected shoreline morphology and increased vulnerability to erosion. Most of these assessments relied on remote sensing data, GIS-based modeling, and DSAS techniques using Landsat imagery, often supplemented by field surveys. The comparative shoreline analysis of Gopalpur and Tuticorin highlights the multifaceted interactions between natural forces and human interventions. Considering the need of the country and its coastal communities, National Coastal Research, Chennai, published a shoreline change assessment report for the Indian coast undertaking long-term shoreline change analysis using the moderate to high resolution satellite images (Kankara et al. 2018). The report reveals that Odisha coast is experienced with erosion, accretion, and stable at 28, 51 and 21% of its total coastline while the Tamil Nadu coast experiences erosion (41%), accretion (23%) and stable (36%) shoreline change of its total coastal line length. These findings reiterate the urgency for Integrated Coastal Zone Management (ICZM) that balances development with ecosystem preservation. Key strategies include erosion control, mangrove restoration, sustainable infrastructure development, and participatory governance involving local communities. While numerous studies have addressed shoreline changes at regional scales, detailed investigations on how different natural-anthropogenic drivers, such as tide, wave, current, river, and coastal infrastructure, individually and collectively alters the shoreline, remain limited. Therefore, the present study aims to bridge this gap by examining shoreline changes in three diverse coastal settings across micro- and mesotidal regimes. This research aims to inform shoreline management plans for effective coastal risk mitigation and long-term resilience.

13.2 Material and Methods

The Indian coastline, characterized by its diverse geomorphological settings and rapid coastal development, is increasingly subjected to intense erosion and shoreline transformation. Shoreline change refers to the temporal variation in the position of the shoreline relative to a defined baseline, and it can be estimated using observational data or remote sensing techniques. In the present study, seasonal shoreline changes rates for 2020 were assessed using Sentinel satellite imagery. Sentinel-2 imagery was acquired for different seasons through the Copernicus Open Access Hub. The images were pre-processed through ortho-rectification, and shoreline delineation was carried out manually following the standard operating procedure (SOP) developed by the National Centre for Coastal Research (Kankara et al. 2018). Special attention was given during pre-processing to minimize both human-induced and system-generated errors. The uncertainties associated with shoreline extraction were evaluated and mitigated through uncertainty estimation protocols, thereby ensuring the accuracy of shoreline positioning.

To quantify shoreline dynamics, three key shorelines change indicators, such as End Point Rate (EPR), Net Shoreline Movement (NSM), and Shoreline Change Envelope (SCE) were computed using the Digital Shoreline Analysis System (DSAS), which processes shorelines from multiple time periods referenced against a defined baseline. The EPR provides the average rate of shoreline change between the earliest and latest shorelines. An increase in EPR indicates accelerated shoreline movement in recent years. NSM represents the net displacement of the shoreline over time, while SCE captures the total variation between the most landward and seaward positions of the shoreline across the observation period. For shoreline change analysis, a baseline was created parallel to the coastline, and seasonal shorelines (pre-monsoon and monsoon) were delineated. Transects were drawn from the baseline, intersecting these seasonal shorelines to quantify their displacement. The resultant shoreline change metrics were evaluated across different coastal typologies, including coasts adjacent to infrastructure, riverine environments, and open coastlines, to understand the spatial variation in shoreline dynamics.

The study focused on two representative locations along the East coast of India, namely (1) Tuticorin, representing a micro-tidal environment influenced by coastal infrastructure and estuarine processes, and (2) Gopalpur, situated in a meso-tidal zone that behaves as a relatively open coast. These sites were selected due to their contrasting tidal regimes, wave climates, and increasing anthropogenic pressures, including planned and ongoing coastal infrastructure projects. The choice of these diverse environments allows for a comparative assessment of how tidal and oceanographic conditions influence shoreline response. The seasonal and site-specific assessment of shoreline change provides critical insights into the behavior of India's dynamic coasts under varying environmental conditions (Fig. 13.1).

The primary objective of this study is to quantify the contribution of shoreline change rates driven by three major driver such as oceanographic forces, anthropogenic activities, and riverine inputs. To achieve this, a comprehensive analysis of shoreline changes was conducted using medium-resolution satellite data derived from Sentinel-2. The satellite data used for the shoreline change analysis are listed in Table 13.1. The methodology follows a structured framework comprising five main steps: (1) data acquisition, (2) preprocessing, (3) shoreline extraction, (4) change rate calculation, and (5) uncertainty analysis.

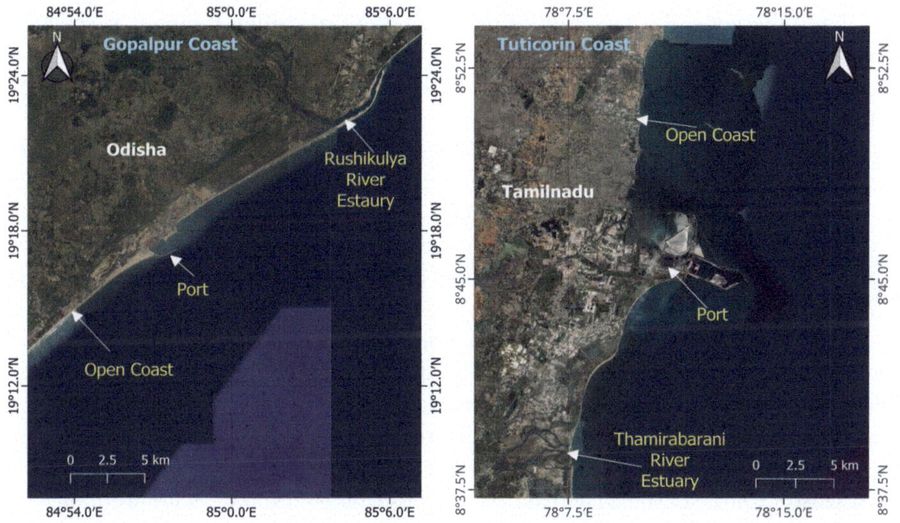

Fig. 13.1 Study area chosen for assessing the shoreline change of natural and anthropogenic impact on the open coast, port and river –induced beach along the micro & meso-tidal coastal environment

Table 13.1 Details of the satellite data used for the shoreline change analysis for the Odisha and Tamil Nadu coast

Sl. No	Satellite data used	Acquisition date	Resolution	Season	State
1	Sentinel 2	03/03/2020	10 m	Pre-monsoon	Odisha
2	Sentinel 2	07/01/2020	10 m	Monsoon	Odisha
3	Sentinel 2	12/28/2020	10 m	Post-monsoon	Odisha
4	Sentinel 2	03/04/2020	10 m	Pre-monsoon	Tamil Nadu
5	Sentinel 2	07/07/2020	10 m	Monsoon	Tamil Nadu
6	Sentinel 2	11/09/2020	10 m	Post-monsoon	Tamil Nadu

Satellite images were selected based on seasonal coverage to represent key periods—pre-monsoon, monsoon, and post-monsoon for the year 2020 for both regions (Table 13.1). This temporal data allows the assessment of inter-seasonal shoreline dynamics and captures the influence of varying hydrodynamic conditions. Selection criteria included cloud-free scenes with appropriate tidal conditions to minimize bias in shoreline position. Shoreline delineation was performed using spectral water indices, particularly the normalized difference water index (NDWI), which enhances the contrast between water and land. The steps included: (1) calculation of NDWI for each image, (2) application of a suitable threshold value to differentiate land and water pixels, and (3) vectorization of the water-land boundary to obtain the shoreline for each seasonal image.

The Digital Shoreline Analysis System (DSAS), an extension of ArcGIS, was used for quantifying shoreline change rates. The process involved: Establishing a baseline parallel to the general orientation of the coast, (2) Generating transects perpendicular to the baseline at regular intervals, (3) Calculating shoreline movement along these transects using standard indicators, such as End Point Rate (EPR)—the average annual rate of change between two shorelines, Net Shoreline Movement (NSM)—total shoreline displacement, Shoreline Change Envelope (SCE)—maximum shoreline variation over time. To account for inherent inaccuracies in shoreline detection and image resolution, uncertainty analysis was conducted following established shoreline mapping error frameworks, which include positional accuracy assessments, estimation of digitization error, tidal correction limits, and sensor resolution constraints.

To distinguish the impact of oceanographic conditions, coastal structure, and river input upon the shoreline change, the shoreline change study was categorized into three coastal typologies: (1) Open coast, (2) anthropogenically influenced coast (e.g., adjacent to ports, seawalls, and infrastructure), and (3) estuarine boundaries (influenced by river discharge). Shoreline change rates are then compared within each season (intra-seasonal) and across seasons (inter-seasonal), with pre-monsoon shoreline serving as the reference for evaluating monsoon and post-monsoon variations. However, to assess the impact of structure and river influence on shoreline change, the shoreline change rate for the open coast was compared to the shoreline change rate at anthropogenic impact and river-influenced coast to determine the structure and river impact on the shoreline, respectively. Similarly, to understand the tidal effect on shoreline change, the study considered two contrasting coasts located a micro-tidal coastal environment—Tuticorin coast (Southern Tamil Nadu) and a meso-tidal coastal environment—Gopalpur coast (Southern Odisha), and detail of the shoreline change statistics are compared with respect to season and environment for better insight into tidal effect on shoreline change.

13.3 Results

Shoreline change caused by natural and anthropogenic drivers along the micro and meso-tidal coast is estimated using high-resolution satellite maps at diverse coastal environment settings during three different seasons. The shoreline changes metrics results and its seasonal and spatial variability are described in detail for an depth understanding and quantification of impact of natural and anthropogenic driver on shoreline change, which is described in detail under the following sub-section:

13.3.1 Shoreline Change Statistics Along Diversified Coastal Environments

The short-term shoreline changes statistics (SSCS) for the monsoon (MON) and post-monsoon (POM) periods were estimated in relation to the shoreline position attained during the pre-monsoon (PRM) period. The SSCS for open coast, estuarine environment and anthropogenically induced coastal structure was studied and the results are presented in the following sub-section.

13.3.1.1 Shoreline Change Statistics Between PRM-MON at Different Coastal Environmental of Odisha and Tamil Nadu

The analysis of shoreline changes indicators (EPR, SCE, and NSM) during the PRM-MON period for Tamil Nadu and Odisha reveals distinct spatial patterns influenced by geomorphological and anthropogenic factors. In Tamil Nadu, the open coast shows moderate erosion with an EPR of 10.38 m/year and negative SCE and NSM values (–4.75 m and –13.88 m), indicating landward shoreline retreat. In contrast, shoreline changes near the port limit and river limit zones exhibit higher accretion trends, with positive NSM values of 18.58 m and 6.64 m, respectively. The derived differences between port limit–open coast (EPR = 3.47 m/year; NSM = 32.46 m) and river limit–open coast (EPR = 4.86 m/year; NSM = 20.52 m) emphasize the protective role of port infrastructure and fluvial sediment inputs in stabilizing the shoreline. Similarly, in Odisha, the open coast is severely erosional, with a negative NSM of –35.18 m and SCE of –11.56 m. However, shoreline segments near the port and river limits show strong, substantial accretion, especially near the port limit, where the EPR reaches 38.19 m/year and NSM is 27.11 m. The contrasts between port limit–open coast (NSM = 62.29 m) and river limit–open coast (NSM = 47.62 m) are notably higher than in Tamil Nadu, indicating a more substantial influence of engineered and natural sediment sources. These findings underscore the importance of ports and rivers as critical buffers that mitigate erosion and enhance sediment deposition, with Odisha exhibiting a more pronounced contrast between engineered and open coastlines compared to Tamil Nadu.

13.3.1.2 Shoreline Change Statistics Between PRM-POM at Different Coastal Environments of Odisha and Tamil Nadu

The shoreline changes indicators between the pre-monsoon and post-monsoon period showed contrasting erosion and accretion patterns at different coastal zones in Tamilnadu and Odisha. In Tamilnadu, the open coast shows moderate accretion (EPR: 8.66 m/year; NSM: 7.40 m), while the port limit zone, despite having a high EPR (25.83 m/year) shows shoreline retreat (NSM: –16.10 m), possibly due to sediment trapping or erosion by the port structures. The river limit area exhibits clear accretion (EPR: 21.73 m/year; NSM: 13.88 m), highlighting the positive influence of fluvial sediment. However, the port limit–open coast segment experiences the highest erosion (NSM: –23.50 m), and the river limit–open coast shows minor net accretion (NSM: 6.49 m). In Odisha, the open coast is erosional (NSM: –6.17 m), though the EPR is relatively high (26.18 m/year), suggesting dynamic shoreline shifts. The port limit is completely stable (NSM: 0.00 m), indicating effective shoreline control through the use of infrastructure. The river limit zone shows accretion (NSM: 10.41 m), and the river limit–open coast segment records the highest shoreline gain (NSM: 16.58 m), emphasizing the strong role of riverine sediment delivery over the shoreline configuration. Meanwhile, the port limit–open coast in Odisha shows moderate accretion (NSM: 6.17 m). Overall results confirm that open coasts are more vulnerable to erosion, while port and river-adjacent zones benefit from sediment accumulation and structural stability. River-fed areas, particularly in Odisha, emerge as the most effective natural buffers against coastal erosion compared to the Tamil Nadu coast (Table 13.3).

13.3.2 Distribution of Shoreline Change Statistics

13.3.2.1 End Point Rate

End point Rate (EPR) is estimated by taking the ratio of the distance of shoreline change movement by the time elapsed between the oldest and the most recent shoreline. The positive values of EPR indicate accretion, while the negative values indicate erosion. The results of EPR for the Odisha and Tamil Nadu Coast during PRM-MON and PRM-POM are computed and presented in Fig. 13.2. The end point rate for the open coast, port limit and river limit along the micro-tidal and meso-tidal coast of Tamil Nadu and Odisha is estimated and results are presented in Fig. 13.2.

Figure 13.2 represents the EPR of diverse coast located in micro and meso-tidal coastal environment of Tamil Nadu and Odisha during PRM-MON and results shows that erosion rate of Open Coast along the Odisha coast (77%) is more than that of Tamil Nadu coast (74%) and the accretion rate of Open Coast along Tamil Nadu coast (27%) is more as compared to Odisha coast (22%) which may be due to the width of the river mouth. Further, it is observed that the magnitude of erosion of the range between 0–50 constitutes 66.46% for Tamil Nadu and Odisha (41.46%). The erosion of the magnitude between 50 to 150 m constitutes 6.21% for Tamil Nadu, while it observed to be high for the Odisha coast (32.75%). The accretion of the magnitude of 0–150 m shows high for Tamil Nadu (24.22%) while it is 18.82% for the Odisha open coast. The accretion magnitude of the order of 50 to 150 m is high for the Odisha coast (3.83%), while it is only 1.86% for the Odisha coast. The EPR at port limit regimes experiences high erosion in Tamil Nadu (45.51%), while is extremely low at Odisha coast (3.03%) and erosion with magnitude of 50 to 150 m is high for Tamil Nadu (6.9.%) while it is 4.65% for Odisha. The accretion of the magnitude of 0 to 50 m is high at Tamil Nadu (33%) while it is 24.14% for the Tamil Nadu coast

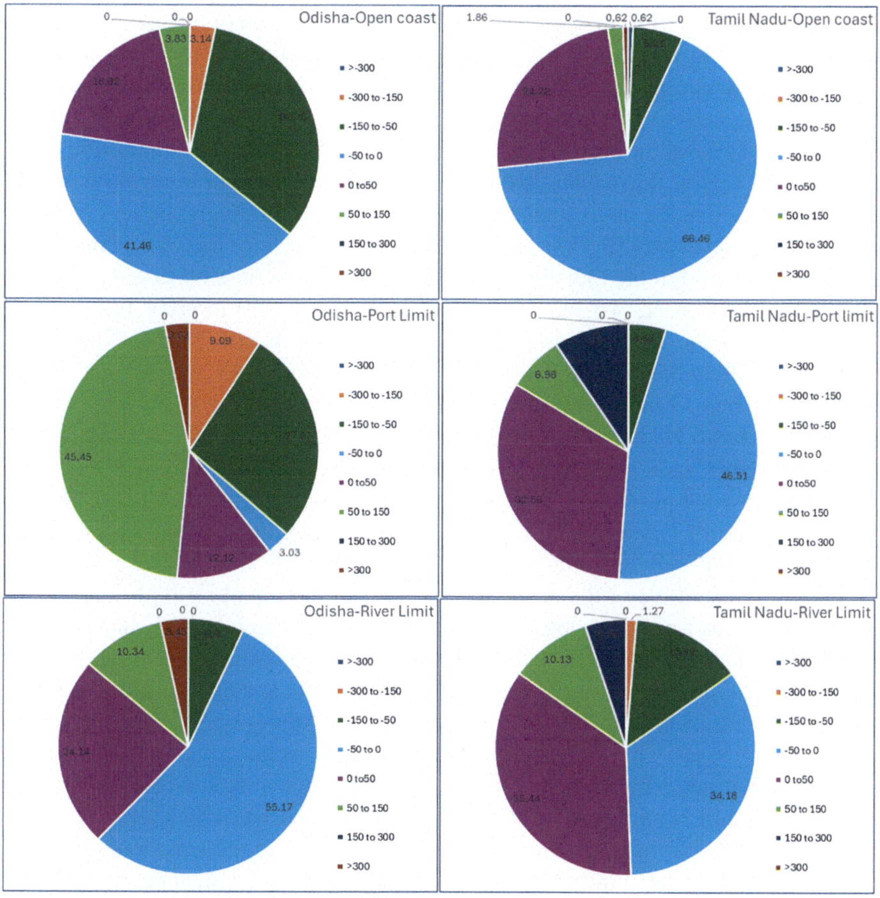

Fig. 13.2 Percentage distribution of End point rate (EPR) along Odisha and Tamil Nadu coast between the PRM and MON with respect to baseline at three different environments

and with the magnitude of 50 to 150 m constitutes of 10.34% for Tamil Nadu and 7% for Odisha. In the river limit, erosion is more for the Odisha compared to the Tamil Nadu. The erosion of the magnitude 0–50 m is high at Odisha (55.17%) and Odisha (34.18%) while erosion of the order of 50 to 100 m is high for Tamil Nadu (14%) and Odisha (7%). The accretion near the river limit is high for Tamil Nadu (35.44%) and Odisha (24.14%) with a magnitude of 0 to 50 m. In comparison, accretion magnitude of 50–150 m constitutes around 10% for both the state and accretion of the order of 150–300 m is high for Tamil Nadu (5.06%), while accretion order of 300 m to 1300 m constitutes 3.45% for the Odisha. The over analysis show that the micro-tidal coast of open coast and port limit area and meso-tidal coast of river limit exhibited maximum percentage of erosion with an order of 0 to 50 m while meso-tidal coast of open coast and port limit and micro-tidal of river limit experiences the intense erosion of the magnitude of 50 m to 150 m. Similarly, the micro-tidal region of the open coast, port limit, and river limit experience maximum accretion of order 0 to 50 m, while the meso-tidal region of the Open coast, port limit, and river limits experience accretion of the magnitude 50 to 150 m, respectively.

Figure 13.3 shows the EPR of diverse coasts located along the Odisha and Tamil Nadu coastal region during the pre-monsoon and post-monsoon periods. Results show that erosion at the open coast of Odisha (72.5%) is relatively higher compared to the Tamil Nadu coast (33.51%), and the accretion is high for the Tamil Nadu coast (65.95%) compared to the Odisha coast (17.96%). Further, it is observed that along the Odisha Open coast, erosion of the magnitude 50 to 150 m constitutes around 4.23% while for the Tamil Nadu coast, such magnitude is not observed during the observation. Along the port limit coastline, the Tamil Nadu region experienced high erosion compared to the Odisha coast, while the accretion rate is high for Odisha compared to Tamil Nadu. It is observed that erosion of the magnitude of 0–50 m is 54% along Tamil Nadu while it is 9.09% for the Odisha coastal region.

The erosion magnitude of 50–150 m is observed to be high for Odisha (30.30%) compared to Tamil Nadu (16.28%). The accretion magnitude of the order of 0 to 50 m is observed high for Odisha (51.52%) and Tamil Nadu (16.28%). However, the accretion magnitude of the order 50–150 m is almost equal for both states (9%). Along the riverine limit, it is observed that erosion rate is high for Odisha compared to the Tamil Nadu coastal region. The erosion of the magnitude between 0–50 m is high for the Tamil Nadu coast (31.65%) while it is (25%) and the erosion magnitude between 50–150 m is high for the Odisha coast (67.86%), while it is low for the Tamil Nadu (5.06%). The accretion rate at river limit is relatively high for the Tamil Nadu coastal water compared to Odisha. The accretion magnitude of 0–50 m is high for the Tamil Nadu coast (54.43%) and low for Odisha (3.57%) and accretion magnitude of 50–150 m is high for the Tamil Nadu coast (5.6%) and magnitude for the 150–300 is high for Tamil Nadu (3.87%) and low for the Odisha coast (3.57%). The overall analysis shows that the micro-tidal port limit and the River limit during PRM-MON, and the micro-tidal open coast during PRM-POM experienced a high percentage of erosion within range of 0–50 m, while mesotidal port and river limit face erosion magnitude of 50–150 m. In contrast, the micro-tidal open coast and river limit followed by port limit area experiences a high percentage of accretion rate of 0–50 m during the PRM-MON period, while the micro and meso-tidal port limit and micro-tidal river limit experience accretion of the order of magnitude of 50–150 m.

13.3.2.2 Net Shoreline Movement

Figure 13.4 shows the NSM of both Odisha and Tamil Nadu coasts along different environmental scenarios between the Pre-monsoon and Monsoon periods. Results show that negative NSM of the order of 0 to 50 m constitutes the maximum percentage (74.22%) along the Odisha Open coast compared to the Tamil Nadu coast (72.67%), followed by negative NSM of the order of −50 to −150 m along the Odisha open coast (3.14%). The positive NSM of the Order of 0 to 50 m is high for the open coast of

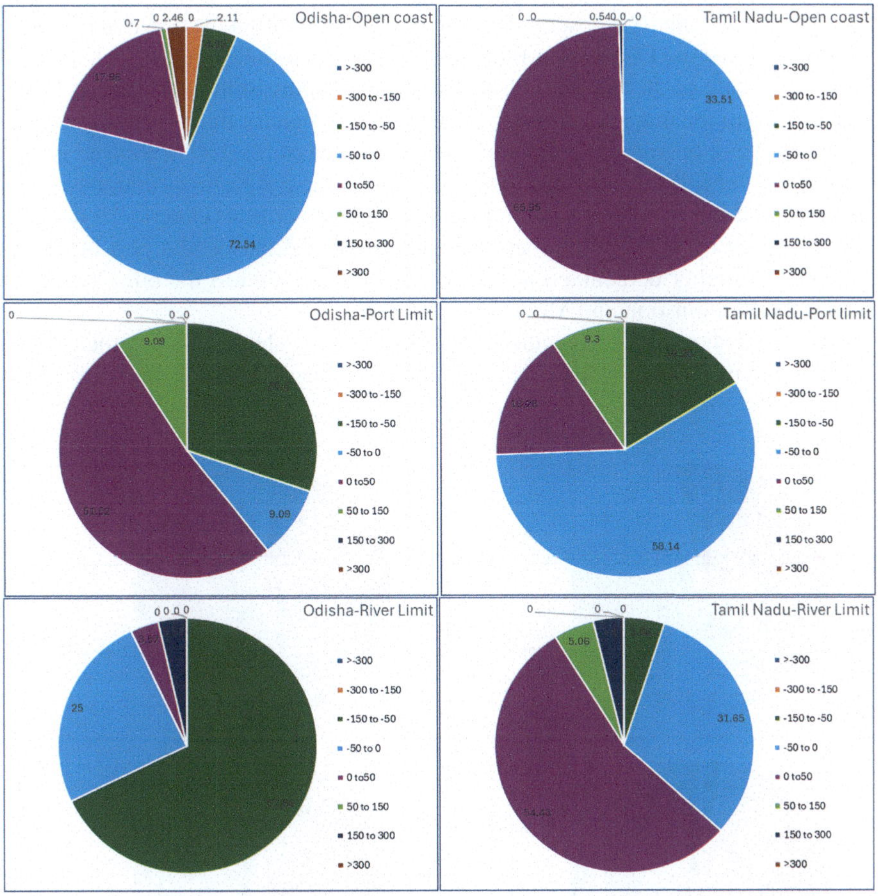

Fig. 13.3 Percentage distribution of End point rate (EPR) along Odisha and Tamilnadu coast between the PRM and POM with respect to baseline at three different environments

Tamil Nadu compared to the Odisha open coast (22.65%). The negative NSM of the Tamil Nadu port limit is more (51.16%) compared to the Odisha port limit (30.30%), while a negative NSM of the order of 50 to 150 m is observed along the Odisha port limit (9.09%), and nothing is observed for Tamil Nadu. The positive NSM within the range of 0 to 50 m is observed high for the Tamil Nadu port limit while for Odisha it is (22.65%) and within the range of 50 to 150 observed 9.30% for Tamil Nadu port limit while no for the Odisha coastal water and NSM of the order of 300 to 1300 m observed for the Odisha coast (3.03%). Similarly, negative NSM of the order of 50–150 m observed high for the Odisha limit area (62.07%) compared to the Tamil Nadu coast (48.10%) and positive NSM within the range of 50 to 150 is high for the Tamil Nadu (15.57%) compared to Odisha (34.48%) and within the range of 50 to 150 m is observed high for the Tamil Nadu coast (5.06%) compared to the Odisha coast (3.45%). Further it is observed that negative NSM within the range of 0 to 50 m observed high for the Open coast followed by river limit and port limit while NSM within magnitude of 50 to 150 m shows high for the port limit and open coast and positive NSM of the magnitude of 0 to 50 m observed high for the port limit followed by river limit and open coast and within the magnitude of 50–150 m showed is high for the port limit followed by river limit. The overall analysis shows that meso-tidal open coast and micro-tidal port limit and meso-tidal river limit experience negative NSM, followed by micro-tidal open and river limit, and meso-tidal port limit area experiences a higher percentage of positive NSM compared to the other scenarios between the Premonsoon and Monsoon season.

Figure 13.5 shows the percentage distribution of net shoreline movement along the micro and meso-tidal open, port limit, and river beach environment between the

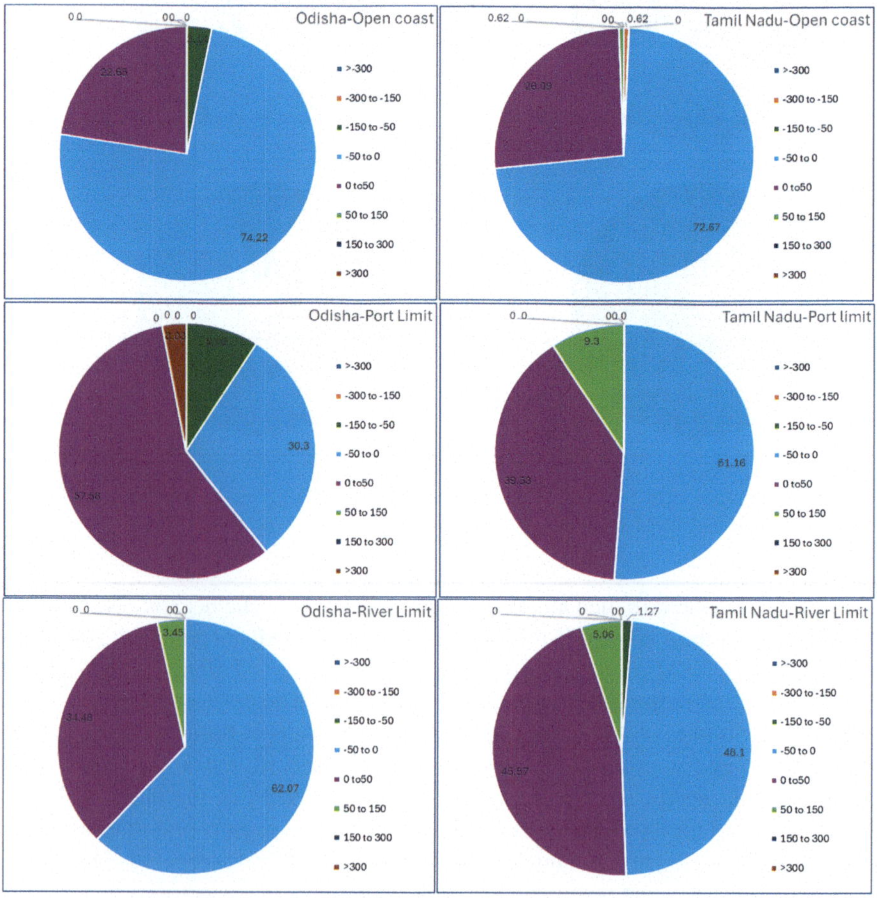

Fig. 13.4 Percentage distribution of Net shoreline movement along Odisha and Tamil Nadu coast between the PRM and MON with respect to baseline at three different environments

pre-monsoon and post-monsoon period. Results show that negative NSM within the 0 to 50 m constitutes high for the Odisha open coast (74.65%) compared to Tamil Nadu (33.51%) and for 50–150 m is high for the Odisha (2.46%) while the positive NSM of the order 0 to 50 m is high for Tamil Nadu (65.95%) compared to Odisha (18.31%). The negative NSM for the port limit of the magnitude of 0 to 50 m is high for the Tamil Nadu coast (69.77%) compared to Odisha (21.21%), and for the magnitude of 50 to 150 m is high for Odisha (18.18%) compared to Tamil Nadu (4.65%). The positive NSM of the order of 0 to 50 m is observed more for Odisha (54.55%) compared to Tamil Nadu (18.60%), and for the magnitude of 50–150 m constitutes a high for Tamil Nadu (6.98%) compared to Odisha (6.06%). Further analysis shows that negative NSM of the order 0 to 50 m observed maximum for the open coast, followed by the port limit and the river limit, while negative NSM within 50 to 150 m observed maximum for the port limit, followed by the river limit and the open coast. The positive NSM of the order of 0 to 50 m shows a maximum for the open coast, followed by the port limit and the river limit. The overall analysis shows that during the post-monsoon period, meso-tidal open coast and river limit and micro-tidal port limit experience predominant negative NSM (0–50 m) while micro-tidal open coast and river limit and meso-tidal port limit experience positive NSM (0–50 m).

13.3.2.3 Shoreline Change Envelope

Figure 13.6 shows the percentage distribution of shoreline change envelope during the monsoon period at various beach environments of micro-tidal and meso-tidal coast. Results show that positive SCE of the magnitude (0–50 m) is high for the open coast of Tamil Nadu (98.76%) compared to Odisha (98.86%), while SCE of the range of 50 to 150 m shows maximum for Odisha (3.14%) compared to Tamil Nadu (0.62%). For Port limit, the positive SCE of order of 0–50 m shows high for Tamil Nadu (90.7%) compared to the Odisha (87.88%). SCE of the magnitude of 50–150 m show high value for Tamil Nadu (9.30%) compared to the Odisha (9%). For the river limit, the positive SCE of the order of 0–50 m show high value for Odisha (96.55%) compared to the Tamil Nadu (93.67%) while for SCE of the magnitude of 50 to 150 m shows a maximum value for the Tamil Nadu (6.33%) compared to Odisha (3.45%). Furthermore, the analysis shows that positive SCE of the magnitude of 0 to 50 m is high for the open coast, followed by the river limit and the port limit, while SCE of the magnitude of 50–150 m shows a high value for the Port limit, followed by the River limit and the open coast, respectively. The over-analysis shows that the micro-tidal open coast, port limit, and meso-tidal coast of river limit experiences positive SCE (0–50 m), and the meso-tidal open coast and micro-tidal port and river limit experiences SCE of the order of 50–150 m during the monsoon season.

Figure 13.7 shows the percentage distribution of SCE occurrence at various beach environments of meso and micro-tidal coast during the post-monsoon period. Results show that positive SCE of the magnitude of 0 to 50 m is high for the micro-tidal open coast (98.92%) compared to the Odisha (92.96%). The SCE of the order 50 to 150 m

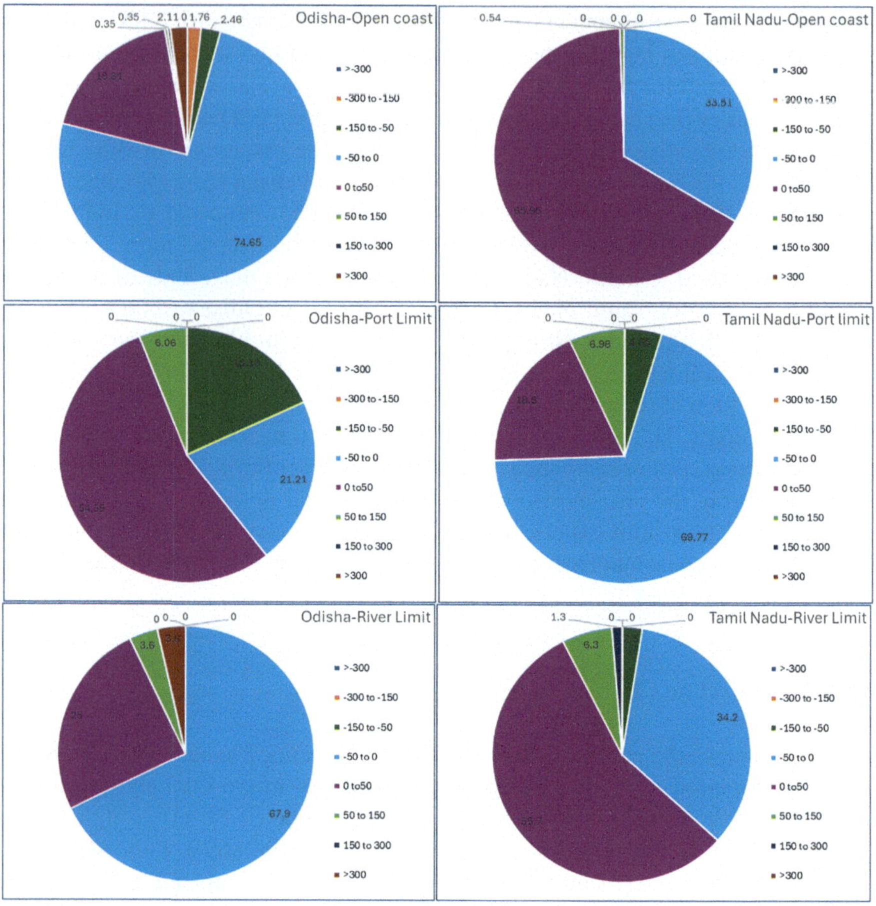

Fig. 13.5 Percentage distribution of Net shoreline movement along Odisha and Tamil Nadu coast between the PRM and POM with respect to baseline at three different environments

is observed high for the meso-tidal open coast (2.82%). For the port limit, the SCE of the magnitude of 0 to 50 m is high for the micro-tidal coast of Tamil Nadu (88.37%) compared to Odisha (75.76%). The SCE of the magnitude of 50–150 m shows high for Odisha (24.24%) compared to Tamil Nadu (11.63%). Near the river limit, the SCE of magnitude of 0 to 50 m show high for the mesotidal coast of Odisha (92.86%) compared to the Tamil Nadu (89.87%) and with SCE magnitude of 50 to 150 m observed maximum in micro-tidal coast of Tamil Nadu (8.86%) compared to Odisha (3.57%). The analysis shows that SCE of the magnitude of 0 to 50 m is experienced maximum near the open coast, followed by the river limit and the port limit. The overall analysis show that micro-tidal open coast and port limit and meso-tidal river limit experienced higher percentage of occurrence of SCE of the magnitude of 0 to 50 m. In comparison, meso-tidal open coast and port limit and micro-tidal river limit

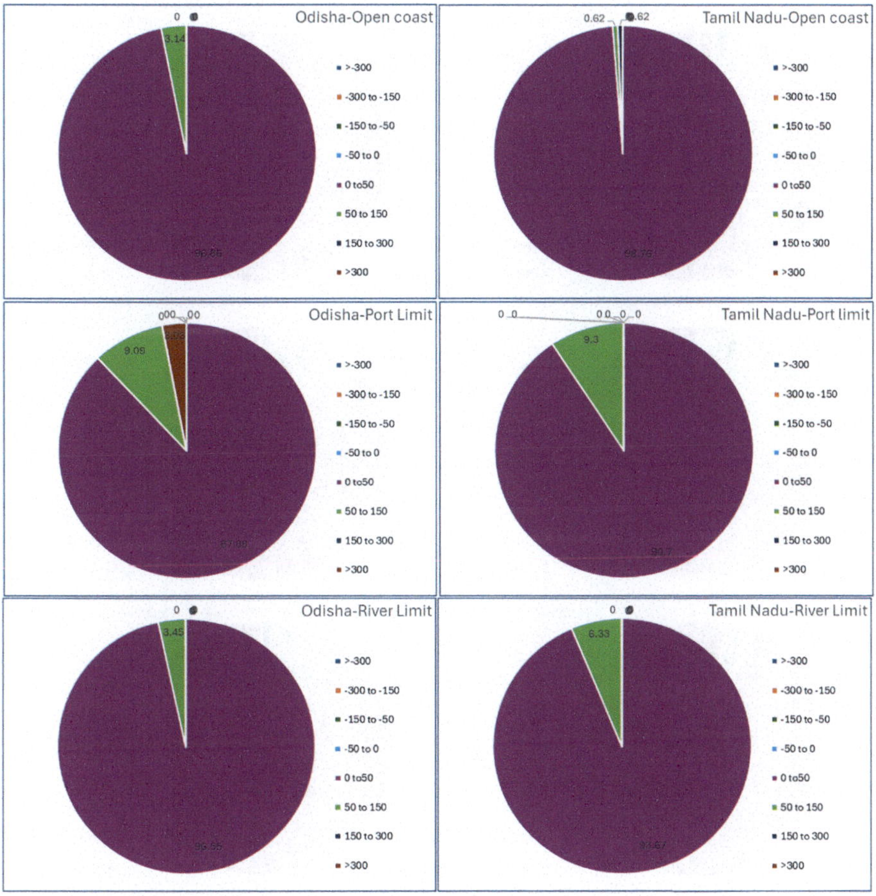

Fig. 13.6 Percentage distribution of Shoreline Change Envelope along Odisha and Tamilnadu coast between the PRM and MON with respect to baseline at three different environments

experienced more percentage of SCE within the range of 50 to 150 m compared to other coastal conditions.

13.3.3 Scaling of the Coastal Erosion and Accretion Process

The analysis of shorelines along the micro-tidal and meso-tidal coast during the south-west and post monsoon season is carried out, and the response of the coast to those is analyzed and presented in Table 13.4. Results show that micro and meso-tidal open coast during monsoon and mesotidal open coast during post-monsoon regime experiences the erosion, while micro tidal open coast during post-monsoon exhibits

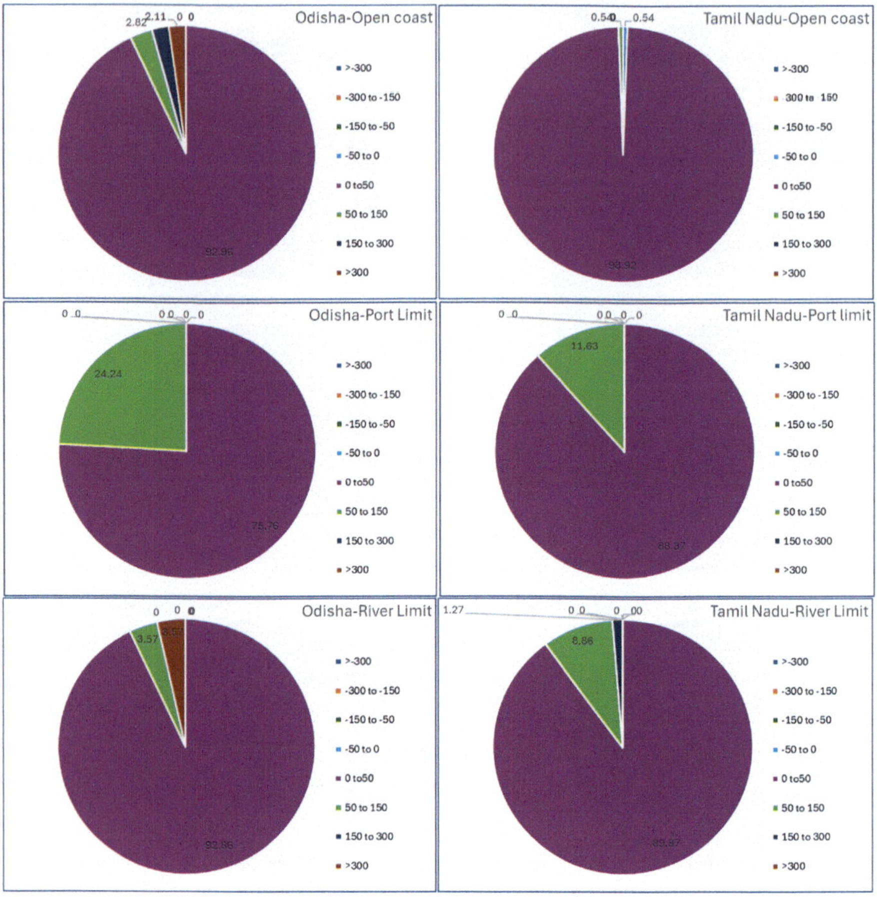

Fig. 13.7 Percentage distribution of Shoreline Change Envelope along Odisha and Tamil Nadu coast between the PRM and POM with respect to baseline at three different environments

the accretion. The coast influenced by the port structure in the micro- and meso-tidal coastal regimes experiences an accretion pattern of the shoreline during the monsoon. In contrast, the micro-tidal coast near the port limit exhibits an erosional trend during the POM, while the meso-tidal coast remains stable to shoreline change. The coast influenced by micro and meso-tidal regime shows the accretion pattern of shoreline during the monsoon and post-monsoon period near the river limit. The analysis of results showed that coastal environments near to the river environment possess a healing mechanism to restore the shoreline, while the open coast at the micro-tidal regime tends to restore the retreated shoreline. Further, we observed that although the POM is regarded as a period of accretion, the hard infrastructure near the port limit coastal environment during the POM is less effective to protect the coast near

Table 13.2 Shoreline change statistics between pre-monsoon and monsoon at three different environmental regimes of Odisha and Tamil Nadu coast

States	Coast type	EPR	SCE	NSM
Tamil Nadu	Open coast	10.38	–4.75	–13.88
	Port limit	13.85	6.36	18.58
	River limit	15.24	2.27	6.64
	Port limit-open coast	3.47	11.11	32.46
	River limit-open coast	4.86	7.02	20.52
Odisha	Open coast	16.09	–11.56	–35.18
	Port limit	38.19	8.91	27.11
	River limit	15.41	4.09	12.44
	Port limit-open coast	22.10	20.47	62.29
	River limit-open coast	–0.68	15.65	47.62

Table 13.3 Shoreline change statistics during PRM_POM at three different environmental regimes of Odisha and Tamil Nadu coast

States	Coast type	EPR	SCE	NSM
Tamil Nadu	Open coast	8.66	5.06	7.40
	Port limit	25.83	−11.02	−16.10
	River limit	21.73	9.50	13.88
	Port limit-Open coast	17.17	−16.08	−23.50
	River limit-open coast	13.07	4.44	6.49
Odisha	Open coast	26.18	−5.07	−6.17
	Port limit	37.78	0.00	0.00
	River limit	22.76	8.55	10.41
	Port limit-Open coast	11.59	5.07	6.17
	River limit-open coast	−3.42	13.62	16.58

the micro-tidal coast, while port limit coastal environment in the meso-tidal coast can able to maintain a stable shoreline.

13.4 Discussion

The analysis of shoreline changes statistics across different coastal environments in Tamil Nadu and Odisha highlight significant spatial and temporal variability driven by both natural processes and anthropogenic interventions. The data derived from short-term shoreline change indicators such as End Point Rate (EPR), Net Shoreline Movement (NSM), and Shoreline Change Envelope (SCE) for the pre-monsoon,

monsoon, and post-monsoon seasons provide critical insights into coastal dynamics influenced by tidal regimes, riverine sediment discharge, and port infrastructure.

13.4.1 Spatial Erosion and Accretion Pattern

The open coasts of both Tamil Nadu and Odisha exhibit pronounced erosional trends, particularly during the monsoon period. Odisha's open coast demonstrates higher erosion intensity compared to Tamil Nadu, as indicated by a significantly negative NSM (–35.18 m versus –13.88 m, respectively). This is consistent with the mesotidal characteristics and higher wave energy exposure of the Odisha coast. In contrast, the riverine and port limit zones in both states show a relative buffering effect, often experiencing accretion or reduced erosion due to fluvial sediment input and the sediment-trapping influence of port infrastructure. It is pertinent to note that Tamil Nadu's port-influenced shoreline shows positive EPR during the monsoon, while during the post-monsoon period, some port-adjacent segments exhibit a retreat trend (NSM: –16.10 m), which suggests port structures may contribute to localized erosion, possibly due to sediment bypassing or wave reflection effects. On the contrary, Odisha's port adjacent coastal zone maintains stability or accretional trends in both seasons, which highlights the efficacy of port-induced sediment trapping in mesotidal environments. Mujabar et al. (2013) observed that southern coast of India along the Kanyakumari and Turticorin coast showed a significant shift of erosional trend during the post-tsunami phase compared to pre-tsunami occurrence and they revealed that headland coastal regions exhibited both the erosion and accretional trend, while the present study along the micro-tidal coast shows the accretion and erosion trend of shoreline near the open, port and river influenced environment and quantified the impact of natural and anthropogenic impact to net shoreline change rate. Along the Odisha coast, shoreline change is mainly controlled by met-oceanographic conditions, coastal structure, and landfall of tropical cyclones, and the erosional process is mostly prevalent along the central Odisha coast (Kendrapara and Jagatsinghpur district) compared to other coastal stretches Kar et al 2021). Their study discerned that shoreline change rate for Gopalpur coast is both negative and positive in different phase of time during 1978 to 2017. Mostly the coast experienced the erosion trend during the landfall of cyclone and the accretion trend during the development of the Port infrastructure. Besides, it observed that shoreline change rate near a river influenced the coast is either negative or positive depending on the site's local conditions and river discharge. The shoreline change rate along the southern Odisha coast (Gopalpur coast) at open, port, and river limits is corroborated with the results of Kar et al. (2021).

13.4.2 Seasonal Variability and Environmental Influence

The Indian coast, being located along a tropical regime, the monsoonal forcing plays a dominant role in shoreline dynamics. Generally, the monsoon season corresponds to erosional processes due to strong wave action and storm surges, while the post-monsoon period is expected to favor accretion as energetic conditions subside. This trend is partially observed in Tamil Nadu, particularly along open and riverine coasts. However, along the Odisha coast, despite post-monsoon prevalence, the open coasts continue to show erosion (NSM: –6.17 m), indicating residual monsoonal effects or limited sediment availability for beach recovery. River limits along both regions consistently exhibit accretion across seasons, with Odisha's River-dominated zone showing particularly high NSM during monsoon (up to + 12.44 m), indicating the significance of sediment supply from rivers in shoreline stabilization. The scaling analysis (Table 13.4) confirms that river-adjacent zones along the coast possess a natural resilience and self-recovery mechanism compared to the port limit coast, especially during the post-monsoon season, while erosion persists in some places. Similar trend of coastal change along the south Odisha coast has been observed by Kar et al (2021) although their study encompasses the short-time shoreline change rate during 1978–2017. Similarly, a study by Mageswaran et al (2015) highlighted the coastal erosion and accretion pattern along the Nagapattinam coast during 1978–2013, with predominant shoreline change skewed towards erosion. Their study observed mean erosion rate as −0.99 m/year along the Nagapattinam coast while preset study observed positive EPR value and negative NSM during monsoon and post-monsoon period respectively indicating erosional trend along the open coast and port limit during the monsoon and post-monsoon period respectively, although their study not revealed the seasonal trend of coastal erosion and accretion and quantification of natural and anthropogenic impact on shoreline change.

Table 13.4 Status of possible coastal problem along the micro and meso-tidal coast during the south-west monsoon and north-east monsoon along the east coast of India Extracted from Tables 13.1 and 13.2

Type of beach	Type of coast	MON	POM
Open coast	Micro-tidal	Erosion	Accretion
	Meso-tidal	Erosion	Erosion
Port limit	Micro-tidal	Accretion	Erosion
	Meso-tidal	Accretion	Stable
River limit	Micro-tidal	Accretion	Accretion
	Meso-tidal	Accretion	Accretion

13.4.3 Tidal Impact on Shoreline Change

The distinction between micro-tidal and meso-tidal regimes clarifies the behavior of shoreline processes due to its gradient in magnitude and associated coastal transformational processes. Micro-tidal open coasts exhibit higher variability in both erosion and accretion, while mesotidal coasts, especially those influenced by river mouths, exhibit more pronounced accretion. The data reveal that erosion within the range 0–50 m is prevalent in both tidal regimes, but erosion of 50–150 m is more common in meso-tidal coasts (e.g., Odisha), especially during the monsoon. In contrast, micro-tidal coasts are more susceptible to shorter-term, lower-magnitude erosion events. Similarly, the accretion pattern within the range of 0–50 is more prevalent along the meso-tidal coast, followed by the accretion within the range of 50–150 m near the open coast and river-influenced coast, while near the port environment, it is quite peculiar. The shoreline accretion percentage is more at the port environment during monsoon along the Turticorin coast while it is less during the post-monsoon period. Similarly, along the meso-tidal coast, the accretion pattern near the port is high during the monsoon season compared to the micro-tidal coast during the post-monsoon season. The above findings are coinciding with the results of the Natesan and Subramanian (2013), Natesan et al. (2015) and studies of PWD, Tamil Nadu (2022). The high accretion environment during the monsoon season along the micro-tidal coast and meso-tidal coast during the post-monsoon period indicates that tidal range has a significant impact on shoreline changes. Further, tide impact on open and river influenced coast shows that accretion process within range of 0–50 m is more prevalent along the micro tidal coast during the monsoon followed by accretion within range of 50–150 m, which could be due to arrest of coastal sediments by the coastal structure, while during the post-monsoon, the accretion within the range of 50–150 m is almost equal magnitude near the port influenced coastal region indicating variation in tidal range has least impact on shoreline changes near the coastal structure.

13.5 Conclusion

The shoreline change study carried out in this piece of work shows a comprehensive assessment of short-term shoreline changes along the East coast of India, considering the differing behaviors of open, port-influenced, and riverine coastal segments subjected to varying tidal conditions. The findings reveal that open coasts, especially along Odisha during the monsoon, are highly susceptible to erosion, with short-term Net Shoreline Movement (NSM) values reaching up to –35 m. In contrast, port and river limit zones often act as sediment buffers, mitigating shoreline retreat. While ports tend to stabilize adjacent coastlines in Odisha, they may contribute to localized erosion in Tamil Nadu, particularly during the post-monsoon period. River-influenced coastal segments in both states show consistent accretion, underscoring the critical role of fluvial sediment supply in counteracting erosion. Tidal regimes also play a pivotal role, with mesotidal environments, notably in Odisha, exhibiting greater resistance to erosion and improved sediment retention compared to microtidal settings. Although post-monsoon shoreline accretion process is common, Odisha's open coasts continue to erode beyond the monsoon season, highlighting limited natural recovery in high-energy, sediment-deficient environments. Furthermore, microtidal port coastal environment may be ineffective at preventing erosion during the post-monsoon without proper sediment management strategies. These insights may call for site-specific coastal management approaches that will integrate tidal regimes, sediment dynamics, and infrastructure impacts for holistic management of the shoreline. The findings of the present work could provide a basis for quantifying natural and anthropogenic impacts on erosion, serving as a framework for similar coastal erosion audits across other parts of the Indian coastline.

Acknowledgements The first and second authors are very thankful to the Director, National Centre for Coastal Research, Chennai, and the Authority of Berhampur University for providing the opportunity and possible support to carry out this work during the dissertation work. On behalf of all the authors, thanks to both reviewers for their valuable suggestions for improvement of this work.

References

Abijith D, Saravanan S, Sundar PKS (2023) Coastal vulnerability assessment for the coast of Tamil Nadu, India—a geospatial approach. Environ Sci Pollut Res 30(30):75610–75628

Bulleri F, Chapman MG (2010) The introduction of coastal infrastructure as a driver of change in marine environments. J Appl Ecol 47(1):26–35

Camfield FE, Morang A (1996) Defining and interpreting shoreline change. Ocean Coast Manag 32(3):129–151

Dwivedi CS, Pampattiwar ST, Pandey AC, Parida BR, Mitra D, Kumar N (2023) Characterization of the coastal vulnerability in different geological settings: a comparative study on kerala and Tamil Nadu coasts using FuzzyAHP. Sustain 15(12):9543

Hapke CJ, Kratzmann MG, Himmelstoss EA (2013) Geomorphic and human influence on large-scale coastal change. Geomorphol 199:160–170

Jana A, Sheena S, Biswas A (2012) Morphological change study of ghoramara island, eastern india using multi-temporal satellite data. Res J Recent Sci 1(10):72–81

Jana A, Biswas A, Maiti S, Bhattacharya AK (2014) Shoreline changes in response to sea level rise along digha coast, eastern india: an analytical approach of remote sensing, GIS and statistical techniques. J Coast Conserv 18(3):145–155

Jana A, Maiti S, Biswas A (2016a) Analysis of short-term shoreline oscillations along midnapur-balasore coast, bay of Bengal, India-A study based on Geospatial technology. Model Earth Syst Environ 2(2):64. https://doi.org/10.1007/s40808-016-0117-7

Jana A, Maiti S, Biswas A (2016b) Seasonal change monitoring and mapping of coastal vegetation types along midnapur-balasore coast, bay of bengal using multi-temporal landsat data. Model Earth Syst Environ 2(1):7

Jana A, Maiti S, Biswas A (2017) Appraisal of long-term shoreline oscillations from a part of coastal zones of sundarban delta, eastern india—a study based on geospatial technology. Spat Inf Res 25:713–723. https://doi.org/10.1007/s41324-017-0139-x

Kankara RS, Ramanmurthy MV, Rajeevan M (2018) National assessment of shoreline changes along Indian coast-A status report for 26 years 1990–2016. NCCR Publication, 2018. Available at http://www.nccr.gov.in

Kar P, Mohanty PK, Pradhan S, Behera B (2021) Padhi S K & Mishra P 2021 Shoreline change along Odisha coast using statistical and geo-spatial techniques, Indian Academy of Sciences. J Earth Syst Sci 130:209. https://doi.org/10.1007/s12040-021-01703-1

Kar PK, Mohanty PK, Barik SK, Behera B (2017) Shoreline change: A study along South Odisha coast using statistical and. geospatial technique; Int Res J Earth Sci 5(1) 1–7, http://www.isca.in/WARTH SCI/Archive/v5/i1/LISCA-IRJES-2017–002.PDF

Kar PK, Mohanty PK, Behera B (2018) December. Shoreline change associated with coastal structures at Gopalpur port, Odisha, East Coast of India. In: Proceedings of the fourth international conference in ocean engineering (ICOE2018) 2 (pp. 535–547). Singapore: Springer Singapore

Kar PK, Mohanty PK, Behera B (2019) Shoreline change associated with coastal structures at Gopalpur port, Odisha, East Coast of India; In: Proceedings of the fourth international conference in ocean engineering (ICOE2018) (eds) Murali K, Sriram V, Samad A and Saha N, Lect Notes Civ Eng 23 535–547, Springer, Singapore, https://doi.org/10.1007/s12665-013-2845-6

Mageswaran T, Mohan RV, Selvan CS, Arumugam T, Usha T, Kankara RS (2015) Assessment of shoreline changes along Nagapattinam coast using geospatial techniques. Int J GeomatS Geosci 5(4):555

Masselink G, Hughes M, Knight J (2014) Introduction to coastal processes and geomorphology. Routledge

Mishra M, Chand P, Pattnaik N, Kattel DB, Panda GK, Mohanti M, Baruah UD, Chandniha SK, Achary S, Mohanty T (2019) Response of long- to short-term changes of the Puri coastline of Odisha (India) to natural and anthropogenic factors: A remote sensing and statistical assessment. Environ Earth Sci 78:338. https://doi.org/10.1007/s12665-019-8336-7

Mishra M, Acharyya T, Chand P, Santos CAG, Kar D, Das PP, Pattnaik N, Silva RMD, Nascimento TVMD (2021) Analyzing shoreline dynamicity and the associated socioecological risk along the southern Odisha Coast of India using remote sensing-based and statistical approaches. Geocarto Int. https://doi.org/10.1080/10106049.2021.1882005

Mishra M, Acharyya T, Chand P, Santos C A G, Kar D, Das P P, Pattnaik N, Silva MR, Nascimento TVM (2023) Analyzing shoreline dynamicity and the associated socioecological risk along the Southern Odisha Coast of India using remote sensing-based and statistical approaches, Geocarto International, 37, 2022, issue 14, https://doi.org/10.1080/10106049.2021.1882005

Mohanty PK, Patra SK, Bramha S, Seth B, Pradhan U, Behera B, Mishra P, Panda US (2012) Impact of groins on beach morphology: A case study near Gopalpur Port, east coast of India. J Coast Res 28(1):132–142. https://doi.org/10.2112/JCOASTRES-D-10-00045.1

Mohanty PK, Barik SK, Kar PK, Behera B, Mishra P (2015) Impact of ports on shoreline change along Odisha coast. Proc Eng 116:647–654. https://doi.org/10.1016/j.proeng.2015.08.339

Mohanty PK, Kar PK, Behera B (2020) Impact of very severe cyclonic storm Phailin on shoreline change along South Odisha Coast; Nat. Hazards 102:633–644. https://doi.org/10.1007/s11069-019-03610-7

Mujabar PS, Chandrasekar N (2013) Shoreline change analysis along the coast between Kanyakumari and Tuticorin of India using remote sensing and GIS. Arabian J Geosci. 6(3):647–664. https://doi.org/10.1007/s12517-011-0394-4

Natesan U, Parthasarathy A, Vishnunath R, Kumar GEJ, Ferrer VA (2015) Monitoring longterm shoreline changes along Tamil Nadu, India using geospatial techniques. Aquatic Procedia 4:325–332

Natesan U, Subramanian SP (2013) Identification of Erosion-Accretion regimes along the Tamil Nadu coast, India. J Coast Res 10(1), 203–205

PWD Tamil Nadu: History of Tamilnadu coastline. IHH Report No. 02/2002

Ranasinghe R (2016) Assessing climate change impacts on open sandy coasts: A review. Earth Sci Rev 160:320–332

Ratheesh R, Rajput P, Bhatti H (2023) Rajawat A S and Rajak D R 2023 Quantification of shoreline changes along the entire Indian coast using Indian Remote Sensing satellite images of 2004–06 and 2014–16. Curr Sci 124(5):578–584

Roy S, Mahapatra M, Chakraborty A (2018) Shoreline change detection along the coast of Odisha, India using digital shoreline analysis system. Spat Inf Res 26(5):563–571

Zhang Z, Liang B, Wang Z, Shi L, Borsje BW (2024) Effects of wave forces on sediment transport patterns in micro-tidal estuaries. Phys Fluids, 36(2)

Chapter 14
Women as Catalysts of Community-Based Flood Resilience and Food Security in Ogbaru LGA, Anambra State

I. E. Okoli, Angela Oyilieze Akanwa, and A. U. Okonkwo

14.1 Introduction

Carbon dioxide emissions have increased significantly due to factors such as fossil fuel use, deforestation, farming practices, poor waste management, solar activity fluctuations, and volcanic eruptions, impacting climate patterns globally (World Meteorological Organization, 2020). In Africa, particularly Nigeria, these climate changes lead to severe effects, including flooding, erosion, drought, and desertification, which threaten agriculture and national food security due to the prevalence of rainfed agriculture. Food security encompasses the availability and quality of safe, nutritious food required for a healthy lifestyle. Women are pivotal in agricultural production and food security, with significant roles in climate resilience (Janna and James, 2021).

Women globally contribute substantially to agricultural and rural economies, relying on agriculture for primary income, especially in rural areas of developing countries. In Africa, approximately 80% of agricultural output comes from smallholders, mostly rural women, with women in Sub-Saharan Africa being more involved in agriculture than men. In Nigeria, women comprise 60–80% of the agricultural labor force, producing two-thirds of food crops (Action Aid, 2015). This underlines the need for gender equality in addressing climate challenges, as involving women in decision-making is crucial for effective climate change mitigation. Women

I. E. Okoli · A. U. Okonkwo
Department of Environmental Management, Chukwuemeka Odumegwu Ojukwu University, Uli, Anambra State, Nigeria

A. O. Akanwa (✉)
Faculty of Environmental Management, Agriculture and Renewable Energy (FEMARE), University of Technology and Arts, Byumba, Rwanda
e-mail: angy.akanwa1@gmail.com

A. Nath et al. (eds.), *Geospatial Application of Coastal Hazards*, Springer Water,
https://doi.org/10.1007/978-3-032-20076-1_14

in agriculture are disproportionately impacted by climate change, affecting their income, health, and livelihoods (UN Women, 2022; Odubo et al. 2019). Historically, the Ogbaru Local Government Area in Anambra State faces flooding issues, aggravated by its flat terrain near the River Niger, where alluvial deposits from the rivers are vital for agriculture. Flooding from heavy rainfall damages crops, displaces communities, endangers livestock, and risks reservoir failures (Ajah et al. 2021). Therefore, understanding women's roles in food security and climate resilience in Ogbaru is essential.

14.2 Concept of Gender Vulnerability

The United Nations defines gender as a social construct involving roles, responsibilities, and opportunities influenced by power dynamics. Vulnerability is multidimensional and context-dependent, influenced by various determinants that may not be immediately apparent. Biological and physiological differences between women and men affect their resilience during disasters; for instance, malnourishment, pregnancy, and age impact women's capacities. Although women significantly contribute to agriculture, they are underrepresented in decision-making and lack access to resources necessary for agribusiness development (Ogato et al. 2009). Empowering rural women with resources like finance and land could reduce vulnerability and enhance food security. Involving women in climate change adaptation is crucial for ensuring food security and community resilience. The geographical traits of Ogbaru LGA heighten disaster susceptibility, making women's empowerment vital for sustainable agricultural practices amidst climate change.

14.3 Materials and Methods

14.3.1 Study Area

The study area is the Ogbaru Local Government Area (LGA) in Anambra State, Nigeria, characterized as a large wetland zone (See Figs 14.1 and 14.2). It is situated between latitudes 5°42′ and 6°08'N and longitudes 6°42′ and 6°50′E, covering four communities: Atani, Ochuche, Ohita, and Ogbakuba. The mean daily temperature is 28 °C, with a range of 24–28 °C throughout the year and nighttime temperatures between 16 and 18 °C (Egbueri, 2019). July relative humidity varies from 60 to 70%. Annual rainfall ranges from 1520 mm to 2020 mm, primarily from March to November (Ejikeme et al. 2017). The dry season lasts from December to February (Nzoiwu et al. 2017). Ogbaru's agricultural focus includes yam, cassava, rice, groundnuts, maize, and fishing, vital to the local economy and contribute significantly to the fish market in Onitsha.

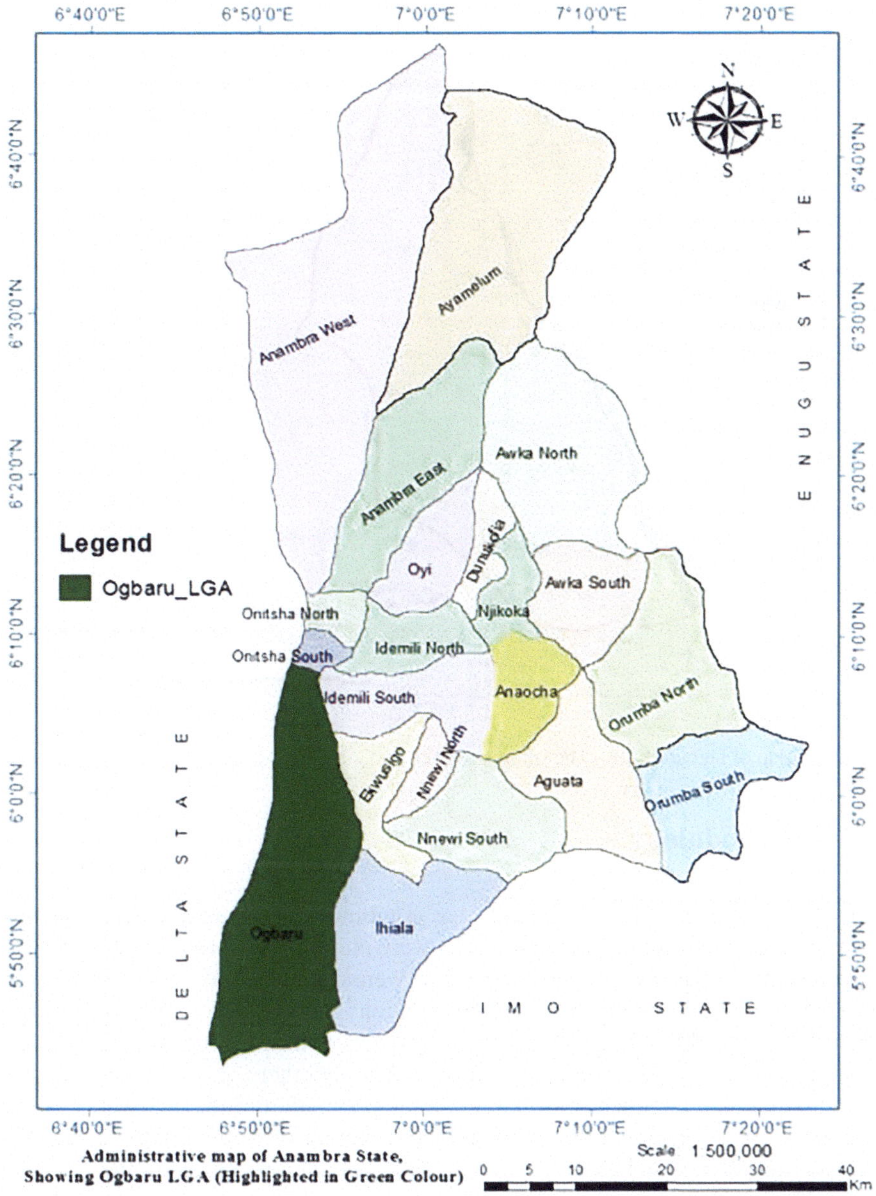

Fig. 14.1 Administrative map of Anambra state showing Ogbaru LGA (Highlighted in green colour) *Source* compiled by authors

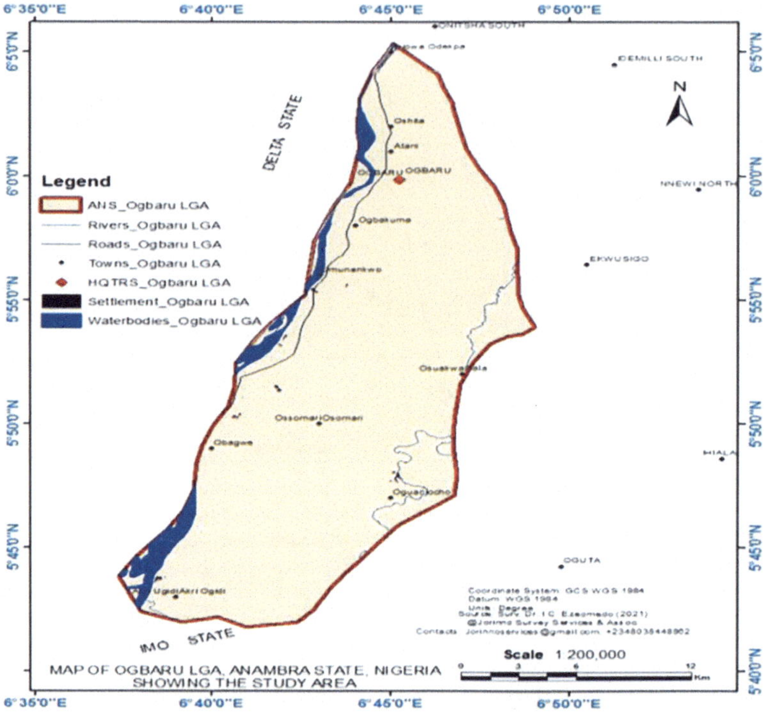

Fig. 14.2 Map of Ogbaru LGA (The study area) *Source* Compiled by authors

14.4 Methodology

Adopting a mixed methods approach that combines quantitative and recognition surveys, four communities in Ogbaru LGA were purposively selected: Atani, Ohita, Ochuche, and Ogbakuba. 400 questionnaires were distributed to women farmers in the study area. Additionally, physical observations were captured on camera. Data were collected on the socioeconomic characteristics of the respondents, such as age, income, level of education, and flood impacts, among others.

Additional data included information on the population of the study area. Since there are no available data records showing the number of women in the communities from the 2006 population census, the 1991 population census by the National Population Commission, which includes the population of women, was estimated and used to establish the number of women in the community. It should be noted that farming is the predominant occupation of the residents. According to NPC, the four communities had a total of 10,883 females. Using the Taro Yamen formula, 400 questionnaires were adopted and distributed. To quantitatively sample and establish the current population, projections were made to 2022 using an annual population growth rate of 2.8%, projecting the population to 25,575 women. This study was conducted over a year and a half from early 2023 to late 2024 to capture the flood season and beyond.

14.5 Findings and Discussion

The findings from the study revealed that women played a significant role in various farming activities. They were involved in tilling the land at the beginning of the farming season, making ridges, planting, weeding, harvesting, processing, marketing, and preparing food for their households. In addition to crop farming, they also participated in fishing and animal husbandry. While women contributed greatly to agricultural production, most of the sales of farm products were often managed by their husbands, even though the women were the primary contributors.

The study also indicated that women farmers produced crops such as yams, cassava, vegetables, maize, and other legumes, as demonstrated in Plate 1. Women were actively engaged in agricultural activities and were productive throughout the farming season and the year. In contrast, the men, who were typically landowners, engaged in planting crops alongside the women but spent less time on farming due to their commitments to other employment. As a result, many men had limited time to devote to agricultural activities (Fig. 14.3)

The age distribution of women farmers in the study area is presented in Table 14.1. It indicates that most women in this region fall within the productive age range, suggesting they are actively engaged in agricultural production to help meet food needs. For example, in Atani, the majority of women farmers are between 41 and 50 years old. A similar age group was observed in Ohita and Ochuche, while in Ogbakuba, women farmers predominantly belong to the 31 to 40-year age group.

Fig. 14.3 Different women farmers preparing harvested vegetables for sale

Table 14.1 Age distribution of respondents

Age	Atani		Ohita		Ochuche		Ogbakuba	
	N/R	%	N/R	%	N/R	%	N/R	%
>20years	32	16.1	6	22.2	5	6	16	20.3
21–30years	36	18.1	3	11.1	11	13.2	12	15.2
31–40years	33	16.6	4	14.8	29	35	19	24.0
41–50years	41	20.6	7	25.9	6	7.2	12	15.3
51–60years	31	15.6	4	14.8	17	20.5	8	10.1
>60years	26	13.1	3	11.1	15	18.1	12	15.2
Total	199	100	27	100	83	100	79	100

Source Compiled by authors

Table 14.2 shows the educational qualifications of women farmers in the four studied communities. 60% of the women farmers had attained a form of education from primary, secondary, and post-secondary education, and about 40% had no formal experience of education. This should show that the farmers could read and write and apply their knowledge concerning climate change, and articulate their challenges and resilience.

Table 14.3 presents the income data, highlighting the variation in earnings among women farmers in Ogbaru LGA, where agriculture serves as their primary source of livelihood. Most residents in the study area likely generate income from the sale of agricultural products, among other sources. The table indicates that the majority of women farmers earn between 30,000 and 90,000 naira (approximately 18.89 to 56.84 USD) per month.

Findings from the study, as shown in Table 14.4, indicated that climate change is the major determinant of flooding. It has a mean score of 4.8, accounting for the highest contributor to flooding in the study area. However, excessive rainfall and overflowing of upstream dams also contribute above 4 points out of a possible 5

Table 14.2 Educational qualification of women farmers in Ogbaru LGA

Educational qualification	Atani		Ohita		Ochuche		Ogbakuba	
	N/R	%	N/R	%	N/R	%	N/R	%
Post-secondary	26	13.1	10	37	17	20.5	12	15.1
Secondary	32	16.1	5	18.5	12	12.5	29	36.7
Primary	42	21.1	5	18.5	21	25.3	14	17.7
None	99	49.7	7	25.9	33	39.8	24	30.4
Total	199	100	27	100	83	100	79	100

Source Compiled by authors

Table 14.3 Monthly INCOME OF the women farmers in Ogbaru LGA

Monthly Income	Atani		Ohita		Ochuche		Ogbakuba	
	N/R	%	N/R	%	N/R	%	N/R	%
<30,000	63	31.7	7	25.9	28	33.7	19	24.1
30,001–N60,000	56	28.1	5	18.5	32	38.6	18	22.8
60,001–90,000	41	20.6	10	37	13	15.7	25	31.7
>90,000	39	19.6	5	18.5	10	12	17	21.5
Total	199	100	27	100	83	100	79	100

Source Compiled by authors

points, which is also very high. The least contributor to flooding in the study area is topography.

Women Farmers in Agricultural Production, Food Security, and Climate Resilience in Ogabru LGA.

Our study highlights the vital role of Ogabru women in agriculture, where they contribute more labor than men to household food production. Although men dominate commercial farming and control resources such as land and mechanized equipment, women represent over half of the agricultural workforce in the area.

The findings of this study are consistent with Doss (2018), where women typically manage smaller, diverse plots essential for household nutrition. In addition to crop production, they are involved in post-harvest processing, food preservation, and small-scale trade, which are crucial for rural livelihoods and nutritional stability, as noted in the works of Meinzen-Dick et al. (2019). Furthermore, women farmers are early adopters of sustainable practices like composting and water harvesting due to their roles in maintaining family well-being (Carr and Thompson, 2014). Despite these contributions, Ogabru women face systemic challenges such as limited land ownership, restricted access to credit, and a lack of decision-making power in agricultural systems. Addressing these barriers by enhancing women's access to resources could potentially boost farm productivity by up to 30%, significantly impacting global hunger reduction (FAO, 2011; UN Women, 2021). This effort aligns with national food security goals and sustainable development objectives.

Table 14.4 Causes of flooding

Variables	5	4	3	2	1	Mean
Excessive rainfall	241	76	39	21	11	4.2
Overflowing of major rivers	216	89	32	11	40	4.0
Insufficient drainages	136	99	68	46	39	3.6
Topography	102	99	88	54	51	3.4
Climate change	301	93		–	–	4.8

Source Compiled by authors

Additionally, the findings resonate with research by Odubo et al. (2019), indicating women's involvement in activities like water collection and trading. In Ogbaru LGA, women's participation in the fishing sector is substantial, attributed to their community's proximity to the Niger River. They handle over half the operational tasks in both fishing and farming. Akanwa et al. found that women primarily engage in setting traps, fishing, and selling their catch. Adepoju and Coker (2018) reported that women farmers contribute about 14–18h of labor to agricultural activities but own only about one percent of farm assets, a gap rooted in cultural and traditional marginalization. Thus, while they are fundamental to food security in Ogbaru LGA and beyond, significant disparities remain in asset ownership and support.

As shown in Table 14.1, women farmers in Ogbaru LGA are predominantly within their productive years, aged between 15 and 49. Their participation is essential for agricultural productivity and food security. Women in this age group possess both the physical capability and the accumulated experience necessary for labor-intensive farming tasks, including land preparation, planting, weeding, harvesting, and post-harvest processing (World Bank, 2022).

In Nigeria, women's involvement in agriculture is closely linked to household food availability and nutrition outcomes. Njoku et al. (2020) highlight that women of productive age in rural communities play a dual role, contributing to both on-farm production and off-farm economic activities. This duality not only sustains food systems but also strengthens rural economies. Despite facing challenges such as limited access to land, inputs, and extension services, women consistently demonstrate resilience and adaptability to climate-related and economic shocks, ultimately enhancing agricultural sustainability (Adebayo and Ogunsumi, 2023).

The education of women farmers in Ogbaru is essential for enhancing agricultural productivity, ensuring food security, and promoting climate resilience. According to Table 14.2, 60% of the women have received education ranging from primary to post-secondary levels, while 40% have no formal education. This educational disparity can lead to significant differences in knowledge acquisition, technology adoption, and resource utilization.

60% of women with education are more likely to adopt climate-smart agricultural practices, access extension services, and interpret early warning systems for climate-related risks (Asfaw and Maggio, 2018). They are also better equipped to manage farm resources efficiently and diversify their livelihoods, both critical for building resilience to climate variability (Njuki et al. 2014).

This educational advantage among women farmers contributes to higher crop yields, improved nutritional outcomes, and sustainable land use practices (Meinzen-Dick et al. 2019). Overall, the education of women farmers in Ogbaru plays a crucial role in enhancing agricultural productivity, ensuring food security, and fostering climate resilience.

Conversely, 40% of women without formal education may encounter barriers to accessing essential information, agricultural innovations, and credit, which limits their capacity to adapt. The lack of education can constrain decision-making, reduce participation in cooperative societies, and perpetuate reliance on traditional farming methods that may not be resilient to climate change (Mutenje et al. 2020). Consequently, disparities in education levels can lead to uneven food production outcomes, increased vulnerability, and widening gender gaps in agricultural development. Bridging this educational divide is crucial for building inclusive and equitable food systems in Nigeria and across Sub-Saharan Africa.

Notably, income level is a critical determinant of women farmers' ability to invest in sustainable agriculture, achieve food security, and build resilience against climate shocks. In this study, Table 14.3 revealed that women farmers earn between ₦30,000 and ₦90,000 monthly (equivalent to approximately $18.89–$56.84). However, their purchasing power and ability to adopt climate-resilient practices remain significantly limited (World Bank, 2021).

Low Income restricts access to essential inputs such as fertilizers, improved seeds, irrigation tools, and farm mechanization, which are crucial for boosting productivity and adapting to erratic weather patterns (FAO, 2017). Additionally, constrained financial resources reduce participation in cooperative savings groups or agricultural training programs, thereby excluding them from value chains that could enhance market access and increase resilience (Njuki et al. 2014). This income insecurity directly impacts household food availability, dietary diversity, and children's nutritional status, particularly in female-headed households.

Furthermore, insufficient income hinders long-term investments in soil conservation, agroforestry, and water management technologies that are vital for climate-smart agriculture (Mutenje et al. 2020). The cumulative effect is a cycle of low productivity, heightened vulnerability to climate risks, and persistent food insecurity. Thus, improving income generation opportunities for rural women farmers is not only essential for gender equity but also a strategic necessity for achieving sustainable agricultural growth and climate resilience.

Findings in Table 14.4 indicate that climate change is the leading cause of flooding, with the highest mean score of 4.8, followed by excessive rainfall with a score of 4.2. Climate-induced floods have become increasingly frequent and severe in Nigeria, posing significant challenges to agricultural production and disproportionately affecting women farmers. In Ogabru LGA, annual floods destroy farmlands, displace rural populations, and erode household food security, with women, who make up the majority of smallholder farmers, bearing the brunt of these impacts, as illustrated in Plate 2, which shows farmlands submerged in floods. Studies conducted by Akanwa et al. indicate that vast farmlands and agricultural products are lost, destroying the income and livelihoods of women farmers (Fig. 14.4)

Women involved in subsistence farming play crucial roles in food preparation and family nutrition, making them particularly vulnerable to climate-related disruptions. Flooding damages crops and erodes soil nutrients, leading to lower yields and income loss. Many women lack secure land rights and access to climate insurance, resulting in prolonged recovery times and increased food insecurity (Mehar et al. 2016). Their caregiving responsibilities also intensify during and after floods, limiting their participation in agricultural recovery efforts. Gender disparities in access to extension services, early warning systems, and adaptive technologies amplify these

Fig. 14.4 Graphics to the women's farmlands were submerged in water

vulnerabilities. Consequently, floods linked to climate change deepen poverty cycles, jeopardizing livelihoods and undermining agricultural sustainability in rural areas.

To address these challenges, it is vital to implement gender-responsive climate adaptation strategies that enhance women's access to information, financial resources, and resilient farming methods. Despite obstacles, women farmers in Ogabru build climate resilience and food security through various techniques. They can preserve cassava for up to a year (See plate 3), although rising flooding limits storage capacity due to a lack of resources, labor for processing, and drilling equipment. They also prepare boats with waterproof materials for navigating floods to access fields. Despite constraints, they employ adaptive strategies such as crop diversification, organic composting, and mixed cropping to mitigate climate risks and floods (Fig. 14.5).

Additionally, women reported utilizing early maturing crop varieties and mulching techniques to address unpredictable rainfall and maintain soil moisture, thereby minimizing their exposure to climate-related challenges. Other research has supported the idea that women are vital to indigenous knowledge systems that promote sustainable practices for soil and water management (Ayanlade and Radeny, 2020; Onyeneke et al. 2020).

Significantly, rural women farmers are essential for ensuring household food security by overseeing post-harvest activities such as storage, processing, and marketing. These efforts guarantee food availability throughout the year and generate income even during times of climate stress. To improve their effectiveness, it is crucial for targeted policies to emphasize gender-responsive agricultural training, climate financing, and access to extension services that cater to the specific needs and contributions of women.

Fig. 14.5 Local means of preserving and storing agricultural products (cassava)

14.6 Conclusion and Recommendations

Ogabru women farmers play a crucial role in sustaining food security, driving agricultural production, and enhancing climate resilience, particularly in light of increasing flood risks. They make significant contributions to the cultivation of staple crops, household nutrition, post-harvest processing, and informal market systems. Despite facing challenges such as limited access to land, credit, and extension services, these women adopt local strategies, including crop diversification, early planting, and traditional storage management techniques, to adapt to climate-induced floods. They act as first responders, safeguarding seeds, food stocks, and children, while also participating in recovery efforts through replanting and soil restoration. Their knowledge of indigenous practices and community-based resilience mechanisms strengthens adaptive capacity at the grassroots level. However, socio-economic barriers and the absence of gender-responsive policies limit their full potential in building resilient food systems. To support these women, it is essential to provide access to flood-tolerant seeds, early warning systems, and irrigation tools tailored to local contexts and managed by women's cooperatives. Furthermore, gender-sensitive training in climate-smart agriculture and flood preparedness should be offered, utilizing local languages and community facilitators. Ensuring equitable land tenure for women and expanding access to microcredit and insurance schemes are necessary to aid recovery from flood losses. It is also important to integrate women's voices into local and national disaster risk reduction and agricultural planning frameworks and encourage literacy and digital training to improve access to climate information, farming apps, and market platforms.

References

Adebayo O, Ogunsumi L (2023) Gender and sustainable agriculture: examining the roles and constraints of Women farmers in Nigeria. Afr J Gend Dev 15(1):45–60

Adepoju AB, Coker OJ (2018) Nutritional status, household food security, and dietary diversity of the elderly residents in Ilaro, Ogun state, Nigeria. Int J Food Sci Nutr. wwwfoodsciencejour nalcom

Aid A (2015) Delivering Women farmers' rights. Action Aid, Johannesburg, South Africa, Policy brief. www.actionaid.org

Ajah SO, Igbokwe JI, Igbokwe EC, Anoke E (2021) Management of flood disaster in Ogbaru L.G.A of Anambra state, Nigeria using remote sensing and GIS techniques. Int J Eng Sci Comput 11:4

Asfaw S, Maggio G (2018) Gender, weather shocks, and welfare: evidence from Malawi. J Dev Stud 54(2):271–291. https://doi.org/10.1080/00220388.2017.1283016

Ayanlade A, Radeny M (2020) Climate variability and household adaptation strategies in rural Nigeria. Clim Risk Manag 27:100210. https://doi.org/10.1016/j.crm.2020.100210

Carr ER, Thompson MC (2014) Gender and climate change adaptation in agrarian settings: current thinking, new directions, and research Frontiers. Geogr Compass 8(3):182–197. https://doi.org/10.1111/gec3.12121

Doss CR (2018) Women and agricultural productivity: reframing the issues. Dev Policy Rev 36(1):35–50. https://doi.org/10.1111/dpr.12243

Egbueri JC (2019) Evaluation and characterization of the groundwater quality and hydrogeochemistry of Ogbaru Farming District in southeastern Nigeria. Appl Sci 1(8):1–16

Ejikeme JO, Ojiako JC, Onwuzuligbo CU, Ezeh FC (2017) Enhancing food security in Anambra state, Nigeria using remote sensing data. Environ Rev 6(1):27–44

FAO (2011) The state of food and agriculture: women in agriculture, closing the gender gap for development. Food and agriculture organization of the United Nations, Rome

FAO (2017) The role of Women in agriculture. Food and agriculture organization of the United Nations, Rome. http://www.fao.org/gender/resources/infographics/en/

IPCC (2022) Climate change: mitigation of climate change contribution of working group III to the Sixth assessment report of the IPCC. https://www.ipcc.org. https://doi.org/10.1017/9781009157926

Janna V, James W (2021) Women's dual centrality in food security solutions: the need for a stronger gender lens in food systems transformation. Curr J Environ Sustain 3(1):100094

Mehar M, Mittal S, Prasad N (2016) Farmers' coping strategies for climate shock: is it differentiated by gender? J Rural Stud 44:123–131. https://doi.org/10.1016/j.jrurstud.2016.01.001

Meinzen-Dick R, Quisumbing A, Doss C, Theis S (2019) Women's land rights as a pathway to poverty reduction: framework and review of available evidence. Agric Syst 172:72–82. https://doi.org/10.1016/j.agsy.2017.10.009

Mutenje MJ, Farnworth CR, Stirling CM (2020) A gender analysis of climate-smart agriculture adoption in southern Africa. World Dev 130:104919. https://doi.org/10.1016/j.worlddev.2020.104919

Njoku JE, Okoye BC, Chukwuone NA (2020) Contributions of women in agriculture to household food security in southeastern Nigeria. Niger J Agric Econ 10(2):71–81

Njuki J, Kaaria S, Chamunorwa A, Chiuri W (2014) Women, livestock ownership and markets: bridging the gender gap in eastern and southern Africa. Int Livest Res Inst. https://hdl.handle.net/10568/51892

Nzoiwu PC, Ezenwaji EE, Enete IC, Igu NI (2017) Analysis of trends in rainfall and water balance characteristics of Awka. Niger J Geogr RegNal Plan 10(7):186–196

Odubo TV, Obafemi AA, Emenike GC (2019) Analysis of Women's participation in agriculture in selected states in Niger Delta, Nigeria. Int J Innov Soc Sci HumIties Res 7(2):51–61

Ogato GS, Boon EK, Boon, Subramani J (2009) Gender roles in CroProuction and management practices: a case study of three rural communities in Ambo District. Ethiop J Hum Ecol 27(1):1–20

Onyeneke RU, Nwajiuba CU, Igberi CO (2020) Climate-smart agriculture in Nigeria: an assessment of farmers' adoption in Abia state. Sci Agric 77(1):e20180150. https://doi.org/10.1590/1678-992x-2018-0150

UN Women (2021) Women and agriculture: empowering women for food security. https://www.unwomen.org

UN Women (2022) Explainer: how gender inequality and climate change are interconnected. https://www.unwomen.org/en/newsstories/explainer/2022/02/explainer

World Bank (2021) Poverty and equity brief: Nigeria. https://www.worldbank.org/en/country/nigeria/publication/poverty-and-equity-brief

World Bank (2022) Gender in agriculture sourcebook. World Bank, Washington, DC

World Meteorological Organization (2020) The state of the climate in Africa: a multi-agency publication. https://hedleyserver.metoffice.gov.uk/wmoic/

Chapter 15
A Geospatial Approach for Flood Risk Evaluation of Nigeria's Most Flood-Affected States

Saurabh Singh

15.1 Introduction

Flooding is one of the most damaging natural disasters worldwide (Jonkman 2005). In Nigeria, major floods have become more frequent in recent years. Because many communities are highly exposed and have limited capacity to cope, climate-driven extremes are putting lives, homes and livelihoods at serious risk (Akatah et al. 2025; Bello et al. 2024). This growing threat makes it essential to adopt smarter, more effective flood-risk management strategies. Floods disrupt roads, delay infrastructure projects and can even affect political processes. It is also damaging buildings, utilities, transport networks and agriculture (Dharmarathne et al. 2024). Data from the Climate Change Knowledge Portal (Fig. 15.1) show that between 1980 and 2020, about one-third (33%) of recorded natural hazards in Nigeria were floods which is the second most common event after epidemics. This pattern shows that why Nigeria needs geospatial technologies such as satellite data and GIS to strengthen flood-risk assessment, early warning and preparedness.

The Nigeria Flood Impact, Recovery and Mitigation Assessment Report (2022–2023) says 64% of households were affected by the 2022 floods (Umar et al. 2025). The effects of this disaster were quite wide-ranging, affecting people's jobs, housing stability, food security, and access to important services like schools and health care.

Figure 15.2 shows how wide the 2022 floods were and what percent of households were hit in each state. The impact was much worse in rural areas 74% of households were affected while about 40% were affected in cities. This is a sign that rural communities are very vulnerable. Kogi and Anambra are especially most affected

S. Singh (✉)
Department of Civil Engineering, Poornima University, Sitapura, Rajasthan, India
e-mail: saurabh.singh@poornima.edu.in; saurabhsingh@ees.hokudai.ac.jp

Faculty of Environmental Earth Science, Hokkaido University, Sapporo, Hokkaido, Japan

A. Nath et al. (eds.), *Geospatial Application of Coastal Hazards*, Springer Water,
https://doi.org/10.1007/978-3-032-20076-1_15

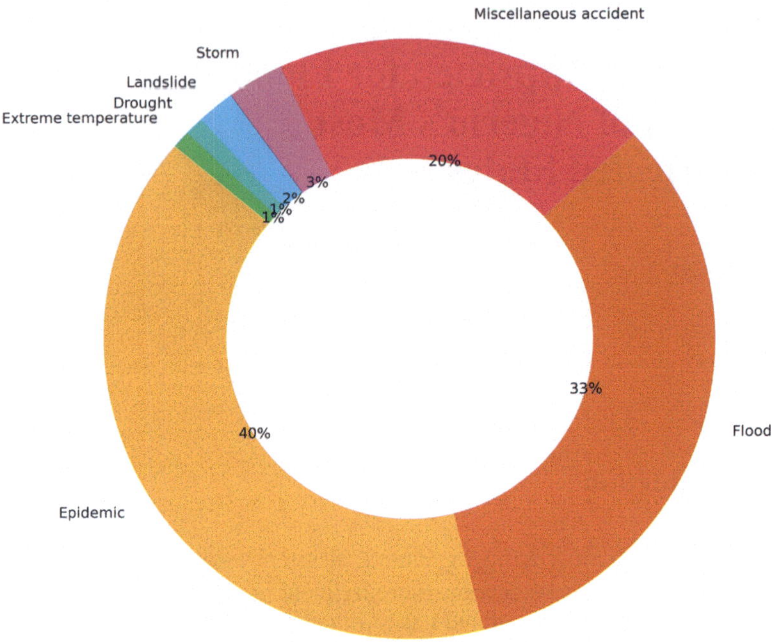

Fig. 15.1 Average annual natural hazard occurrence from 1980–2020 of Nigeria

state in Nigeria, so urgent action and better preparedness are needed in that places (Fig. 15.3).

Anambra and Kogi carefully measure the damage in terms of lost property, transportation delays, power and water supply problems, lost food stocks, lost lives, and job losses. The table's detailed list shows the different vulnerabilities and the many ways floods can affect people, which are important areas of concern for disaster management and efforts to build resilience. Nigeria is growing the fastest of the ten biggest countries in the world. As a result, Nigeria's population, which is now the seventh largest in the world, is expected to grow larger than that of the United States and become the third largest country in the world just before 2050. Nkwunonwo et al. (2016) reviewed urban flood risk management in Lagos, identifying challenges such as inadequate data, limited public awareness, and insufficient mitigation strategies, emphasizing the need for improved flood modeling and communication (Nkwunonwo, Whitworth, and Baily 2016). Adelekan (2011) assessed vulnerability in Abeokuta, finding that while many residents had experienced floods, most were unprepared for the 2007 flood, highlighting the need for better early warning systems (Adelekan 2011). Olanrewaju et al. (2019) examined the health impacts of flooding, identifying waterborne disease outbreaks and advocating for comprehensive disaster management that includes health considerations (Olanrewaju et al. 2019). Nkeki et al. (2022) introduced GIS and HEC-RAS for flood risk mapping in the Ona River Basin, revealing 48.2% of the area at moderate to high flood risk, important for prioritizing

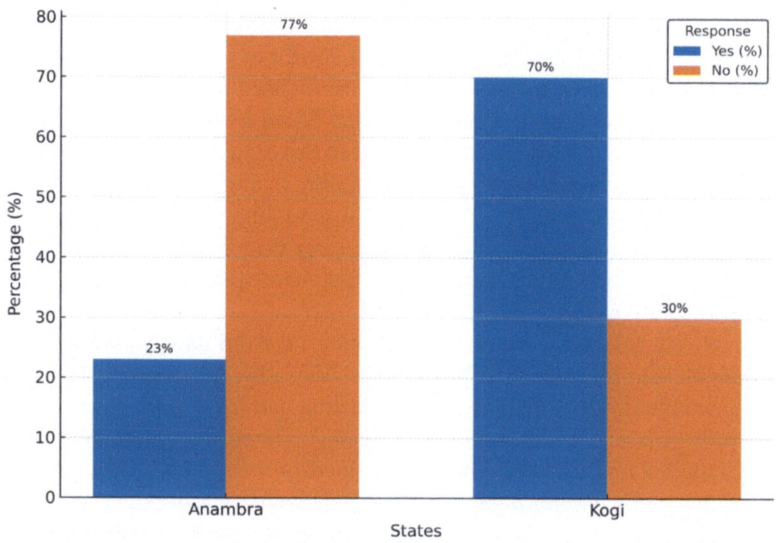

Fig. 15.2 Households affected by the 2022 floods

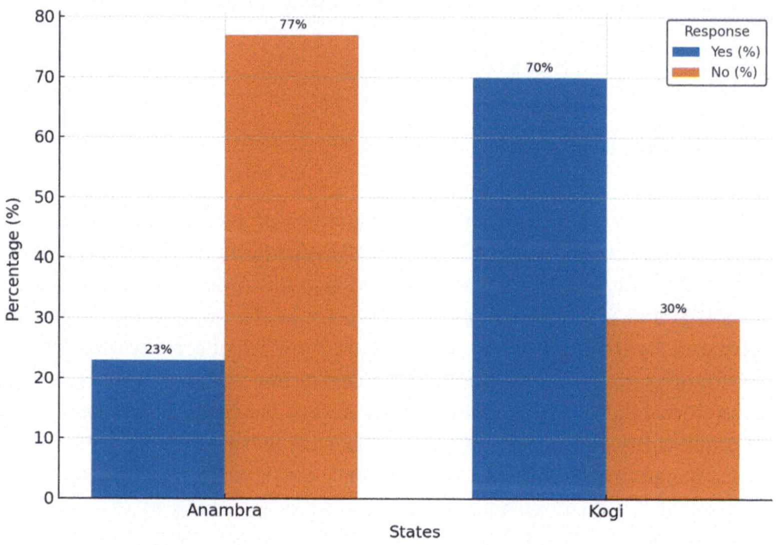

Fig. 15.3 Severity of impact on households

mitigation efforts (Nkeki et al. 2022). Musa and Shabu (2019) studied the impact of flooding on land use along the Benue River, suggesting the relocation of agriculture from flood-prone areas to reduce food insecurity (Musa and Shabu 2019).

Though there have been marked improvements in flood risk management, however, there is a scarce record of detailed region-specific flood risk assessments within Nigeria particularly in States such as Kogi and Anambra. Previous research has been conducted in urban centers such as Lagos, but rural and semi-urban areas with different flood risk dynamics require more focused analysis. Furthermore, the fully possible usage of remote sensing and GIS-based hydrological modeling have not been implemented in these areas. The goal of this study is to assess and delineate flood risk in two Nigerian states Kogi and Anambra, using a suite of geospatial tools such as remote sensing and GIS, hydrological modeling, including applications in assessing flood prone areas and classifying risks of flooding. The focus is on delivering actionable recommendations and applicable strategies for dealing with natural disasters in such areas.

The novelty in this study is the concentration on Kogi and Anambra states and use of sophisticated geospatial techniques for flood risk mapping at fine-scale. By integrating satellite images, hydrological modeling and local demographic information, this study provides a more detailed regionalized flood risk assessment in Nigeria and offers a model applicable to other global flood-prone regions. This study is relevant given that it fills essential gaps in flood risk evaluation in Nigerian vulnerable zones. Results will inform the development of local-specific flood risk reduction plans, community level resilience building and policy making, thus contributing to SDG 13: climate action. Through innovative geospatial approaches, the article offers a strong basis for addressing flood hazards and promoting sustainable resilience in flood-prone areas.

15.2 Study Area

The research focuses on Kogi and Anambra states of Nigeria that were hit much significantly by the 2022 floods in Nigeria and considered to be more vulnerable to flooding post-2022 events. Delta State is included in this report simply to offer context it was affected, and remains vulnerable (Fig. 15.4). The two states have different topography and hydro-logies that gave a wide view of observing the differen-tial effects of flooding in diverse ecological and eco-nomic zones in the country.

The selection of these states presents a transparent and equitable means of focusing on flood vulnerability in Nigeria. The study seeks to generate lessons that are valuable for Kogi and Anambra as well as other regions experiencing similar problems from such analyses of varied environments and conditions. The aim of the method is to assist in creating locally relevant strategies to manage flood risks and prevent disasters, not just in Nigeria but elsewhere on Earth.

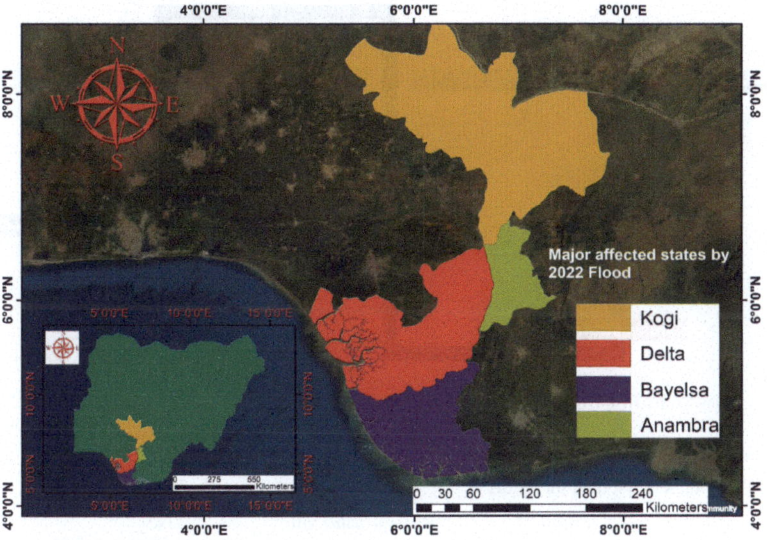

Fig. 15.4 Map of the study area

15.3 Methodology

The present study applies an integrated geo-spatial analysis to assess and map flood risk across the Nigerian most susceptible states. The strategy is comprised of three components: (1) monitoring system to look at water and land conditions from satellites, (2) hydrological modeling using GIS technology to determine the elevation, slope, drainage features and how water might flow or accumulate and (3) human impacts analysis by using a household questionnaire to understand who is affected and how communities are adapting. Together, these steps are used to create clear maps of flood-prone areas and risk levels and to link physical risk with people's experience and needs (Fig. 15.5).

15.3.1 Data Collection

Satellite images and GIS datasets were used to capture conditions before and after the floods (Table 15.1). Sentinel-1 images were used because they are not blocked by clouds and are good at showing open water during storms. Landsat-8 optical images were used to calculate NDWI (for water) and NDVI (for vegetation), so changes in water and vegetation could be seen across the study area. The SRTM digital elevation model was used to provide elevation, from which slope, flow direction, and flow accumulation were derived; these layers are important for understanding how water is likely to move and where it may collect. Global Surface Water (GSW) data

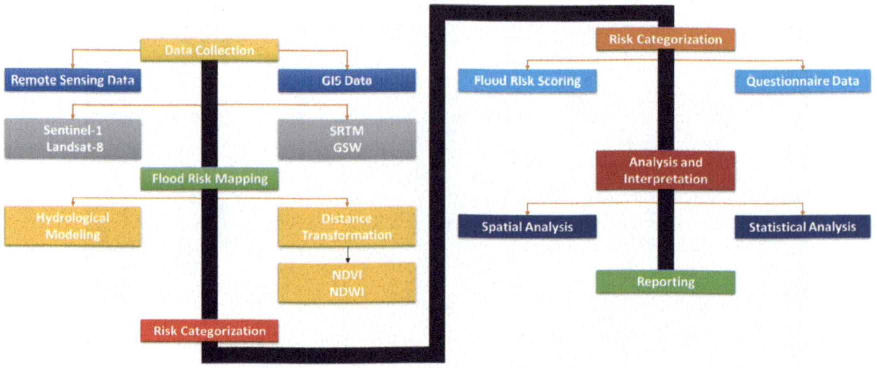

Fig. 15.5 Flowchart illustrating the methodology for flood risk assessment

Table 15.1 Datasets used in the study

Dataset	Source	Resolution/Date	Use
SRTM DEM (30 m)	NASA	30 m; 2000	Elevation, slope, flow direction/accumulation
Global surface water (GSW) v1.x	JRC	30 m; 1984–2023	Permanent water mask and water occurrence
Landsat-8 OLI	USGS	30 m; 2022–2023 (pre/post flood)	NDWI (water), NDVI (vegetation)
Sentinel-1 (optional)	ESA	10 m; 2022–2023	Open water under clouds
Administrative boundaries	GADM/FGN	1:250 k	State masks and area stats

were used to identify where permanent and seasonal water bodies are, which helps measure proximity to water and water persistence. Administrative boundary data were used to clip all layers to each state and to compute area statistics. All datasets were brought to a common projection and a 30-m resolution (unless noted), so the layers are consistent and can be compared across states.

15.3.2 Flood Risk Mapping

We use GIS to build hydrology layers from the SRTM DEM, so we can see slope, flow direction and where water is likely to collect. We also use the JRC Global Surface Water (GSW) maps to show where water has been present over many years; this helps us measure how close each place is to permanent water and where water tends to return. GSW is a long-term, high-resolution record of surface water presence and seasonality, so it is well suited for flood work (Pekel et al. 2016). To map water on optical images, we calculate two standard indices on Landsat-8. The NDWI is used to highlight open water, and the NDVI is used to track vegetation condition before and after floods. These indices are simple and well known, and a modified NDWI with SWIR can be used when needed to reduce confusion with built-up areas (McFEETERS 1996). Because clouds are common during floods, we also consider Sentinel-1 SAR. Radar is not blocked by clouds and can detect floodwater over large areas. Recent studies show that Sentinel-1 can improve near-real-time flood mapping and that machine-learning methods on SAR are increasingly effective (Garg et al. 2024). Our mapping follows the multi-criteria approach widely used in Nigeria. Prior work has combined elevation, slope, drainage/flow accumulation, land use, rainfall and distance to rivers to produce flood-susceptibility maps with weighted overlays or AHP. For example, studies in Kaduna and Ilorin West use GIS-MCDA/AHP with terrain and proximity factors to identify high-risk zones for planning. In Anambra, earlier work reports that most of the state falls from medium to very high vulnerability, with river corridors being the most exposed. These findings support our choice to weight terrain and proximity to water strongly in the risk score (Abubakar et al. 2025).

15.3.3 Risk Categorization and Validation

Based on the outcomes of hydrological modeling, distance analysis and vegetation/water indices, a scoring system was devised. This system categorized regions into very low, low, medium, high and very high-risk zones.

Although not the main focus, data from questionnaires filled out by affected households gave us more information about how flooding affects society. This information helped to check and confirm the results of the geospatial analysis, making sure that we fully understood the effects of the flood.

15.3.4 Analysis and Interpretation

Using GIS, spatial analysis methods were used to put demographic data on top of flood risk maps. This made it possible to find important areas where a high risk of flooding meets a high population density or important infrastructure.

Statistical tools were used to look at the relationship between the level of flood risk and the effects on different groups of people. This analysis yielded quantitative insights regarding the severity of flooding in communities situated within the designated high-risk zones.

15.3.5 Reporting

We produced maps that show flood-risk levels for each state, and we added overlays that show where people and key infrastructure are most exposed. These maps are useful for planning because they help officials see priority areas and decide where action is needed first an approach that is recommended in international guidance on flood-risk mapping and disaster risk assessment.

15.4 Result and Disscussion

The results of our multi-scale geospatial analysis including remote sensing information and GIS-mediated hydrological modeling, emphasize complex and variable flood risk in Kogi and Anambra states. We look at the numbers that indicate how risky these zones are. With the more detailed mapping and risk categorisation, it is easier to select where resources should be concentrated—this enables better strategic planning for comprehensive flood management and resilience building.

The map shows the geospatial flood risk assessment for Kogi State, Nigeria. It divides the area's risk of flooding into four levels: very low, low, medium, high, and very high (Fig. 15.6). The assessment shows that a large part of Kogi State is in the medium flood risk category, covering 22,767.99 km^2. The areas with high risk cover 5398.47 km^2, and the areas with very high risk cover 157.41 km^2. 218.53 km^2 of land have a low risk of flooding. This detailed, spatial measurement is important for deciding which flood prevention and resilience-building projects should get the most attention in the state. It helps direct resources to the areas that are most likely to be affected by flooding events.

The map reflected the flood risk zones in Anambra State and portions were marked according to the susceptibility of flooding. High-risk areas are 1238.936 km^2 as it is the region required special attention in flood control and management (Fig. 15.7). And by the way, there's 3233.248 km^2 of medium-risk areas, so we must spend a lot of attention on how to reduce it all this risk." Then there are 45.894 km^2 high-risk areas, which are even smaller areas that "there is a need for immediate action". The low and very low risk areas are both 12.365 km^2, and 14.352 km^2 respectively. This makes them relatively less at risk, but they are still crucial. This delineation serves as an inventory for flood hazard mapping and land use planning in Anambra State (Fig. 15.8).

Our multi-tiered geospatial analysis, which combines remote sensing data and GIS-based hydrological modelling, shows that Kogi and Anambra states have a wide range of flood risks. The numbers reveal just how risky these areas are. The study revealed that 22,767.99 km^2 of landmass in Kogi State is under medium threat followed by 5398.47 km^2 at high threat and 157.41 km^2 at very high threat. That's an indication that we need to zero in on mitigation. Anambra State has 1238.936 km^2

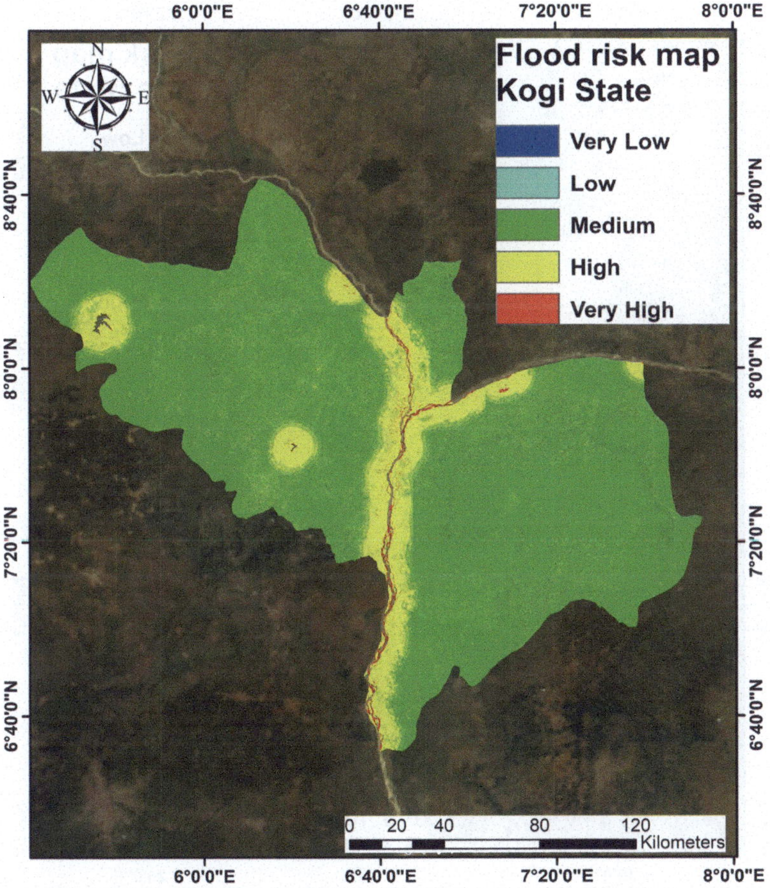

Fig. 15.6 Comprehensive flood risk assessment for Kogi State, Nigeria State

of high-risk areas and 45.894 km^2 of very-high-risk areas respectively. These are crucial areas on which to be prioritizing for flood prevention and preparedness.

It is to achieve the efficient resource distribution, which will help policy makers, urban planners and disaster management experts in mind bottleneck places. The risk classification and mapping facilitate the identification of priority areas for intervention so that flood management and resilience can be more effectively programmed. Furthermore, these results demonstrate the benefit of using geospatial analysis to locate flood-prone areas and provide a source of information for decision makers to respond strategically so that community resilience and sustainable socio-economic development will be assured when impacted by climate-induced flooding.

To complement these geospatial findings, we conducted a public opinion survey on government-led flood prevention and maintenance activities. Of the 1000 people questioned, 44.1% said the government should act to prevent flooding with 36.9%

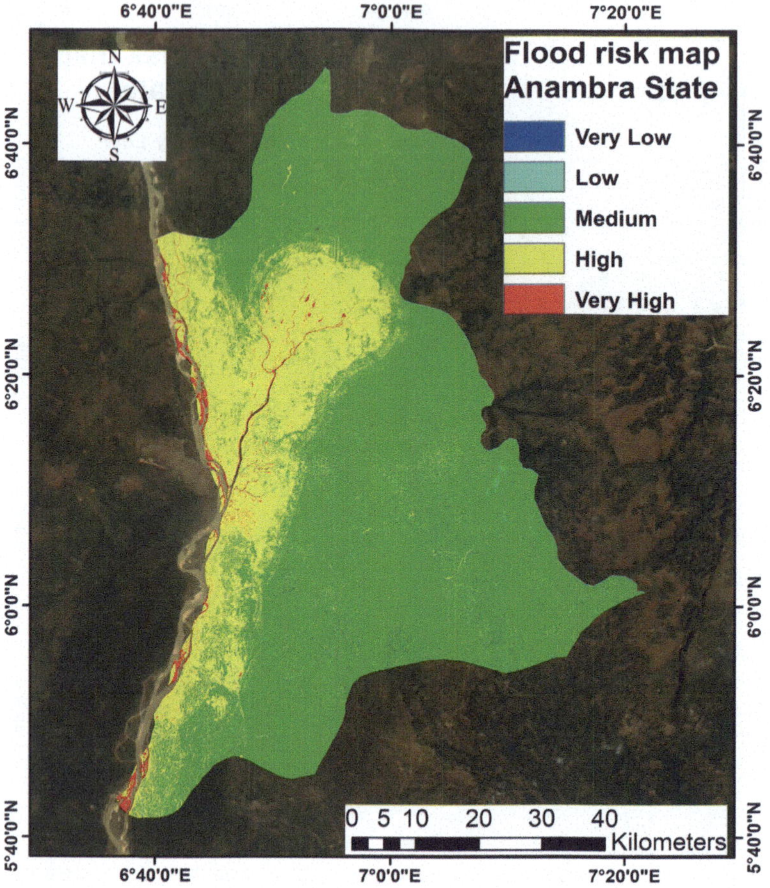

Fig. 15.7 Comprehensive flood risk assessment for Anambra State, Nigeria

saying no and 18.9% unsure. Regarding regular maintenance of drainage systems as one of the most important ways to prevent flooded, 65.8% claimed they did not see any kind upkeep, whereas 15.3% stated that they had seen it and 18.9% were indecisive (Fig. 15.9).

There's a significant catch in that what people think my government does for flooding and what is actually done, particularly to keep our drainage systems maintained: Compare these numbers as well to the patterns of responses collected before Katrina. Whether the inverse environmental policy would form part of twenty first century strategies for flood risk management remains an open question, but this contrast highlights the need for governments to get better not just at their communications but what they do—so that perceptions match more closely with what strategies actually are for flood risk management. Africa's most populous country is thus more exposed to the negative impacts of climate change, in particular floods. And it's not

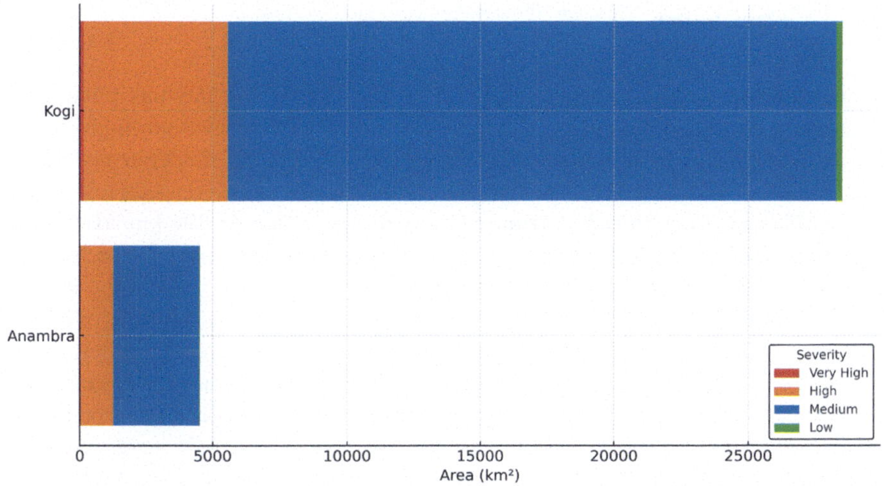

Fig. 15.8 Bar chart illustrating the distribution of flood risk severity across the study area

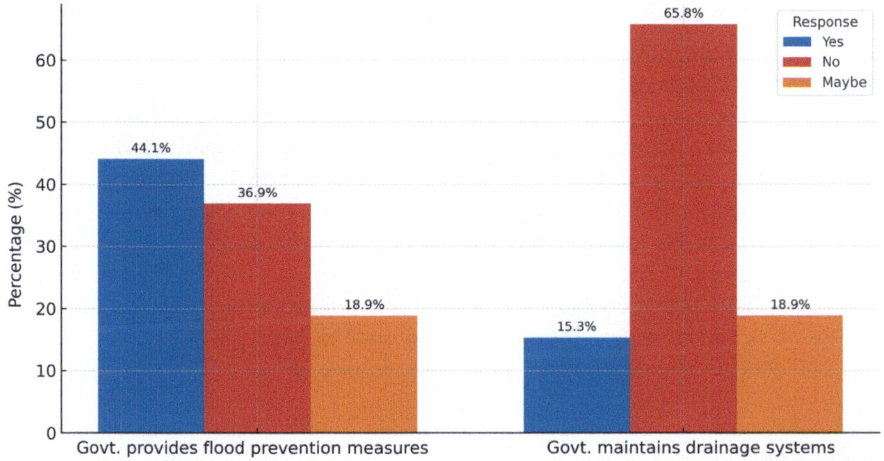

Fig. 15.9 Bar chart showing public responses to government-led flood prevention and maintenance initiatives

helped by another significant weakness in our early action arsenal: Nigeria lacks a dedicated flood reporting tool and effective means of disseminating early warning information. Therefore, societies are less ready to face them and more at risk of floods. It is clear that the Nigerian government has worked to mitigate the risk of flooding, but more work needs to be done in terms of early warning systems and establishment of real-time flood reportage system.

15.5 Conclusion

In this study the satellite imagery and GIS are used to delimit the maximum flood risk areas in Kogi and Anambra. The number of places in the medium-high to very high risk categories is numerous. Kogi has 22,767.99 km^2 at moderate risk level, 5398.47 km^2 at high and cc157.41 km^2 at very high levels of risk respectively. There is 1238.936 km^2 of high risk Anambra land while a further 45.894 km^2 (0,63% of projected Anambra State Land) is considered to be very high risk. These findings indicate where the first steps could be made.

Why it matters: Nigeria has many frequent and severe flooding. Millions of people were affected, and hundreds died in the flood of 2022, for instance. Families, roads, schools, hospitals and farms require better planning, early warning systems and maintenance to remain safe. High resolution satellite images, elevation models and GIS are cheap and useful geospatial tools for this task. They guide governments to hotspots, when and where to deploy money first and how well they make progress over time. This is consistent with international principles of disaster risk reduction, and country-level risk assessment. UNDRR+1 And equipment like Sentinel-1, but also other radar satellites, are able to map floods even if there are clouds in the way—a huge help when it comes to monitoring events while they unfold in storm conditions. These maps would support a range of specific actions, from regular cleaning and maintenance of drainage systems to land use rules to keep new buildings out of the riskiest areas, preserving or restoring wetlands for nature-based flood retention when possible, and early warning systems that are people-focused and include last-mile communication. This would support SDG 13 (Climate Action) by pushing us towards more climate history-risk sensitive planning. This process should be further refined in the future with satellite images being taken each year, expanded to cover more states and directly linked to river models and local flood reports. This will provide more accurate predictions and enable Nigeria and other countries at risk from flooding to plan, lose less and be better prepared, for now and the future.

References

Abubakar ML, Abdussalam AF, Mohammed HI, Ahmed MS, Musa AS (2025) Flood susceptibility assessment via GIS and the analytical hierarchy process in Kaduna. Niger Discov Geosci 3(1):94. https://doi.org/10.1007/s44288-025-00204-8

Adelekan IO (2011) Vulnerability assessment of an urban flood in Nigeria: Abeokuta flood 2007. Nat Hazards 56(1):215–231. https://doi.org/10.1007/s11069-010-9564-z

Akatah BM, Onyeaka H, Onungwe I, Akpan PP, Igulu KT, Elenwa EP (2025) Climate change-induced heatwaves in Nigeria: causes, challenges, and adaptive strategies. J Environ Manag 394:127433. https://doi.org/10.1016/j.jenvman.2025.127433

Bello M, Singh S, Singh SK, Pandey V, Kumar P, Meraj G, Kanga S, Sajan B (2024) Geospatial analysis of flood susceptibility in Nigeria's vulnerable coastal states: a detailed assessment and mitigation strategy proposal. Climate 12(7):93. https://doi.org/10.3390/cli12070093

Dharmarathne G, Waduge AO, Bogahawaththa M, Rathnayake U, Meddage DPP (2024) Adapting cities to the surge: a comprehensive review of climate-induced urban flooding. Results Eng 22:102123. https://doi.org/10.1016/j.rineng.2024.102123

Garg S, Dasgupta A, Motagh M, Martinis S, Selvakumaran S (2024) Unlocking the full potential of Sentinel-1 for flood detection in arid regions. Remote Sens Environ 315:114417. https://doi.org/10.1016/j.rse.2024.114417

Jonkman SN (2005) Global perspectives on loss of human life caused by floods. Nat Hazards 34(2):151–175. https://doi.org/10.1007/s11069-004-8891-3

McFeeters SK (1996) The use of the normalized difference water index (NDWI) in the delineation of open water features. Int J Remote Sens 17(7):1425–1432. https://doi.org/10.1080/01431169608948714

Musa SD, Shabu T (2019) Using geographic information system to evaluate land use and land cover affected by flooding in Adamawa state, Nigeria. Jàmbá J Disaster Risk Stud 11(1):11. https://doi.org/10.4102/jamba.v11i1.494

Nkeki FN, Bello EL, Agbaje IG (2022) Flood risk mapping and urban infrastructural susceptibility assessment using a GIS and analytic hierarchical raster fusion approach in the Ona River basin, Nigeria. Int J Disaster Risk Reduct 77:103097. https://doi.org/10.1016/j.ijdrr.2022.103097

Nkwunonwo UC, Whitworth M, Baily B (2016) Review article: a review and critical analysis of the efforts towards urban flood risk management in the Lagos region of Nigeria. Nat Hazards Earth Syst Sci 16(2):349–369. https://doi.org/10.5194/nhess-16-349-2016

Olanrewaju CC, Chitakira M, Olanrewaju OO, Louw E (2019) Impacts of flood disasters in Nigeria: a critical evaluation of health implications and management. Jàmbá J Disaster Risk Stud 11(1):9. https://doi.org/10.4102/jamba.v11i1.557

Pekel JF, Cottam A, Gorelick N, Belward AS (2016) High-resolution mapping of global surface water and its long-term changes. Nature 540(7633):418–422. https://doi.org/10.1038/nature20584

Umar IA, Bukar WM, Mustapha A, Kyari A, Bukar MA, Ibrahim HH, Jimme MA, Kachalla U (2025) Flood risk and resilience: evidence from the 2024 flood in Maiduguri, Nigeria. Int J Environ Climate Chang 15(1):490–505. https://doi.org/10.9734/ijecc/2025/v15i14708

Chapter 16
Assessing Coastal Erosion and Shoreline Shifts Due to Cyclone Fani (2019) in Puri District, Odisha

Indrakant Behera and Sathi Mondal

16.1 Introduction

Cyclones have a significant impact on coastal dynamics due to their strong winds, intense precipitation, and storm surges. The intensity, course, and proximity of these cyclonic storms are intimately related to the severity of coastal erosion and inundation, as noted by Mohanty et al. (2019). According to Li et al. (2001), landforms along the coast are constantly changing, making the shoreline one of Earth's most significant and distinctive features. This is caused by a variety of coastal activities, including storms, wind direction, and wave motion.

The two main components of coastal dynamics that shape the world's coastlines and have a significant impact on coastal communities and ecosystems are coastal erosion and shoreline changes. The term "shoreline change" refers to the gradual transformation of a coastline over time, encompassing changes in its location, form, or features. The shoreline undergoes various alterations due to both human and natural influences. Man-made factors include infrastructure development and the impacts of climate change, while natural processes such as erosion, accretion, morphological transformation, and sediment transport also play major roles (Jana et al., 2012, 2015). Coastal erosion refers to the gradual wearing away or displacement of land or dunes along the coast caused by wave action, tidal forces, currents, and storm surges. This process results in the loss of valuable coastal land, infrastructure, and habitats, thereby threatening human settlements, the economy, and biodiversity (Jana et al., 2013, 2016, 2017).

I. Behera (✉)
Department of Civil Engineering, Indian Institute of Technology Roorkee, Roorkee, Uttarakhand, India
e-mail: indrakant.ce@sric.iitr.ac.in

S. Mondal
Department of Geography, Hemvati Nandan Bahuguna Garhwal University, Srinagar (A Central University), Srinagar, Uttarakhand, India

A. Nath et al. (eds.), *Geospatial Application of Coastal Hazards*, Springer Water,
https://doi.org/10.1007/978-3-032-20076-1_16

Worldwide, coastal erosion brought on by cyclones is a serious worry, particularly in areas that are vulnerable to cyclonic storms, such as the Gulf of Mexico, the Caribbean, South-East Asia, and the Indian Ocean. According to UN estimates, approximately 40% of the world's population lives within 100 km of the shore, making them especially vulnerable to the problems associated with erosion. This vulnerability highlights the urgency of understanding and addressing the complex impacts of coastal erosion.

Research on coastal erosion and related topics has been conducted by numerous researchers and scientists. AlRuheili (2022) found that, as a result of the Shaheen Cyclone in Al Khaboura City, Oman, the coastline eroded at a rate of 16.33 m per year, representing 7.82% of the total coastline, with a maximum erosion distance of 1428.5 m. Abushandi and Abualkishik (2020) reported that, due to Cyclone Gonu in Oman, the coastline experienced severe negative volumetric shoreline erosion, averaging 5.2 m per year, with some variations near catchment outlets. In the Indian subcontinent, Mishra et al. (2021) found that Cyclone Nisarga in the coastal zone of Maharashtra caused erosion along 56.32% of the shoreline, with the highest levels observed near Mumbai. Noujas et al. (2023) studied the impact of Cyclone Phethai on the eastern coast of India and reported that storm surges of 0.5–1 meter caused flooding ranging from 40 to 120 m along the affected coastline. Herlekar et al. (2023) used Survey of India toposheets and remote sensing data in GIS to analyze coastline changes along Maharashtra's West Coast between 1980 and 2013. Their results indicated significant erosion of 45.28%, deposition of 40.0%, and stability of 14.70% between Alibag and Shrivardhan.

Due to the significant impacts of cyclone-induced shoreline changes and coastal erosion on coastal communities and ecosystems, research on these phenomena is crucial. A key location for studying the intricate relationship between coastal erosion and cyclone effects is Puri, in the eastern Indian state of Odisha, particularly in light of the destructive consequences of Cyclone Fani in 2019. As Beura (2016) and Kar and Banerjee (2021) have shown, the persistent pressures of ocean waves and tides modify the coastal erosion process in this region, reflecting a dynamic equilibrium of erosion and deposition. The aftermath of Cyclone Fani highlighted Puri's vulnerability to such storms and the significant coastline erosion they cause, underscoring the urgent need for a thorough examination of these interactions. Several studies have utilized remote sensing, GIS, and statistical assessments to map shoreline changes and understand coastal dynamics (Ahammed & Pandey, 2018; Mishra et al., 2019; Markose et al., 2016; Jose et al., 2018).

This research mainly focuses on revealing the complex relationships that exist between cyclones and coastal erosion processes. A thorough evaluation and documentation of shoreline changes pre- and post-Cyclone Fani using GIS and remote sensing technologies has been carried out. Precise quantification of erosion and accretion processes along the coast and identification and delineation of vulnerable zones prone to significant coastal erosion, recognizing areas most susceptible to environmental shifts and cyclonic impacts, are the primary goals of the research. Sophisticated procedures are being used in this study, such as statistical techniques, data interpretation from remote sensing, and Geographic Information System (GIS) analysis.

16.2 Study Area

Puri District is a coastal District in the eastern part of Odisha, India. This District derives its name from the heritage city of Puri, one of the four pilgrimage centers (Char Dham) of India. Covering an area of 3479 sq./km, it is situated along the Bay of Bengal at approximately 19.8177° N latitude and 85.8315° E longitude as shown in Fig. 16.1. It boasts a unique blend of geographical, climatic, & topographic features. Paddy, Wheat, Mung, bori, Kulthi, and groundnut are some of the major crops grown in the District. The number of rivers flowing through the District helps in sustaining agriculture. Being in proximity to the Bay of Bengal, this District has a tropical climate. The summers are from March to June, the monsoons from June to September, and the winters are from October to February. The minimum temperature of the District is approximately 16.45 °C, and the maximum 33.9 °C, with an average annual rainfall of 1449.1 mm (Mullick and Das, 2014).

Located on India's east coast, Puri is frequently hit by storms that originate in the Bay of Bengal and make landfall there, as shown in Fig. 16.2. Due to its unique physical structure–namely, the bulging landmass between Puri and Bhadrak–the area

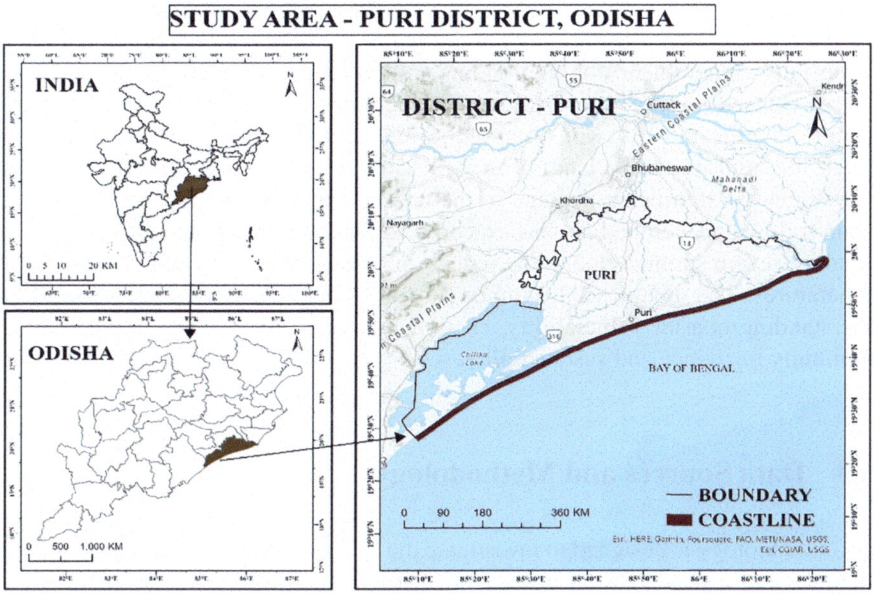

Fig. 16.1 Study area showing the coastline of Puri district of Odisha, India

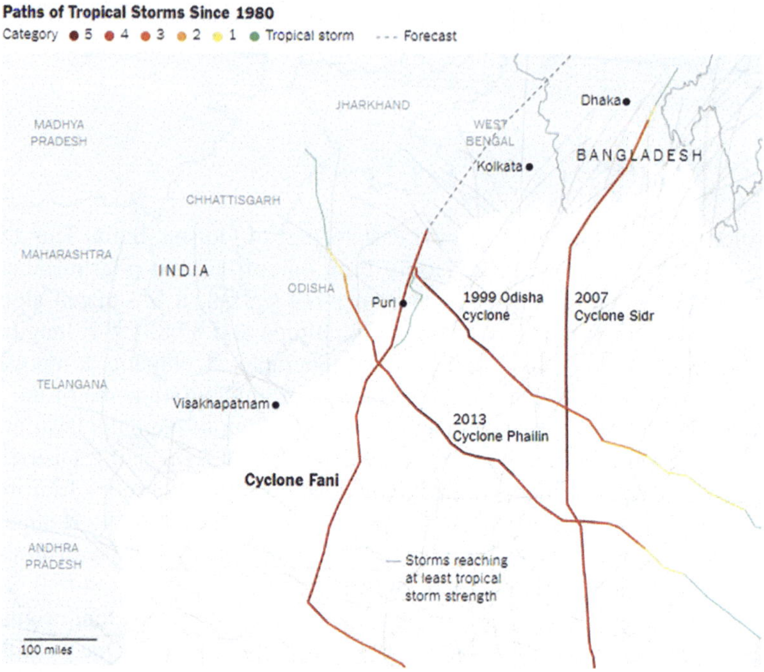

Fig. 16.2 Path of major cyclones that hit on Puri district. *Source* New York Times, NOAA (Bloch et al., 2019)

is especially vulnerable to coastal flooding and cyclonic activity. In the past, the city has experienced multiple powerful storms, including the mega storm in 1999 and Cyclone Phailin in 2013, which caused immense destruction and fatalities. Even though they did not make landfall in Puri, storms like Hudhud in 2014 and Titli in 2018 caused significant damage to the coastal villages. In addition, the altered terrain following Cyclone Fani has exacerbated regional weather patterns, resulting in a more severe summer due to the reduction of coastal canopy and elevated local temperatures. The area is vitally important for a comprehensive study focused on understanding coastal vulnerability, erosion dynamics, and their implications for community resilience and sustainability.

16.3 Data Sources and Methodology

This methodology is designed to investigate the impact of Cyclone Fani on shoreline dynamics and coastal erosion in the Puri district. The project objectives are well-defined, focusing on assessing the effects of Cyclone Fani. Figure 16.3 represents the systematic approach of the methodology opted to assess the shoreline change and coastal erosion dynamics. Starting with extracting shoreline and baseline data from satellite imagery, then by creating transects, which help in the calculation of shoreline statistics and identifying vulnerable zones of coastal erosion and accretion. The datasets related to cyclone Fani, such as wind direction, wind speed, and atmospheric pressure, are used to understand the impact of cyclone Fani on shoreline change and coastal Erosion (Table 16.1). With the help of various tools and plugins available in QGIS, the results obtained were visualised through a map to yield more enhanced results.

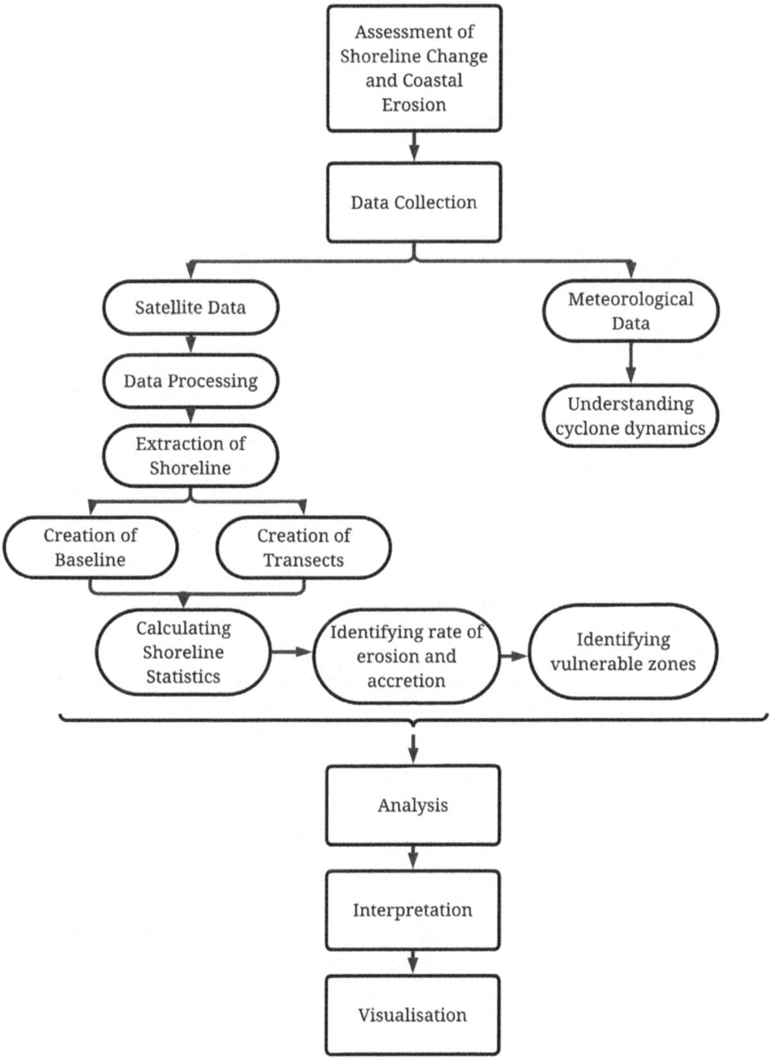

Fig. 16.3 Flow chart of the adopted methodology in the present study

Table 16.1 Datasets used and their sources

Datasets	Purpose	Source	URL
Landsat 8 OLI/ TIRS (30 m resolution)	Extraction of shoreline and baseline	USGS earth explorer	https://earthexplorer.usgs.gov/
Village boundary	Identifying vulnerable zones	NASA (.gov)	https://cmr.earthdata.nasa.gov/search/concepts/C1513253831-SEDAC.html
Wind speed	Assessing cyclone dynamics	Zoom earth	https://zoom.earth/storms/fani-2019/#map=satellite-hd
Wind direction	Assessing cyclone dynamics	Research article	https://doi.org/10.1007/s11069-020-04033-5
Atmospheric pressure	Assessing cyclone dynamics	Zoom earth	https://zoom.earth/storms/fani-2019/#map=satellite-hd
Precipitation	Assessing cyclone dynamics	Research article	https://doi.org/10.1007/s11069-020-04033-5
Government report	Impact assessment	RMSC, New Delhi	https://rsmcnewdelhi.imd.gov.in/
		UNICEF	https://www.unicef.org/
		The World Bank	https://www.worldbank.org/en/home
		NDM, India	https://ndma.gov.in/

16.3.1 Data Collection and Acquisition

The research utilizes a diverse array of datasets, including satellite imagery, meteorological data, and GIS information, to comprehensively analyse coastal dynamics.

16.3.2 Datasets Related to Cyclone Fani

Data related to Cyclone Fani were carefully collected from various sources for the research study to comprehend the cyclone's effects fully. The primary datasets came from the "Zoom Earth" platform, which is available to the general public (URL: https://zoom.earth/storms/fani-2019/#map=satellite-hd). These statistics included information on wind speed and pressure variations recorded between May 26 and May 4, 2019, as well as real-time tracking of the cyclone's progress (Fig. 16.4

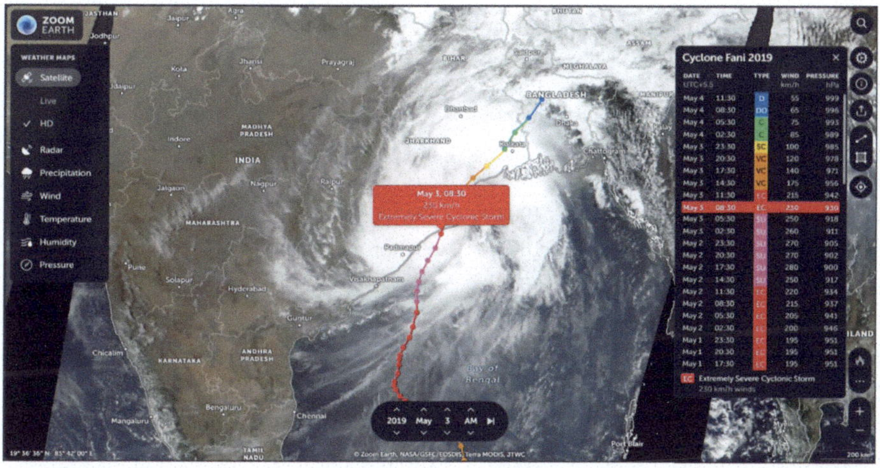

Fig. 16.4 Track of Cyclone Fani. *Source* Zoom Earth, https://zoom.earth/storms/fani-2019/#map=satellite-hd

and Table 16.2). Furthermore, supplementary datasets regarding wind direction and precipitation patterns during this time period were obtained from academic research publications (Figs. 16.5 and 16.6). Combining these datasets provides a comprehensive perspective of Cyclone Fani's path, dynamic features, and climatic parameters during the designated period of time. This overview is essential for establishing a connection between the cyclone's behavior and observed erosion patterns and coastline alterations.

16.3.3 Extraction of Shoreline Data

Interpreting or identifying the shoreline's position is the most important issue when specifying any changes to the shoreline. This is because identifying a coastline position is typically a challenging undertaking due to the changing character of the coasts (Kankara et al., 2014). The primary focus of this research was coastline extraction using QGIS, an open-source platform, performing manual digitization followed by preprocessing, to extract the shoreline. A cross-validation was carried out with Google Earth Pro to validate the accuracy of the extracted data, proving the dependability of the acquired coastline data. To ensure a comprehensive and accurate assessment of coastal erosion and shoreline change dynamics following Cyclone Fani, the procedures included image acquisition, preprocessing involving mosaicking and clipping, NDWI classification, reclassification into binary values, and final manual digitization.

Table 16.2 Characteristics of Cyclone Fani before, during, and after landfall. Full forms of the codes are provided in Table 16.3

Date	Time	Type	Wind (km/h)	Pressure (h pa)	Date	Time	Type	Wind (km/h)	Pressure (h pa)
25-Apr-19	23:30	D	45	1004	30-Apr-19	5:30	VC	145	971
26-Apr-19	2:30	D	45	1003	30-Apr-19	8:30	VC	155	969
26-Apr-19	5:30	D	45	1003	30-Apr-19	11:30	VC	160	967
26-Apr-19	8:30	D	45	1003	30-Apr-19	14:30	VC	165	965
26-Apr-19	11:30	D	45	1003	30-Apr-19	17:30	VC	170	964
26-Apr-19	14:30	D	45	1003	30-Apr-19	20:30	EC	175	959
26-Apr-19	17:30	D	45	1003	30-Apr-19	23:30	EC	185	955
26-Apr-19	20:30	D	50	1003	1-May-19	2:30	EC	195	953
26-Apr-19	23:30	D	55	1004	1-May-19	5:30	EC	195	952
27-Apr-19	2:30	D	60	1003	1-May-19	8:30	EC	195	951
27-Apr-19	5:30	DD	65	1003	1-May-19	11:30	EC	195	951
27-Apr-19	8:30	C	75	1000	1-May-19	14:30	EC	195	951
27-Apr-19	11:30	C	85	998	1-May-19	17:30	EC	195	951
27-Apr-19	14:30	C	85	997	1-May-19	20:30	EC	195	951
27-Apr-19	17:30	C	85	996	1-May-19	23:30	EC	195	951
27-Apr-19	20:30	C	85	996	2-May-19	2:30	EC	200	946
27-Apr-19	23:30	C	85	996	2-May-19	5:30	EC	205	941
28-Apr-19	2:30	C	85	997	2-May-19	8:30	EC	215	937
28-Apr-19	5:30	C	85	998	2-May-19	11:30	EC	220	934
28-Apr-19	8:30	C	85	998	2-May-19	14:30	SU	250	917
28-Apr-19	11:30	C	85	998	2-May-19	17:30	SU	280	900
28-Apr-19	14:30	C	85	997	2-May-19	20:30	SU	270	902
28-Apr-19	17:30	C	85	996	2-May-19	23:30	SU	270	905
28-Apr-19	20:30	C	85	995	3-May-19	2:30	SU	260	911
28-Apr-19	23:30	C	85	995	3-May-19	5:30	SU	250	918
29-Apr-19	2:30	C	85	992	3-May-19	8:30	EC	230	930
29-Apr-19	5:30	C	85	989	3-May-19	11:30	EC	215	942
29-Apr-19	8:30	C	85	987	3-May-19	14:30	VC	175	956
29-Apr-19	11:30	C	95	985	3-May-19	17:30	VC	140	971
29-Apr-19	14:30	SC	95	986	3-May-19	20:30	VC	120	978
29-Apr-19	17:30	SC	100	987	3-May-19	23:30	SC	100	985
29-Apr-19	20:30	VC	110	984	4-May-19	2:30	C	84	989
29-Apr-19	23:30	VC	120	981	4-May-19	5:30	C	75	993
30-Apr-19	2:30	VC	130	976	4-May-19	8:30	DD	65	996

(continued)

Table 16.2 (continued)

Date	Time	Type	Wind (km/h)	Pressure (h pa)	Date	Time	Type	Wind (km/h)	Pressure (h pa)
30-Apr-19	5:30	VC	145	971	4-May-19	11:30	D	55	999
30-Apr-19	8:30	VC	155	969	–	–	–	–	–

Source Zoom Earth, https://zoom.earth/storms/fani-2019/#map=satellite-hd

Table 16.3 Abbreviations used in Table 16.2

Variables	Categories
D	Depression
DD	Deep Depression
C	Cyclonic Storm
SC	Severe Cyclonic Storm
VC	Very Severe Cyclonic Storm
EC	Extremely Severe Cyclonic Storm
SU	Super Cyclonic Storm

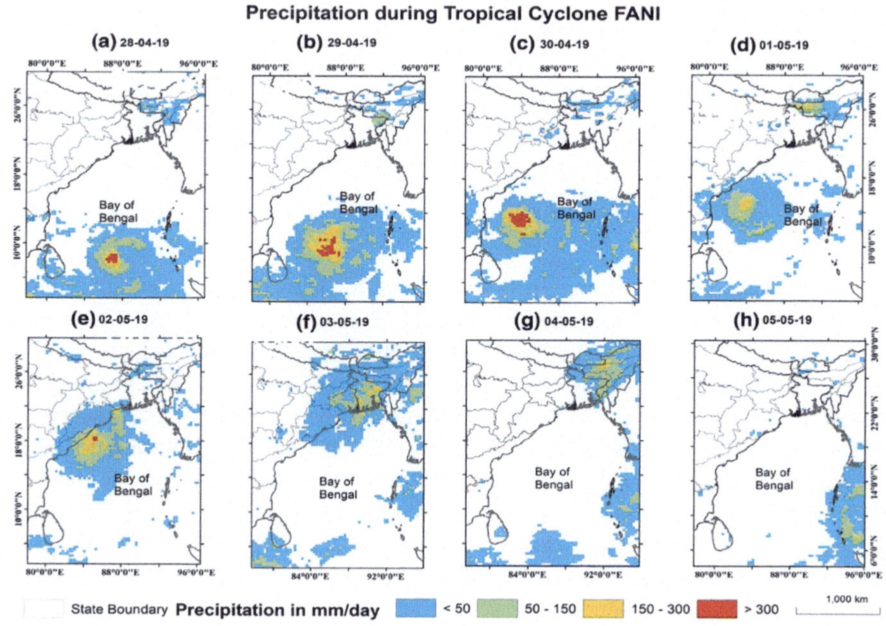

Fig. 16.5 Precipitation during Cyclone Fani. Adapted from Kumar et al. (2020)

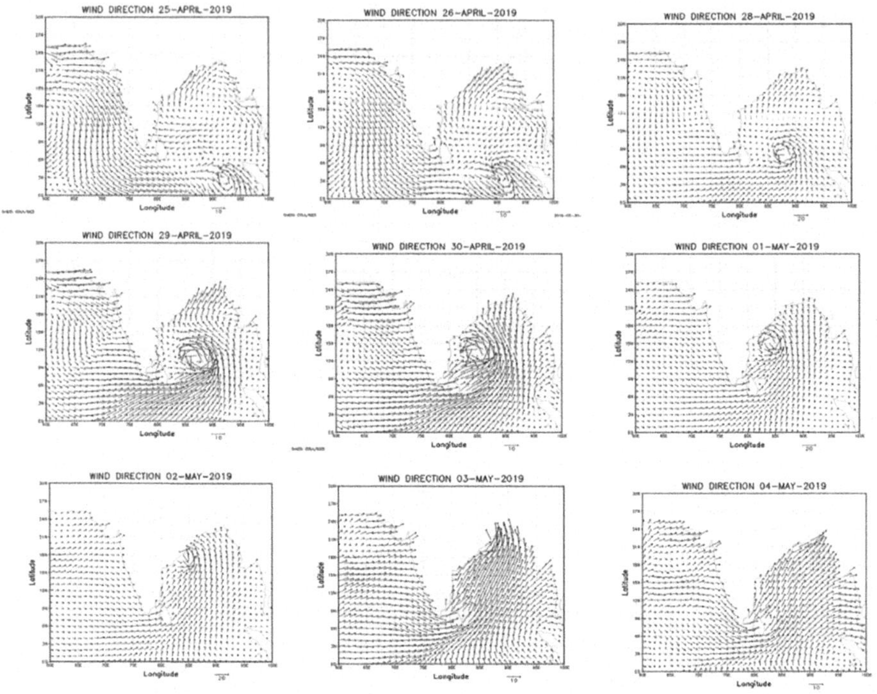

Fig. 16.6 Wind direction during cyclonic events. Adapted from Kumar et al. (2020)

16.3.4 Creation of Base Line

A baseline is an imaginary line or reference point that is formed perpendicular to the shoreline in coastal geomorphology and shoreline management. It acts as a constant point of reference for calculating how the shoreline has changed throughout time.

The next step was to create the baseline. First, the pre- and post-Cyclone Fani shorelines were merged into a single dataset, and then a 500-m buffer was built around the merged line. This buffer was converted to a line by using the polygon-to-line conversion. Following that, the baseline was manually digitized to ensure a precise depiction of the coastline situation before the cyclone. This rigorous approach aimed to provide an extensive baseline for evaluating the degree of shoreline changes and coastal erosion caused by Cyclone Fani's impact.

16.3.5 Creation of Transects

A transect is a straight line or linear path made perpendicular to the shore in shoreline change studies. Transects are used to quantify and evaluate changes in the shoreline over time. These lines offer a standardized method for gathering information about the location and changes to the coastline, enabling reliable assessments of variations in erosion or accretion along particular intervals. Using QGIS, transect generation was completed for the analysis of coastal dynamism. First, two fundamental layers were created: the baseline layer and the layer that represented the shoreline. Transects were created using the Processing Toolbox's "Geometry by expression" tool. An important step was to produce lines perpendicular to the baseline layer by using the "Geometry generator" option and choosing the shoreline layer. In particular, the procedure utilized the baseline layer to create transects necessary for analyzing and researching changes and variances in coastal features. By carrying out this process, transects were produced that served as essential reference lines for thorough examination and precise mapping of shoreline dynamics and possible erosion patterns within the study region. A total of 1398 transects were obtained within the shoreline and baseline, with 100 m distance between each.

16.3.6 Calculation of Shoreline Change Rates

The methodology for determining the Shoreline Change Rate involved a systematic approach within QGIS tools and attribute table modifications. Initially, the dataset organization included three crucial layers: Baseline, Pre-Cyclone Shoreline, and Post-Cyclone Shoreline, alongside distinct attributions for each transect to differentiate between pre- and post-cyclone data. Intersection points were generated using the 'Intersection' tool, marking locations where transects crossed both shoreline layers, pivotal for subsequent analyses. The attribute table was structured to include transect IDs and lengths for pre- and post-cyclone shorelines, facilitating a clear distinction between temporal states. Employing the Field Calculator, a new attribute field captured the differences in shoreline lengths between the pre- and post-cyclone phases.

Subsequently, the shoreline change rate was calculated:

$$\frac{\text{Shorelength in post cyclone} - \text{Shorelength in pre cyclone}}{\text{Time interval}} \tag{16.1}$$

These calculations offer insights into variations–positive values indicating accretion, negative values signifying erosion, and zeros denoting no change.

16.3.7 Calculation of Net Shoreline Movement

The term "Net Shoreline Movement" (NSM) refers to the overall shift in the shoreline's position over a specified period of time. To evaluate the dynamic character of coastal settings, coastal geomorphology utilizes this essential metric. NSM can show how a shoreline has changed over time, whether it has retreated, progressed, or mainly stayed stable. NSM, which is typically expressed in linear units such as meters or feet, essentially measures the net movement of the shoreline along a baseline. The lengths between comparable points on two shoreline features (such as pre- and post-cyclone shorelines) can be measured along a baseline in QGIS to compute NSM.

16.3.8 Identification of Vulnerable Zones

The computed net shoreline movement from transects is used to identify susceptible zones, particularly villages that are experiencing severe erosion and accretion. Villages that exhibit a marked retreat or advancement are identified as being seriously impacted based on an analysis of the extent of shoreline alterations. Setting these areas as priorities enables focused interventions and mitigation plans to mitigate the effects of coastal accretion and erosion, thereby strengthening the ability of coastal communities to recover from Cyclone Fani.

16.3.9 Visualization of Results

The final phase is visualizing the outcomes of the process. Through this technique, vulnerable zones, including villages that are badly affected, are identified, and the geographical distribution of shoreline changes and the processes of coastal erosion are portrayed in a visual style.

16.4 Results

This section aims to clarify the varied patterns of erosion and accretion observed before and after the cyclone's occurrence through a thorough analysis of coastline data, meteorological parameters, and vulnerability reports.

16.4.1 Trajectory of Cyclone Fani

On April 25, 2019, a depression in the southern Bay of Bengal gave rise to cyclone Fani, which on April 26, 2019, strengthened into a deep depression while traveling at a high wind speed of 50–60 km/h. The deep depression moved to 9° N 87° E toward the northwest Bay of Bengal on April 28 and 29, 2019, when it developed into a cyclonic storm with a wind speed of 120–140 km/h. The cyclonic storm moved northwest of the Bay of Bengal and was approaching the Indian coast on April 30, 2019. Fani became a powerful cyclonic storm on May 2, 2019, with a maximum wind speed of 210 km/h. It made landfall at 16° N, 82° E (the coast of Odisha). The Fani cyclone, which was impacting northern Odisha on May 3, 2019, lost some of its intensity (175–185 km/h) and shifted northeast (Fig. 16.7).

Following May 3, 2019, the Fani cyclone moved over southern and central West Bengal and northern Odisha, weakening into a very severe cyclonic storm with winds of 130–140 km/h. On May 4, the intensity of the Fani cyclone decreased, primarily in the West Bengal and Bangladesh region, becoming a severe cyclonic storm with winds of 90–100 km/h. It finally vanished on May fifth, reaching Assam with very little wind (50–60 km/h) (Fig. 16.8).

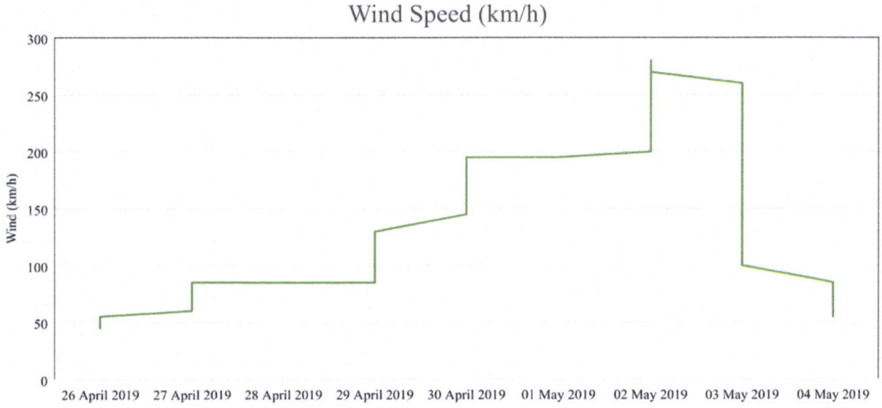

Fig. 16.7 Wind speed during cyclonic events

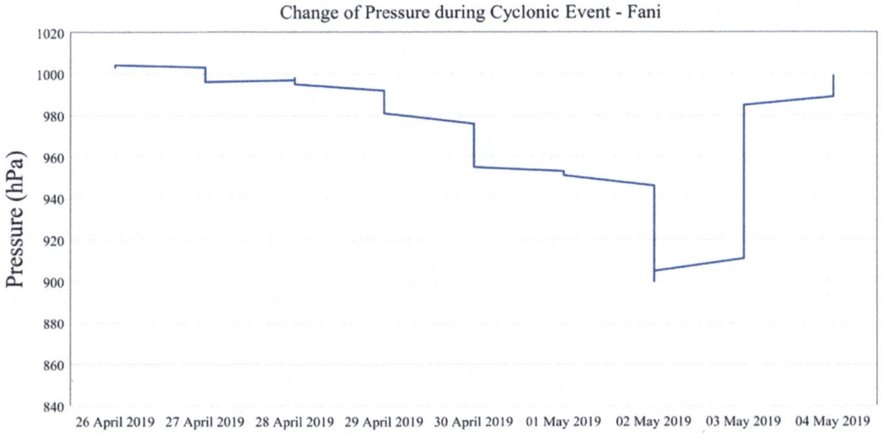

Fig. 16.8 Atmospheric pressure during cyclonic events

16.4.2 Shoreline Changes Pre- and Post-cyclone Fani

Utilizing Landsat 8 OLI/TIRS imagery with a resolution of 30 m obtained from the USGS Earth Explorer platform of two dates, 11/04/2019 (Pre-Cyclone) and 29/05/2019 (Post Cyclone). The shoreline extraction process begins by identifying the interface between land and water bodies. Next, extensive image pre-processing was carried out, and the shoreline was determined by using the Normalized Difference Water Index (NDWI).

A baseline was created perpendicular to the water after the shoreline had been defined, usually at a predefined range inland. The shoreline variations can be measured using this baseline as a reference (Fig. 16.9). Differential patterns of erosion and accretion are revealed by contrasting shoreline data obtained before the cyclone event with data obtained after it. The analysis reveals that while certain areas of the shoreline experienced accretion (indicated by a positive shoreline change rate), a substantial portion of the beach underwent erosion (indicated by a negative shoreline change rate).

16.4.3 Rate of Shoreline Change

A total of 1398 Transects were created, out of which 771 transects have a negative shoreline change rate, which represents Erosion, and 627 transects have a positive Shoreline change rate, which represents Accretion. In total, the coastline of Puri district has witnessed more erosion than accretion. About 55% of the total coastline has witnessed erosion, and 45% has witnessed accretion (Fig. 16.10).

On average, during the time period of 48 days of pre and post cyclone, 46 m of accretion has taken place, and 41 m of erosion has taken place during this time period, in which there was a significant amount of wind speed, atmospheric pressure, and tides were generated.

16.4.4 Net Shoreline Movement

There has been significant coastline movement throughout the 48 days under investigation, with notable examples of erosion and accretion. The largest erosion-related coastal retreat measured was 201 m, indicating a significant shift of the coastline toward the land (Fig. 16.11).

Several villages in the Puri district have experienced significant erosion following Cyclone Fani. Rahakhandol, Keutajanga, Gadabangor, Beldal, Mohanipur, Puri, Narasinghpatana, Anandapur, Naba, Bajrakote, Fatepur, Udayagiri, Baliapokhari, Hunjan Forest, Nandala, and Paikarapur are among the areas that have experienced significant shoreline retreat.

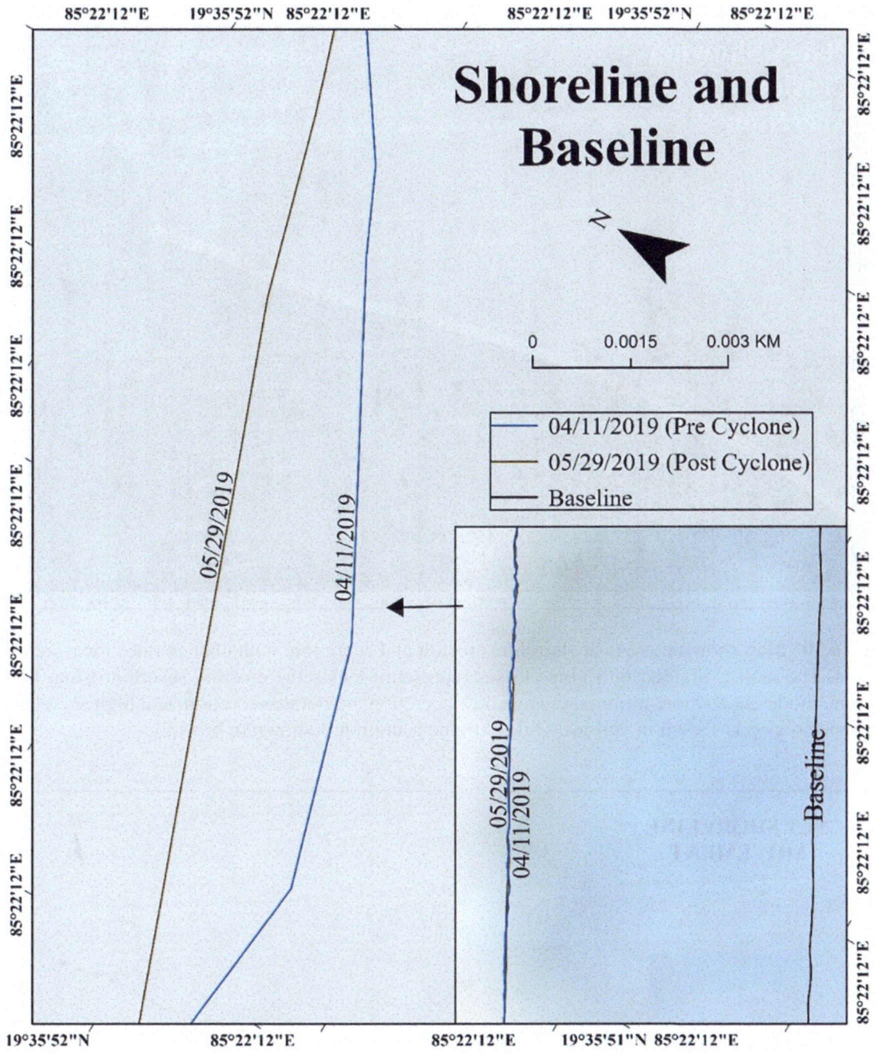

Fig. 16.9 Shorelines of pre-cyclone and post-cyclone, and baseline

On the other hand, there was significant accretion in the Puri district village of Manik Patna, where the shoreline moved up to 449.94 m seaward. This important development suggests that silt has been deposited along the coastline, expanding the coastal land.

With a frequency of 17 transects reflecting the prevalence of erosion processes, Zone 1 is distinguished by extremely high erosion and very low accretion. Zone 5, on the other hand, shows extremely high accretion and very low erosion, with a frequency of 23 transects indicating a considerable landward migration of the

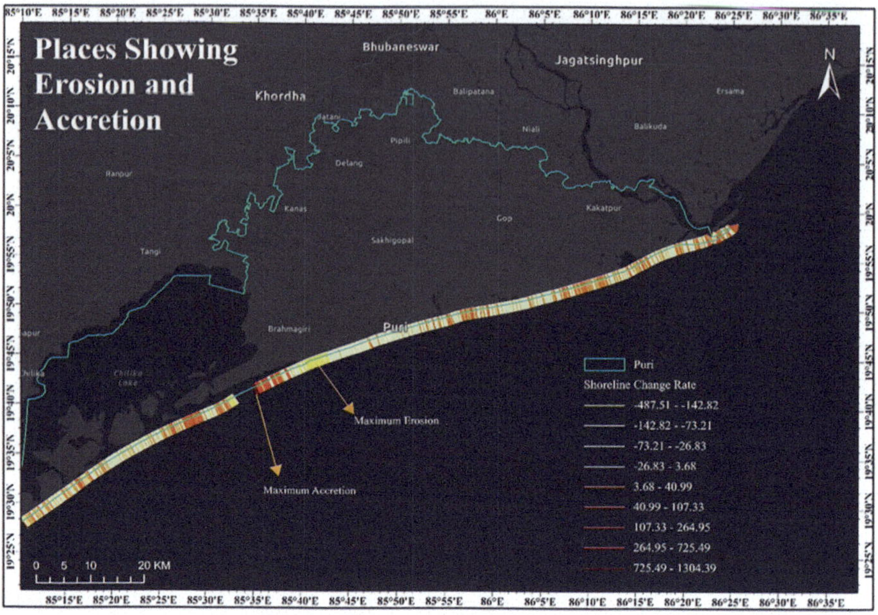

Fig. 16.10 Map showing areas of shoreline erosion and accretion, with change rates measured in meters. The map is divided into nine classes representing extreme erosion, severe erosion, high erosion, moderate erosion, minimal change, low accretion, moderate accretion, and high accretion. Extreme erosion is shown in yellow, while extreme accretion is shown in brown

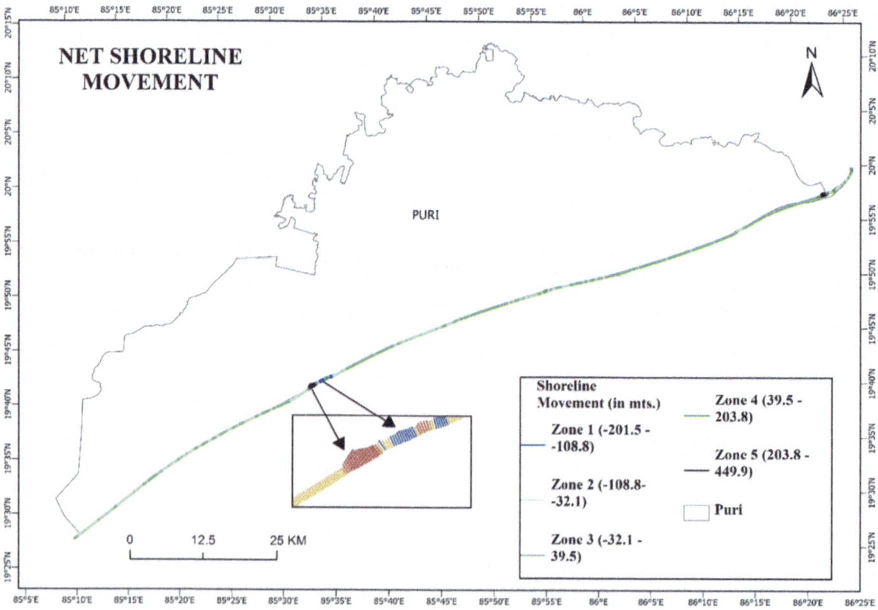

Fig. 16.11 Net shoreline movement during the period of the cyclone

coastline. With high frequencies of 561 and 836 transects, respectively, zones 2 and 3 exhibit a strong trend of erosion. Zone 4, on the other hand, shows comparatively less intense erosion than Zones 1, 2, and 3. Zone 4 has a frequency of 35 transects.

16.4.5 Identification of Vulnerable Villages

To identify vulnerable zones, a detailed analysis was conducted to pinpoint villages that are particularly susceptible to erosion and accretion within the study area. Through meticulous examination of shoreline data and transect analysis, several villages have been identified as being more prone to erosion, while others exhibit a higher susceptibility to accretion. Among these vulnerable villages, those experiencing severe erosion, such as Rahakhandol, Keutajanga, Gadabangor, and Beldal, stand out as areas of heightened concern. These villages have witnessed significant landward movement of the shoreline, resulting in the loss of coastal land and potential damage to infrastructure and habitats (Fig. 16.12). Conversely, villages exhibiting pronounced accretion, such as Manik Patna, highlight areas where coastal land is expanding due to sediment deposition (Fig. 16.13). Understanding the vulnerability of these villages to erosion and accretion is crucial for informing targeted coastal management strategies and adaptation measures that aim to mitigate the impacts of future cyclones and enhance the resilience of coastal communities.

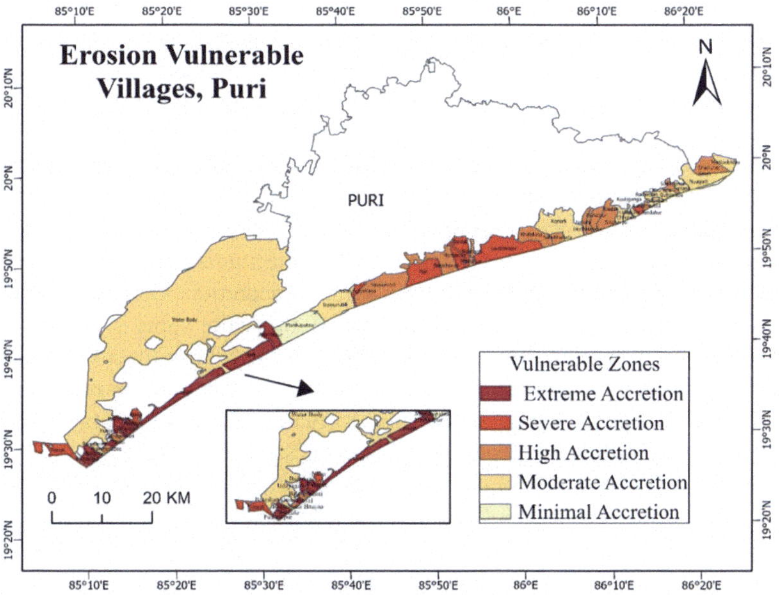

Fig. 16.12 Map showing villages prone to coastal erosion

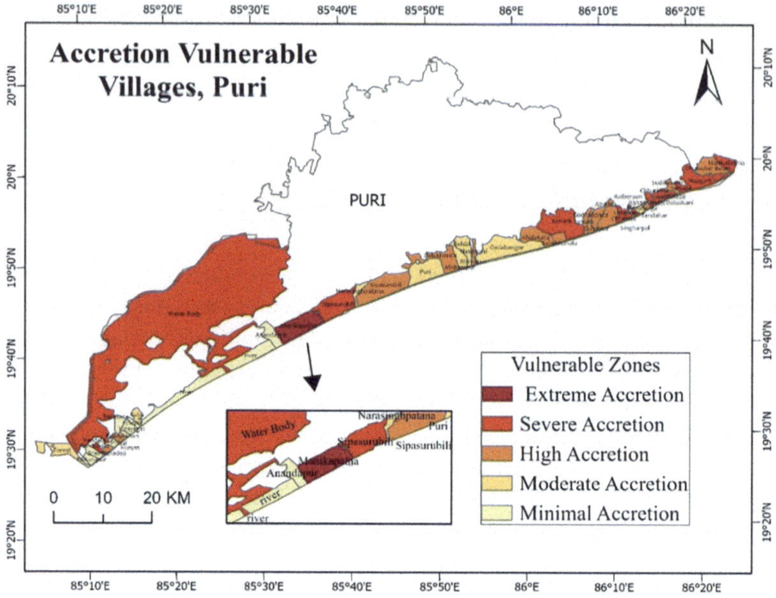

Fig. 16.13 Map showing villages prone to coastal accretion

16.5 Discussions

The findings from this study demonstrate the havoc caused by Cyclone Fani on the coastal regions of Puri District. Characterized by exceptionally high-speed winds and low atmospheric pressure, during the landfall process, it has triggered significant geomorphic alterations along the shoreline. Use of Landsat-8 imagery and NDWI-based shoreline and baseline extraction has enabled the precise identification of erosion and accretion patterns after the cyclone, revealing short-term effects of extreme weather events on coastal areas.

The results indicate that approximately 55% of the studied coastline experienced erosion, while the remaining 45% showed signs of accretion. On average, over the 48-day study period, the shoreline eroded by 41 m, while some segments gained land through 46 m of accretion. These rates reflect the high-impact effects associated with Cyclone Fani, particularly due to sustained wind speeds reaching up to 210 km/h, significant atmospheric pressure fluctuations, and tidal surges generated during and after landfall. Erosion was most witnessed in Zones 1,2, and 3, where the frequency and magnitude of shoreline retreat were highest, with some transects indicating landward movement of 201 m. Villages such as Rahakhandol, Keutajanga, Gadabangor, and Beldal were particularly vulnerable with extensive loss of coastal land. Substantial accretion representing Zones 4 and 5, notably in Manik Patna, was found, where shoreline advancement reached nearly 450 m seaward. These unusual changes helped carry and deposit sand and other sediments in areas where the water was calmer. As a result, the coastal area experienced temporary growth. However, this new land is not very stable and can easily be washed away during the next storm or high tide.

In summary, this study demonstrates how high-resolution remote sensing data, when combined with cyclone characteristics, can effectively quantify and interpret the short-term impact of cyclones on coastlines.

16.6 Conclusion and Recommendations

The research has yielded significant knowledge regarding the effects of Cyclone Fani on the dynamics of the shoreline and coastal erosion in the Puri district. The results underscore the critical need for efficient coastal management and adaptation plans, highlighting the vulnerability of coastal populations to cyclonic disturbances. Several recommendations are made to direct future efforts and interventions targeted at improving coastal resilience and reducing the effects of cyclonic disturbances, such as investing in Early Warning Systems, which will entail building strong communication networks and providing vulnerable groups with accurate and readily available information. Monitoring activities should be carried out long after Cyclone Fani has passed to track the resilience and recovery of coastal ecosystems and communities. Research on how projected sea level rise, modifications to storm frequency and intensity, and adjustments to precipitation patterns may affect coastal resilience and susceptibility. Artificial Intelligence and Machine Learning (ML) are being used to analyse massive datasets and find trends in coastline changes, cyclone activity, and community susceptibility.

Statements and Declarations Ethical Approval Not Applicable.

Availability of Supporting Data Not Applicable.

Funding This research did not receive a specific grant from public, commercial,or not-for-profit funding agencies.

Competing Interests The authors declare no competing interests.

Data Availability All the datasets are publicly accessible.

Author Contributions Author A carried out the research, conceptualized the study, developed the methodology, utilized software for data analysis and visualization, and took the lead in drafting, reviewing, and editing the manuscript.Author B provided significant contributions to conceptualizing the research,supervising the methodology, and extensively drafting, reviewing, and editing the manuscript.

References

Abushandi E, Abualkishik A (2020) Shoreline erosion assessment modelling for Sohar region: measurements, analysis, and scenario. Sci Rep 10(1). https://doi.org/10.1038/s41598-020-61033-y

Ahammed KKB, Pandey AC (2018) Shoreline morphology changes along the eastern coast of India, Andhra Pradesh, by using geospatial technology. J Coast Conserv 23(2):331–353. https://doi.org/10.1007/s11852-018-0662-5

AlRuheili AM (2022) A tale of Shaheen's cyclone consequences in Al Khaboura City. Oman Water 14(3):340. https://doi.org/10.3390/w14030340

Beura D (2016) Coastal erosion in Odisha: causes and consequences with special reference to Puri beach erosion

Bloch M, Cai W, Migliozzi B (2019) Live tracking map: Cyclone Fani batters India. The New York Times. https://www.nytimes.com/interactive/2019/05/02/world/asia/india-cyclone-fani-map.html

Herlekar MA, Kamble PB, Gawali PB, Hanamgond PT, Aher SP (2023) Quantitative assessment of shoreline changes along the tropical west coast, Maharashtra, India: a remote sensing and GIS approach. J Earth Syst Sci 132(1). https://doi.org/10.1007/s12040-023-02047-8

Jana A, Biswas A, Maiti S, Bhattacharya AK (2013) Shoreline changes in response to sea level rise along Digha coast, eastern India: an analytical approach of remote sensing, GIS and statistical techniques. J Coast Conserv 18(3):145–155. https://doi.org/10.1007/s11852-013-0297-5

Jana A, Maiti S, Biswas A (2015) Seasonal change monitoring and mapping of coastal vegetation types along Midnapur-Balasore coast, Bay of Bengal using multi-temporal Landsat data. Model Earth Syst Environ 2(1). https://doi.org/10.1007/s40808-015-0062-x

Jana A, Maiti S, Biswas A (2016) Analysis of short-term shoreline oscillations along Midnapur-Balasore coast, Bay of Bengal, India: a study based on geospatial technology. Model Earth Syst Environ 2(2). https://doi.org/10.1007/s40808-016-0117-7

Jana A, Maiti S, Biswas A (2017) Appraisal of long-term shoreline oscillations from a part of coastal zones of Sundarban delta, eastern India: a study based on geospatial technology. Spat Inf Res 25(5):713–723. https://doi.org/10.1007/s41324-017-0139-x

Jana A, Sheena S, Biswas A (2012) Morphological change study of Ghoramara Island, eastern India using multi-temporal satellite data. Res J Recent Sci 1(10):72–81. https://www.isca.me/rjrs/archive/v1/i10/13.ISCA-RJRS-2012-319.pdf

Jose M, Pereira GF, Bairavi S (2018) Shoreline mapping and coastal change studies along Chennai coast using remote sensing and GIS. Int J Sci Res Sci Technol 4(5):1394–1398. https://doi.org/10.32628/ijsrst1845360

Kankara RS, Selvan S, Bhoopathy R, Samykannu A (2014) An adaptive approach to monitor the shoreline changes in ICZM framework: a case study of Chennai coast. 43:1266–1271

Kar C, Banerjee S (2021) Tropical cyclone intensity classification from infrared images of clouds over Bay of Bengal and Arabian Sea using machine learning classifiers. Arab J Geosci 14(8). https://doi.org/10.1007/s12517-021-06997-5

Kumar S, Lal P, Kumar A (2020) Turbulence of tropical cyclone 'Fani' in the Bay of Bengal and Indian subcontinent. Nat Hazards 103(1):1613–1622. https://doi.org/10.1007/s11069-020-04033-5

Li R, Liu J-K, Felus Y (2001) Spatial modeling and analysis for shoreline change detection and coastal erosion monitoring. Mar Geod 24(1):1–12. https://doi.org/10.1080/01490410121502

Markose VJ, Rajan B, Kankara RS, Selvan SC, Dhanalakshmi S (2016) Quantitative analysis of temporal variations on shoreline change pattern along Ganjam district, Odisha, east coast of India. Environ Earth Sci 75(10). https://doi.org/10.1007/s12665-016-5723-1

Mishra M, Chand P, Pattnaik N, Kattel DB, Panda GK, Mohanti M, Ba-ruah UD, Chandniha SK, Achary S, Mohanty T (2019) Response of long–to short-term changes of the Puri coastline of Odisha (India) to natural and anthropogenic factors: a remote sensing and statistical assessment. Environ Earth Sci 78(11). https://doi.org/10.1007/s12665-019-8336-7

Mishra M, Kar D, Santos CAG, Da Silva RM, Das PP (2021) Assessment of impacts to the sequence of the tropical cyclone Nisarga and monsoon events in shoreline changes and vegetation damage in the coastal zone of Maharashtra. India Mar Pollut Bull 174:113262. https://doi.org/10.1016/j.marpolbul.2021.113262

Mohanty PK, Kar PK, Behera B (2019) Impact of very severe cyclonic storm Phailin on shoreline change along South Odisha coast. Nat Hazards 102(2):633–644. https://doi.org/10.1007/s11069-019-03610-7

Mullick PD, Das A (2014) Puri: an overview on the space, potential, and possibilities for social research. Radix Int J Res Soc Sci 3(11):1–7

Noujas V, Kankara RS, Rajan B (2023) Assessment of beach morphological changes in the east coast of India during cyclone Phethai, through field data and numerical modeling. Mar Geod 46(4):305–332. https://doi.org/10.1080/01490419.2023.2200043

Chapter 17
Impact of Tropical Cyclones on Coastal Mangroves Using Earth Observation Systems

M. Murugan, V. Mahalakshmi, Gautam Saha, and B. Santhoshkumar

17.1 Introduction

Mangrove forests are highly valuable in coastal ecosystems situated in the intertidal zones of tropical and subtropical regions of the world. There are approximately 70 species in the world which form an important linkage between continental and oceanic environments. Mangrove forest is also referred as Blue Forests 'as they are often growing in and under the water or 'Wetland Ecosystem Engineers' due to the ability to maintain their environment without much degrading. They provide important ecological and social benefits. They are very important for maintaining and acting as a habitat for a variety of marine and terrestrial species. It also improves the livelihood of millions of people by providing valuable food, medicine, fuel, building materials and so on. And most importantly mangroves act as a primary defense in the coastal ecosystem against most of the natural disasters such as flood, tsunami etc.

Tropical cyclone is a phenomenon that occurs when the sea surface temperature crosses over 27 °C, the warm ocean water heats the air above it, causing it to rise and create a low-pressure area, this air continuously rise and condense causing thunderstorm, as the storm organizes, it develops a spiral structure and strong winds, with a central low-pressure area. It can cause extreme winds, heavy rainfall, storm surge, and even tornadoes pose a significant threat to coastal ecosystems, particularly mangrove forests. The rotation of the storm is driven by low atmospheric pressure and the Earth's rotation, causing a counterclockwise rotation in the Northern Hemisphere and clockwise in the Southern Hemisphere. Due to climate change and global

M. Murugan (✉) · G. Saha · B. Santhoshkumar
Department of Agricultural Meteorology and Physics, Bidhan Chandra Krishi Viswavidhyalaya, Haringhata, West Bengal, India
e-mail: murugankarthick796@gmail.com

V. Mahalakshmi
Division of Environmental Sciences, ICAR—Indian Agricultural Research Institute, New Delhi, India

A. Nath et al. (eds.), *Geospatial Application of Coastal Hazards*, Springer Water,
https://doi.org/10.1007/978-3-032-20076-1_17

warming, the SST is increasing day by day which in turn increases the frequency and intensity of tropical cyclones formed in a year have placed mangrove ecosystems under severe stress.

Mangroves are resilient to occasional disturbance but due to the repeated and high intensity of these events the ecosystem is highly harmed and damaged. More specifically, the frequency of the intense tropical cyclone is projected to increase in the upcoming decades as the world is undergoing more climate change. Climate change is expected to shift polewards especially to cyclones and migrates to coastlines on poles. From this we might experience the unaffected mangroves may get affected in the future in case of improper action towards climate change and lacking formulating the policies for conservation of mangroves.

Earth Observation System (EOS), includes remote sensing and geospatial technologies like Geographical Information system (GIS) gives us valuable information on monitoring and analysing the surface from space. Remote Sensing—the technology that gets information about the study area without any physical contact, they use sensors mounted on satellites, hot air balloons, drones and aircrafts. Previously the spaceborne sensors were highly used like Sentinel, Landsat and so on but nowadays modern technologies like Unmanned Aerial Vehicle for Ultra High-resolution data. Platforms like Google Earth Engine (GEE) facilitate processing vast amounts of geospatial data. Using EOS for impact assessment, particularly on ecosystems like mangrove forests affected by tropical cyclones, can be justified for multiple reasons. RS offers synoptic coverage over large areas, which is an important advantage of it. It allows monitoring and assessing the area which are inaccessible after the intense damage due to the calamity. RS allows to get the main ecological parameters for environmental assessment such as canopy cover, vegetation health index like Normalized Difference Vegetation Index (NDVI) and Enhanced Vegetation index (EVI) and structural attributes like biomass and canopy height which are important for assessment of cyclone impact and tracking post cyclone recovery. The data availability over the long term enables us to monitor and to conduct time series study of the particular zone. EOS is highly efficient when compared to traditional field surveys which is highly laborious and costly and consumes more time and energy. These technologies are essential for moving beyond assessments focused solely on casualties and property loss to include impacts on biodiversity and ecosystem services. They can help quantify damage extent and severity, track recovery, and analyse structural changes in ecosystems like mangroves. In this chapter, we will see how Earth Observation systems enhance our understanding of cyclone impacts on mangroves. The integration of remote sensing data with analytical techniques aids in monitoring mangrove health and resilience. This EOS approach supports the development of evidence-based policies to safeguard mangrove forests against increasing cyclone activity.

17.2 Tropical Cyclones and Mangrove Ecosystems

Tropical cyclones make significant and impactful large-scale disturbances to the mangrove ecosystem globally. Mangrove forest is a unique type of forest of woody plants and trees that grows in the intertidal regions of coastal lagoons and estuaries, positioning them at the terrestrial–marine ecotone. It makes them highly prone to cyclones. Though they have their inherent resilience and adaptations to these kinds of events the increasing frequency and intensity of tropical cyclones, driven by global climate change, possess threat to the ecosystem and people who all are getting important resources from them.

17.2.1 Cyclonic Hazards and Its Characteristics

Tropical cyclones cause damage through the combination of multiple extreme environmental events, primarily high-speed winds, storm surges, and heavy rainfall leading to extensive flooding etc.

17.2.1.1 Wind

Tropical cyclones are typically characterised by high velocity surface wind, strong wind gusts, and intense wind speeds that cause immediate and widespread physical damage to coastal environments, including coastal mangroves forest. The extent of damage caused by this wind is often classified using scales like the Saffir–Simpson Hurricane Wind Scale, where Category 3 winds ($\geq$178 km h^{-1}) and higher are known to cause extreme visible effects on mangroves almost without exception. Even the lower wind velocities can lead to substantial crown alteration. The magnitude of canopy damage is directly proportional to the maximum wind speed a stand is exposed to, and the distance from the path of cyclone movement, with closer cyclones inducing much more canopy loss. Particularly tall mangrove trees are susceptible to being removed by cyclones due to defoliation, branch breaking, and windthrow.

17.2.1.2 Storm Surge

A critical destructive element of tropical cyclones is the massive overflow caused by storm surges. It leads to changes in hydrology and can alter sedimentological conditions within mangroves. For example, a Category 1 cyclone can cause storm surges of 1.2–1.5 m above normal, while a Category 4 cyclone can produce surges ranging from 3.9–5.5 m. Such surges can cause severe inundation and water ponding, which is identified as an important factor in mangrove dieback.

17.2.1.3 Flooding and Rainfall

Tropical cyclones are often accompanied by heavy rainfall and extreme precipitation, which, accompanied with high velocity winds, which altogether contribute to a massive loss in the coastal mangrove ecosystem. Heavy rainfall can result in flooding and embankment breaches, allowing saltwater from the sea to enter mangrove regions, which can lead to increased soil salinity level. The inundation from flooding can extend far inland, and if floodwaters are impounded by natural or anthropogenic barriers, can create anaerobic conditions, high sulfide concentrations, or hypersaline soils, causing stress and mortality. The climate change projections indicate an expected increase in the amount of rainfall in the future caused by tropical cyclones near the equatorial region.

17.2.2 Biophysical Impacts on Mangroves

The combined impact of high velocity wind, storm surge, and rainfall results in a range of biophysical impacts on mangrove ecosystems affecting their physical structure, soil conditions, and overall health.

17.2.2.1 Defoliation and Canopy Damage

Defoliation caused by the heavy wind and storm surge and rainfall is the primary impact on mangrove trees by the tropical cyclone. It removes the leaves and branches, which will cause significant loss in the canopy cover. Which in turn causes reduction in the rate of photosynthesis. This defoliation also disturbs the typical seasonal growth phenology of surviving mangrove trees. In many cases, when the entire crown of the tree is lost, death of the tree can occur even in species with the capacity for regrowth. The reduced canopy density due to defoliation may lead to lower canopy CO_2 assimilation rates, affecting the ecosystem's carbon balance.

17.2.2.2 Uprooting, Breakage and Structural Changes

Branch breaking, tree windthrow (uprooting), and stem breakage are some of the direct physical impacts of cyclones. A category 4 cyclone can cause most trees and shrubs to be blown down or broken, with saplings often leading to defoliation. Studies show massive tree death and extensive forest structural change. The severity also depends on the species of tree that are found in that region not only on cyclone intensity. Some species like *Avicennia* are known for prolific re-sprouting after canopy damage, others like *Rhizophora* and *Ceriops* may struggle to recover due to their inability to re-sprout or regrow.

17.2.2.3 Soil Salinity and Hydrological Alterations

These events also cause changes in soil conditions, notably the increase of soil salinity. It can be caused by the infiltration of sea water during flooding and embankment breaches, followed by subsequent evaporation and prolonged inundation (Azad et al. 2019).

The disturbance of natural tidal connectivity, either by storm-transported sediments or anthropogenic barriers, can lead to flood water infiltration, creating anaerobic conditions and high sulphide concentrations in soils. The salinity of soil causes abiotic stress to plants which also impact the plant health. These changes in hydrological regimes are critical as they can reduce the recovery and growth of new plants or seeds.

17.2.2.4 Biodiversity Loss and Ecosystem Degradation

Tropical cyclones markedly change the floristic pattern and structure of mangrove forest. They cause shrinkage of dense and moderately dense tree cover, leading to an expansion of less dense and sparse vegetation cover over the affected area. The cyclone not only affects the trees but also the fauna presents there leading to a reduction in zooplankton biomass, density, and species composition, which in turn impacts juvenile fish—which feeds on the zooplankton. This has a cascading effect on the food chain in that particular biodiversity. The destructive force of cyclones can also alter the behaviour and life of regional fauna, such as crocodiles, and prevent populations from recovering, increasing human-wildlife interactions.

17.2.3 Ecological and Socioeconomic Consequences

The biophysical impacts of cyclones continue to affect the ecological and socioeconomic problems by degrading the many advantage that mangrove ecosystem to provide to humans for good livelihood.

17.2.3.1 Degradation of Ecosystem Services

These types of forests are essential for giving the necessary ecosystem services which include:

Coastal Protection: This specialized ecosystem acts as natural barriers against high-speed winds, heavy rains, storm surges, and tsunamis, by reducing their damaging effects and safeguarding coastal communities and infrastructure of people who live near the sea shore or mangrove forest areas. It destroys this crucial protective function.

Carbon Sequestration and Storage: Mangroves are the most carbon-rich forest ecosystems, which play an important role in carbon storage and sequestration (often referred to as 'blue carbon'). Canopy damage leads to lower CO_2 assimilation rates, impacting the ecosystem's carbon balance. While recovery can restore carbon stocks in the long term, severe disturbances can alter the substrate, potentially negating short-term sequestration.

Sediment Trapping and Erosion Control: Mangroves are very good trees in controlling coastal erosion and it acts as sediment traps and helps in exchanging sediment and nutrients with nearby ecosystems. This type of cyclone used to cause significant erosion, create new channels, and remove mangrove roots, and affecting these important processes which is crucial for the environment.

Impacts on Fisheries and Livelihoods: The degradation of mangrove forest can directly affect fisheries, because these forests act as coastal nurseries for a number of fish species. The loss of juvenile fish, due to impacts on zooplankton which is affected by salinity intrusion in the ecosystem and direct habitat destruction, which can reduce the stock available for local fishermen, severely affecting their livelihoods. Beyond fisheries, cyclones inflict socio economic losses across coastal regions, impacting agricultural lands through sea water intrusion in agricultural fields, storage areas and crop damage, and disrupting the overall socio-economic conditions of dependent communities. The loss of ecosystem services, such as the provision of food, fuel, and timber, also directly affects local lives.

Long-term Trajectories and Resilience: While mangroves are considered disturbance-adapted and ecologically resilient, capable of recovering from periodic storms this resilience is not without limits. Recovery times can vary from years to decades, depending on the damage occurred and condition of the site. Climate change is expected to increase the frequency and intensity of the most damaging cyclones, lowering regeneration and recovery time for these forests and resulting in cascading consequences on biodiversity. According to some study, mangroves may be less resilient to long-term trends in cyclones. Additionally, factors such as long-term drought, human infrastructure, geomorphology, and pre-cyclone forest structure can influence a mangrove's vulnerability and resilience, making it necessary to address compounding effects. Therefore, regular monitoring, particularly using Earth Observation (EO) systems and remote sensing techniques, is crucial for swiftly and reliably assessing cyclone impacts and understanding the long-term recovery dynamics of these vital ecosystems across wide areas.

17.3 Earth Observation Systems in Mangrove Monitoring

Earth Observation Systems (EOS) play an important role in the monitoring and assessing of mangrove ecosystems, provides a best alternative to challenging field surveys which are highly difficult due to logistical constraints associated with these environments (Vizcaya-Martínez et al. 2022). This is very important for assessing the susceptibility of mangrove forests to the impact of tropical cyclones and other disturbances when direct field survey by the humans is not possible due to the calamities (Krauss and Osland 2020; Vizcaya-Martínez et al. 2022).

17.3.1 Types of Remote Sensing Data Used

The data sources through remote sensing for mangrove monitoring and assessing are diverse and vast, which includes various parts of the electromagnetic spectrum (EMS) to capture the data from different aspects of forest health, structure and other parameters which are required for various studies.

17.3.1.1 Optical Data

Optical remote sensing depends on sensors that detect reflected part of sunlight. These types of sensors are widely used because to its broad temporal coverage and open accessibility (Roy et al. 2024).

Landsat Missions

Thematic Mapper (TM), Enhanced Thematic Mapper Plus (ETM +), Operational Land Imager (OLI)) are the some of the common sources of remote sensing data which have medium to high spatial resolution, which are commonly used for various applications (Vizcaya-Martínez et al. 2022). Landsat imagery has been important tool in long-term assessments of mangrove changes, such as monitoring hurricane-induced impacts (Roy et al. 2024) and mapping land cover dynamics and time series based analysis of various ecosystem (Kingsley Kanjin and Alam 2024) Some of the important examples are, the Sundarbans mangrove forest has been monitored using multi-temporal Landsat data from 1973 to 2000 (Giri et al. 2007) and from 2000 to 2020 (Samanta et al. 2021), providing information on canopy changes (Krauss and Osland 2020).

Sentinel-2 Satellites

Sentinel-2's high resolution and revisits make it a best option for mapping and forest monitoring (Roy et al. 2024). In order to classify mangrove species (Roy et al. 2025), study mangrove phenology (Vizcaya-Martínez et al. 2022), and study forest fragmentation (Roy et al. 2025), this satellite's Multi Spectral Instrument (MSI) sensor (Vizcaya-Martínez et al. 2022) provides ultra-high spatial resolution data.

Moderate Resolution Imaging Spectroradiometer (MODIS)

MODIS data are commonly used for rapid assessments and monitoring due to their short revisit time (Islam and Assal 2023). MODIS products are used to monitor mangrove health indicators (Samanta et al. 2021) and assess ecological disturbances caused by cyclones. They have also been used to analyse the relationship between vegetation health and land surface temperature in the Sundarbans (Kingsley Kanjin and Alam 2024). Other optical sources include QuickBird (Kovacs et al. 2009), WorldView-2 (Dutta Roy et al. 202,420), SPOT 6 (Dutta Roy et al. 2024), and Advanced Very-High-Resolution Radiometer (AVHRR).

17.3.1.2 Radar Data

Radar sensors importantly Synthetic Aperture Radar (SAR) are highly advantageous as they can penetrate into cloud (Mondal et al. 2022) It works in all weather conditions, making them suitable for monitoring during and immediately after and during natural disaster such as cyclonic events, flood etc. (Dutta Roy et al. 2024).

Sentinel-1 SAR

Sentinel-1 SAR data are open source and free available, have been used for inundation analysis (Mondal et al. 2022; Mishra et al. 2024) which includes detailed flood mapping (Mondal et al. 2022).

ALOS (Advanced Land Observing Satellite) PALSAR (L-band SAR Data)

ALOS, PALSAR are much useful in detecting flooded vegetation due to its higher canopy penetration capability (Mondal et al. 2022). SAR data are broadly used for mangrove extent and biomass estimation (Dutta Roy et al. 2024).

17.3.1.3 LiDAR and Hyperspectral Data

LiDAR and hyperspectral data are advanced remote sensing technologies provides information which are highly detailed and structural about mangrove forests.

Light Detection and Ranging (LiDAR) Sensors

Light Detection and Ranging (LiDAR) sensors provides data with high accuracy in capturing 3 Dimensional (3D) structural details (Dutta Roy et al. 2024; Chavez et al. 2023), which is an important tool for estimating canopy height and biomass (Dutta Roy et al. 2024). LiDAR has been extensively used to assess hurricane-induced damage, quantify defoliation (Chavez et al. 2023), and analyse structural changes in mangrove forests (Dutta Roy et al. 2024). The NASA Goddard's LiDAR, Hyperspectral and Thermal Imager (G-LiHT) system is an important airborne imaging system that maps the structure, composition and function of terrestrial ecosystems (Chavez et al. 2023).

Hyperspectral Sensors

Hyperspectral sensors provide high-resolution spectral data, enabling detailed analysis for applications like object-based mangrove classification (Vizcaya-Martínez et al. 2022). G-LiHT also includes hyperspectral sensors for comprehensive data collection (Chavez et al. 2023).

17.3.2 Data Platforms and Tools

Modern advancements in cloud computing and geospatial software have revolutionized the processing and analysis of remote sensing data for mangrove monitoring.

17.3.2.1 Google Earth Engine

GEE is a widely accessible, multi-petabyte open-source catalogue of satellite imagery and geospatial datasets (Islam and Assal 2023). It offers extensive data handling and processing capabilities for planetary-scale geospatial analysis. GEE is frequently used for time series analysis (Vizcaya-Martínez et al. 2022), extracting vegetation indices (Vizcaya-Martínez et al. 2022), cloud masking (Samanta et al. 2021), flood detection, and monitoring annual forest health (Roy et al. 2025).

17.3.2.2 Geographic Information System

(GIS) Software such as ArcGIS Pro, QGIS, and ENVI are essential tools for spatial analysis (Mishra et al. 2021). They are used to process Landsat images (Kanjin and Alam 2024), perform raster-to-vector conversions (Mishra et al. 2021c38), conduct supervised classification (Samanta et al. 2021), and implement decision tree analysis for classifying mangrove species and features (Vizcaya-Martínez et al. 2022). Time Series Analysis is a crucial approach for understanding long-term changes and recovery patterns in mangrove ecosystems following disturbances (Asbridge et al. 2024). It involves analysing sequences of satellite data over time (Vizcaya-Martínez et al. 2022) to observe trends in vegetation indices (Vizcaya-Martínez et al. 2022) and assess post-storm recovery (Islam and Assal 2023). Techniques like continuous change detection and classification are employed to investigate long-term mangrove dynamics.

17.3.3 Metrics and Indices

Remote sensing method uses different metrics and indices derived from satellite imagery and other aerial data sources to quantify changes in vegetation health, structure, and composition. Some of the commonly used metrices to assess the mangrove ecosystem is discussed in the Table 17.1.

Table 17.1 Various indices that are commonly used in remote sensing to assess the mangrove ecosystem

Index/ Metric	Full Form (Abbreviation)	Formula	What it measures	Primary uses
NDVI	Normalized Difference Vegetation Index (NDVI)	(NIR−R)/(NIR + R) {(IR–R)/ (IR + R)} × 100 + 100 (ArcGIS)	Vegetation health, greenness, chlorophyll (indirectly)	Vegetation mapping, cyclone impact, canopy cover monitoring
EVI	Enhanced Vegetation Index (EVI)	2.5 × (NIR−Red)/ (NIR + 6 × Red−7.5 × Blue + 1)	Canopy vigour, leaf physiology	Dense canopy detection, cyclone impacts, GPP
NDII	Normalized Difference Infrared Index (NDII)	(NIR−SWIR)/(NIR + SWIR)	Leaf chlorophyll and water content	Monitoring canopy damage and recovery
NDWI	Normalized Difference Water Index (NDWI)	(NIR−SWIR)/(NIR + SWIR)	Surface water presence	Flood mapping, shoreline detection
NDSI	Normalized Difference Salinity Index (NDSI)	(Green–SWIR)/(Green + SWIR)	Soil salinity in mangrove zones	Mapping salinity from floods and salt intrusion
CHLc	Chlorophyll Content (CHLc)	(127 × NDVI)−46.61	Chlorophyll concentration, photosynthesis	Tracking cyclone impact on chlorophyll
GPP	Gross Primary Productivity (GPP)	(0.0983 × EVI) + 0.0161	Net productivity, carbon absorption	Cyclone/land use impact on productivity
CSIF/ SIF	Solar-Induced Chlorophyll Fluorescence (CSIF/SIF)	Neural network-based from OCO-2 and MODIS	Chlorophyll absorption energy	Productivity trends, climate response
VARI	Visible Atmospherically Resistant Index (VARI)	(G−R)/(G + R−B)	Vegetation greenness (RGB)	Mangrove species mapping, RGB analysis
NDVIre	Red Edge Normalized Difference Vegetation Index (NDVIre)	(NIR−RE1)/(NIR + RE1)	Vegetation greenness via red edge	Improved greenness detection
CIg-re1/ 2/3	Chlorophyll Index Red Edge (CIg-re1/2/3)	RE/G−1	Chlorophyll via red edge	Estimating plant health/ chlorophyll

(continued)

Table 17.1 (continued)

Index/ Metric	Full Form (Abbreviation)	Formula	What it measures	Primary uses
CMRI	Combined Mangrove Recognition Index (CMRI)	NDVI−((G−NIR)/(G + NIR))	Mangrove vs non-mangrove areas	Accurate mangrove classification
NDMI	Normalized Difference Mangrove Index (NDMI)	(SWIR2−G)/(SWIR2 + G)	Mangrove condition	Monitoring mangrove condition
TGI	Triangular Greenness Index (TGI)	(G−0.39)(R−0.61)B	Greenness via RGB	Matches NIR-based vegetation indices
ExG	Excess Green Index (ExG)	2G−R−B	Vegetation via RGB	General RGB-based vegetation index
VEG	Vegetation Index (VEG)	G/(R^0.67 × B^0.33)	Vegetation density	Simple RGB-based plant cover measure
CIVE	Color Index of Vegetation (CIVE)	0.44R−0.88G + 0.39B + 18.79	Greenness by colour	Visual vegetation analysis
NGRDI	Normalized Green–Red Difference Index (NGRDI)	(G−R)/(G + R)	RGB-based greenness	Tracks vegetation cover changes
NGBDI	Normalized Green–Blue Difference Index (NGBDI)	(G−B)/(G + B)	RGB-based greenness	Alternate vegetation greenness index
VDVI	Visible-Band Difference Vegetation Index (VDVI)	(2G−R−B)/(2G + R + B)	RGB greenness	Alternative to multispectral VIs
RGRI	Red-Green Ratio Index (RGRI)	R/G	Redness-greenness ratio	Vegetation redness detection
BGRI	Blue-Green Ratio Index (BGRI)	B/G	Blueness-greenness ratio	Vegetation stress via RGB
WI	Woebbecke Index (WI)	(G−B)/(R−G)	Vegetation colour characteristics	General vegetation analysis

(continued)

Table 17.1 (continued)

Index/ Metric	Full Form (Abbreviation)	Formula	What it measures	Primary uses
RGBRI	Red–Green–Blue Ratio Index (RGBRI)	(R + B)/2G	RGB balance	Differentiates vegetation types
RGBVI	Red–Green–Blue Vegetation Index (RGBVI)	(G2−R × B)/(G2 + R × B)	RGB-based greenness	Comparable to NDVI trends

17.4 Methodological Approaches for Assessing Cyclone Impacts

17.4.1 Pre- and Post-Event Imagery Analysis

Pre- and post-event imagery analysis is one of the basic methods for assessing cyclone damage to an ecosystem which involves comparing satellite images captured before and after a cyclonic event to quantify the changes in vegetation cover and health of the vegetation. Many studies are based on Landsat 8 and Sentinel-2 satellite data to detect changes in mangrove forest canopy density (FCD) calculating and comparing advanced vegetation index (AVI) bareness index (BI), shadow index (SI) and thermal index (TI) (Samanta et al. 2021). The Enhanced Vegetation Index (EVI) and Normalized Difference Vegetation Index (NDVI) are commonly used in this type of analysis. A threshold change response of -0.2 NDVI is often used to identify significant damage or loss in mangrove areas (Vizcaya-Martínez et al. 2022). With the help of this technique, mangrove canopy declines can be more easily and rapidly quantified. Change detection analyses are used to classify and quantify immediate impacts and also reclassifying canopy cover based on variations in pre-cyclone and 1-year post-cyclone images which is crucial for quick assessment following a natural calamity to understand immediate impacts like inundation and vegetation damage by the cyclone.

17.4.2 Time Series Analysis and Change Point Detection

Time series analysis is a method which involves using a sequence of multiple satellite images over time to detect mangrove dynamics, assess damage and track recovery (Chowdhury and Hafsa 2022; Islam and Assal 2023; Kingsley Kanjin and Alam 2024; Vizcaya-Martínez et al. 2022) The Landsat archive and Moderate Resolution Imaging Spectroradiometer (MODIS) data are highly utilized for their wide temporal coverage which have the ability to capture long-term changes. It allows for the calculation of mangrove canopy loss with respect to cyclones and the assessment of recovery dynamics over time ranging from months to years or even several decades (Islam and Assal 2023; Sharma et al. 2022). A study from Sundarbans using MODIS time series data (2001–2010) have been used to quantify monthly plant productivity anomalies to assess post-storm vegetation change and recovery time, showing a near-full recovery within 26 months after Cyclone Sidr in some areas (Islam and Assal 2023). Likewise, Landsat time series (1985 to 2017) analysis found significant decreases in mangrove cover with post-disturbance recovery taking 3–4 years in the Everglades (Chavez et al. 2023). This method also helps in understanding how this cyclone frequency relates to canopy cover changes and whether more frequently disturbed mangroves recover faster. It also enables the assessment of repeated cyclone impacts on mangrove forests over immediate and longer period.

17.4.3 Machine Learning and AI Applications

Machine learning (ML) and Artificial Intelligence (AI) applications have made a major advancement in the cyclone impact assessment on mangroves by offering improving accuracy and flexibility in data analysis (Amaral et al. 2023a, b, c). Random Forest (RF) and Support Vector Machines (SVMs) are the most common ML techniques which is used for classifying satellite images and predicting mangrove damage. It can effectively handle highly complex and high-dimensional remote sensing data which pave the way to improved accuracy in vegetation classification and biomass estimation. RF have achieved a correlation coefficient of 0.84 for mangrove damage assessment (Roy et al. 2024). ML techniques can also be used to find the soil fertility by many parameters which offers more efficiency, scalability and helps in real-time decision-making (Muthulakshmi et al. 2024). Supervised classification techniques, such as Maximum Likelihood Classification (MLC) and Classification and Regression Tree (CART) methods are nowadays widely used for mapping and studying land use/land cover (LULC) changes (Hossain et al. 2022). Advanced Deep learning tools such as Fully Convolutional Network (FCN) are also helpful in quantifying the mangrove degradation by classifying satellite images (Mishra et al. 2024; Roy et al. 2024).

17.4.4 Field Data Integration and Validation

Field data integration and validation are important for verifying and supporting remote sensing assessments of cyclone impacts on mangroves. While EO system provide large scale insight, field study is much more important for studying the specific information which are not possible to obtain from imagery (Vizcaya-Martínez et al. 2022) Field survey can be used to record specific damage type such as canopy defoliation, broken stem/trees and species damaged. Despite logistic difficulties and time consuming, field survey has been a traditional approach for examining mangrove damage and recovery rates. The combination of field study along with remote sensing provide us with more insightful information (Chavez et al. 2023).

17.5 Case Study from Global Hotspot—The Sundarbans

The Sundarbans, the world's largest mangrove forest and a UNESCO heritage Site is shared between India and Bangladesh, covering an estimated 10,000 km^2 (Giri et al. 2007). This important ecosystem acts as a natural bio-shield protecting millions of people and properties of the Bay of Bengal from cyclones, tidal surges and coastal erosion in the densely populated coastal areas. Despite its role, the ecosystem currently faces many intensifying stresses from climate change, sea-level rise and frequent devastating cyclonic events. The Bay of Bengal is one of the prominent cyclone hotspots which experiences a heightened susceptibility to tropical cyclones due to factors like rise in sea surface temperature and many other factors (Halder et al. 2021). While the worldwide frequency of tropical cyclones may decrease or remain stable, projections indicate an increase in the frequency of the strongest cyclones, greater precipitation rates and a poleward shift in their distribution (Gupta 2021). The Northern Bay of Bengal has seen a 26% increase in very severe cyclonic storms for the past 120 years, with notable increases in both intensity and frequency over the last two decades (Samanta et al. 2021). These changes pose major challenges to the resilience of the Sundarbans, as discussed in the study area overview Fig. 17.1, which highlights zones frequently exposed to cyclonic impacts and their ecological significance.

Mangroves are considered as ecosystem that are well adapted to disturbances, showing ecological robustness and the ability to recover from structural damages caused by cyclonic events (Krauss and Osland 2020). However, environmental changes, particularly those that impact tidal hydrological boundaries, can weaken its resistance. Tropical cyclone impacts include defoliation, bole breakage, uprooting, and alterations to sediment and hydrological conditions, all of which influence biogeochemical processess and forest recovery. Cyclone Aila, which attacked in May 2009 with wind speeds of 120.38 km/h, have shown a major impact on the Sundarbans (Mohammed et al. 2024). Its large land areas unproductive—sometimes permanently—forcing local people to shift their livelihoods from land-based activities to forest resource dependency which is increasing pressure to the ecosystem. Aila caused a 20–50% reduction in mangrove forest cover across the Sundarbans, with particular devastation in the Oligohaline zone of the Bangladesh Sundarbans, where 475.75 km^2 was heavily damaged. Species of trees such as Gewa (*Excoecaria agallocha*) were uprooted and damaged which resulted in a notable decrease in individual trees (Mohammed et al. 2024). Though widespread damage, some places have showed natural vegetation recovery, with approximately 5.31% of the vegetation area regenerating over time (Mohammed et al. 2024).

Cyclone Amphan which affected in May 2020, was a Category 5 tropical cyclonic storm that affected both West Bengal (India) and parts of Bangladesh. It was the third

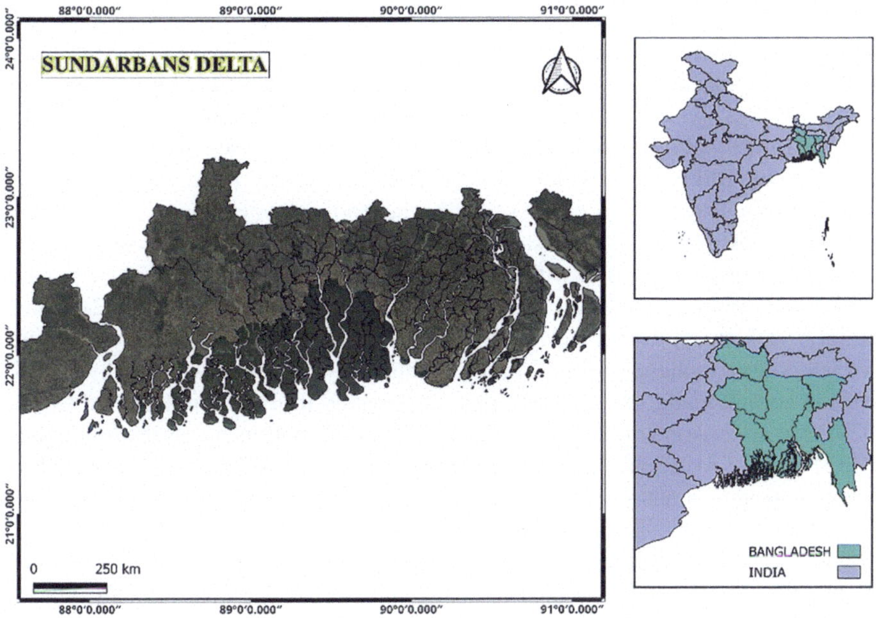

Fig. 17.1 Map showing the location of the Sundarbans

most severe cyclone to affect the Sundarbans in three decades and caused 11.5% damage to the mangrove vegetation (Sharma et al. 2022). During the COVID-19 pandemic, the cyclone worsened disaster response and may have affected recovery efforts by limiting mobility and straining resources. It led to major degradation and fragmentation of mangrove vegetation as evidenced by reduction in NDVI and EVI values significantly. Dense and moderately dense mangrove cover decreased from 77 to 34%, while less dense and sparse cover increased from 20 to 63% (Mishra et al. 2021) The greatest fragmentation and loss occurred in areas with high pre-cyclone NDVI values, showing explicitly that denser mangrove areas suffered the most. There was also a lot of Visible damage which includes widespread defoliation and a shift in canopy color to reddish-brown (Sharma et al. 2022). Shoreline erosion was very high over 68% of transects showing signs of erosion (Mishra et al. 2021). This was more affected on the Indian side (89% of transects) than on the Bangladeshi side (63%), primarily due to proximity to the cyclone path. Particularly the untouched islands with dense mangrove forests experienced severe erosion. Additionally, land use/ land cover (LULC) changes were observed, including the inundation of agricultural lands and aquaculture ponds, leading to crop damage. Salt-sensitive species like Sundari (*Heritiera fomes*) were especially affected, with tall trees being knocked down (Mishra et al. 2021). The Polyhaline zone, closer to the cyclone paths and more extensive in coverage, suffered the most damage (Mohammed et al. 2024).

Despite the severe impacts of Amphan, immediate post-cyclone image monitoring suggested a relatively rapid recovery in the Sundarbans mangrove vegetation. This observation aligns with broader ecological studies showing that mangrove productivity always declines after extreme weather but can begin to recover within one to two weeks, even under conditions of altered water chemistry, which is further supported with the NDVI trend graph Fig. 17.2. Tropical Cyclone Yaas, which occurred in May 2021, further highlighted the fragility of coastal mangroves. It caused extensive damage to the Bhitarkanika National Park (BNP) in Odisha, India—a region known for its rich mangrove biodiversity. Following the storm, NDVI values fell from 0.55 to 0.45, particularly on the seaward side that was subjected to strong winds (Mishra et al. 2024). Soil salinity also rose, with NDSI values decreasing from 0.24 to 0.15, while flood inundation increased (NDWI from 0.75 to 0.81) (Mishra et al. 2024). Interestingly, such storm events can sometimes introduce nutrients and freshwater that may temporarily enhance Gross Primary Productivity (GPP) and Chlorophyll (CHL) levels in certain mangrove communities, as observed in Bhitarkanika (Mishra et al. 2024). A comparison with the most recent NDVI distribution across the Sundarbans in 2024 (Fig. 17.3) reveals that while overall vegetation greenness has improved, fragmented and edge zones still show lower recovery, suggesting incomplete restoration in some areas.

Under repeated cyclonic events, the ecosystem exhibits complex dynamics but mangroves show high resistance—meaning relatively few areas are initially damaged—their short-term resilience or ability to recover quickly is limited. Even though physiological stress can resolve within weeks, full restoration may take much longer duration. The time of recovery can vary from as little as 3–4 years to nearly

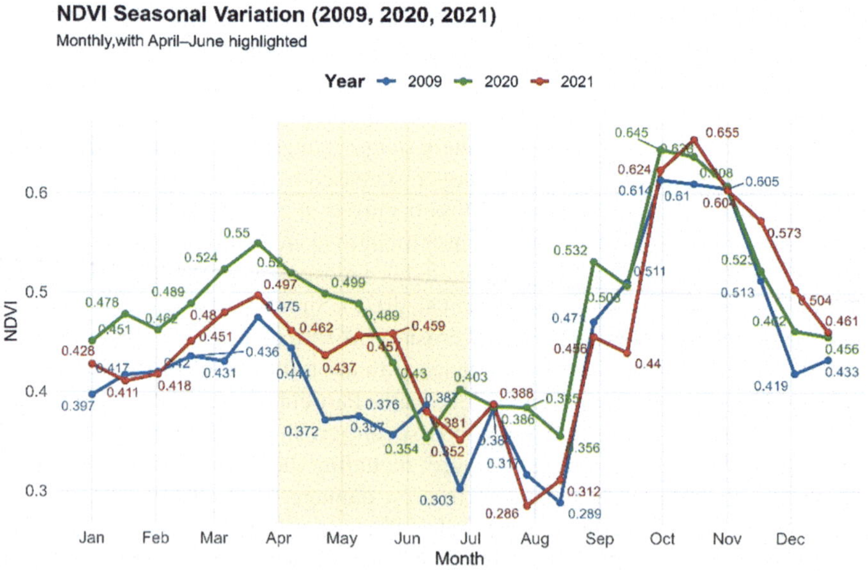

Fig. 17.2 NDVI trends from 2009, 2020 and 2021 showing vegetation stress and partial recovery patterns across cyclonic events

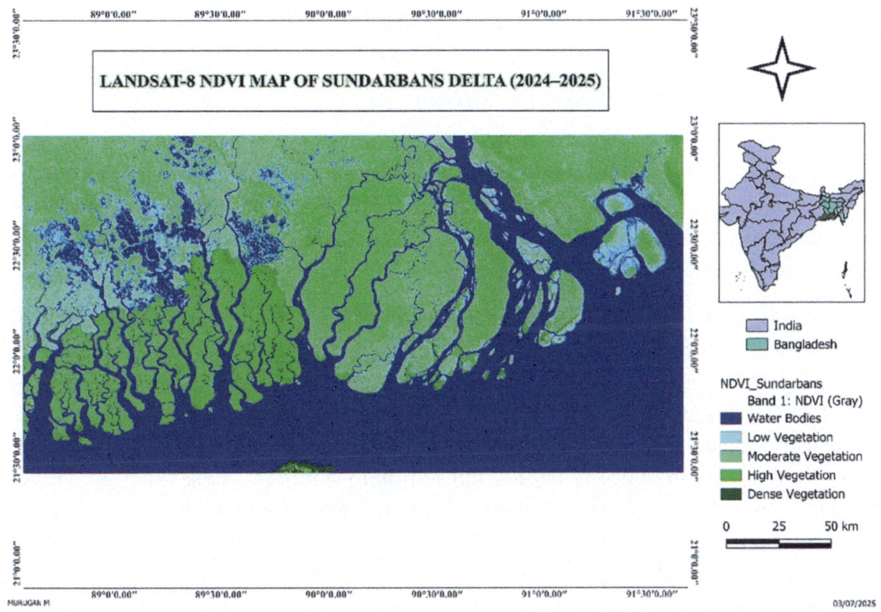

Fig. 17.3 Current NDVI map of the Sundarbans (2024) showing present vegetation condition

23 years which is highly depends on site conditions, forest structure and species composition. Cyclone Sidr's recovery took around three years in many parts, though other reports suggest 6–7 years for full regrowth (Samanta et al. 2021). Increasing cyclone frequency always shortens recovery windows, leading to continuous ecological stress which is very dangerous to this ecosystem making it more vulnerable. Species vulnerability also plays a role: *Heritiera fomes* tends to suffer more damage as tree diameter increases, while *Excoecaria agallocha* often fares better (Halder et al. 2021). Impacts are heterogeneous—seaward edges face greater damage, while interior forests may be shielded by local topography. Anthropogenic stressors—urban expansion, aquaculture, deforestation—further weaken resilience, especially in fragmented forests. There is a marked reduction has been observed in the 'core' forest class of the Sundarbans due to human development. Furthermore, drought-prone areas exhibit lower resilience and local geomorphological features such as ponding, can exacerbate vulnerability. With increasing climate change driving both more frequent and more intense cyclones the natural recovery potential of mangroves risks being overwhelmed. This clearly shows that the urgent need for adaptive environmental management and community-based resilience planning across the Sundarbans landscape is very much crucial to save the ecosystem.

17.6 Resilience, Recovery, and Vulnerability

17.6.1 Drivers of Mangrove Vulnerability

Mangrove ecosystem is vulnerable by several interconnected factors, notably forest structure, past disturbances, anthropogenic stress, and climate change factors. These are influencing the response of mangroves ecosystem towards the environmental pressures and recovery from disturbances.

Forest Structure: Height and biomass of mangrove, significantly influence the vulnerability. Taller mangroves (>10 m) are exposed to heavy wind and storm surges, leading to increased canopy loss and structural damages. Lagoon mangroves grow in sheltered lagoon environments, are more susceptible to cyclone damage than mangroves located on open coasts, estuaries and deltas. Lagoonal mangroves are differentiated from in hydrodynamics and sediment stability. These are more vulnerable due to high wind speeds and intense cyclonic rainfall which abruptly affects the complex structure of forests and leads to greater canopy losses (Peereman et al. 2021). Mangrove forest height is a key predictor to estimate the extent of damage severity. Taller forests tend to suffer more extensive canopy loss during cyclones when the wind speeds exceeding 100 km/h and heavy associated rainfall around 1050 mm are exacerbate flooding and mechanical stress on trees (Amaral et al. 2023a, b, c).

Past Disturbance: The frequency and intensity of tropical cyclones are significantly influencing the mangrove ecosystem. Frequent cyclone prone regions experienced a cumulative damage which reduced the forest resilience. For example, regions like the Bahamian subregion shows that occurrence of frequent rainfall and wind speed increase the damage risk of mangrove system. Long-term drought conditions also exacerbate the vulnerability by weakens the physiological health of mangrove and reduces their ability to regenerate after the disturbance. Mangroves exposed to frequent and intense cyclones with the wind speeds of $\geq$100 km/h associated with heavy rainfall are more likely to suffer severe damage and slower recovery (Amaral al. 2023a, b, c).

Anthropogenic Stress: Human activities such as coastal development, land restoration, pollution and expansion of aquaculture area are degrading habitat quality and altering natural processes of the ecosystem. Coastal infrastructure and human settlements increase the exposure of stress and reduce the mangrove resilience by habitat fragmentation and hampering natural regeneration. In China, Guangdong and Fujian provinces shows increased land-based pollution and coastal aquaculture expansion are accelerate the susceptibly in mangrove system. Pollution hotspots like Hainan Province shows a sea-level rise pressure concentrated over the mangrove ecosystem. Coastal artificial structures like seawalls hinder the mangrove migration towards the land and sea-level rise threatens the ecosystem. Chronic pollution from industrial and agricultural runoff damages the mangrove health which can leads to ecosystem collapse if it is not concerned. These anthropogenic pressures reduce the resistance to natural disturbances such as storms and biological invasions, its further triggers to degradation and carbon release (Hagger et al. 2022).

Climate Change Factors: Climate change compounds are altering the disturbance and environmental conditions of the system. Increasing frequency cyclone and intensity are coupled with accelerated sea-level rise, pose a compounded risks to mangrove forests. Major disturbances expected to cause due to climate change are structural changes, degradation, and loss of mangrove extent, threatening to their critical ecosystem services including coastal protection, carbon sequestration and fisheries support. Long-term climate trends such as drought and warming also reduce soil organic carbon stocks and impair the mangrove recovery capacity. The combined effects of climate change and anthropogenic stress can create loss of hotspots resilience. Prolonged drought area is facing heavy human pressure which emphasizes the need for mangrove conservation and management (Hagger et al. 2022).

17.6.2 Indicators of Recovery and Resilience

Recovery and resilience of mangroves after the disturbance are effectively monitored by using the combination of regeneration rates, vegetation indices, remote sensing techniques, and composite indices. Each methods provide critical insights of ecosystem health and success of restoration.

Regeneration Rates: The increase in mangrove area over time after the disturbance is a direct and tangible indicator for successful regeneration and recovery. Studies have shown that increase in mangrove cover reflect the effective natural and assisted restoration. Decreases or stagnation in area shows a poor recovery or remain degradation. For example, in the Guangxi Shankou Mangrove National Natural Reserve in China, mangrove area expanded from 235 hectares in 1990 to over 870 hectares by 2022, indicating that substantial increase in regeneration and afforestation efforts (Wang et al. 2023). Tracking the affected area over a period of time provides a quantifiable measure of ecosystem recovery and provides management decisions.

Vegetation Indices: Normalized Difference Vegetation Index (NDVI) is a remote sensing-based vegetation index provides the green cover of the earth which is widely used to assess the proxies of mangrove health, canopy density, and productivity. NDVI correlates the leaf area index, species richness, and aboveground net primary productivity. It delivers reliable an indication of ecosystem resilience and vitality. Multispectral and hyperspectral images capture the differences in spectral signatures between healthy and degraded mangroves which detects variations in leaf pigments, water content, and canopy structure (Tran et al. 2022). Hurricane disturbance in Mexico effectively quantified by using NDVI and other vegetation indices to assess the canopy defoliation and early recovery trends. It shows canopy defoliation reaching 79% and gradual recovery was projected that it takes at least eight years to reach pre-impact conditions. These indices enable us to monitor large-scale mangrove condition over time.

Remote Sensing and monitoring: The integration of satellite images (SPOT-5 and Sentinel 2A) with vegetation indices and Geographic Information Systems (GIS) allows us to get accurate mapping and monitoring of changes in mangrove cover. Recent studies have demonstrated that medium-resolution satellite data are coupled with machine learning algorithms (e.g., Random Trees) to produce highly accurate mangrove distribution maps also facilitating the decadal assessments of mangrove expansion or decline (Reshma Sunkur et al. 2024). Remote sensing approaches are cost-effective, repeatable and capable of overcoming logistical challenges associated with field surveys. It provides critical data collection from inaccessible mangrove areas for conservation and restoration planning. These technologies also support the evaluation of carbon stocks assessment in mangrove region and vulnerability assessments. It helps us to identify the changing patten in mangrove ecosystem and allows to necessary actions to be taken.

Composite Indices: Utilization of composite indices serves better evaluation and delivers successful restoration changes in mangrove areas. Mangrove Restoration Effectiveness Index (MREI) quantifies the restoration trajectories by integrating spatial and spectral data over multiple evaluation phases (Wang et al. 2023). The MREI has been applied to assess restoration effectiveness in both landscape and village scales. It reveals the spatial heterogeneity in restoration outcomes and persistence of recovery. The zones with high MREI values indicate that active afforestation and natural regeneration, while zones with low values indicated poor or decline in restoration effectiveness. Such composite indices provide exact and scalable tools for monitoring restoration progress and guiding adaptive measures.

Together, these indicators form a robust framework for assessing mangrove recovery and resilience. Leveraging the advancement of remote sensing technology and ecological monitoring are support conservation, restoration, and climate adaptation effectively.

17.6.3 Cumulative and Long-Term Impacts

Repeated disturbances and long-term environmental changes exert cumulative impacts on mangrove ecosystems. It resulting in significant loss of ecosystem function and resilience.

Repeated Disturbances: Successive tropical cyclones and other natural hazards cause progressive damage to mangroves. It reduces their capacity to recover and leads to persistent degradation of the system. In North Atlantic Basin, approximately 48% of mangrove areas affected by post-disturbance loss without any recovery trend even after eight months of cyclone impact. It indicates that substantial loss of resilience and potential for permanent ecosystem decline. This cumulative damage from repeated storms weakens the structural integrity of mangrove forest, regeneration processes, degrade the ecosystem, which are difficult to recover. The frequency and intensity of such disturbances are expected to increase under climate change and further exacerbating these impacts (Gomes et al. 2021).

Loss of Ecosystem Function: Mangrove structural degradation and loss of forests cover disturb their potential to provide ecosystem services. It affects the coastal protection against storm surges, carbon storage, which mitigates climate change and habitat support for fisheries. Climate change increases the frequency of cyclones, sea-level rise and irregular precipitation patterns. The near future impact of sea-level rise leads to salt water intrusion, coastal soil erosion and habitat losses. These fluctuations affect the mangrove productivity, resilience and also diminishing their capacity to buffer the coastal communities.

Long-Term Trends: Prolonged drought alter the rainfall pattern and sustained climate warming. It reduces the soil organic carbon stocks and overall forest health. Simultaneous exposure to drought, pollution, and human disturbance affects the hotspots of mangrove and resilience, which leads to permanent degradation of ecosystem. Reduced freshwater input elevates salinity stress, which hamper the seedling establishment and growth. In addition, mangrove migration towards inland was obstructed by expansion of coastal aquaculture and seawalls construction. Proper management plan was needed to avoid the ecosystem collapse from the consecutive environmental threats (Fan et al. 2024).

Socioeconomic Implications: Loss of mangroves and its degradation have significant impact on socioeconomic status. It affects millions of people who depends on mangrove ecosystem services for their livelihoods and food security. Degradation of mangrove forests exaggerates the risk of coastal communities to climate-related impacts like flooding and storm surges. It leads to potential economic loss and livelihood loss. In the Caribbean and Southeast Asia, decline in mangrove forests causes risk to fisheries, reduces carbon storage and weakens the climate adaptation and mitigation strategies. Instantaneous and effective measures for mangrove conservation and restoration are essential to protect the ecosystems and livelihoods (Gomes et al. 2021).

17.7 Emerging Trends and Technologies

17.7.1 Real-Time and Near-Real-Time Monitoring

Rapid assessment of mangrove health and disturbance are monitored by real-time monitoring by integration of radar remote sensing and weather (Reshma Sunkur et al. 2024). Sentinel-1 satellites use Synthetic Aperture Radar (SAR) sensors to penetrate the cloud cover and operate in all weather conditions to make them useful for monitoring tropical mangrove regions (Reshma Sunkur et al. 2024). Radar images helps to identify the changes in mangrove canopy structure and biomass. It enables to quick evaluations of cyclone impacts, illegal logging and shifts in hydrology (Saoum and Sarkar 2024). Accuracy of the radar improved by combined analysis with optical data gives high spatial and temporal resolution (Reshma Sunkur et al. 2024). These integrated data sources are used in MangMap to provide frequent updates for every few days on mangrove coverage and ecosystem health (Saoum and Sarkar 2024). These advanced technologies ease the conservation efforts and tracking carbon stocks (Reshma Sunkur et al. 2024).

17.7.2 Integration of Multi-Source Data: Combining Optical, Radar, LiDAR, and Socio-Ecological Data

Mangrove monitoring and its management by integrating multiple data sources delivers wide range of reliable information (Saoum and Sarkar 2024). Optical sensors (e.g., Sentinel-2, Landsat) provide spectral information about vegetation health and species composition. Radar sensors can penetrate the cloud and offer structural and moisture-related data (Reshma Sunkur et al. 2024). To assess the carbon stock, LiDAR contributes three-dimensional canopy structure and biomass estimating with high precision (Abu et al. 2024). Integration of these datasets give comprehensive characterization of mangrove ecosystems from species identification to biomass and hydrological dynamics (Saoum and Sarkar 2024). Incorporating socio-ecological data (e.g., community knowledge and land use) enriches the ecological data, enriches management decisions and restoration planning. The citizen science biodiversity data integrated with satellite and LiDAR information enlightens ecosystem health indicators and restoration success. This holistic data fusion approach supports the adaptive management, enhance successful restoration and strengthens climate mitigation strategies (Abu et al. 2024).

17.7.3 Cloud Computing and Big Data Analytics: Use of GEE, AWS, and AI Frameworks

Cloud computing platforms like, Google Earth Engine (GEE) and Amazon Web Services (AWS) have eased the mangrove monitoring by enabling large-scale remote sensing data processing with high computational efficiency (Alexey Valero-Jorge et al. 2024; José et al. 2024). These platforms support the application of artificial intelligence (AI) and machine learning algorithms to automate the mangrove classification, growth trend and biomass estimation across large spatial and temporal scales (José et al. 2024). Hybrid deep learning models like U-Net have been developed to extract the mangrove area cover and species related information from multispectral and hyperspectral data with high accuracy (José et al. 2024). In Philippine "Baka-Juan Project" integrates the citizen science data with AI based analysis on cloud platforms by combining volunteer-collected field data with Sentinel-2 images are processed in GEE to update national mangrove maps. These technologies provide near-real-time alert on deforestation, effectiveness of restoration and carbon stock monitoring to researchers, policymakers and local communities (Alexey Valero-Jorge et al. 2024). The scalability and accessibility of cloud computing reduce the cost and improve the timeliness of mangrove management interventions (José et al. 2024).

17.7.4 Community-Based and Citizen Science Approaches

Nowadays, Community and citizen science-based approaches are recognized as a vital complement to technological monitoring and it empowers the local stakeholders to actively participate in mangrove conservation. These approaches are leveraging local knowledge, foster stewardship and provide ground-truth data that enhance the accuracy and relevance of remote sensing products. Australia's MangroveWatch program engage Traditional Owners and volunteers to collect the biodiversity, health data for integrate with satellite-derived indices to prepare ecosystem reports and management plans. In Philippines, "BakaJuan Project" mobilizes citizen scientists to validate satellite mangrove maps using mobile apps and drones. It bridges the gaps between government monitoring and enabling rapid response to illegal activities. In India and Southeast Asian train local communities in sustainable resource use by Community-Based Mangrove Management (CBMM) and Community-Based Ecological Mangrove Restoration (CBEMR) initiatives for ensuring long-term resilience and socio-economic benefits. These participatory models align with conservation goals and local livelihoods to enhance the adaptive capacity to climate change.

17.8 Policy, Conservation, and Future Directions

Integration of these advanced Earth observation tools into policy, conservation, and future research is important for improving the resilience of both mangrove ecosystems and the human communities they protect.

17.8.1 Use of EO in Disaster Response and Mitigation

Remote sensing plays multiple roles in disaster response and mitigation, encompassing risk zoning, early warning systems, and post-disaster assessments.

Risk Zoning: EOS effectively identify areas that are vulnerable to the effects of cyclones. For example, studies conducted in the Sundarbans have assessed the susceptibility to cyclones in three different salinity zones: oligohaline, mesohaline, and polyhaline. The results indicate that the Polyhaline zone which is the biggest and lowest in elevation and suffered the most damage due to its proximity to cyclone paths (Mohammed et al. 2024). Factors such as elevation, proximity to the sea and height of the canopy are influencing mangrove recovery rates and vulnerability.

Early Warning: The technological advancement in the field of remote sensing and cloud computing platforms like Google Earth Engine (GEE) allow us to early forecast of cyclonic activities, real-time tracking of cyclone evolution, and projections of impacted areas more rapid and precisely. These systems have reduced human casualties (Mondal et al. 2022). Satellite imagery and Unmanned Aerial Vehicles (UAVs) should be used to develop crucial early warning metrics.

Post-Disaster Assessment: EOS offers cost effective, quick and efficient method for monitoring the post disaster recovery after the natural calamity. UAVs (Unmanned Aerial Vehicles) ultra-high spatial resolution (cm-level) which is highly useful for detailed canopy damage assessment and vertical canopy structure analysis using VARI index and decision tree classification and reducing manual efforts (Vizcaya-Martínez et al. 2022). High-resolution airborne lidar (e.g., G-LiHT) which will accurately model and estimate hurricane-induced structural damages and provides the detailed information on Aboveground Necromass (AGN) distribution which is linked to factors like distance from the coast or forest edge, tree height, and the surrounding landscape (Chavez et al. 2023). Timeseries data from satellites such as MODIS and Landsat also provide us a comprehensive view of damage and recovery pattern time to time. RADAR, geospatial and GIS technologies are also very much crucial for studying the post disaster activity (Mishra et al. 2021; Mondal et al. 2022).

17.8.2 Policy Integration and Coastal Management

Earth observation system is an important and effective tool for policy integration and coastal management of mangrove forests, which offers cost-effective and efficient monitoring capabilities and reduce the time of operation manyfold. It helps by providing synoptic coverage, historical data and repeated and continuous observation from satellites like Landsat and sentinel-2, UAVs which enables the detection of spatial and temporal changes in the mangrove ecosystem which is very important for informed decision making and sustainable practices. Especially for disaster management remote sensing is important for rapidly assessing damage from tropical cyclones and storm surges, guiding conservation efforts and advancing disaster preparedness and coastal resilience. It helps to find the vulnerable and resilient areas, supports targeted mitigation strategies, and aids in selecting sites for mangrove rehabilitation. This informative data makes the policymakers to formulate effective regulations for the sustainable management of these mangrove ecosystems amidst climate change and urbanization.

17.8.3 Future Research Needs

Future studies that make use of remote sensing are essential to enhancing planning and policy in the management of coastal mangroves are enhanced data and monitoring, holistic impact and driver analysis and predictive modelling and policy tools.

Enhanced Data and Monitoring—Collecting higher quality spatial and temporal data through satellites and UAVs which is crucial for mapping the damage assessment, shoreline dynamics and fragmentation. Long term and continuous monitoring are required to study the recovery pattern and resilience.

Holistic Impact and Driver Analysis—researcher must broaden to holistic assessment of biological, physical, chemical and ecological impacts from the cyclone to the ecosystem including understudied belowground carbon dynamics and sedimentation effects of the cyclone. Identifying the drivers of the cyclone like anthropogenic activity, climatic effects like El Nino and La Nina or other environmental stresses which is crucial for modelling.

Predictive Modelling and Policy Tools—The developer should make more accurate predictive models for mangroves response to cyclone and carbon dynamics. Targeted conservation, rehabilitation, and evidence-based policy for sustainable management can be informed by remote sensing ability to facilitate early warning systems and quick post-disaster evaluations. It is also necessary to evaluate the efficacy of transboundary freshwater management.

17.9 Conclusion

The above sections have shown the intricate dynamics of mangrove ecosystem revealing them as remarkably resilient ecosystems that are nonetheless highly vulnerable to both natural and anthropogenic stressors, particularly tropical cyclones. Many findings shows that cyclone characteristics such as wind speed and closeness to cyclone path are primary drivers of canopy damage. While mangroves exhibit rapid recovery from physiological stress (within one to two weeks for some areas like the Sundarbans) the complete recovery of vertical canopy structure can take much more longer periods, exceeding eight years. Damage magnitude is greater in taller stands, but higher aboveground biomass can enhance resilience. However, repeated cyclone impacts along with long-term climate trends like drought and heat and also the pervasive effects of anthropogenic activities can significantly compromise this resilience which leads to increased vulnerability and delayed recovery. In the Sundarbans, case studies have revealed that the loss of 'Core' mangrove areas, as well as localized

'Patch' regeneration, is a cause for concern. These losses are influenced by factors such as salinity, temperature rainfall instability and many other factors.

The integration of Earth Observation (EO) is very much crucial for effective mangrove resilience strategies. Remote sensing techniques, including satellite-derived vegetation indices and high-resolution Unmanned Aerial Vehicle (UAV) imagery offers cost-effective, efficient and comprehensive monitoring capabilities over large and inaccessible areas. The information from EOS is vital for rapidly assessing post-disaster damage from cyclones informing conservation efforts and advancing disaster preparedness and coastal resilicnce. Cloud-computing further enhance the processing of large-scale datasets and provide timely insights for immediate post-cyclone inundation assessment and long-term trend study. Remote sensing empowers policymakers to formulate evidence-based regulations and targeted mitigation strategies for the sustainable management of these critical ecosystems by helping to identify vulnerable hotspots and providing baselines for monitoring ecosystem services. The recognition of its role in presenting mangroves as Nature-based Solutions (NbS) for coastal protection against climate change and severe events is growing.

In conclusion, the inherent complexity of mangrove ecosystems, as well as the multifaceted nature of the threats they face, necessitate a strong call for multidisciplinary and transboundary approaches. Future research must move beyond simple cover assessments to holistic evaluations of ecological, biological and physical indicators including belowground carbon dynamics and sedimentation effects which are remains understudied. Understanding the interplay of environmental and other drivers requires integrating diverse scientific disciplines. Ultimately, collaborative efforts involving the people from the local communities, governmental bodies, and environmental organizations are essential for effective policy implementation at the grassroots level for protecting the long-term future of these vital coastal habitats. Such integrated approaches will be pivotal in improving our capacity for planning to sustainably manage mangroves, especially in the face of increasing climate change impacts.

17.10 Competing Interests

The authors declare no competing interests.

Acknowledgements The authors thank Bidhan Chandra Krishi Viswavidhyalaya (BCKV), West Bengal, for institutional support during the course of this work. We are also grateful to the anonymous reviewers for their valuable feedback, which helped improve the quality of this chapter. Appreciation is extended to the global scientific community for their continuous contributions and insights that enriched our understanding of the subject. We further acknowledge the efforts of the editors and publishers for their guidance and for facilitating the publication of this chapter.

References

Abu A, Shafrina W, Maulud A, Aisyah Marliza Muhmad Kamarulzaman Nor S, Mohan M (2024) Monitoring mangrove-based blue carbon ecosystems using UAVs: a review. Geocarto Int 39(1). https://doi.org/10.1080/10106049.2024.2405123

Amaral C, Poulter B, Lagomasino D, Fatoyinbo T, Taillie P, Lizcano G, Canty S, Alfredo J, TeutliHernández C, CifuentesJara M, Charles SP, Moreno CS, GonzálezTrujillo JD, RomanCuesta RM (2023a) Drivers of mangrove vulnerability and resilience to tropical cyclones in the North Atlantic Basin. Sci Total Environ 898:165413. https://doi.org/10.1016/j.scitotenv.2023.165413

Amaral C, Poulter B, Lagomasino D, Fatoyinbo T, Taillie P, Lizcano G, Canty S, Silveira JAH, Teutli-Hernández C, Cifuentes-Jara M, Charles SP, Moreno CS, González-Trujillo JD, Roman-Cuesta RM (2023b) Drivers of mangrove vulnerability and resilience to tropical cyclones in the North Atlantic Basin. Sci Total Environ 898:165413. https://doi.org/10.1016/j.scitotenv.2023.165413

Amaral C, Poulter B, Lagomasino D, Temilola Fatoyinbo Taillie P, Lizcano G, Canty S, Alfredo J, Teutli-Hernández C, Cifuentes-Jara M, Charles SP, Moreno CS, González-Trujillo JD, Roman-Cuesta RM (2023) Drivers of mangrove vulnerability and resilience to tropical cyclones in the North Atlantic Basin. Sci Total Environ 898(1). https://doi.org/10.1016/j.scitotenv.2023.165413

Asbridge E, Krause C, Lucas R, Owers CJ, Rogers K, Lymburner L, Mueller N, Ai E. Wong S (2024) Characterising the short- and long-term impacts of tropical cyclones on mangroves using the Landsat archive. Cambridge Prisms: Coastal Futures 3(4). https://doi.org/10.1017/cft.2024.19

Azad MS, Kamruzzaman M, Osawa A (2019) The influences of cyclone on abundance, species diversity and floristic composition in mangrove ecosystem in the Sundarbans. Bangladesh. Regional Studies Mar Sci 28. https://doi.org/10.1016/j.rsma.2019.100621

Chavez S, Wdowinski S, Lagomasino D, Castañeda-Moya E, Fatoyinbo T, Moyer RP, Smoak JM (2023) Estimating structural damage to mangrove forests using airborne lidar imagery: Case study of damage induced by the 2017 hurricane irma to mangroves in the florida everglades, USA. Sensors 23(15). https://doi.org/10.3390/s23156669

Chowdhury MS, Hafsa B (2022) Multi-decadal land cover change analysis over Sundarbans mangrove forest of Bangladesh: a GIS and remote sensing based approach. Global Ecology Conserv 37. https://doi.org/10.1016/j.gecco.2022.e02151

Fan B, Li Y, Zhang Z, Yang Y, Li Y (2024) Exploring cumulative vulnerability of mangrove forests to intensive coastal anthropogenic threats. Ecosyst Health Sustain, 10. https://doi.org/10.34133/ehs.0153

Giri C, Pengra B, Zhu Z, Singh A, Tieszen LL (2007) Monitoring mangrove forest dynamics of the Sundarbans in Bangladesh and India using multi-temporal satellite data from 1973 to 2000. Estuar Coast Shelf Sci 73(1–2):91–100. https://doi.org/10.1016/j.ecss.2006.12.019

Gomes LE de O, Sanders CJ, Nobrega GN, Vescovi LC, Queiroz HM, Kauffman JB, Ferreira TO, Bernardino AF (2021) Ecosystem carbon losses following a climate-induced mangrove mortality in Brazil. J Environ Manag 297:113381. https://doi.org/10.1016/j.jenvman.2021.113381

Gupta S (2021) Disturbance of mangrove forest due to climate change: the prospects of sundarban. Int J Multidiscip Appl Bus Educ Res 2(12):1306–1313. https://doi.org/10.11594/ijmaber.02.12.02

Hagger V, Worthington TA, Lovelock CE, Adame MF, Amano T, Brown BM, Friess DA, Landis E, Mumby PJ, Morrison TH, O'Brien KR, Wilson KA, Zganjar C, Saunders MI (2022) Drivers of global mangrove loss and gain in social-ecological systems. Nat Commun 13(1):6373. https://doi.org/10.1038/s41467-022-33962-x

Halder NK, Merchant A, Misbahuzzaman K, Wagner S, Mukul SA (2021) Why some trees are more vulnerable during catastrophic cyclone events in the Sundarbans mangrove forest of Bangladesh? For Ecol Manage 490. https://doi.org/10.1016/j.foreco.2021.119117

Hossain KA, Masiero M, Pirotti F (2022) Land cover change across 45 years in the world's largest mangrove forest (Sundarbans): the contribution of remote sensing in forest monitoring. Europ J Remote Sensing 57(1). https://doi.org/10.1080/22797254.2022.2097450

Islam AM, Assal TJ (2023) Rapid, landscape-scale assessment of cyclonic impacts on mangrove forests using MODIS imagery. Coasts 3(3):280–293. https://doi.org/10.3390/coasts3030017

José L, Valverde I, Salem-Sermanet A, Seghouani N, Pourchier L (2024).A deep learning-based approach for mangrove monitoring. ArXiv.org. https://arxiv.org/abs/2410.05443

Kanjin K, Alam BM (2024) Assessing changes in land cover, NDVI, and LST in the Sundarbans mangrove forest in Bangladesh and India: a GIS and remote sensing approach. Remote Sens Appl Soc Environ 36. https://doi.org/10.1016/j.rsase.2024.101289

Krauss KW, Osland MJ (2020) Tropical cyclones and the organization of mangrove forests: a review. Ann Bot 125(2):213–234. https://doi.org/10.1093/aob/mcz161

Mishra M, Acharyya T, Halder B, Santos S, Rout NR, Bhattacharyya D (2024) Impact assessment of Cyclone Yaas on the mangrove forest area in the Bhitarkanika National Park (India). J Mar Syst 242. https://doi.org/10.1016/j.jmarsys.2023.103947

Mishra M, Acharyya T, Santos, Silva, Kar D, Abu, Raulo S (2021) Geo-ecological impact assessment of severe cyclonic storm Amphan on Sundarban mangrove forest using geospatial technology. Estuarine Coastal Shelf Sci, 260. https://doi.org/10.1016/j.ecss.2021.107486

Mohammed, Sultana F, Khan A, Ahammed S, Shamim M, Bhuiyan MS, Srivastava SK, Mukul SA, Arfin-Khan MAS (2024) Assessing vulnerability to cyclone hazards in the world's largest mangrove forest, the sundarbans: a geospatial analysis. Forests 15(10). https://doi.org/10.3390/f15101722

Mondal P, Dutta T, Qadir A, Sharma S (2022) Radar and optical remote sensing for near real-time assessments of cyclone impacts on coastal ecosystems. Remote Sensing Ecol Conserv 8(4):506–520. https://doi.org/10.1002/rse2.257

Muthulakshmi S, Backiyavathy MR, Gopalakrishnan M, Kalpana R, Thangamani C, Pavithra SI (2024) Insights of machine learning approach for soil fertility assessment and management strategy. Commun Soil Sci Plant Anal 56(3):436–457. https://doi.org/10.1080/00103624.2024.2416920

Peereman J, Hogan JA, Lin T, Michaletz S (2021) Disturbance frequency, intensity and forest structure modulate cyclone-induced changes in mangrove forest canopy cover. Glob Ecol Biogeogr 31(1):37–50. https://doi.org/10.1111/geb.13407

Roy SK, Mojumder P, Arman A, Hasan MM (2025) Evaluating mangrove forest dynamics and fragmentation in Sundarbans, Bangladesh using high-resolution Sentinel-2 satellite images. Global Ecology and Conserv 58. https://doi.org/10.1016/j.gecco.2025.e03493

Roy AD, Karpowicz DA, Hendy I, Rog SM, Watt MS, Reef R, Broadbent EN, Asbridge EF, Gebrie A, Ali T, Mohan M (2024) Current status of remote sensing for studying the impacts of hurricanes on mangrove forests in the coastal united states. Remote Sensing 16(19). https://doi.org/10.3390/rs16193596

Samanta S, Hazra S, Mondal PP, Chanda A, Giri S, French JR, Nicholls RJ (2021) Assessment and attribution of mangrove forest changes in the indian sundarbans from 2000 to 2020. Remote Sensing 13(24). https://doi.org/10.3390/rs13244957

Saoum MR, Sarkar SK (2024) Monitoring mangrove forest change and its impacts on the environment. Ecol Ind 159:111666. https://doi.org/10.1016/j.ecolind.2024.111666

Sharma S, Suwa R, Ray R, Shamim M (2022) Successive cyclones attacked the world's largest mangrove forest located in the bay of bengal under pandemic. Sustainability (Switzerland) 14(9). https://doi.org/10.3390/su14095130

Reshma S, Kantamaneni K, Bokhoree C, Rathnayake U, Fernando M (2024) Mangrove mapping and monitoring using remote sensing techniques towards climate change resilience. Scientific Reports 14(1). https://doi.org/10.1038/s41598-024-57563-4

Tran TV, Reef R, Zhu X (2022) A review of spectral indices for mangrove remote sensing. Remote Sensing 14(19):4868. https://doi.org/10.3390/rs14194868

Valero-Jorge A, González-Lozano R, Zayas RG-D, Matos-Pupo F, Sorí R, Stojanovic M (2024) An innovative tool for monitoring mangrove forest dynamics in Cuba using remote sensing and WebGIS technologies: SIGMEM. Remote Sensing 16(20):3802–3802. https://doi.org/10.3390/rs16203802

Vizcaya-Martínez DA, Flores-de-Santiago F, Valderrama-Landeros L, Serrano D, Rodríguez-Sobreyra R, Álvarez-Sánchez LF, Flores-Verdugo F (2022) Monitoring detailed mangrove hurricane damage and early recovery using multisource remote sensing data. J Environ Manage 320. https://doi.org/10.1016/j.jenvman.2022.115830

Wang Y, Wang X, Khan S, Zhou D, Ke Y (2023) Evaluation of mangrove restoration effectiveness using remote sensing indices—a case study in Guangxi Shankou Mangrove National Natural Reserve. China. Front Mar Sci 10. https://doi.org/10.3389/fmars.2023.1280373

Chapter 18
Assessment of the Status of Coastal Green Belt and Casuarina Plantations Along the Shoreline of New Chandipur

Suman Chatterjee, Dhananjoy Mahapatra, Bivash Dhali, Indrani Haldar, Rakesh Samanta, Sayan Samanta, Sanjoy Mahapatra, Amit Dey, Arup Sarkar, Hrikta Dey, Abhishikta Biswas, Sreelekha Manna, Priyanshu Mukherjee, Sourav Manna, Chitra Bhowmick, Rashmi Roy, Disha Maity, Camelia Nandi, Abhisek Santra, Supratim Karmakar, and Manisha baral

18.1 Introduction

Coastal Green Belt and vegetation cover greatly reduce the impacts of climate change-induced sea-level rise and subsequent coastal erosion, change in shorelines, and coastal inundations, also protect against the storm surges (Griggs and Reguero 2021). There has been a consistent effort to conserve the natural vegetation over time, along with the growing awareness of climate change. New Plantation along the shoreline has also been considered as one of the integral parts of these conservation endeavors (Powell 2022). Different species have been used by government and non-government agencies to strategically plant them, maximizing mitigation efforts and efficiency. Plantation is not the sole factor; rather, regular monitoring of the planted vegetation cover and maintenance is also required. Among different halophytic species, such as *Casuarina Equisetifolia* acts as a protective measure for coastal marine flora and fauna. These species are strategically planted along the coastal dunes to block the effects of storm surges and to stabilize and solidify the shorelines. Green belts help fight biodiversity loss, desertification, and carbon sequestration in addition to

S. Chatterjee (✉)
Department of Geography, Loreto College, Kolkata, West Bengal, India
e-mail: sumanchatterjee@loretocollege.in

D. Mahapatra · B. Dhali · I. Haldar · R. Samanta · S. Samanta · S. Mahapatra · A. Dey · A. Sarkar · H. Dey · A. Biswas · S. Manna · P. Mukherjee · S. Manna · C. Bhowmick · R. Roy · D. Maity · C. Nandi · A. Santra · S. Karmakar · M. baral
Department of Geography, Adamas University, Barasat, West Bengal, India

A. Nath et al. (eds.), *Geospatial Application of Coastal Hazards*, Springer Water,
https://doi.org/10.1007/978-3-032-20076-1_18

protecting the land physically. Casuarina plantations are widely utilized in India to stabilize the coast and establish new trees (Feagin et al. 2010).

The Chandipur coast in Balasore district of Odisha (Fig. 18.1) is a flat, marshy area that becomes visible when the tide is low and goes underwater when the tide rises. This is known as the intertidal zone. The land above the high-tide line is called the supratidal zone, while the area below the low-tide line is the subtidal zone. But in reality, it's not easy to tell where the intertidal zone ends and the supratidal zone begins (Majumder 2025). To the north, there is a wide stretch of tidal flat with a narrow beach around it. Behind the beach, sand dunes formed by wind (called Aeolian dunes) can be seen, and further inland, there is newer soil deposited by rivers. On the eastern side, the Buribalam River flows, along with its marshy terrace that has risen above water level. In the south, the land beyond the tidal flat and shoreface is mostly flat and level. As the sea gradually advances, the coast is facing serious erosion. The shape and form of the land here don't match the way waves behave today. The shoreline runs in a northeast to southwest (NE–SW) direction (Saha et al. 2023).

This research has assessed the ecological status and protective function of the green belt using field observation, remote sensing, and machine learning. The focus is on estimating biomass and species numbers through satellite imagery and predictive models to better understand changes in coastal vegetation over time. Tree and shrub species, sizes, health conditions, and mowing data were collected through direct observation within selected sample plots for comprehensive analysis. The structural parameters and perceived health informed the development of these methods.

Above Ground Biomass (AGB) is quantified mainly using four approaches: empirical methods incorporating visible and infrared data along with field observations, the allometric equation method, RADAR and LiDAR-based techniques, and carbon sequestration models such as the InVEST model and CBM-CFS3 (Angkahad et al. 2024; Tian et al. 2023). The AGB calculation based on actual field measurements for a small forest stand is more accurate (Li et al. 2020). Forest AGB calculation through the field measurement is impossible for the regional scale, as it is too costly, time- consuming, and labor-intensive. Past work has shown that remote sensing has the capability to estimate and monitor forest biomass at the regional scale reliably; consequently, various types of remote sensors, utilizing passive and active sensors, have been used to de-rive AGB. In its capacity as an active type of remote sensing, radar can penetrate the canopy layer and interact with the main biomass components, namely the branches and trunks of the trees. The strength of the backscattering of the radar grows as the forest biomass increases. Sensitivity of radar to biomass estimation is wavelength dependent. With a longer wavelength, the value of the scattering saturation value increases, and the radar backscatter-biomass relationship gets enhanced as well; therefore, long wavelength bands are more suitable for biomass estimation (Khan et al. 2024; Tian et al. 2023).

Additionally, it is necessary to select an appropriate algorithm for building AGB estimation models. Classical statistical regression is easy to calculate and straightforward (Prakash et al. 2022). The linear regression (LR) method has been the most widely used approach for estimating AGB in previous studies. Remote

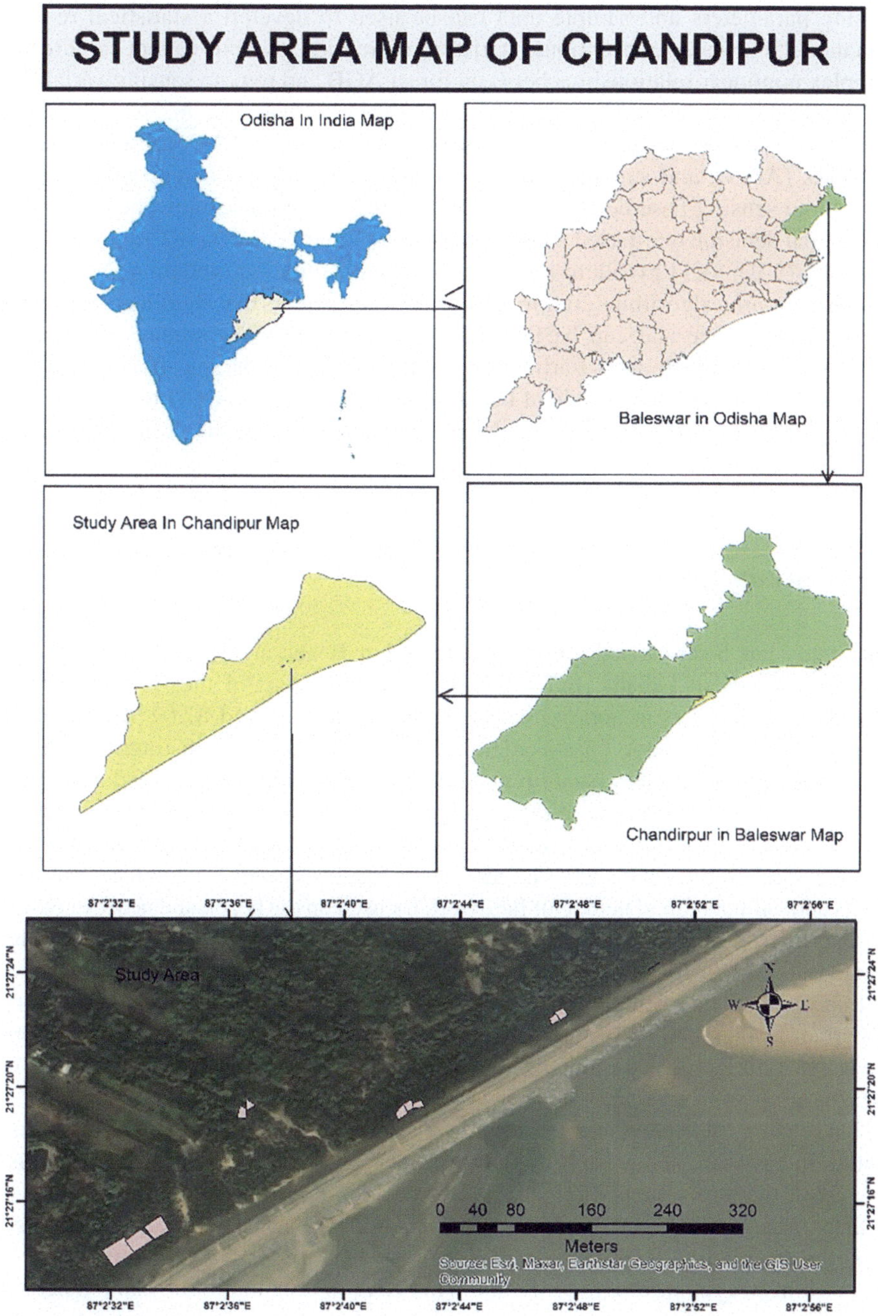

Fig. 18.1 Study area location map

sensing parameters and sample data can be used to develop a statistical regression model. However, conventional statistical regression methods cannot capture the complex nonlinear relationships between forest AGB and remote sensing variables. To improve the model's ability to estimate nonlinear patterns, machine learning methods—such as decision trees, K-nearest neighbors (KNN), artificial neural networks (ANN), and support vector machines (SVM)—are increasingly used in the remote sensing-based estimation of forest AGB. Previous studies have demonstrated that decision tree-based algorithms, such as Random Forest (RF) and Gradient Boosting (GB), yield excellent results in biomass estimation modeling. Moreover, machine learning algorithms comprise numerous parameters that significantly impact model performance and can be fine-tuned. However, in some studies, the tuning process has been overlooked. Earlier research has shown that the sensitivity of parameters in RF and stochastic gradient boosting models differs, and the tuning process plays a key role in determining model performance (Tamiminia et al. 2022; Tian et al. 2023).

18.2 Study Area

This study has been carried out in the Chandipur Beach area, which is located in Balasore district of Odisha, India, along the eastern coast of the Bay of Bengal. Chandipur Beach lies at approximately 21.47° N latitude and 87.02° E longitude (Fig. 18.1). It is a unique coastal region, known globally for a distinguished tidal event where the sea moves back up to 5 km during low tide and returns during high tide. This unusual movement of the sea creates a dynamic coastal environment with wide, muddy flats, sandy areas, shallow water pools, and small, river-like channels. These natural features have a big impact on the local environment and the variety of plants and animals found here. Our fieldwork focused on the beach and nearby coastal areas, primarily concentrating on studying and documenting the local vegetation. The region is home to a diverse array of plants, including salt-tolerant grasses, halophytes, and shrubs that thrive in salty, wet, and sandy conditions. These plants are very important as they help hold the soil together, reduce erosion, filter pollutants from the water by trapping sediments, and support a wide range of other forms of life in the area.

These plants also play a key role in the coastal food chain. They produce organic matter and provide shelter and breeding spaces for many land and water animals. As part of the larger vegetation study, a team focused on measuring plant biomass in the Casuarina Research Plantation at Chandipur. The Jashipur Research Range manages this plantation under the Odisha Forest Department. It was set up to protect the coast from cyclones, sea waves, and wind erosion. The main tree planted here is *Casuarina equisetifolia*, a fast-growing and salt-tolerant species. It is useful not only for its ability to grow well in tough conditions but also for the many environmental benefits it provides. These trees grow in thick lines, forming green belts that block strong winds and offer homes to birds, small animals, and reptiles. The biomass study aimed to measure the amount of plant material above and below the ground in this plantation. This helps us understand how much carbon is stored, how healthy the plants are, and what benefits these man-made forests give to the environment. Apart from this plantation, the nearby tidal areas and open sandy zones of Chandipur Beach also have some vegetation. Although not dense, plants such as Suaeda maritima, Salicornia brachiata, and various grasses grow here. These are salt-tolerant plants, capable of surviving in salty water, regular flooding, and shifting sand. They help hold the soil together, recycle nutrients, and offer homes for insects and birds. Overall, this survey of the Chandipur Beach area and its Casuarina plantation gave us useful information about the types of plants growing there, where they grow, and what role they play in the coastal ecosystem. It demonstrates the importance of both natural and planted greenery in protecting the coast, maintaining biodiversity, and stabilizing the environment along India's eastern coastline. The main objectives of the study are:

i. This study aims to estimate the above-ground biomass (AGB) of *Casuarina equisetifolia* and other coastal vegetation in the chosen site;
ii. This study aims to map the spatial distribution of species across the selected plantation site;
iii. This study aims to assess the floristic composition and diversity index of the green belt.

18.3 Data and Methods

In this study, we utilized secondary remotely sensed data from ISRO's ResourceSat-2, specifically the LISS-IV sensor, which offers a spatial resolution of 5 m. Additionally, we derived ancillary datasets, including NDVI, GNDVI, and SAVI. Software libraries such as Dask and Xarray in Python were utilized for efficient data processing. Most importantly, ground-based field data—including geotagged measurements of DBH (Diameter at Breast Height), tree height, and species type—were utilized.

18.3.1 Plot Selection and Sampling

This field study was carried out along Chandipur Beach in the Balasore district of Odisha. The main goal was to survey the vegetation and measure plant biomass. The study area encompassed the beach, the intertidal mudflats, and the Casuarina Research Plantation, which is managed by the Jashipur Research Range. Because of the changing tides and uneven ground, it wasn't possible to reach every part of the area for detailed measurement. To handle this, the team employed a plot-based sampling method, selecting specific locations that represented the various types of vegetation in the accessible parts of the study area. In total, 11 sample plots were marked out. These plots were labeled A1, A2, A3, B1, B2, C1, C2, C3, D1, D2, and E1. Among them, A1, A2, and A3 were larger plots, each measuring 10 m by 10 m (Griggs and

Reguero 2024; Lu et al. 2022). These were placed within the Casuarina Plantation, where the trees grow in a uniform pattern, allowing for accurate measurement of biomass and tree density. The other eight plots—B1 to E1—were smaller, each 10 m by 10 m. They were spread across different parts of Chandipur Beach that were easier to access, including areas with salt-tolerant grasses, halophytes, and scattered coastal shrubs. Some places couldn't be surveyed due to high tides, muddy ground, or thick vegetation. The plots were chosen based on where the team could safely reach, while still attempting to cover the diverse types of plant life found in the area. The research team used both big and small plots to study the plants in Chandipur. This helped them see how plant types, structure, and biomass changed in different parts of the area. Their method made it possible to gather useful information without needing too many resources. In each plot, they counted all the trees and noted their species. They also measured the trunk size (DBH) and height of each tree using an Abney level and measuring tape. To keep track of the locations, they marked all the plots using GPS.

Tree height was measured using an Abney level, and the trunk size (GBH/DBH) was measured with a tape at a height of 1.37 m (Fig. 18.2). Tree species were identified on the spot using plant guidebooks. Our team also recorded biodiversity details like how many of each species were present (abundance), how many different species there were (richness), and how evenly they were spread (evenness).

Fig. 18.2 Data collection of the present study

We have used satellite images from ISRO's ResourceSat-2 LISS-IV, which has a resolution of 5 m. The bands used for analysis were Red, Green, and Near-Infrared (NIR). To study the health and density of the plants, they calculated vegetation indices like NDVI, GNDVI, and SAVI (Huete 1988; Tucker 1979).

The stratified random sampling method has been used. First, we divided the area into different vegetation or habitat types based on secondary data derived maps. Then, we randomly selected a certain number of sample plots from each type.

Each sample plot was marked out as a square plot e.g. 10 m × 10 m plots, depending on the type of vegetation i.e. **Trees and shrubs or Herbaceous plants**. We marked the corners of each plot with stakes and flagging tape, and we recorded the exact location using a GPS device (with about ±3 m accuracy). Each plot was assigned a unique ID, allowing us to track it throughout the study.

We measured plant biomass using both **destructive** (cutting and weighing) and **non-destructive** (measuring and calculating) methods.

For trees, we recorded:

- **Diameter at Breast Height (DBH)**: Measured at 1.3 m from the ground
- **Tree Height**: Measured with an abney level or laser range finder (for hard-to-reach trees)

We used a standard formula (Chave et al. 2014) to estimate the **Above Ground Biomass (AGB)** of each tree (Eq. 18.1):

$$\mathrm{AGB} = 0.0673 \times \left(\mathrm{D}^2 \times \mathrm{H} \times \mathrm{WD}\right)^{0.976} \tag{18.1}$$

where:

D = diameter at breast height (cm), **H** = height (m), **WD** = wood density (g/cm^3).

In August 2024, our team spent two full days in the field, collecting data from 11 sites using this technique. For bushes, we either measured height and basal diameter and estimated biomass using formulae. Collected and weighed biomass in a 10 m × 10 m sample area, using approximation methods.

Plant species within each sampling plot were identified through direct observation in the field using taxonomic keys, field guides, and expert knowledge, and some smartphone-based applications i.e. iNaturalist, Plantix, Google Lense, Leafsnap. If we weren't sure about a plant, we collected a sample (like leaves or flowers), dried and pressed it, and sent it to a herbarium for expert identification. We also kept a voucher specimen of each species for future reference, and noted down the scientific name (genus, species, and authority).

We recorded the GPS location (latitude and longitude) of each plot using a hand-held device. We also noted the **date, time**, and **environmental features** like elevation, slope, and direction of the land (aspect), if they were important for the study.

All field data was first written down in prepared datasheets and logbook, and later typed into a digital database. We double-checked the entries and flagged any unusual values to ensure everything was accurate. The database connects **biomass data**, **species info**, and **location details** for each plot.

18.3.2 Data Preprocessing and Analysis

Remotely sensed data were preprocessed using atmospheric correction with inbuilt models available in tools such as 6S, Sen2Cor, or LEDAPS. The necessary spectral bands were stacked, and vegetation masking was performed to remove non-vegetative land cover types (e.g., dunes, water bodies). The datasets were inspected for outlier detection and removal to clean the data and improve model training quality. Field plot locations were projected and matched with corresponding satellite pixels to integrate geotagged field data, including DBH, tree height, and species type. Biomass estimation modeling was conducted using the processed dataset to train machine learning models.

18.3.3 Biomass Modelling

In this study, **two machine learning models** were applied to estimate the **above-ground biomass** in the study area. The models that were used included **XGBoost** and **Random Forest** (Ramdani and Furqon 2022).

Among the two, the **XGBoost model** was found to perform exceptionally well. It achieved a **coefficient of determination (R^2) of 0.95**, which indicates a very strong relationship between the predicted and actual biomass values. On the other hand, the **Random Forest model** yielded a significantly lower R^2 of 0.54, indicating that it was less effective for this purpose.

To make the models more accurate and reduce the time and computing power needed, a method called **Bayesian optimization** was used to fine-tune their parameters. During this tuning process, several parameters were adjusted, including colsample_bytree, subsample, n_estimators, gamma, eta, min_child_weight, max_depth, and verbosity.

Both models were trained using a set of input features that included satellite image bands like **Near-Infrared (NIR), Red, and Green**, along with vegetation indices such as **NDVI (Normalized Difference Vegetation Index), GNDVI (Green NDVI), and SAVI (Soil Adjusted Vegetation Index)**.

18.3.4 Species Mapping

A Random Forest classification model was used to map the spatial distribution of the dominant plant species in the area. The model's overall accuracy of 0.79, which is deemed suitable for ecological studies, was attained through the use of satellite imagery and vegetation classes based on shapefiles.

18.4 Results and Discussions

The XGBoost model was tuned using Bayesian optimization, which was preferred over traditional grid search to save on computational time and cost. A range of hyper-parameters—such as learning rate, tree depth, and sampling rates—was adjusted during the tuning process (Fig. 18.3). It was observed that no overfitting occurred even as the number of boosting rounds increased. Instead, the model's RMSE steadily decreased and then stabilized, suggesting strong predictive consistency. When tree depth was increased, the performance (measured by R^2 and RMSE) remained largely unchanged, although deeper trees were more prone to overfitting. It was found that a lower learning rate led to better performance. With smaller learning rates, the model became more cautious in its learning, requiring more iterations but offering better generalization. In contrast, higher learning rates (such as 0.3 and 0.5) caused the model to overfit after a certain number of rounds—although the RMSE initially dropped, it then started to rise. The best results were achieved when the learning rate was set to 0.23. Performance was also influenced by the subsample rate, which determines the proportion of the training data used for each tree. As the subsample value decreased, more iterations were needed, but the model's performance improved. Similarly, column sampling by tree was applied to prevent the model from becoming overly dependent on specific features. When a higher proportion of features was used (0.86 in this case), the model performed better, likely due to the richness of informative variables in the dataset.

The final AGB map was classified into three categories based on estimated biomass density, expressed in megagrams per hectare (Mg/ha). These categories allowed for

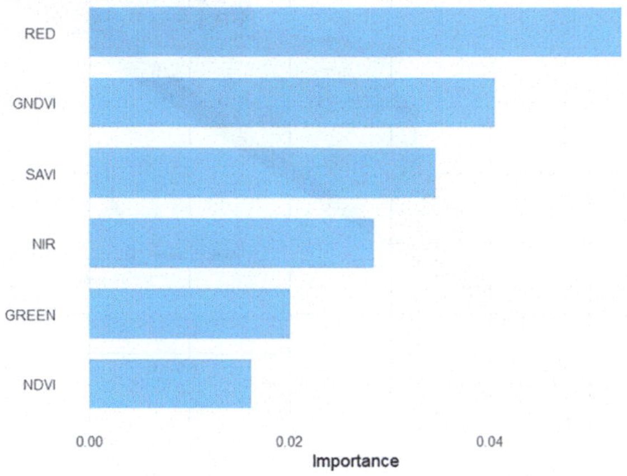

Fig. 18.3 Feature importance plot of XGBoost model

a clearer understanding of biomass distribution across the study area (Fig. 18.4). A Bayesian optimizer was used to fine-tune XGBoost parameters, resulting in the best model for biomass modeling with an R-squared value of 0.95. We have categorized the map index into three classes. Class 1 with low biomass density (<200 Mg/ha) indicated in purple color. These areas were mainly located along the coastal region and represent sparse vegetation or bare ground, likely due to the influence of sandy beaches and limited plant cover. The second class or category we classified as moderate biomass levels (200–300 Mg/ha) indicated by yellow color. This category appeared in a scattered pattern, indicating regions with moderate vegetation density. The third category, represented in green, marks high biomass density areas (>300 Mg/ha). These zones were characterized by dense vegetation, often found in association with plantations or conservation areas. High-biomass areas are particularly important for carbon storage and long-term ecological stability. Three dominant plant species were identified based on the spatial distribution of the region's dominant plant species.

The largest area was occupied by *Casuarina Equisetifolia*, which was mostly found along the beach and in nearby coastal areas. Due to its robust root system, ability to withstand salt, and capacity to mitigate coastal erosion, it is widely distributed (Fig. 18.5).

Smaller patches of Ficus benghalensis are usually found close to semi-urban or green, isolated areas. This species is well-known for both its ecological function of stabilizing microenvironments and its wide canopy. Due to its seasonal growth cycle

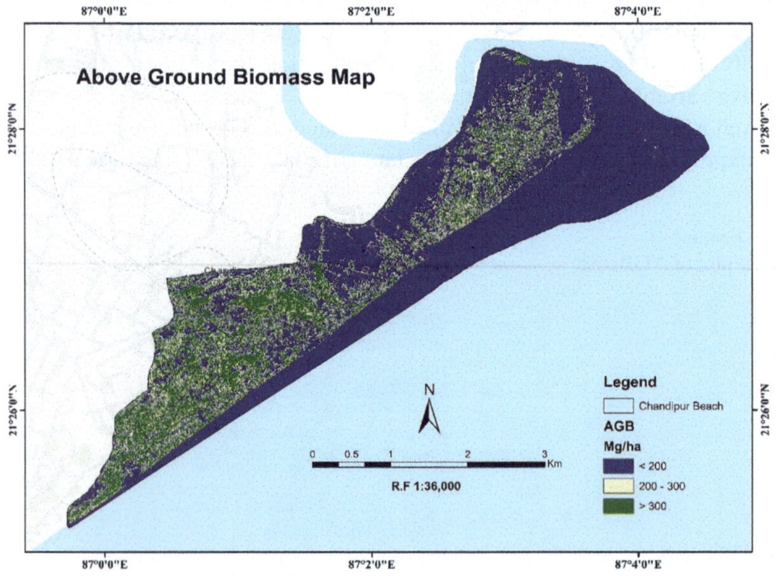

Fig. 18.4 Above-ground biomass map

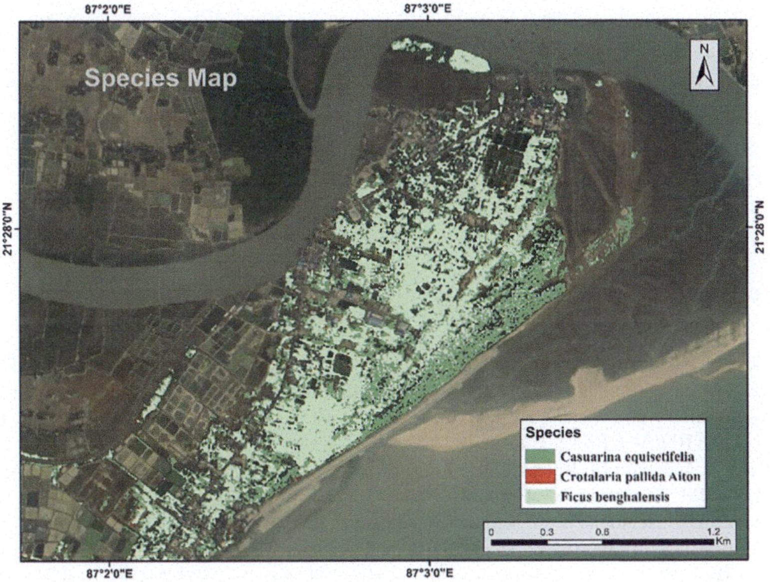

Fig. 18.5 Species distribution map

and particular soil or moisture requirements, Crotalaria pallida Aiton had a sparse and dispersed distribution.

18.5 Conclusion

This study demonstrates how well machine learning techniques, such as Random Forest, can be combined with geospatial tools to create detailed vegetation maps. Planning restoration projects, promoting biodiversity conservation, and managing coastal landscapes sustainably all depend on such spatially explicit analyses. Future ecological assessments and policy decisions aimed at increasing coastal green cover in response to anthropogenic and environmental pressures will greatly benefit from the findings. The above-ground biomass (AGB) map that was produced offered valuable information for tracking the health of the forest ecosystem and vegetation dynamics in the coastal area of Chandipur. Tidal influences and a variety of vegetation types have shaped this ecologically sensitive landscape, necessitating careful monitoring for effective conservation initiatives. This study emphasises how useful machine learning, high-resolution remote sensing, and integrated field data are for evaluating coastal vegetation. *Casuarina equisetifolia* plots' high biomass levels support their function in carbon sequestration and shoreline protection. Nonetheless, the ecological hazards of monocultures emphasize the necessity of species diversity in coastal afforestation initiatives. XGBoost's superior performance highlights the promise of advanced machine learning in ecological modelling. Important new information about the spatial variability of biomass and species composition was made possible by the utilisation of ResourceSat-2 LISS-IV data and vegetation indices such as NDVI, GNDVI, and SAVI. To improve sustainability, mixed-species plantations should be given priority in future coastal resilience plans, and data-driven monitoring should be used.

References

Angkahad T, Laosuwan T, Sangpradid S, Prasertsri N, Uttaruk Y, Phoophathong T, Nuchthapho J (2024) An empirical analysis of above-ground biomass and carbon sequestration using UAV photogrammetry and machine learning techniques. IEEE Access 12:186740–186752. https://doi.org/10.1109/ACCESS.2024.3514074

Chave J, Réjou-Méchain M, Búrquez A, Chidumayo E, Colgan MS, Delitti WBC, Duque A, Eid T, Fearnside PM, Goodman RC, Henry M, Martínez-Yrízar A, Mugasha WA, Muller-Landau HC, Mencuccini M, Nelson BW, Ngomanda A, Nogueira EM, Ortiz-Malavassi E, Vieilledent G et al (2014) Improved allometric models to estimate the aboveground biomass of tropical trees. Glob Change Biol 20(10):3177–3190. https://doi.org/10.1111/GCB.12629;PAGEGROUP:STRING:PUBLICATION

Feagin RA, Mukherjee N, Shanker K, Baird AH, Cinner J, Kerr AM, Koedam N, Sridhar A, Arthur R, Jayatissa LP, Lo Seen D, Menon M, Rodriguez S, Shamsuddoha M, Dahdouh-Guebas F (2010) Shelter from the storm? Use and misuse of coastal vegetation bioshields for managing natural disasters. Conserv Lett 3(1):1–11. https://doi.org/10.1111/J.1755-263X.2009.00087.X

Koppad AG, Gowda GB, Das A (2024) Assessment of forest above ground biomass with SAR data using linear modeling in Joida Taluk, Uttara Kannada, India. Int J Environ Clim Change 14(12):241–259. https://doi.org/10.34894/VQ1DJA

Griggs G, Reguero BG (2021) Coastal adaptation to climate change and sea-level rise. Water (Switzerland) 13(16). https://doi.org/10.3390/W13162151

Huete AR (1988) A soil-adjusted vegetation index (SAVI). Remote Sens Environ 25(3):295–309. https://doi.org/10.1016/0034-4257(88)90106-X

Khan MN, Tan Y, Gul AA, Abbas S, Wang J (2024) Forest aboveground biomass estimation and inventory: evaluating remote sensing-based approaches. Forests 15(6). https://doi.org/10.3390/F15061055

Li Y, Li M, Li C, Liu Z (2020) Forest aboveground biomass estimation using Landsat 8 and Sentinel-1A data with machine learning algorithms. Sci Rep 10(1):1–12. https://doi.org/10.1038/S41598-020-67024-3

Lu C, Xu H, Zhang J, Wang A, Wu H, Bao R, Ou G (2022) A method for estimating forest aboveground biomass at the plot scale combining the horizontal distribution model of biomass and sampling technique. Forests 13(10):1612. https://doi.org/10.3390/F13101612

Majumder R (2025) A study of intertidal macroinvertebrate assemblages in the Chandipur Coast, Odisha, Eastern India. Thalassas 41(1):1–15. https://doi.org/10.1007/S41208-024-00764-5

Powell K (2022) Building a living shoreline to help combat climate change. Nature 612(7941):806. https://doi.org/10.1038/D41586-022-04468-9

Prakash AJ, Behera MD, Ghosh SM, Das A, Mishra DR (2022) A new synergistic approach for Sentinel-1 and PALSAR-2 in a machine learning framework to predict aboveground biomass of a dense mangrove forest. Eco Inform 72:101900. https://doi.org/10.1016/J.ECOINF.2022.101900

Ramdani F, Furqon MT (2022) The simplicity of XGBoost algorithm versus the complexity of Random Forest, Support Vector Machine, and Neural Networks algorithms in urban forest classification. F1000Research 11:1069. https://doi.org/10.12688/f1000research.124604.1

Saha B, Samanta P, Mukhopadhyay S, Eriksson PG (2023) Effects of transgression on sedimentation system vis-à-vis coastal erosion on the Chandipur coast, Odisha, India. J Indian Assoc Sedimentol (Peer Reviewed) 40(I):5–19. https://doi.org/10.51710/JIAS.V40II.276

Tamiminia H, Salehi B, Mahdianpari M, Beier CM, Johnson L, Phoenix DB, Mahoney M (2022) Decision tree-based machine learning models for above-ground biomass estimation using multi-source remote sensing data and object-based image analysis. Geocarto Int 37(26):12763–12791. https://doi.org/10.1080/10106049.2022.2071475

Tian L, Wu X, Tao Y, Li M, Qian C, Liao L, Fu W (2023) Review of remote sensing-based methods for forest aboveground biomass estimation: progress, challenges, and prospects. Forests 14(6). https://doi.org/10.3390/F14061086

Tucker CJ (1979) Red and photographic infrared linear combinations for monitoring vegetation. Remote Sens Environ 8(2):127–150. https://doi.org/10.1016/0034-4257(79)90013-0

Chapter 19
Anticipatory Actions to Climate Crisis: Integrating Community-Based Adaptive Responses and GIS Analysis to Socio-economic Impacts of Flooding in Coastal Communities in Niger Delta, Rivers State, Nigeria

B. C. Oburu, Akanwa Angela Oyilieze, Bappaditya Koely, Alom Anselem Chukwuma, and Evariste Nshimyimana

19.1 Introduction

Flood frequency and intensity are rising globally due to climate change, which amplifies moisture and extreme rainfall events. This trend is evident from deadly recent floods, from Texas and China to Nigeria and Bolivia, demonstrating how warming fuels more frequent and severe inundations across diverse regions (IPCC 2022; Reuters 2025; Time 2025; Associated Press 2025).

Additionally, urbanization has intensified, bringing about changes in land cover and increasing impervious surfaces that reduce natural infiltration and increase surface runoff (Ogunrinde et al. 2024). This trend has heightened the frequency and

B. C. Oburu · A. A. Oyilieze (✉)
Department of Environmental Management, Chukwuemeka Odumegwu Ojukwu University (COOU), Anambra, Nigeria
e-mail: angy.akanwa1@gmail.com

A. A. Oyilieze · E. Nshimyimana
Faculty of Environmental Management, Agriculture and Renewable Energy (FEMARE), University of Technology and Arts (UTAB), Byumba, Rwanda

B. Koely
Department of Geography, Bankim Sardar College, Tangrakhali, West Bengal, India

A. A. Chukwuma
Faculty of Environmental Sciences, Department of Urban and Regional Planning, Nnamdi Azikwe University (NAU), Awka, Anambra, Nigeria
e-mail: ac.alom@unzikedu.ng

A. Nath et al. (eds.), *Geospatial Application of Coastal Hazards*, Springer Water,
https://doi.org/10.1007/978-3-032-20076-1_19

severity of floods, causing widespread socio-economic disruptions and underscoring the urgent need for effective flood management and adaptive strategies (Adedeji et al. 2021).

Flooding has increasingly become one of the most devastating natural hazards worldwide, disproportionately affecting vulnerable coastal and riverine communities, particularly in developing countries (Akanwa et al. 2022). In Nigeria, recurring flood events have caused significant socio-economic disruption, destroying homes, infrastructure, and livelihoods, while exacerbating poverty and inequality in flood-prone areas. The Niger Delta region, characterized by low-lying terrain and dense population, remains especially susceptible to flood hazards intensified by climate change, poor urban planning, and environmental degradation (Nwilo and Badejo 2006; Olajuyigbe and Ogunyemi 2022).

The growing frequency and severity of floods demand not only reactive emergency responses but also proactive, anticipatory actions that enable communities to prepare and adapt before disasters occur (Kelman et al. 2022). Given these factors, the Ogba/Egbema/Ndoni Local Government Area (ONELGA) (LGA) of Rivers State represents a typical Niger Delta community grappling with the compounded effects of climate variability, environmental degradation, and socio-economic vulnerability.

ONELGA is among the most catastrophic affected region in recent historical flood accounts, impacting over 200 communities and resulting in considerable socio-economic upheaval. This disastrous event resulted in the submersion of 51 communities within the LGA, displacing thousands of inhabitants and inflicting extensive damage upon properties and infrastructure (Godwin 2024).

Responses from governmental authorities to these flooding incidents in the Ogba/Egbema/Ndoni LGA have primarily been characterized by a reactive orientation, concentrating on post-disaster relief initiatives rather than adopting proactive strategies that anticipate and ameliorate the impacts of flooding. This reactive approach has been demonstrably inadequate in addressing the underlying causes of flooding and in mitigating its recurrent effects. The resultant loss of housing and the displacement of families have intensified the socio-economic adversities confronting the populace, thereby elucidating the inadequacies inherent in the current management frameworks.

There exists an urgent imperative for the formulation of a comprehensive flood management strategy that integrates infrastructure development, community education, and sustainable land use practices to effectively confront and mitigate the challenges engendered by flooding. This thus offers a compelling context for studying anticipatory actions and community-based adaptive responses to flooding within the Omoku and Okansu communities.

Anticipatory actions encompass early warning systems, risk communication, capacity building, and locally driven adaptive strategies aimed at minimizing flood impacts (Stein et al. 2021). In this context, community-based adaptive responses play a critical role in enhancing resilience by leveraging indigenous knowledge, social networks, and place-based coping mechanisms (Birkmann et al. 2017).

While previous studies in Nigerian flood-prone areas have predominantly focused on post-disaster impacts and policy recommendations (Nwankwoala and Udom

2019), there is a need for integrated approaches that combine socio-economic data with geospatial analysis to better understand spatial patterns of vulnerability and adaptive capacity. Geographic Information Systems (GIS) and remote sensing technologies provide powerful tools for mapping flood-prone zones and analyzing environmental factors such as elevation and land use, thereby supporting targeted interventions (Ogunrinde et al. 2024).

This study adopts a mixed-methods approach that integrates household questionnaires, photographic documentation, and GIS spatial analysis to investigate anticipatory actions and community-based adaptive responses to flooding in Omoku and Okansu communities in Rivers State. By linking local perceptions and adaptive behaviours with spatial flood risk indicators, this research aims to provide nuanced insights to inform policy and practice for climate-resilient flood management in vulnerable Nigerian communities.

19.1.1 The Concept of Anticipatory Actions

Anticipatory actions represent proactive measures taken in advance of predictable climate-related hazards to minimize adverse impacts on vulnerable communities (Birkmann et al. 2017). This approach shifts the focus from reactive disaster response to preparedness and risk reduction, enabling communities to act before a crisis fully unfolds (Stein et al. 2021). In the context of flooding, anticipatory actions may include early warning dissemination, pre-positioning of resources, temporary relocation, and community mobilization to reduce exposure and loss (Kelman et al. 2022).

The concept aligns closely with the Sendai Framework for Disaster Risk Reduction (2015–2030), which emphasizes preparedness and anticipatory risk management as critical to building resilience in the face of increasing climate variability (UNDRR 2015). By adopting anticipatory actions, communities can reduce reliance on post-disaster humanitarian aid and better safeguard livelihoods, particularly in flood-prone areas like Rivers State, Niger Delta Region.

In this study, anticipatory actions provide a theoretical lens to understand how local populations in coastal communities engage in flood preparedness and adaptive responses. Research shows that community-based anticipatory actions are often shaped by local knowledge, social networks, and the capacity to interpret early warnings (Birkmann et al. 2017). However, challenges such as inadequate infrastructure, limited access to timely information, and governance gaps can constrain effective anticipatory measures (Lindsey et al. 2020).

Integrating anticipatory actions with geospatial analysis of elevation and land use offers a more comprehensive understanding of flood risk and community vulnerability. GIS mapping can identify high-risk zones, enabling targeted dissemination of early warnings and resource allocation (Paton and Johnston 2017). This combined approach enhances both the accuracy of flood risk assessments and the relevance of community adaptive strategies.

Ultimately, anticipatory actions underscore the importance of proactive, locally grounded, and spatially informed interventions to mitigate flooding impacts. Their incorporation in this study supports the development of tailored policies and capacity-building efforts that empower communities to anticipate and effectively respond to the increasing threats posed by climate-induced flooding.

19.2 Study Area

The Ogba/Egbema/Ndoni Local Government Area (ONELGA), located in Rivers State, Nigeria (Fig. 19.1). It was established on September 21, 1991, by then Head of State Ibrahim Badamasi Babangida. It is one of the 774 local government areas recognized by Nigeria's 1999 constitution. The headquarters of ONELGA is situated in Omoku, which is strategically important as the second largest industrial and commercial hub in Rivers State after Port Harcourt, playing a critical role in the regional economy (Guide for Indigenous People of Nigeria 2021; OENW 2018). One of the major centers of economic activities in Rivers State, and one of the major cities in the Niger Delta, is located 93 kilometers from the state capital, Port Harcourt. The ONELGA local government area covers a land mass of 1621 km^2 (National Boundary Commission 1991). And at the 2006 census, it had a population of 284,010 with a 2013 projected figure of 400,000 residing in 96 communities (NPC 2006, 2013).

However, based on the 2006 national census and subsequent projections (National Population Commission 2006, 2013). Using a 3.2% annual growth rate, ONELGA's 2013 estimated population of 400,000 was projected to be approximately 565,636 in 2024. Based on available demographic sources, Omoku, the LGA headquarters and largest urban settlement, was assigned an estimated 45% share (254,536 people), while Okansu, a smaller rural community, was conservatively estimated at 2% (11,313 people) (Wikipedia 2024; National Network 2024; Places in the World 2024). The original indigenous occupants of the area are the Ogba people (Guide for Indigenous People of Nigeria 2021).

Ogba/Egbema/Ndoni LGA is bounded on the North by Ogbaru Local Government Area of Anambra State, on the North-East by Uguta and Ohiaji/Egbema LGA of Imo State, at the West by the Sagbama/Yenagoa LGA of Bayelsa State and Ndokwa–East LGA of Delta State, South by Ahoada –West LGA and Emohua LGA of Rivers State on the East located the eastern bank of the River Niger and in the heart of the Niger Delta region. It is located between latitudes 5′5, 25N and longitudes 60, 40–60,50E and is located at the North-Western fringe of Rivers State (NBC 1991). Ogba/Egbema/Ndoni LGA has a topography of flat plain netted in a web of rivers- the Niger, Sombreiro (Nkisa), Orashi, and their tributaries, as well as dotted creeks, Idia, Omoku, and Onitsha among others, rendering it susceptible to inundation during heavy rainfall events and upstream river overflow. It is generally a lowland area with an average elevation below 30 meters above sea level; its geology comprises basically of alluvial sedimentary basin and a basement complex. Its climate is tropical

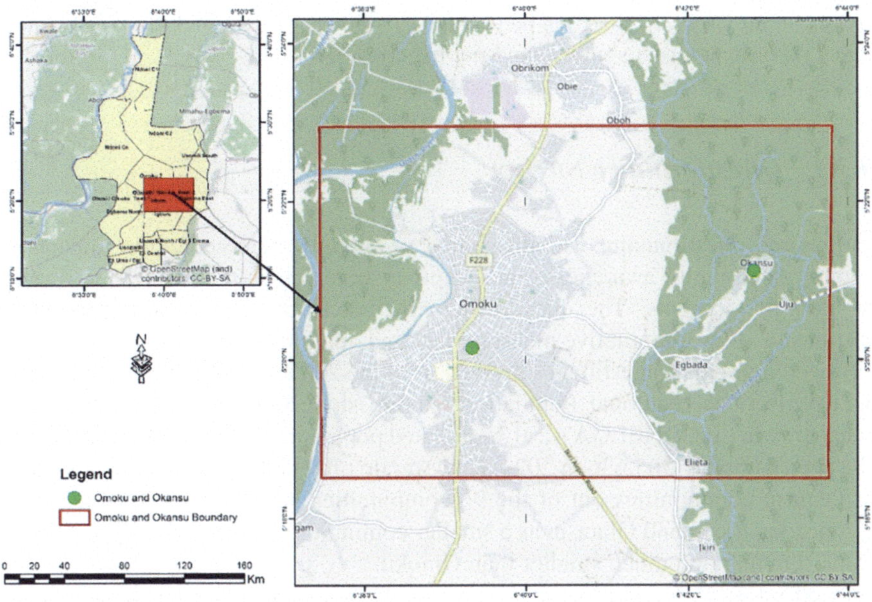

Fig. 19.1 Map of rivers state showing ONELGA and Omoku, and Okansu communities

with two distinct seasons: a wet season from April to October and a dry season from November to March. The average annual rainfall ranges between 1,500 mm and 2,000 mm, contributing to periodic flooding, especially during peak rainfall months (Akanwa et al. 2022). However, increasing anthropogenic activities such as unregulated construction, poor drainage infrastructure, and wetland encroachment have exacerbated flood risk (Obasi and Chukwu 2020).

Land use is predominantly comprised of residential settlements, commercial centers, agricultural lands, and patches of remnant natural vegetation. Agriculture remains a major livelihood source, with crops like cassava, yam, and maize cultivated on smallholder farms that are vulnerable to seasonal flooding, which often leads to crop loss and soil degradation, threatening food security and economic stability (Ezenwa and Ofor 2019).

19.3 Data Sources and Methodology

This study employed a mixed-methods approach combining quantitative data from questionnaires, qualitative visual documentation through photography, and spatial analysis using Geographic Information Systems (GIS) to comprehensively assess community-based adaptive responses to flooding in ONELGA, Rivers State, Nigeria

(Creswell and Creswell 2018). The secondary sources of data include Government agency records, internet sources, and published works.

19.3.1 Questionnaire Survey

A structured questionnaire was developed to capture socio-economic impacts of flooding, community awareness, and adaptive strategies employed before, during, and after flood events. The questionnaire included both closed and open-ended questions to allow quantitative measurement and qualitative insights.

Based on population estimates derived from the 2006 national census and subsequent projections (National Population Commission 2006, 2013). Using a 3.2% annual growth rate, ONELGA's 2013 estimated population of 400,000 was projected to be approximately 565,636 in 2024. However, to project the populations of Omoku and Okansu communities out of the 96 communities in ONELGA, Omoku is the LGA headquarters, and Okansu, as a smaller community within ONELGA, is likely to have a population much smaller than Omoku.

Based on available demographic sources, Omoku, the LGA headquarters and largest urban settlement, was assigned an estimated 45% share (254,536 people), while Okansu, a smaller rural community, was conservatively estimated at 2% (11,313 people) (Wikipedia 2024; National Network Online 2024; Places in the World 2024). A sample size of 400 questionnaires was determined using Taro Yamane's (1967). However, out of the 400 questionnaires distributed, 320 were correctly completed and returned, while 48 questionnaires were not returned. This indicated a high response rate of 87% highlighting the significance and urgency of flooding issues in Omoku and Okansu communities, providing a strong dataset for analysis (Creswell and Creswell 2018). The questionnaires were administered face-to-face by trained enumerators, ensuring clarity and completeness.

19.3.2 Photographic Documentation

Photographic methods were employed to visually capture flood impacts, adaptive responses, and key landscape features relevant to flood risk. Photography served both as a qualitative data source and a means to triangulate information gathered from questionnaires (Rose 2016). Field photographs documented damaged infrastructure, waterlogged areas, and environmental conditions. These images provided contextual understanding and were analyzed thematically alongside survey data.

19.3.3 GIS and Remote Sensing Analysis

Spatial analysis was conducted using GIS software (ArcGIS 10.8) and high-resolution satellite imagery. The GIS component involved mapping elevation, land use/land cover, and hydrological features to identify flood-prone zones and analyze spatial relationships between environmental factors and flood vulnerability. A Digital Elevation Model (DEM) was used to determine low-lying areas susceptible to inundation, while land use maps highlighted urbanized zones, agricultural lands, and wetlands. These layers were overlaid with community locations and questionnaire data to spatially correlate socio-economic impacts and adaptive strategies.

19.3.4 Data Analysis

Quantitative data from questionnaires were coded and analyzed using SPSS version 26. Descriptive statistics (frequencies, percentages) and inferential statistics (Chi-square tests) were applied to test hypotheses regarding flood impacts and community responses (Field 2018). Photographic data were subjected to qualitative content analysis to identify recurring themes related to vulnerability and adaptation (Rose 2016). GIS outputs were interpreted visually and quantitatively to support spatial understanding of flood risk.

19.4 Results and Discussion

The results from the 320 questionnaires, along with visual documentation through photography and spatial analysis using Geographic Information Systems (GIS), are presented to comprehensively examine the socio-economic impacts of flooding in the Omoku and Okansu communities in ONELGA, Rivers State, Nigeria. Additionally, we assessed community-based adaptive responses before, during, and after floods, while proposing anticipatory actions to address the climate crisis in these communities.

This combination of approaches provided a solid foundation for analyzing localized community adaptive responses and formulating policy-relevant recommendations for flood adaptation. According to the analyzed questionnaires, as shown in Table 19.1, out of 320 participants, 219 were males (68.4%) and 101 were females (31.6%). This indicates that there were significantly more males than females participating in the study.

Table 19.2 shows the age of the participants, where 18–24 (10.6%), 25–31 years (18.4%), 32–38 years (19.7%), 39–45 years (17.5%), 46–52 years (18.1%), 53–59 years (9.4%), 60–66 years (3.8%) and 67 years (2.5%) and above is the lowest age group, while the highest age group is 32–38 years (19.7).

Table 19.1 Gender distribution of participants

Sex	Frequency	Percentages (%)
Male	219	68.4
Female	101	31.6
Total	320	100

Source Compiled by authors, 2024

Table 19.2 Age distribution of participants

Age	Frequency	Percentage (%)
18–24 years	34	10.6
25–31 years	54	18.4
32–38 years	63	19.7
39–45 years	56	17.5
46–52 years	58	18.1
53–59 years	35	9.4
60–66 years	12	3.8
67 years and above	8	2.5
Total	320	100

Source Compiled by authors, Work 2024

Table 19.3 reveals the marital status where 34.7% are single, 58.1% are married, 3.4% are divorced, 2.5% are widows, and 1.3% are widowers. This implies that the largest population is married.

Table 19.4 shows the educational attainment of the participants in the study areas. Where 16% had no formal education, 43.4% have attained primary education, 39.6% have obtained their WAEC certificate, 7.5% attended a school of technology, and 4.3% have a BSc. This indicates that the majority of the participants completed their first school leaving certificate (FSLC) and can read and write.

Table 19.5 shows the occupational status of the participants. 3.4% of the participants are involved in Teaching/Workers, 38.1% are into business/Trading, 26.9% are Civil Servants (working at the LGA), 21.9% are involved in Farming and Fishing,

Table 19.3 Marital status of participants

Marital Status	Frequency	Percentage (%)
Single	111	34.7
Married	186	58.1
Divorced	11	3.4
Widow	8	2.5
Widower	4	1.3
Total	320	100

Source Compiled by authors, 2024

Table 19.4 Educational status of the respondent

Education	Frequency	Percentage (%)
No formal education	16	5
Primary school completed (FSLC)	139	43.4
Secondary School (WAEC)	127	39.6
School of Technology	24	7.5
BSc	14	4.3
Total	320	100

Source Compiled by authors, 2024

Table 19.5 Occupation of participants

Occupation	Frequency	Percentage%
Workers/Teachers	11	3.4
Trader/Business	122	38.1
Civil Servants	86	26.9
Fishing/Farming	70	21.9
Artisans	18	5.6
Professionals	6	1.9
Religious leader	7	2.2
Total	320	100

Source Compiled by authors, 2024

5.6% are Artisans (Tilers, Plumbers, Electrician etc), 1.9% are Professionals and 2.2% are Religious leaders. Business and Trading make up the highest occupation, followed by Civil Servants.

Table 19.6 shows the income range of the participants, were the highest income category was 27.2%, earning below 100,000naira (65.33USD) monthly, 24.1% earn between ₦110,000–₦130,000 (84.97USD) monthly, 20.6% earn between ₦131,000–₦150,000 (97.7 USD) monthly, 18.4% earn between ₦151,000–₦160,000 (103.18 USD) monthly, 5.9% earn between ₦161,000 -180,000 (116.6 USD) monthly and 1.9% earn above 200,000 (130.35 USD). This indicates that the majority of the participants earn below 100,000 naira (65.33 USD).

Table 19.7 shows that the highest number of years spent by participants living in their communities is 20 years, accounting for 29.6%. This is followed by 11–20 years (28.1%), with others having lived there for 6–10 years (21.8%), and 20.3% of the participants having lived less than 5 years. With nearly 30% indicating they have lived over 20 years, this suggests they have substantial experience regarding the implications of flooding in their communities.

Table 19.8 displays the factors contributing to flooding in Okansu and Omoku communities, where 92 participants, making up 28.75%, identified obstruction and the absence of drainage systems as the primary cause, representing the highest factor.

Table 19.6 Monthly income of respondent

Income Level (N)	Frequency	Percentage (%)
Below 100,000	87	27.2
110,000–130,000	77	24.1
131,000–150,000	66	20.6
151,000–160,000	59	18..4
161,000–180,000	19	5.9
181,000–200,000	6	1.9
210,000 and above	4	1.3
Total	320	100

Source Compiled by authors, 2024

Table 19.7 Number of years lived in the community

	Years	Frequency	Percentage (%)
1	Less than 5 years	65	20.3
2	6–10 years	70	21.8
3	11–20 years	90	28.1
4	20 years and beyond	95	29.6
	Total	320	100

Source Compiled by authors, 2024

Seventy-five and 38 participants, accounting for 17.1% and 11.8% respectively, cited rainfall and low-lying areas as contributing factors. Additionally, 56 and 26 participants, comprising 17.5% and 8.1% respectively, pointed to the presence of illegal structures on flood plains and swamp cover as significant contributors to flooding, along with heavy rainfall. Furthermore, 53 responses, representing 16.5%, indicated that climate change plays a role in flooding within these communities.

Table 19.8 Causes of flooding in the community

Response	Frequency	Percentages (%)
Blockage/absence of drainage systems	92	28.75
Heavy rainfall	55	17.1
Low land	38	11.8
Illegal Structure/Building on Flood plains/Swamp cover	56	17.5
Release of excess water from Kainji Dam	26	8.1
Climate Change	53	16.5
Total	320	100

Source Compiled by authors, 2024

In Table 19.9, shows 91.3% of the participants indicated that they were severely affected by the flood, while 8.7% reported that they were not affected by the flooding menace.

Table 19.10 indicates that 34.1% of participants experienced the loss of houses and infrastructure, 0.9% faced deaths, and 3.8% and 21.2% faced an increase in loss of farms and disruption of economic activities, respectively. 18.1% and 5% experienced restricted movement (Human and vehicular) and disruption in school attendance, as students faced significant challenges traversing floodwaters respectively. However, 2.8% and 5.3% reported that flooding had adversely affected them through disruptions in power supply, and significantly disrupted social activities, including funerals, weddings, and other gatherings, respectively. Additionally, 6.3% of respondents identified an increase in transportation costs as one of the challenges encountered, while 2.5% noted an increase in food expenses as a related challenge.

Table 19.11 shows the financial losses participants experienced due to flooding in Naira, and the equivalent USD is provided. 12.8% and 12.2% of participants reported losing between N 250,000 (N350) (approximately $227.60) and N 450,000 (N360), respectively. About 9.7% and 9.7% lost between N 60,000 and N 650,000 ($227.6) and N 560,000 and N 450,000 (approximately $424.85), respectively. Additionally,

Table 19.9 The effect of flooding on the respondent

Response	Frequency	Percentages (%)
Yes	292	91.3
No	28	8.7
Total	320	100

Source Compiled by authors, 2024

Table 19.10 Effects of flooding in the community

Effects	Respondent	Percentages (%)
Destruction of houses/infrastructure	109	34.1
Increase in deaths	3	0.9
Increase in loss of farmlands	12	3.8
Disruption of economic activities.	68	21.2
Restriction of human and vehicular movement	58	18.1
Disruption of school attendance	16	5
Loss of power supply	9	2.8
Disruption of social activities	17	5.3
Increased cost of transportation	20	6.3
Increase Cost of food.	8	2.5
Total	320	100

Source Compiled by authors, 2024

Table 19.11 Estimated financial losses

Financial losses	Frequency	Percentages (%)
Below N 50,000	21	6.5
N 60,000–N 150,000	31	9.7
N 160,000–N 200,000	30	9.4
N 250,000– N 350,000	41	12.8
N 360,000–N 450,000	39	12.2
N 460,000–N 550,000	23	7.2
N 560,000–N 650,000	31	9.7
N 660,000–N 750,000	27	8.4
N 760,000–N 850,000	17	5.3
N 860,000–N 950,000	20	6.3
N 960,000–N 1,200,000	18	5.6
Above N 1,500,000	22	6.9
Total	**320**	**100**

Source Compiled by authors, 2024

9.4% and 9.7% lost between N 200,000 (N160) and N 650,000 ($130.35), and N 660,000 and N 750,000 (around $490.11), respectively.

Moreover, 7.2% and 6.9% reported their highest financial losses between N 460,000 and N 550,000 (about $359.83) and N 1,500,000 (roughly $979.75), respectively. 6.5% and 6.3% lost between N 50,000 ($32.72) and N 860,000 and N 950,000 (approximately $620.94), respectively. Finally, 5.6% and 5.3 % experienced losses between N 960,000 and N 1,200,000 (about $784.80) and N 760,000 and N 850,000 (around $556.17), respectively. The highest financial losses ranged from N 250,000 to N 350,000, accounting for 12.8%, which is about $227.60 according to the current exchange rate as of July 2025.

19.4.1 Socio-economic impacts of flooding in the Omoku and Okansu communities in ONELGA, Rivers State, Nigeria

Our findings form this study reveal that 29.6% of respondents have lived in their communities for over 20 years, with 28.1% residing there for 11–20 years, indicating deep-rooted local knowledge and lived experience with flooding events. Notably, long-term residents are often more aware of historical flood patterns and more likely to provide reliable information, participate and lead adaptive responses (Sinthumule and Mudau 2019). Such tenure provides valuable context for understanding anticipatory actions and risk perception at the grassroots level. Their insights are critical for designing locally relevant flood mitigation strategies (Buba et al. 2021).

In our study, we had 68.4% identifying as male and 31.6% as female, this indicates a potential male dominance in household decision-making. In many Nigerian homes, men are typically the heads of households, which likely leads to their perspectives being more prominently represented in discussions about flooding issues in the study area. This trend has also been observed in similar flood risk studies conducted in Nigeria (Nwankwoala and Udom 2019). Also, the age distribution of respondents provides crucial insight into the demographic most exposed to and involved in flood risk and adaptation. According to the survey data, a significant proportion of participants fell within the 30–49 age group, representing the economically active and socially responsible population segment. This age bracket is often most engaged in livelihood activities such as farming, trade, or civil service, which are highly susceptible to flooding impacts (Adedeji et al. 2021). Moreover, this demographic typically assumes decision-making roles in households and communities, making them central to anticipatory and adaptive responses to flooding (Adelekan 2016). The relative underrepresentation of older adults and teenagers may suggest either reduced exposure or limited participation in public discourse around climate risk. Similar trends in age-related vulnerability and participation have been documented in flood-prone Nigerian communities (Olajuyigbe and Ogunyemi 2022), highlighting the need for age-sensitive adaptation planning.

Further, our findings on the marital status distribution reveal that 58.1% of respondents were married, making them the predominant group in the study area. This demographic pattern implies that most participants are likely responsible for family welfare, economic stability, and housing decisions, factors highly relevant to flood preparedness and adaptation (Ndimele et al. 2024). Married individuals often exhibit stronger ties to place and property, which may influence their willingness to invest in anticipatory actions or recovery efforts (Rowland et al. 2024). The presence of widows (2.5%) and widowers (1.3%) also highlights a socially vulnerable subgroup, especially in rural Nigerian contexts where widowhood may limit access to resources and decision-making roles (Adedeji et al. 2021). Meanwhile, the 34.7% single population may reflect younger or unmarried adults who may be less engaged in long-term community resilience planning. These dynamics underscore the need for marital-status-sensitive approaches in flood adaptation interventions.

Findings show that 43.4% of participants achieved primary education, and 39.6% finished secondary school (WAEC), while merely 4.3% obtained a Bachelor's degree. This suggests that most participants possess basic literacy and can engage effectively with flood awareness messages, early warning systems, and adaptive information (Mitheu 2022). The 16% lacking formal education represent a group that may be at greater risk due to restricted access to disaster communication or involvement in policy (Hussainzad and Gou 2025). Educational background plays a crucial role in individuals' ability to prepare for and respond to climate-related disasters, as higher education levels positively correlate with risk awareness and proactive actions (Olajuyigbe and Ogunyemi 2022). The relatively low rate of tertiary education indicates that community-based awareness and risk communication should be designed to fit the literacy levels of the wider populace, focusing on visual aids, local languages, and interactive learning.

The data indicates that business and trading (38.1%) make up the largest portion of participants, followed by civil servants (26.9%) and those in farming/fishing (21.9%). This demonstrates that the local economy primarily relies on informal and commerce-driven activities, a common characteristic of many semi-urban and rural areas in Nigeria (Akanwa et al. 2022). Traders and civil servants typically have better access to adaptive resources such as income and information, enabling them to respond more effectively to flood alerts (Osuteye et al. 2019). However, individuals involved in farming and fishing face heightened vulnerability to flooding, as their livelihoods are dependent on land and water conditions (Akanwa et al. 2022; Okoli et al. 2025). Although the presence of artisans, professionals, and religious leaders is smaller in scale, it reflects a varied community structure that can be engaged for proactive flood measures. These occupational trends underscore the necessity for flood risk communication strategies to be customized according to economic roles and exposure to livelihood risks.

The income distribution among participants indicates that a significant portion (27.2%) earns below ₦100,000 (approximately USD 65.33) monthly, reflecting widespread low-income levels. Furthermore, 68.9% of participants earn less than ₦150,000 (about USD 97.70) monthly, placing the majority in the low-income bracket. This pattern aligns with broader trends observed in rural and peri-urban Nigerian communities, where income instability limits access to flood-resistant infrastructure, insurance, or emergency savings (Buba and Ajayi 2024). Low income is a critical factor that increases vulnerability to climate-related disasters, as it restricts both anticipatory capacity and adaptive choices (Olajuyigbe and Ogunyemi 2022). The small percentage of individuals (1.9%) earning above ₦200,000 (approximately USD 130.79) indicates a significant economic gap, with very few having the resources to invest in proactive flood resilience. This economic fragility highlights the need for targeted financial interventions, community-based early warning systems, and affordable adaptation strategies for low-income populations.

Our Findings also show that 28.75% of respondents identified obstruction and the absence of drainage systems as the leading cause of flooding in Okansu and Omoku. This finding aligns with previous studies that emphasize poor urban infrastructure as a major driver of flooding in Nigerian cities (NIMFRU Climate Centre 2023). Additionally, rainfall (17.1%) and low-lying topography (11.8%) were recognized as significant contributors, consistent with the climate-induced intensification of precipitation and the vulnerabilities posed by natural terrain (Okolie et al. 2025). Illegal structures on floodplains (17.5%) and swamp encroachment (8.1%), including buildings like Chicken Hill Eatery and Everyday Supermarket, further exacerbate flood risks by disrupting natural drainage and water absorption zones. Finally, 16.5% of participants explicitly recognized climate change as a contributing factor, suggesting a growing awareness of broader environmental influences. These findings underscore the pressing need for enhanced land-use enforcement, sustainable urban planning, and community-based drainage interventions to mitigate flood exposure and foster anticipatory resilience (Okonkwo and Akanwa 2017).

The data in Table 19.10 illustrate the multi-dimensional impacts of flooding on the residents of Okansu and Omoku. The most reported consequence was the loss of homes and infrastructure (34.1%), reflecting how poorly planned or low-lying settlements are highly susceptible to flood damage (Ajayi et al.2024). Additionally, our photographic documentation in Plates 19.1 and 19.2 captures the impact of the flood disrupting human and vehicular movements, general restriction of economic activities, and breakdown of the transportation system resulting from the collapse of the bridge. This is also reflected in the results, where 21.2% experienced disruptions to economic activities, underscoring the vulnerability of informal and daily-income earners to environmental shocks (Olajuyigbe and Ogunyemi 2022). Loss of farmlands (3.8%), restricted movement (18.1%), and school disruption (5%) also signal critical impacts on food security, education, and mobility, factors that deeply affect long-term community resilience. Although deaths (0.9%) were reported at a lower rate, the significance of life-threatening situations cannot be understated. The disruption of social events (5.3%) and power outages (2.8%) reflect the broader social dislocation that floods can cause. Rising transportation (6.3%) and food costs (2.5%) suggest that the aftermath of floods continues to strain household income long after the waters recede. These findings are consistent with broader literature on flooding in Nigeria, where physical, social, and economic vulnerabilities intersect to create compounding impacts (Buba and Ajayi 2024). This calls for anticipatory strategies that address both infrastructure and socio-economic resilience.

Our findings from the study indicate that flooding has a significant and overwhelming effect, where 91.3% of the participants reported being severely affected by flooding in Okansu and Omoku, while only 8.7% claimed not to have been affected. This high percentage reflects the widespread vulnerability of residents in flood-prone,

Plate 19.1 Road Submerged by flood restricting human and vehicular access and movement

Plate 19.2 Trans-Okansu bridge damaged by flood

poorly drained, and low-lying areas typical of the Niger Delta (Okolie et al. 2025). The extent of impact highlights how inadequate infrastructure, unregulated development, as captured in Plate 19.3, a gas station built on a flood plain, exacerbates flood impacts, added to the increasingly erratic rainfall linked to climate change, adds to heightens exposure (Akanwa et al. 2022). The minority unaffected may reside in less vulnerable zones or possess better adaptive capacities. This finding confirms patterns observed in other studies where socio-economic and environmental conditions exacerbate flood impacts across Delta and Rivers States (Akanwa et al. 2022). Findings from this study indicate that flooding in the Omoku and Okansu communities affect socio-economic activities particularly human movement is restricted, making it difficult for people to go work, children attend school, or access to healthcare (Balomenos et al. 2020). Transportation systems are often paralyzed, as roads become impassable, public transit is suspended, and vehicles are damaged (Zeng et al. 2024). This leads to delays, increased transport costs, and reduced mobility. Further, economic activities suffer, particularly the informal sectors like trading, business, farming and artisanal work, where daily income depends on physical presence (Olasimbo 2024).

Evidently, our findings reveal that social and cultural life events such as weddings, burials, church services and community meetings are postponed or cancelled. Religious gatherings may experience low attendance or total disruption due to unsafe travel conditions or flooded worship venues (Buba, and Ajayi 2024). These disturbances reduce community cohesion, increase stress, and further weaken recovery efforts. Ultimately, flooding erodes both the physical and social infrastructure that communities rely on for resilience and continuity. It underscores the urgent need for anticipatory action frameworks, community-based risk reduction strategies, and investments in resilient infrastructure to mitigate the socio-economic burdens of recurrent flooding. Further, the financial loss data in Table 19.11 highlights the

Plate 19.3 Showing an authorized Structure on Flood floodplain at the study Area

widespread economic vulnerability of households in Okansu and Omoku as a result of recurrent flooding. The largest proportion of respondents (12.8%) reported losing between ₦250,000 and ₦350,000, which equals approximately $227.60, while another 12.2% lost ₦450,000, emphasizing the substantial burden even moderate flooding imposes on residents with limited income. Additionally, 9.7% of participants reported losses of ₦650,000, equivalent to $424.85, while 6.9% faced extreme losses of up to ₦1.5 million (about $979.75), underscoring the scale of asset destruction in high-exposure households. These losses, considering that many participants earn below ₦150,000 monthly, represent not just a setback but potentially long-term impoverishment. Households without access to insurance or credit often struggle to recover, deepening cycles of vulnerability and inequality (Olasimbo 2024; Nkwunonwo 2020). The high variation in reported losses, ranging from ₦50,000 to over ₦1 million, also reflects unequal flood exposure based on location, asset ownership, and housing conditions (Zhu et al. 2018). direct damage, such as destruction of homes and goods, financial losses, also likely include indirect costs, such as business interruption, emergency expenses, and post-disaster inflation, especially in transportation and food (Zeng et al. 2024). This economic impact affirms the necessity for anticipatory action frameworks that offer risk financing mechanisms, community insurance schemes, and livelihood diversification programs. Flood adaptation efforts must prioritize financial resilience, especially in low-income and informal economies where such shocks have cascading effects on welfare, education, and health outcomes.

19.4.2 Community-Based Flood Response in Omoku and Okansu Communities, ONELGA, Rivers State

Our findings on community-based flood responses in Omoku and Okansu communities in Ogba/Egbema/Ndoni Local Government Area (ONELGA), Rivers State, are presented in Table 19.12, which reflects the local knowledge, resilience, and adaptive innovation across the three phases of flooding: before, during, and after. These strategies, although often informal, demonstrate an understanding of environmental hazards and communal efforts to reduce vulnerability. However, their adequacy and success vary widely based on resource availability, institutional support, and environmental intensity (Ndimele et al.2024).

Before floods, communities relied on early warning systems via informal channels such as local radio, churches, and town criers. Also, there were clean-up exercises and drainage desilting efforts by the community, especially among youth and women's groups. These proactive actions help to reduce blockage of waterways and are often more consistent in Okansu, where communal ties are stronger. Nevertheless, the absence of technologically supported early warning systems limits the precision and timeliness of such warnings (Osuteye et al. 2019).

During floods, the use of elevated buildings, community shelters (such as schools and churches), and traditional boats to transport goods and people has proven useful. However, they were overwhelmed because the flood was severe displacing a large number of people. Similar experience of flood crisis was also experienced in other communities in Nigeria, as reported by Okoli et al. (2025), that floods in Ogbaru LGA, Anambra state restricted socio-economic activities, and transportation both by vehicles and boats. Health vulnerabilities also increase due to low access to clean water and adequate first aid services (Nkwunonwo et al. 2020). Women's groups and youth volunteers often serve as first responders, but lack of formal training limits the scope of emergency care.

However, the post-flood recovery involves self-help and communal reconstruction of roads, pathways, cleaning of flood debris from drainage channels, water purification efforts, and small-scale farming. These initiatives are largely driven by households and cooperative groups but remain underfunded and poorly coordinated without significant government or NGO intervention (Buba, and Ajayi 2024). Notably, there is need for community-based post trauma support and resilience building for long-term impact, however, it is constrained by systemic gaps in planning and funding.

In conclusion, the community-based responses in Omoku and Okansu reveal adaptability and grassroots resilience but are limited in adequacy and scope. These efforts require stronger institutional support, integration of ICT-based early warning systems, and targeted investments in infrastructure and health services to enhance their effectiveness.

Table 19.12 Community-based adaptive strategies for flooding in Omoku and Okansu communities

Flood stage	Adaptive strategies	Description of community practice
Before flood	Risk mapping and early warning systems	Use of local knowledge to identify flood hotspots and share early warning signs via churches, radio, town criers.
	Elevated construction (e.g., raised foundations)	Community members raise buildings and store food on upper floors or shelves.
	Community clean-up and drainage clearing	Regular desilting of gutters and waterways by youths and community groups to ensure free water flow.
	Flood preparedness education and sensitization	Local awareness campaigns by NGOs, health workers, and community leaders on flood risks and response.
	Formation of community emergency task groups	Volunteer groups trained to mobilize resources and assist vulnerable households.
	Household sandbagging and embankments	Placement of sandbags and makeshift barriers around homes and key roads.
During flood	Temporary evacuation to schools, churches, or elevated areas	Use of public spaces on higher ground as informal shelters.
	Protection of water sources and food stores	Use of covered containers and elevated storage to reduce contamination.
	First aid and community health response	Local health volunteers respond to minor injuries and waterborne disease outbreaks.
	Use of canoes, rafts, and local boats for transport	Community improvises with traditional boats to move people and supplies.
	Communication via local radio or mobile networks	Updates on water levels, relief distribution, and safety tips shared through mobile phones or radio.
After flood	Rehabilitation of homes and roads	Community-led repairs using local materials and communal labor.
	Disinfection of wells and water sources	Boiling water, chlorination, and community hygiene initiatives.
	Livelihood recovery initiatives	Communal farming, cooperative savings.

19.4.3 Spatial Analysis of Digital Elevation Model (DEM), and land use/land cover of Omoku and Okansu Communities

Findings from our study on the flood vulnerability mapping of the Okansu and Omoku communities reveals significant spatial variations in flood risk, largely influenced by elevation, land use, drainage patterns, and hydrological connectivity. The identification of flood hotspots within these communities highlights critical areas where vulnerability to flooding is most pronounced as shown in Fig. 19.2. Notably, Omoku

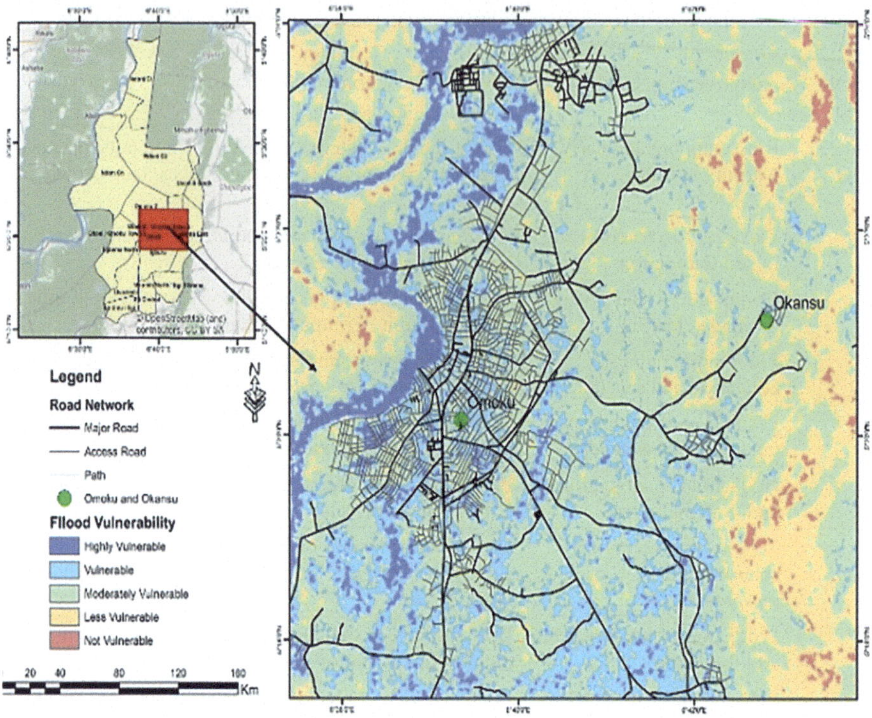

Fig. 19.2 Digital Elevation Model (DEM) delineating Omoku and Okansu as Flood hotspot zones. *Source* Compiled by authors 2024

community is positioned in a highly flood prone area as shown by the purple colour compared to Okansu community which is designated by an orange colour indicating it is less vulnerable to flood as shown in Fig. 19.2.

However, both communities are positioned in the low-lying zones, particularly those below 15meters above sea level, dominate the flood-prone regions in southern Okansu and western Omoku. These areas coincide with floodplains and riparian wetlands adjacent to the Orashi River tributaries, confirming earlier studies that link low elevation and flat terrain to heightened flood exposure in Niger Delta communities (Fagunloye 2024). The Digital Elevation Model (DEM) used in delineating these zones (Fig. 19.2) aligns with findings by Shekhar et al. (2022a, b), underscoring the critical role of topography in surface water accumulation and flood risk.

Land use within these flood hotspots further compounds vulnerability. In Omoku, urban expansion with extensive impervious surfaces reduces natural infiltration, thereby increasing surface runoff and localized flooding (Ayanlade et al. 2020). Similarly, in Okansu, the conversion of wetlands and natural floodplains into agricultural land diminishes the landscape's natural water retention capacity, exacerbating flood severity during heavy rainfall events (Ogundele et al. 2023).

This pattern reflects a broader regional challenge in balancing economic development with ecological preservation (Numbere 2021). Figure 19.2 also highlights areas where drainage networks are disrupted or insufficient. Blocked or poorly maintained channels in densely settled parts of Omoku and central Okansu impede water flow, leading to frequent inundation even during moderate rains. This finding corresponds with documented impacts of anthropogenic interference on flood dynamics in similar Nigerian watersheds (Nwankwoala and Udom 2019; Fagunloye 2024). The spatial overlay of drainage deficiencies and high population density intensifies flood risk and underscores the urgency for infrastructure upgrades and urban planning reforms.

Flood risk classification in Fig. 19.2 stratifies our study communities into high, moderate, and low-risk zones. High-risk zones, located primarily in densely populated lowlands and encroached wetlands, require prioritized intervention. Strategies such as early warning systems, flood-resilient building codes, and wetland restoration are essential to mitigate impacts in these areas (Ogunrinde et al. 2024). Moderate-risk zones, typically transitional terraces, call for controlled development with integrated drainage solutions. Low-risk zones on elevated ridges offer safer locations for future expansion and infrastructure development.

Overall, the GIS-based flood vulnerability map (Fig. 19.2) offers a robust tool for evidence-based planning and disaster risk reduction in Okansu and Omoku. By integrating hydrological data with land use and topographic variables, stakeholders can formulate targeted adaptation strategies that balance community growth with environmental sustainability. This approach aligns with best practices in flood risk management for vulnerable riverine communities in the Niger Delta (Shekhar et al. 2022a, b; Numbere 2021).

Findings in Fig. 19.3 visually represents the Omoku and Okansu watershed, focusing on elevation values, contour lines, and water body distribution. The elevation values range from -13 meters to 268 meters above sea level, divided into distinct intervals represented by the following colour gradients.

Elevation Ranges and Values

i. **−13 to 6.9 m** (Dark Blue):

   These areas represent the lowest elevations, likely comprising river valleys, wetlands, or flood-prone zones. They are concentrated in the western and southern parts of the watershed.

ii. **7 to 18.2 m** (Light Blue):

   Slightly elevated regions, potentially gentle plains or areas transitioning from lowlands to slightly higher terrain.

iii. **18.3 to 24.7 m** (Greenish Blue):

   This range may include areas along tributaries or flatlands suitable for agriculture or settlements.

iv. **24.8 to 28.4 m** (Light Green):

   Representing moderately low elevations, often part of transitional zones between valleys and hills.

v. **28.5 to 34.9 m** (Yellowish Green):

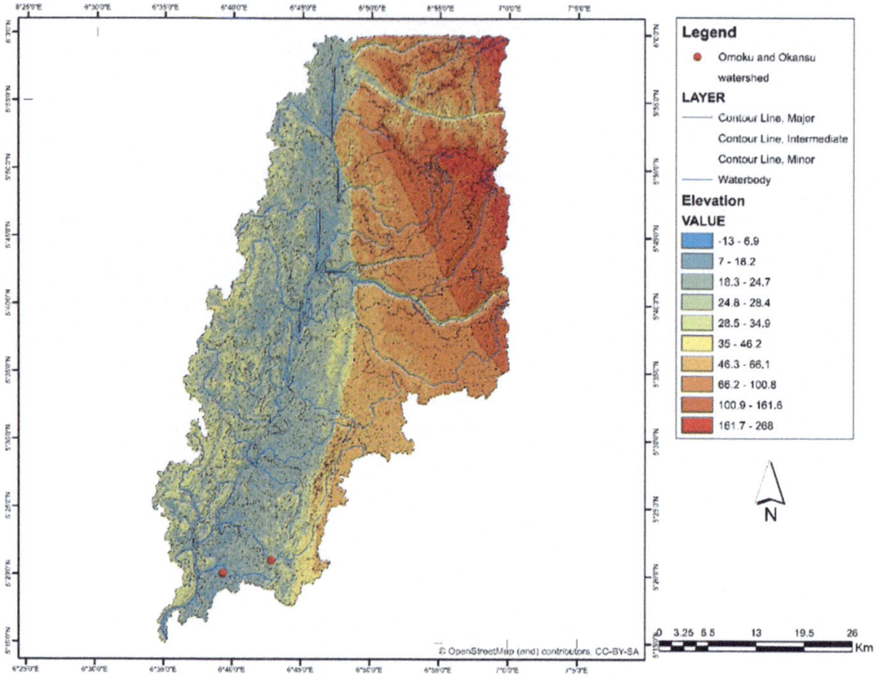

Fig. 19.3 Showing the Elevation Ranges of Omoku and Okansu watershed region. *Source* Compiled by authors, 2024

These areas are higher than typical lowlands but are still relatively close to water systems.

vi. **35 to 46.2 m** (Yellow):
Indicative of foothills or gradual inclines, showing a mix of landforms.

vii. **46.3 to 66.1 m** (Orange):
Represents more elevated regions, likely including hills or sloping terrains.

viii. **66.2 to 100.8 m** (Light Orange):
Signifies higher terrain, possibly with sparse vegetation and less human activity.

ix. **100.9 to 161.6 m** (Red):
Denotes significant elevations, likely associated with uplands or hilly terrains. These areas may have limited accessibility and play a critical role in watershed hydrology.

x. **161.7 to 268 m** (Dark Red):
The highest elevations in the watershed, likely representing steep hills or ridges. These regions are crucial for water flow and may be erosion-prone.

The distribution of these elevation ranges reveals how the watershed is structured, with lower elevations concentrating along rivers and higher values forming the backdrop of ridges and uplands. The contour lines accentuate these patterns, showing densely packed lines in high-elevation zones, indicative of steep slopes, and widely spaced lines in flat or gradual incline areas. The presence of water bodies, depicted in blue, highlights the hydrological network, with rivers flowing from higher elevations to the lower-lying areas. The two red dots marks are Omoku and Okansu communities, serving as our significant study locations.

19.4.4 Elevation of Omoku and Okansu Communities to Flood Susceptibility

In Fig. 19.3, the elevation gradient, ranging from 7 meters (low, green) to 39 meters (high, yellow and brown), is a key factor in flood dynamics. The low-lying green areas are particularly vulnerable to flooding, noting it where our study communities-Omoku and Okansu, are located, as water naturally flows from the higher elevations toward these regions. These areas may act as natural floodplains, experiencing periodic inundation during heavy rainfall or river overflow.

The higher elevations around the watershed's edges, shown in yellow and brown, are less prone to flooding but play a significant role in directing runoff to lower areas. However, these regions could contribute to flash flooding in downstream zones if the runoff is rapid and intense due to steep slopes, as indicated by closely spaced contour lines.

The elevation profile of the Omoku–Okansu watershed provides critical insight into the region's topographic variability and its implications for flood risk, land use, and watershed management. The elevation ranges from −13 meters to 268 meters above sea level, stratified into ten intervals. These gradients, when analyzed through a GIS lens, highlight both ecological vulnerability and potential for sustainable development.

19.4.5 Hydrological and Environmental Implications

Lower elevation zones, particularly those between −13 and 18.2 meters (dark blue to light blue), are concentrated in the western and southern parts of the watershed. These areas correspond to river valleys, swamps, and wetlands that are highly prone to seasonal flooding, sediment deposition, and poor drainage (Fagunloye 2024). Similar GIS-based flood risk assessments in Southern Nigeria have emphasized that low-lying terrain remains the most vulnerable to flash floods, especially when urban expansion lacks proper stormwater infrastructure (Ogunrinde et al. 2024).

As the terrain transitions into the greenish-blue to yellow-green zones (18.3 to 34.9 meters), there is a noticeable shift toward agriculturally viable land and semi-permanent settlements. These elevations are typically better drained and suitable for both arable farming and rural development. However, they still require flood preparedness plans, particularly during the peak of the rainy season.

The mid-range elevations (35 to 66.1 meters), represented by yellow and orange shades, denote foothills and moderate slopes. These areas often mark the drainage divides within the watershed, where surface runoff patterns begin to converge or diverge. Slope stability in these regions is relatively higher, but they are susceptible to sheet erosion, especially where vegetation has been cleared for farming or construction (Konadu and Fosu 2009). These foothills also serve as recharge zones for lower areas and contribute significantly to stream flow regulation.

Elevations between 66.2 and 100.8 meters (light orange) and 100.9 to 161.6 meters (red) signify the upper watershed, characterized by sparse human settlement, undulating terrain, and forest patches. These uplands, while less disturbed, are hydrologically significant. They host stream headwaters and control the rate of downstream water flow. These features align with observations in similar studies across Nigeria, where highland zones are identified as key regulators of runoff and sediment transfer.

The highest elevation class (161.7 to 268 meters), shown in dark red, comprises steep ridges and hills. Though limited in areal coverage, these zones are ecologically important. They influence microclimatic conditions, serve as natural barriers, and are erosion-prone due to the steep gradient and loose surface materials. Such areas are generally unsuitable for intensive land use and are best preserved through afforestation or conservation zoning (Fagunloye 2024).

Our findings from Fig. 19.4, showing the flood map for the Omoku and Okansu watershed, provides critical insights into areas susceptible to flooding and the factors influencing water flow and accumulation. By combining watershed boundaries, elevation data, hydrological networks, and topographic features, it serves as a valuable tool for flood risk assessment and management.

19.4.6 Watershed Boundary/Hydrology and Flood Risk

The red boundary defines the watershed's extent, showing the area contributing surface runoff to rivers and streams. This delineation is essential in understanding flood dynamics, as it highlights the regions where rainfall will eventually flow into the hydrological system. Managing land use within this boundary is crucial to minimize excess runoff that could exacerbate flooding in downstream areas, particularly the low-lying central zones.

The blue lines representing rivers and water bodies indicate natural drainage pathways. These rivers likely serve as conduits for floodwaters during heavy rainfall, with the risk of overflow heightened in low-lying green areas where Omoku and Okansu communities are located. Figure 19.4 shows that our study communities adjacent to these rivers, particularly in the central and southern parts of the watershed, are more prone to flooding due to their proximity to major watercourses and lower elevation.

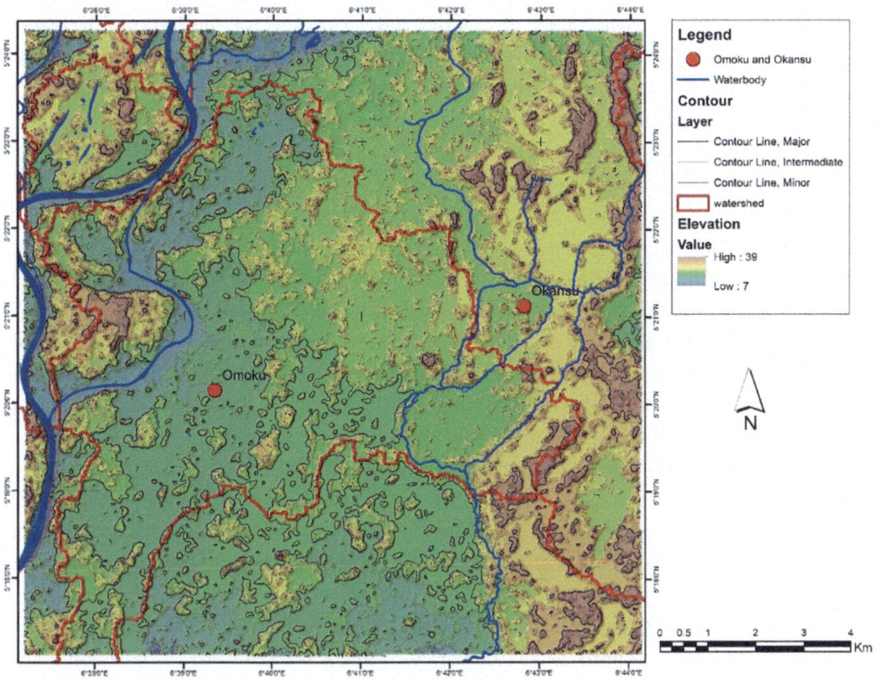

Fig. 19.4 Showing the Flood Map of Omoku and Okansu Communities

Flood risk is further influenced by the density and connectivity of the river network. Omoku and Okansu communities have dense hydrological networks, this makes them likely to experience localized flooding during intense rainfall, as water accumulates faster than it can be drained. Additionally, the convergence of tributaries into major rivers increases flood potential in downstream areas, particularly near settlements like Omoku and Okansu.

19.4.7 GIS-Based Land Use and Land Cover (LULC) Map for Okansu and Omoku Communities, Rivers State (2024)

Our findings from the GIS-based Land Use and Land Cover (LULC) map for our study communities for the year 2024, offers an insightful spatial understanding of how land resources are distributed and utilized across the Okansu and Omoku communities in Rivers State, Nigeria. By analyzing land cover patterns using remote sensing and GIS techniques, the map highlights dominant land use classes including built-up areas, agricultural land, forested zones, water bodies, wetlands, and bare surfaces.

The Built-up areas comprises of residential, commercial, and road infrastructure which are more concentrated in Omoku, which serves as a major urban center in Ogba/Egbema/Ndoni Local Government Area. This expansion is indicative of population growth, urbanization, and socio-economic development, but also exerts pressure on natural vegetation and agricultural land (Ayanlade et al. 2020). The clustering of built-up zones near road networks suggests transport-oriented development, while their encroachment into natural habitats raises concerns about urban sprawl and land degradation.

Agricultural land is prominent in both communities, especially Okansu, where subsistence farming and oil palm plantations dominate. This land use category reflects the communities' reliance on agriculture for food and livelihood, consistent with findings from rural land assessments in the Niger Delta (Ogundele et al. 2023). However, the expansion of farmlands into wetlands and fallow lands may lead to biodiversity loss and soil depletion, especially where shifting cultivation is practiced without sustainable land management. The remnant natural and secondary forests are scattered in the peripheries of both communities, particularly in less accessible upland and wetland areas. These forests are critical for ecosystem services, including carbon sequestration, biodiversity conservation, and watershed protection (Numbere 2021).

However, deforestation driven by farming, fuelwood collection, and infrastructure projects has fragmented forest cover, which can accelerate erosion and alter local microclimates. Further, Fig. 19.5, identifies riparian wetlands and small water bodies, especially along tributaries of the Orashi River system. These features are ecologically sensitive and play vital roles in flood control, aquifer recharge, and fishery support (Nwankwoala and Udom 2019). Their proximity to human activities, however, makes them vulnerable to pollution, encroachment, and seasonal flooding, particularly in low-lying parts of Okansu. Also, Bare land areas, including exposed soils and degraded patches, are found in areas undergoing construction, sand mining, or vegetation clearing. These surfaces, if left untreated, can become hotspots for soil erosion and sediment runoff, especially during the rainy season. The presence of such areas reflects the dynamic interaction between development and land degradation, necessitating targeted land reclamation or revegetation efforts (Ezekiel et al. 2022).

19.4.8 Spatial Implications and Management

The LULC dynamics in Okansu and Omoku show a clear tension between human development and environmental conservation. As Omoku urbanizes and Okansu expands its agricultural frontier, there is a growing need for integrated land use planning that balances livelihoods with landscape sustainability. GIS mapping, when updated annually, provides a decision-support system for local authorities to monitor land changes, predict urban expansion, and implement climate-smart land policies (Ayanlade et al. 2020). Moreover, recognizing the interconnectivity between land

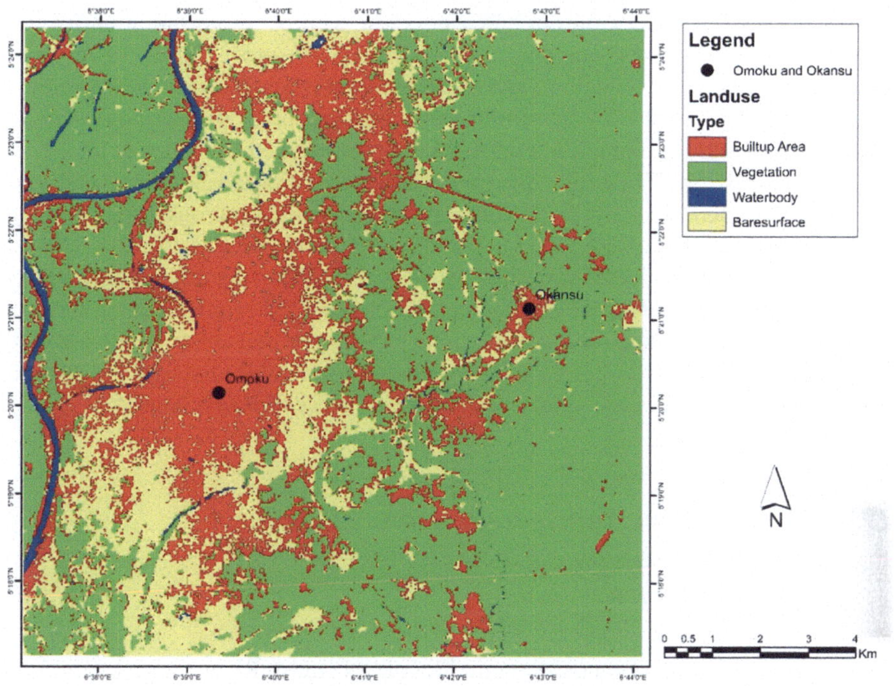

Fig. 19.5 Land Use/Cover Map 2024 for Omoku and Okansu Communities

use and hydrology, especially in wetland zones, can guide efforts in flood mitigation, wastewater control, and sustainable agriculture. Figure 19.5 also supports spatial zoning for infrastructure placement, protected area designation, and resilience building in climate-vulnerable zones.

The 2024 LULC map of Okansu and Omoku communities serves as a spatial baseline for understanding land cover distribution and the drivers of land transformation in this part of Rivers State. It highlights the interplay between urban growth, agricultural expansion, ecological stress, and natural resource use. Going forward, this spatial intelligence should inform evidence-based planning, community land governance, and sustainable development strategies across the watershed.

19.4.9 Land Use Planning and Risk Management

This GIS-derived elevation stratification reveals zones that require different land management strategies. Lower elevations should be designated for flood buffer zones, wetland protection, and adaptive agriculture. Middle elevations may accommodate controlled development, provided slope stabilization and erosion control measures are enforced. Uplands, due to their ecological sensitivity, should be managed through conservation-oriented policies, such as reforestation and watershed protection programs. The use of DEMs in GIS has proven effective in watershed delineation, flood modeling, and environmental risk assessment. For example, studies have shown that DEM resolution impacts the accuracy of hydrological simulations, particularly in complex terrains like this watershed (Shekhar et al. 2022a; Shekhar et al. 2022b). Therefore, integrating high-resolution elevation data enhances the precision of terrain interpretation and supports proactive environmental planning.

The elevation map of the Omoku–Okansu watershed offers a strategic tool for understanding the spatial relationship between terrain and hydrological behavior. From floodplains to upland ridges, each elevation band holds unique environmental and development implications. Incorporating elevation-based analysis into planning frameworks can guide risk-sensitive development, sustainable land use, and climate-resilient infrastructure, particularly in flood-prone areas of Rivers State.

19.4.10 Expedient Anticipatory Actions to Prevent Socio-Economic Flood Damage in Omoku and Okansu

Flooding in Omoku and Okansu communities of ONELGA, Rivers State, has become a recurrent hazard with escalating socio-economic consequences, including destruction of homes, disruption of livelihoods, displacement, food insecurity, and public health crises. In the context of climate change and rising rainfall variability, and poor land use planning, anticipatory actions, proactive steps taken before disaster strikes, offer a more sustainable alternative to reactive emergency response (IFRC 2020). Reducing flood impacts requires a coordinated approach by government institutions, communities, and non-governmental organizations (NGOs).

First, the government must strengthen flood risk governance through early warning systems, improved hydro-meteorological forecasting, and real-time data dissemination tailored to local languages and technology access levels. The Nigeria Hydrological Services Agency (NIHSA) and Nigerian Meteorological Agency (NiMet) must collaborate with local authorities to translate forecasts into actionable local alerts. In addition, investments in urban planning, such as enforcing building codes, upgrading drainage infrastructure, and protecting floodplains from development, are critical (Nkwunonwo et al. 2020). At the community level, anticipatory actions should involve risk education, disaster preparedness training, and the formation of local emergency task forces. These groups can be trained to lead evacuation, first aid, and resource mobilization. Community-led mapping of flood-prone areas using participatory GIS can enhance local preparedness and planning (Adelekan 2016). Cultural institutions such as churches and traditional councils also play key roles in trust-building and mobilizing early action.

NGOs can bridge institutional gaps by supporting capacity building programs and piloting forecast-based financing (FbF), where resources are pre-positioned based on predicted events. They can also facilitate access to micro-insurance schemes and cash transfer programs, which buffer poor households against income shocks. Health-focused NGOs should pre-deploy sanitation kits, mobile clinics, and hygiene education materials to prevent disease outbreaks after floods (Adedeji et al. 2021). Collectively, anticipatory strategies must be inclusive, gender-sensitive, and embedded in long-term climate adaptation frameworks. Reducing the socio-economic toll of flooding in Omoku and Okansu requires shifting from reactive to proactive governance, where timely, locally tailored interventions protect lives, assets, and livelihoods before disaster strikes.

19.5 Conclusion

This study explored the anticipatory actions and adaptive strategies employed by Omoku and Okansu communities in ONELGA, Rivers State, in response to recurrent flood crises, using an integrated methodological approach. Drawing from qualitative interviews, quantitative survey data from 320 respondents, photographic evidence, and GIS-based elevation and land use analysis, the research provides a multi-dimensional understanding of flood vulnerability, socio-economic impacts, and community resilience.

Findings reveal that flooding significantly disrupts livelihoods, housing, agriculture, education and social-cultural activities in both communities. Inadequate urban planning, blocked drainage systems, settlement on low-lying floodplains, and limited early warning systems exacerbate exposure. While communities demonstrate commendable adaptive behaviours, such as using sandbags, elevated structures, evacuation to public buildings, and community cleanup, these efforts are reactive, fragmented, and often unsupported by formal institutions. GIS elevation and land use maps further confirm that flood-prone hotspots coincide with densely settled low-lying areas, intensifying the risk and potential impact on human life and assets.

The study concludes that anticipatory actions must be formalized through coordinated efforts among government, civil society, and local communities. This includes establishing localized early warning systems, enforcing land-use regulations, improving stormwater infrastructure, and integrating spatial planning with community risk knowledge. Moreover, long-term resilience depends on inclusive planning, education, and investment in sustainable livelihoods and infrastructure. Looking ahead, proactive governance that leverages local capacity, spatial data, and predictive analytics can shift communities from reactive survival to strategic adaptation. The integration of GIS with participatory methods presents a valuable tool for mapping risk, guiding policy, and empowering at-risk populations to prepare before floods occur. Such anticipatory frameworks are essential not only for Omoku and Okansu but also for broader regional planning under Nigeria's climate adaptation strategy.

References

Adedeji OH, Ajayi SO, Balogun EA (2021) Assessing flood vulnerability and risk in Lagos State, Nigeria. Int J Disaster Risk Reduct 53:102014. https://doi.org/10.1016/j.ijdrr.2021.102014

Adelekan IO (2016) Vulnerability of poor urban coastal communities to flooding in Lagos. Nigeria. Environ Urbanization 28(1):235–254. https://doi.org/10.1177/0956247815615759

Akanwa AO, Joe-Ikechebelu N, Enweruzor AC, Okafor KJ, Omoruyi FA, Oranu CB, Umeh UC(2022) Changing climate, flood footprints, and climate-related actions: effects on ecosocial and health risks along Ugbowo- Benin Road, Edo State, Nigeria In: Chatterjee U, Akanwa AO, Kumar S, Singh SK, Roy AD (eds), Ecological footprints of climate change: adaptive approaches and sustainability. Published by Springer International Publishing, Springer Nature Switzerland AG(Publisher). https://doi.org/10.1007/978-3-031-15501-1

Ajayi FO et al (2024) Socio-meteorology: flood prediction, social preparedness, and cry-wolf effects. arXiv. https://arxiv.org/abs/2104.11422

Associated Press (2025) Floods kill at least 111 as northern Nigeria battles climate change, dry spells and heavy rainfall. https://apnews.com/article/69dcc0b8c06a052e83643edb20838db6

Ayanlade A, Jegede OO, Babatimehin O (2020) Land use and land cover change in urban Nigeria: remote sensing and social assessment approach. Afr J Environ Sci Technol 14(3):78–89. https://doi.org/10.5897/AJEST2020.2824

Balomenos E, Samitas A, Bithas K (2020) Flood impacts on socio-economic activities and community mobility: evidence from flood-prone regions. Int J Disaster Risk Reduction 46:101495. https://doi.org/10.1016/j.ijdrr.2020.101495

Birkmann J, Mechler R, Nadim F (2017) Enhancing disaster preparedness for effective response through anticipatory action. Int J Disaster Risk Reduction 25:1–9. https://doi.org/10.1016/j.ijdrr.2017.05.004

Buba FN, Ajayi SO (2024) Socioeconomic impacts of flooding and its coping strategies in Nigeria: evidence from Dagiri community. Abuja. Nat Hazards Res 4(3):374–386. https://doi.org/10.1016/j.nhres.2023.09.010

Creswell JW, Creswell JD (2018) Research design: qualitative, quantitative, and mixed methods approaches (5th ed.). SAGE Publications

Ezenwa V, Ofor U (2019) Seasonal flooding impacts on crop production and food security: a study of smallholder farms in Nigeria. J Agric Environ Sustain 7(3):112–125

Ezekiel CC, Nwachukwu AN, Ezenwa VI (2022) Environmental impacts of sand mining in parts of Rivers State, Nigeria. J Environ Manag Tourism 13(4):1061–1073. https://doi.org/10.14505/jemt.v13.4(60).10

Fagunloye OC (2024) Mapping of flood risk zones using a multi-criteria approach: a case study of Ala and Akure-Ofosu communities, Ondo State, Nigeria. Int J Geosci 15:605–631. https://doi.org/10.4236/ijg.2024.158035

Godwin A (2024) Floods sack more communities in Rivers. The Guardian Nigeria News. https://guardian.ng/news/nigeria/metro/floods-sack-more-communities-in-rivers/

Guide for Indigenous People of Nigeria (2021) Overview of Omoku and its economic significance [Report]. Guide for Indigenous People of Nigeria

Hussainzad EA, Gou Z (2025) A review study on community-based flood adaptation in informal settlements in the Global South. Discover Sustain.https://doi.org/10.1007/s43621-025-01449-6

IFRC (2020) Anticipatory action: the new normal in humanitarian response. https://www.ifrc.org/document/anticipatory-action-new-normal

Intergovernmental Panel on Climate Change (IPCC) (2022) Climate change 2022: impacts, adaptation and vulnerability—Chapter 4: water. In: Pörtner H-O et al (Eds), Sixth Assessment Report of the IPCC. https://www.ipcc.ch/report/ar6/wg2/chapter/chapter-4/

Kelman I, Gaillard JC, Mercer J (2022) Early warning and anticipatory action in disaster risk reduction. Disasters 46(S1):S4–S22. https://doi.org/10.1111/disa.12485

Konadu DD, Fosu C (2009) Digital Elevation Models and GIS for watershed modelling and flood prediction: a case study of Accra, Ghana. In: Yanful EK (Ed), Appropriate technologies for

environmental protection in the developing world, pp 325–332. Springer. https://doi.org/10.1007/978-1-4020-9139-1_31
Lindsey A, Etkin D, Morchain D (2020) Barriers and opportunities for anticipatory action: lessons from pilot projects. Clim Risk Manag 28:100224. https://doi.org/10.1016/j.crm.2020.100224
Mitheu F (2022) Community-led approaches towards flood disaster risk reduction. Walker Institute. Retrieved from https://walker.reading.ac.uk/community-led-approaches-towards-flood-disaster-risk-reduction/
National Network (2024) A publication of patriotic citizens of ONELGA. Retrieved from https://nationalnetworkonline.com/2024/03/20/a-publication-of-patriotic-citizens-of-onelga/
National Boundary Commission (1991) Geographical description: Ogba/Egbema/Ndoni Local Government Area. National Boundary Commission
National Population Commission (NPC) (2006) Population and housing census of the Federal Republic of Nigeria: analytical report at the national level. Abuja, Nigeria: NPC
National Population Commission (NPC) (2013) Nigeria demographic and health survey 2013. Abuja, Nigeria: NPC & ICF International
Ndimele PE et al (2024) Vulnerability, resilience and adaptation of Lagos coastal communities to flooding. Earth Syst Soc.https://doi.org/10.3389/esss.2024.10087
NIMFRU Climate Centre (2023) Impact-based flood early warning for rural livelihoods in Uganda. Weather Climate Soc 15(3), Article WCAS-D-22-0089. https://doi.org/10.1175/WCAS-D-22-0089.1
Nkwunonwo UC, Malcolm C, Brian B (2020) Flooding and flood risk reduction in Nigeria: a review of the perceptions of flood risk and strategies for sustainable management. J Environ Manag 276:111295
Numbere AO (2021) Forest loss and fragmentation in the Niger Delta: patterns, causes and consequences. Trop Ecol 62(2):245–255. https://doi.org/10.1007/s42965-021-00137-4
Nwankwoala HO, Udom GJ (2019) Wetlands and water resources in the Niger Delta: understanding values, threats and management. Hydrology Curr Res 10(3):1–9. https://doi.org/10.4172/2157-7587.1000326
Nwilo PC, Badejo OT (2006) Impacts and management of oil spill pollution along the Nigerian coastal areas. International Federation of Surveyors Congress. https://fig.net/resources/proceedings/fig_proceedings/fig2006/papers/ts47/ts47_03_nwilo_0814.pdf
Obasi IJ, Chukwu FA (2020) Anthropogenic factors and flood risk exacerbation: a study of wetland encroachment and drainage infrastructure in Nigeria. Nigerian J Environ Manag 12(2):45–59
Ogba/Egbema/Ndoni Watch (OENW) (2018) Strategic Profile: Omoku—Industrial and Commercial Hub in Rivers State [Brief]. OENW
Ogundele AT, Oyeleye D, Bakare T (2023) Agricultural land conversion and food security implications in southern Nigeria. J Agric Extension Rural Dev 15(2):26–35. https://doi.org/10.5897/JAERD2023.1405
Ogunrinde O, Adebayo S, Ajayi F (2024) Comprehensive GIS-based flood risk assessment of vulnerable areas. World J Innov Modern Technol 8(4):1–23. https://iiardjournals.org/get/WJIMT/VOL.%208%20NO.%204%202024/Comprehensive%20GIS-Based%20Flood%201-23.pdf
Okoli E, Okonkwo AU, Akanwa A (2025) Gendered impacts of climate change: flooding, livelihoods, and food security in Ogbaru LGA, Anambra State, Nigeria. A Paper presented at the 5th International Conference (FESCON 2025), Charting a Resilient Future: Climate Change as a Catalyst for Sustainable National Development at Chukwuemeka Odumegwu Ojukwu University, Uli Campus, Uli, Nigeria 18–20 June 2025.Vol. 5(1). Page: 139–149) https://ajer.org.ng/index.php/fescon-proceedings/index
Okonkwo AU, Akanwa AO (2017) Sustainable urban design: a tool for reducing flood vulnerability and building resilient cities in Nigeria. On transforming Nigeria through the sustainable development goals: options for environmental management.; a paper presented at the annual conference for environmental management association of Nigeria (EMAN), Nigeria, 6–8 Dec 2017

Olajuyigbe A, Ogunyemi O (2022) Climate change impacts and adaptation strategies in Nigeria: a systematic review. Climate 10(2):24. https://doi.org/10.3390/cli10020024

Olasimbo OT (2024) Economic disruptions due to flood events: effects on informal sector livelihoods in Nigeria. J Environ Econ Policy 13(1):65–80

Osuteye EO et al (2019) Information for adaptation and response to flooding: multi-stakeholder perspectives in Nigeria. Climate Serv 7(4), Article 46

Paton D, Johnston D (2017) Disaster resilience: an integrated approach (2nd ed.). Charles C Thomas Publisher

Places in the World (2024) Okansu, Nigeria. Retrieved July 6, 2025, from https://www.places-in-the-world.com/nigeria/okansu/7101353/

Reuters (2025) China's north and west on alert after sweeping rains trigger deadly floods. https://www.reuters.com/sustainability/climate-energy/chinas-north-west-alert-after-sweeping-rains-trigger-deadly-floods-2025-07-03/

Rose G (2016) Visual methodologies: an introduction to researching with visual materials (4th ed.). SAGE Publications

Rowland Eteh D, Egobueze FE, Paaru M (2024) The impact of dam management and rainfall patterns on flooding in the Niger Delta: using Sentinel-1 SAR data. Discover Water, 4(1), Article 1

Shekhar AS, Kumar V, Kumari S (2022a) To study the hydrological system and watershed using remote sensing and DEM. J Water Res 2022:6094160. https://doi.org/10.1155/2022/6094160

Shekhar B, Jain M, Sharma R (2022b) Effects of DEM resolution in hydrological modeling. Water 14(15):2363. https://doi.org/10.3390/w14152363

Sinthumule NI, Mudau NI (2019) Participatory approach to flood disaster management in Thohoyandou. Jàmbá: J Disaster Risk Studies 11(3), Article a711. https://doi.org/10.4102/jamba.v11i3.711

Stein D, Dahal P, Herweg K (2021) Anticipatory humanitarian action in practice: evidence from pilot projects. Glob Environ Chang 69:102306. https://doi.org/10.1016/j.gloenvcha.2021.102306

Time (2025) Texas suffers deadly flash floods on July 4. https://time.com/7300153/texas-flash-flooding-climate-change/

United Nations Office for Disaster Risk Reduction (UNDRR) (2015) Sendai framework for disaster risk reduction 2015–2030. https://www.undrr.org/publication/sendai-framework-disaster-risk-reduction-2015-2030

Wikipedia contributors (2024) Omoku. In Wikipedia. Retrieved July 6, 2025, from https://en.wikipedia.org/wiki/Omoku

Yamane T (1967) Statistics: an introductory analysis, 2nd edn. Harper & Row, New York

Zeng L, Huang Y, Wang X (2024) Impacts of flooding on transportation infrastructure and mobility in urban areas. Transp Res Part D: Transp Environ 116:103815. https://doi.org/10.1016/j.trd.2023.103815

Zhu X, Li Y, Chen J (2018) Socio-economic disparities and flood loss variability: an empirical study in flood-prone regions. Nat Hazards 94(1):103–120. https://doi.org/10.1007/s11069-018-3256-7

Chapter 20
Eyes on the Coast: Advancing Coastal Zone Management with Earth Observation Tools

Saroj Senapati and Gouri Sankar Bhunia

20.1 Introduction

Coastal zones are among the most ecologically vibrant, economically valuable, and densely populated regions globally. However, they are under increasing stress from rapid urbanization, industrial expansion, growing maritime trade, mass tourism, and significant human migration. These factors have contributed to the deterioration of coastal water quality, habitat fragmentation, and an increase in environmental risks. Currently, coastal zones support over 40% of the global population within 100 km of the sea (UNEP 2021). They are expected to face further demographic pressure due to climate-induced displacement and economic centralization in port cities. Ecologically, these areas are highly productive and play a vital role in the global biosphere. According to Rodriguez et al. (2005) and Neumann et al. (2017), coastal areas provide all four primary ecosystem services: Supporting, Provisioning, Regulating, and Cultural.

Despite covering only approximately 356,000 km^2 of coastline globally (CIA World Factbook 2016), these regions act as natural buffers against climate variability and support significant economic sectors, including shipping (90% of world trade), fisheries (employing over 60 million people globally), and coastal tourism (valued at over USD 1 trillion annually) (UNCTAD 2023; WTTC 2022). Coastal morphology—the structure and form of coastal landforms—is constantly re-shaped by both natural processes (e.g., tectonics, tides, wave action, sediment transport) and anthropogenic influences such as land reclamation, port construction, and shoreline hardening. Rising sea levels, frequent storm surges, and coastal erosion now pose a

S. Senapati
Annapurna Memorial College of Education, Kasinagar, West Bengal, India
e-mail: senapati.saroj123@gmail.com

G. S. Bhunia (✉)
Kharagpur, Paschim Medinipur, West Bengal, India
e-mail: rsgis_gouri@rediff.com

A. Nath et al. (eds.), *Geospatial Application of Coastal Hazards*, Springer Water,
https://doi.org/10.1007/978-3-032-20076-1_20

threat to both human and ecological systems. As Spalding et al. (2014) note, coastal hazard protection, along with carbon sequestration (Chmura et al. 2003) and livelihood security (Visbeck et al. 2014), are critical to the resilience of coastal populations. However, despite global policy frameworks such as the Integrated Coastal Zone Management (ICZM) approach, implementation remains fragmented and hindered by a lack of standardized, real-time spatial data (Ali et al. 2018).

Remote Sensing (RS) and Earth Observation (EO) systems offer transformative capabilities for coastal research and management. Satellite-based observations enable the collection of synoptic, multi-spectral, and time-series data at scales that are unfeasible with traditional field surveys. With advances in sensor technologies, high-resolution imagery, and cloud-based processing platforms like Google Earth Engine, remote sensing has become a cornerstone of coastal mapping, monitoring, and early warning systems. Recent studies have demonstrated the utility of RS/EO in mapping shoreline change (Luijendijk et al. 2018), tracking land use–land cover (LULC) dynamics, monitoring coastal vegetation such as mangroves (Giri et al. 2011), assessing coastal vulnerability (Ramachandran et al. 2020), and estimating suspended sediment and chlorophyll concentrations through hyperspectral and multispectral sensors.

Machine learning and AI-driven classification of EO datasets now enable automated feature extraction and predictive modeling for climate adaptation and resilience planning. However, a significant knowledge gap still exists in terms of integrated frameworks for EO-based Integrated Coastal Zone Management (ICZM), especially in low- and middle-income countries where ground truthing and infrastructure capacity are limited. This paper reviews the evolution and current applications of RS and EO in the context of coastal zone management. It details the platforms (e.g., Landsat, Sentinel, MODIS, PlanetScope), sensors (optical, SAR, LiDAR), spectral indices (NDWI, MNDWI, LST, NDTI), and processing techniques. It critically examines case studies to identify best practices and future potential.

20.2 Materials and Methods

The bibliometric analysis in this study was conducted to evaluate the global research trends and intellectual structure surrounding the application of remote sensing, GIS, and AI/ML in Integrated Coastal Zone Management (ICZM) over the period from 1995 to 2025. The analysis was performed using data sourced from the Dimensions AI research database, one of the most comprehensive bibliometric platforms. A total of 28,166 publications were retrieved using a targeted keyword strategy that included terms such as "*Coastal Zone Management*," "*Remote Sensing*," "*GIS*," "*AI/ML*," "*Spatial Technologies*," "*Earth Observation*," "*Integrated Coastal Zone*," and "*Mapping*." The analysis was visualized and interpreted using VOSviewer, a tool designed to construct and visualize bibliometric networks. The bibliometric evaluation encompassed various dimensions, including publication volume per year, total citation counts, Field Citation Ratio (FCR), and Relative Citation Ratio (RCR), which helped assess the academic impact and relevance of the retrieved articles. Author-wise analysis identified prolific contributors, with metrics such as h-index, publication frequency, and citation impact highlighting key opinion leaders in the field. Similarly, a journal-wise analysis revealed that leading journals actively publish in this domain, such as the 'Journal of Coastal Research', 'Remote Sensing of Environment', and 'Marine Policy', which together account for a significant proportion of influential studies (Fig. 20.1).

Co-authorship analysis has revealed patterns of international collaboration and knowledge exchange, indicating that research in ICZM is highly interdisciplinary and globally connected. Co-citation analysis further mapped the intellectual structure of the field by identifying clusters of frequently cited references, which included foundational studies on coastal vulnerability, ecosystem services, and applications of Earth observation technologies. The network visualizations from VOSviewer revealed dense clusters around key thematic areas, including climate resilience, coastal morphology, land-use change, and technological integration in monitoring. It also indicates a gradual transition from descriptive coastal mapping to more predictive, decision-support systems leveraging advanced analytics and machine learning. The significant rise in publication volume post-2010 highlights the growing

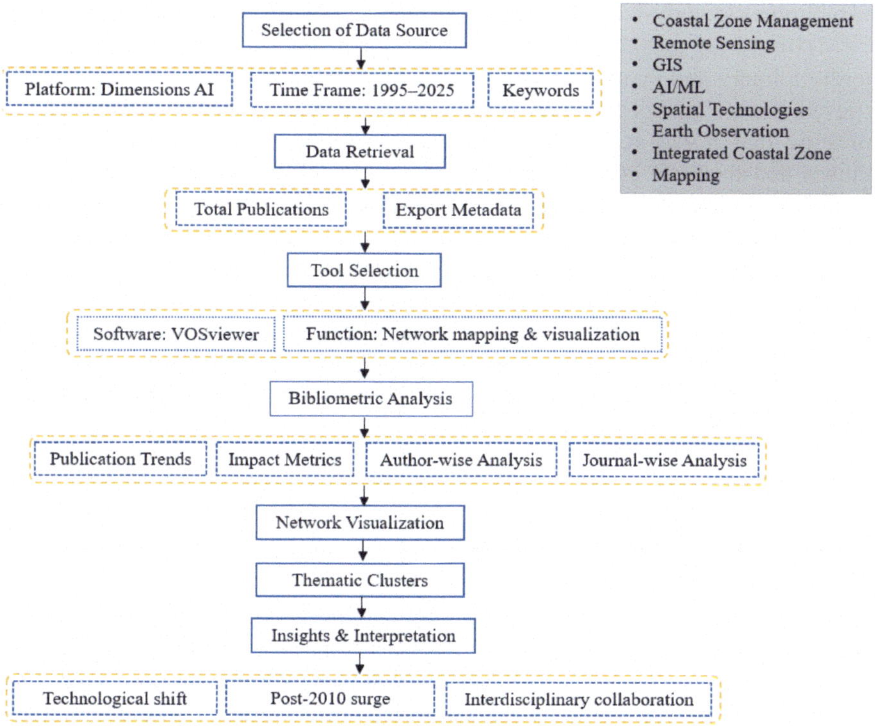

Fig. 20.1 Flow chat of the proposed methodology

Table 20.1 Publication type for integrated coastal zone management (ICZM)

Publication type	Number	Percentage
Chapter	11,576	41.10
Edited book	9,491	33.70
Article	4,981	17.68
Monograph	1,250	4.44
Proceeding	776	2.76
Preprint	92	0.33

urgency to address coastal challenges through science-based, technology-enabled interventions.

20.3 Results and Discussion

20.3.1 Distribution of Publication Types

The bibliometric analysis of global research on Integrated Coastal Zone Management (ICZM) from 1995 to 2025 reveals a rich diversity in publication types, reflecting the interdisciplinary and evolving nature of this field. Among the 28,166 publications retrieved, book chapters constitute the largest share, with 11,576 records (41.10%), indicating that ICZM research is frequently included as part of broader edited volumes and thematic compilations. Edited books are the second most common format, accounting for 9,491 publications (33.70%), underscoring the importance of curated, multi-author contributions in framing complex coastal management issues. Journal articles, typically considered the gold standard in academic publishing, represent 4,981 entries (17.68%). Monographs, which represent in-depth explorations by individual authors or small teams, contribute 1,250 records (4.44%). Conference proceedings, totaling 776 publications (2.76%), play a vital role in presenting preliminary research, pilot studies, and technological innovations in coastal monitoring and planning (Table 20.1).

20.3.2 Journal-wise Publications

Among the top sources, 'HortScience' leads in publication count with 1,753 papers, though its citation impact is relatively modest (mean citation: 0.8), suggesting its content may be more technical or localized. Similarly, 'Epidemiology' contributed 1,239 publications with a slightly lower average of 0.78 citations per article, likely pointing to the linkage of Integrated Coastal Zone Management (ICZM) with public health in coastal regions. In contrast, specialized and encyclopedic series like the

Table 20.2 Journal-wise publications and citations during the period between 1995 and 2025

Name	Publications	Citations	Mean
HortScience	1753	1,405	0.8
Epidemiology	1239	967	0.78
Encyclopedia of earth sciences series	835	3,677	4.4
Lecture notes in computer science	545	3,545	6.5
Lecture notes in civil engineering	343	774	2.26
Lecture notes in networks and systems	333	934	2.8
Encyclopedia of the UN sustainable development goals	280	610	2.18
Lecture notes in electrical engineering	261	1,028	3.94
Communications in computer and information science	247	692	2.8
Advances in intelligent systems and computing	206	2,108	10.23

'Encyclopedia of Earth Sciences Series' (835 publications, mean: 4.4) and the 'Lecture Notes in Computer Science' (545 publications, mean: 6.5) have higher per-article impact, indicating the growing importance of computational methods in coastal studies (Table 20.2).

Notably, the 'Ecological Studies' series, despite having only 70 publications, achieved a remarkable 6,937 citations, averaging 99.1 citations per paper. This underscores the foundational influence of ecological research in ICZM. Similarly, 'Remote Sensing' (69 publications, mean: 31.06) and 'Water Science and Technology Library' (54 publications, mean: 50.8) demonstrate high research visibility, highlighting the centrality of spatial monitoring and hydrological modeling. The 'NATO ASI Series' (45 publications, mean: 60.62) and 'The Handbook of Environmental Chemistry' (57 publications, mean: 15.72) also reflect highly cited contributions, possibly due to their theoretical and policy-oriented content. On the emerging front, platforms like 'Advances in Intelligent Systems and Computing' (206 publications, mean: 10.23) and 'Springer Water' (44 publications, mean: 14.02) indicate the integration of AI and water-centric studies into ICZM frameworks.

20.3.3 Author-wise Publications

Leading the list in publication volume is Polona Šprajc from the University of Maribor, Slovenia, with 104 publications but a relatively low citation count of 21, indicating an average of 0.2 citations per publication. Similar trends are observed for Iztok Podbregar (101 publications, 24 citations) and Olja Arsenijević (100 publications, 20 citations), suggesting a focus on volume rather than citation influence. In contrast, Dorian Q. Fuller of University College London demonstrates a significant research impact, with only 22 publications but 185 citations, averaging 8.41 citations per paper, which highlights the quality and influence of his work. Abu Reza

Table 20.3 Top ten author published work on integrated coastal zone management during the period between 1995 and 2025

Name	Organization, country	Publications	Citations	Mean
Polona Šprajc	University of Maribor, Slovenia	104	21	0.2
Iztok Podbregar	University of Maribor, Slovenia	101	24	0.24
Olja Arsenijević	–	100	20	0.2
Yvonne Ziegler	–	98	20	0.2
Dragan Trivan	–	98	20	0.2
Yun-Chul Hong	Seoul National University, South Korea	38	11	0.29
Mi Na Ha	Dankook University, South Korea	25	19	0.76
Dorian Q Fuller	University College London, United Kingdom	22	185	8.41
Krystyna Konstankiewicz	Polish Academy of Sciences, Poland	21	60	2.86
William Joseph Edmonds	–	18	53	2.94

Md Towfiqul Islam from Begum Rokeya University, Bangladesh also stands out with 14 publications and 258 citations, yielding the highest mean of 18.43, followed by Swapan Talukdar of University of Calcutta, India, with 14 publications and 197 citations (mean 14.07). This disparity between quantity and citation quality underscores a diverse landscape in ICZM research, where some authors produce highly influential work, while others contribute more extensively in terms of volume (Table 20.3).

20.3.4 Co-authorship Analysis

An analysis identified a total of 79 researchers who actively collaborated in Integrated Coastal Zone Management (ICZM)-related publications. These collaborations formed 220 co-authorship links, representing the total number of documented co-authored works between these researchers. The authors were grouped into 11 distinct clusters, each reflecting a subgroup of closely collaborating scholars, often centered around specific thematic areas or geographic affiliations (Fig. 20.2). The clustering suggests a fragmented but interconnected research landscape, where regional and institutional collaborations dominate, with a few bridging figures acting as nodes between multiple groups. This type of analysis is instrumental in identifying collaborative patterns, influential research communities, and potential gaps or isolated researchers in the field. The visual map generated by VOSviewer effectively showcases the strength of co-authorship ties (represented by link thickness) and proximity

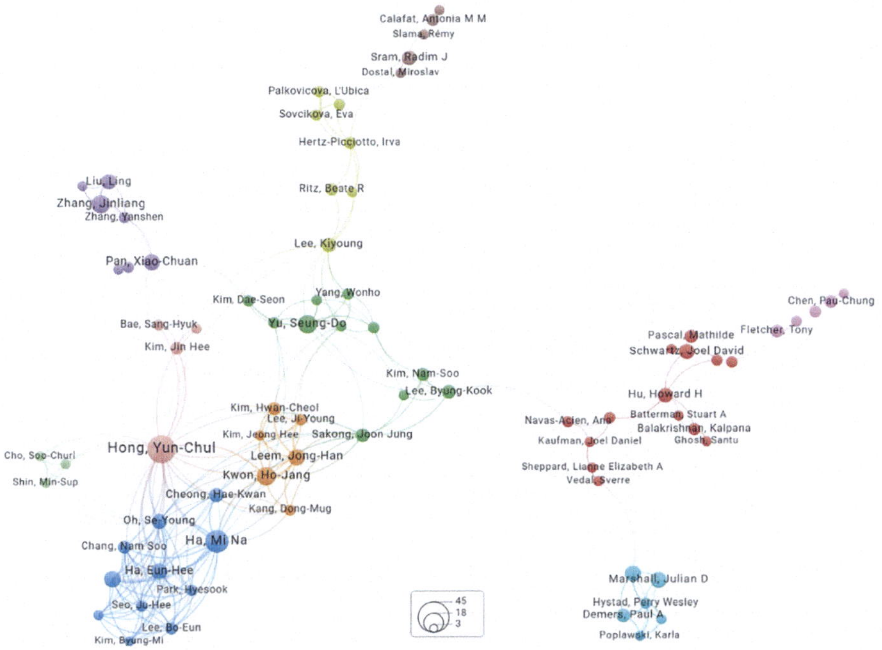

Fig. 20.2 Co-authorship analysis of integrated coastal zone management during the period between 1995 and 2025

between authors, helping highlight key influencers and emerging collaboration hubs within the global ICZM research ecosystem.

20.3.5 Co-citation Analysis

The analysis revealed a total of 86 frequently co-cited researchers, indicating that their works are consistently referenced together in the same publications, which signifies a thematic or conceptual alignment. A total of 1,278 co-citations were mapped across 725 citation links, reflecting how these authors are interlinked through shared academic influence (Fig. 20.3). The network was organized into 10 distinct clusters, each representing a knowledge domain or thematic area within ICZM, such as coastal ecosystem services, climate change adaptation, spatial planning, GIS-based modeling, and remote sensing applications. These clusters highlight core scholars who serve as intellectual anchors and whose work has significantly shaped the evolution of ICZM thought. Through VOSviewer's visualization capabilities, the map clearly illustrates the relative prominence of authors (based on citation strength) and the interconnectivity of scholarly ideas across disciplines. This co-citation structure is essential for understanding the theoretical and methodological foundations of

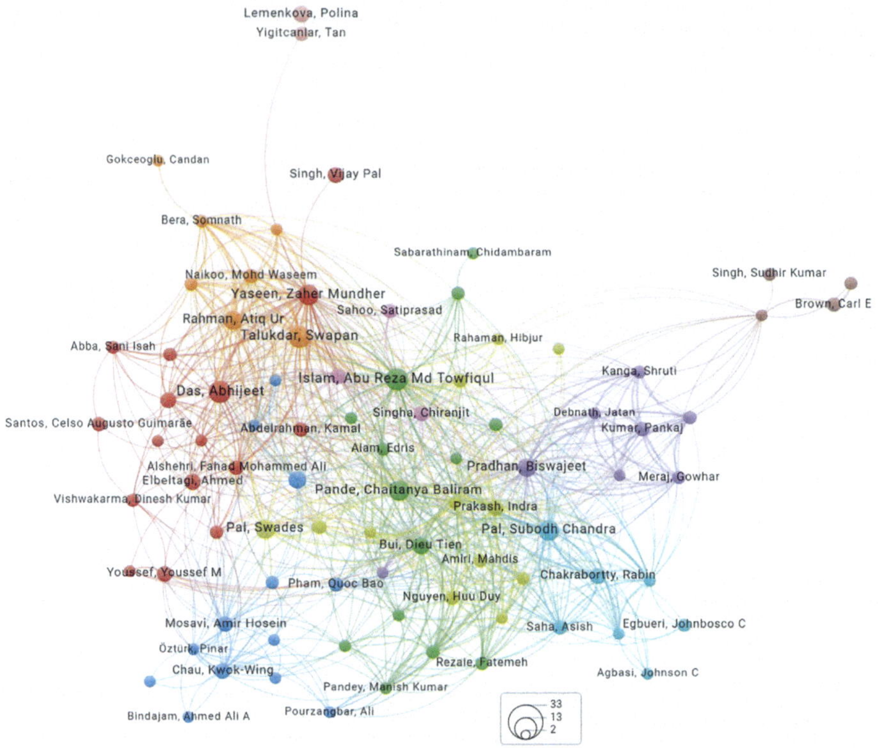

Fig. 20.3 Co-citation analysis of integrated coastal zone management during the period between 1995 and 2025

ICZM and identifying key researchers who have provided enduring contributions to the field.

20.4 Application Areas of EO in Coastal Zone Management

The use of EO tools has significantly enhanced our ability to monitor, assess, and respond to changes in coastal environments. Satellite-based shoreline and coastal erosion mapping enables the detection of spatial and temporal variations in coastlines, providing critical insights into sediment transport and identifying erosion-prone areas. For instance, multi-temporal satellite datasets such as Landsat, Sentinel-2, and ASTER have been widely utilized for detecting shoreline retreat and progradation (Table 20.4). Furthermore, mangrove and wetland monitoring using EO data—particularly from sensors like Sentinel-1 (SAR) and Sentinel-2 (optical)—has provided accurate assessments of vegetation health, species distribution, and degradation caused by anthropogenic and climatic stressors. Urban sprawl and land use/land cover

(LULC) changes in coastal zones have also been effectively tracked using medium- to high-resolution imagery (e.g., Landsat, SPOT, and MODIS), which helps planners understand the encroachment of built-up areas into ecologically sensitive zones.

In disaster risk reduction and early warning, EO plays a vital role in identifying flood zones, storm surge-prone areas, and cyclone tracks. Real-time imagery from geostationary satellites such as INSAT and Meteosat, along with radar data from Sentinel-1 and RADARSAT, support disaster preparedness and response systems. Marine spatial planning and habitat mapping rely on multispectral and hyperspectral remote sensing data to classify coral reefs, sea grass beds, and benthic habitats, providing essential input for zoning and conservation decisions. Furthermore, satellite-based water quality monitoring utilizes EO-derived parameters such as chlorophyll-a, turbidity, and suspended sediment concentration to assess pollution levels and detect harmful algal blooms. These EO applications, when integrated into a cohesive decision support system, offer robust, cost-effective, and scalable solutions for adaptive and sustainable coastal zone management.

20.5 Way Forward and Recommendation

Coastal zones are experiencing growing pressures from rapid urbanization, climate change-induced sea level rise, extreme weather events, and increased human activities. Given these challenges, the integration of EO tools into CZM is not just advantageous but essential. The path forward involves institutionalizing EO-based practices within the planning, policy-making, and regulatory frameworks at both national and sub-national levels.

- Mainstreaming EO data into decision-support systems allows for real-time monitoring of dynamic coastal parameters. This includes tracking shoreline changes, land use transformations, shifts in mangrove cover, and patterns of marine pollution, which enables proactive management rather than reactive interventions. To support this, governments should invest in interoperable, open-access coastal data platforms. These platforms should integrate EO data with socio-economic, ecological, and infrastructure datasets to provide comprehensive insights into coastal vulnerabilities and resource allocations.
- Capacity building and training for local governance institutions, coastal zone authorities, and disaster response agencies in EO data interpretation and geospatial analytics is essential. This will significantly enhance the utility of satellite products in community-level planning, particularly for early warning systems, hazard zoning, and conservation monitoring. Furthermore, incorporating participatory mapping and community-sourced observations can further enrich EO-derived datasets.
- The integration of multi-sensor and multi-temporal satellite data is crucial for more nuanced analyses of coastal processes. This includes data from medium-resolution sources like Landsat and Sentinel, as well as high-resolution and radar

Table 20.4 Summarizing the application of satellite data for integrated coastal zone management

Objectives	Scale	Satellite data used	Methods	Outcomes	References
To assess historical and recent shoreline changes	Regional to local scale	Landsat (TM, ETM +, OLI) (30 m), Sentinel-2 (10 m), IRS LISS-III (23.5 m), MODIS (250 m), High-res (WorldView, SPOT)	Image classification, NDWI, Digital Shoreline Analysis System (DSAS), Transect-based shoreline extraction, Change detection	Identification of erosion-prone hotspots, average erosion/ accretion rates, spatiotemporal shoreline movement	Maiti and Bhattacharya (2009), Bagli et al. (2004)
To model the impact of sea-level rise and human interventions on shoreline retreat	National scale	Landsat, ASTER, SAR (RADARSAT, Sentinel-1)	Time-series analysis, SLR modeling, GIS overlay with infrastructure	Forecasted shoreline positions under RCP scenarios, risk zoning of coastal assets	Appeaning Addo et al. (2011), Gornitz et al. (1997)
To evaluate policy impacts (e.g., buffer zone regulations) on erosion trends	Local scale	IKONOS, QuickBird, Drone-based imagery, Sentinel-2	High-res shoreline delineation, policy overlay in GIS, pre/post regulation comparison	Evidence-based evaluation of coastal management strategies	Thieler et al. (2005), Chen and Chang (2020)
To develop community-level erosion monitoring tools	Community/ village scale	Drone/UAV, Sentinel-2, Landsat	Participatory GIS, community validation of shoreline changes, training local stakeholders	Empowered communities with spatial knowledge, low-cost early warning systems	Jiang et al. (2020), Addo (2013)
To compare natural vs. anthropogenic drivers of shoreline change	Regional scale	MODIS, Sentinel-1 SAR, Landsat	Regression and correlation analysis between shoreline change and variables (e.g., wave energy, population density)	Quantification of anthropogenic contribution to erosion	Luijendijk et al. (2018), Ozyurt and Ergin (2010)

(continued)

Table 20.4 (continued)

Objectives	Scale	Satellite data used	Methods	Outcomes	References
Mapping extent and spatial distribution of mangroves and wetlands	Regional/ National scale	Landsat (TM, ETM + , OLI) (30 m) IRS LISS-III (23.5 m) MODIS (250 m)	Supervised/unsupervised classification (e.g., MLC), NDVI, NDMI, object-based classification (OBIA)	Baseline maps of mangrove/wetland cover, national inventories (e.g., India, Bangladesh)	Giri et al. (2011), Blasco et al. (1998a, b)
Monitoring degradation, deforestation and regrowth	Local/ Watershed scale	Landsat, Sentinel-2 (10 m), ASTER, SPOT, PlanetScope (3 m)	Time-series change detection (e.g., post-classification comparison, image differencing), trend analysis	Identification of hotspots of mangrove degradation and gain, temporal statistics	Hamilton and Casey (2016), Nandy and Kushwaha (2011)
Classifying wetland types (salt marsh, mudflats, mangrove, etc.)	National to sub-regional	Landsat, Sentinel-2, Sentinel-1 (SAR), ALOS PALSAR (L-band SAR)	Hybrid classification (optical + SAR), texture analysis, seasonal compositing	Wetland typology layers, floodplain-wetland interfaces mapped	Gupta et al. (2005), Van Beijma et al. (2014)
Carbon stock and biomass estimation in mangroves	Plot to regional scale	Landsat, Sentinel-2, SAR (Sentinel-1, ALOS), LiDAR (GEDI)	Allometric models + NDVI/EVI + InSAR + canopy height models	Estimations of above-ground biomass, blue carbon potential, carbon loss maps	Simard et al. (2019), Fatoyinbo et al. (2008)
Tracking wetland hydrological connectivity and inundation dynamics	Watershed to river delta scale	MODIS, Sentinel-1 SAR, Radarsat, SMAP, Envisat	SAR-based water detection, time-series NDWI, hydrodynamic modeling	Seasonal wetland flooding maps, delineation of perennial vs temporary wetlands	Karim et al. (2020), Liu et al. (2022)
Detecting aquaculture encroachment and land use pressures	Local to regional scale	Sentinel-2, Planet, Landsat	Change detection, high-resolution time series, OBIA, visual interpretation	Quantification of mangrove conversion to aquaculture/farms; regulatory inputs	Richards and Friess (2016), Mcleod et al. (2011)

(continued)

Table 20.4 (continued)

Objectives	Scale	Satellite data used	Methods	Outcomes	References
Support for Ramsar site monitoring and compliance	Site-level	High-res (WorldView-2, QuickBird), Landsat, Sentinel-2	Periodic EO-based mapping, comparison with Ramsar condition criteria	Support for wetland management, Ramsar documentation, and early warnings	Davidson (2014), Murray et al. (2018)
To assess the temporal and spatial growth of urban areas in coastal zones	Regional to city-scale	Landsat (TM, ETM + , OLI), Sentinel-2, IRS-LISS III, SPOT	Supervised/Unsupervised Classification, NDVI, Built-up Index (NDBI), Change Detection, Object-Based Image Analysis (OBIA), Machine Learning (Random Forest, SVM)	Quantified urban expansion, Land Use/ Land Cover (LULC) maps, detection of peri-urban growth, and loss of agricultural and green spaces	Sudhira et al. (2004), Seto et al. (2011), Schneider et al. (2010), Bhatti and Tripathi (2014), Dewan and Yamaguchi (2009)
To monitor urban encroachment into ecologically sensitive coastal wetlands and mangroves	Local scale	High-resolution imagery (QuickBird, IKONOS, WorldView-2), Landsat series	Multitemporal LULC classification, accuracy assessment, urban footprint extraction	Identification of encroachment patterns, degradation of buffer zones, impacts on biodiversity	Rahman et al. (2011), Rajan et al. (2013), Islam et al. (2015)
To integrate urban planning with coastal ecosystem conservation	City-scale	Sentinel-2A/B, PlanetScope, Google Earth Engine datasets	GIS-based Urban Growth Modeling (CA-Markov, SLEUTH), Transition Matrices	Predictive modeling of future sprawl, suggestion for smart zoning, critical zone delineation	Bhatta et al. (2010), Taubenböck et al. (2012), Hasan et al. (2020)

(continued)

Table 20.4 (continued)

Objectives	Scale	Satellite data used	Methods	Outcomes	References
To understand land consumption pattern and its impact on stormwater and drainage in coastal zones	Urban corridor scale	Landsat series, IRS Resourcesat, TerraSAR-X	LULC change analysis, impervious surface extraction, hydrological modeling integration	Evidence of urban-induced hydrological stress, flooding hotspots identified	Jain et al. (2012)
To link population growth with land use dynamics for sustainable coastal development	Regional scale	Landsat OLI, MODIS, VIIRS Nighttime Lights	Spatiotemporal analysis, population-LULC overlay, correlation analysis	Urban footprint trends aligned with demographic shifts; recommendation for low-impact development	Dibba et al. (2025), Polyakov and Zhang (2008)
To monitor and assess cyclone path, intensity, and landfall location for early warning	Global & regional scale	NOAA AVHRR, INSAT, MODIS, Himawari, METEOSAT, Sentinel-3 SLSTR	Multi-spectral image analysis, Cloud top temperature, Sea Surface Temperature (SST), Wind vectors from scatterometers (e.g., ASCAT), AI/ML-based cyclone prediction	Improved lead time for cyclone warnings, Better path prediction and mitigation planning	Pattnaik and Ramarao (2010), Mukherjee and Mitra (2022)
To detect storm surge extent and post-event inundation mapping	Local to regional	Synthetic Aperture Radar (SAR)—RADARSAT, Sentinel-1, RISAT-1/2, ALOS PALSAR	Change detection, water mask classification, backscatter intensity analysis in SAR, thresholding	Accurate storm surge maps even during cloud cover, estimation of damage extent, and emergency response planning	Kumar et al. (2016)
To monitor and model coastal flooding due to heavy rainfall and riverine overflow	Regional	Landsat TM/ETM + / OLI, Sentinel-2 MSI, MODIS, CARTOSAT, Google Earth Engine	NDWI, SAR-based water mapping, DEM-integrated flood modeling (HEC-RAS, MIKE-FLOOD)	Rapid flood mapping, vulnerable zone delineation, improved preparedness planning	Bhatt et al. (2020), Alam and Dominey-Howes (2015), Singh et al. (2021)

(continued)

Table 20.4 (continued)

Objectives	Scale	Satellite data used	Methods	Outcomes	References
To develop multi-hazard risk zonation for cyclones, storm surges, and sea-level rise	National & state-level	Multi-source EO (Landsat, Sentinel-1/2, SRTM/ALOS DEM, VIIRS Nightlights)	GIS-based overlay analysis, hazard-exposure-vulnerability matrix, weighted overlay methods	Cyclone-prone zone mapping, high-risk coastal strip identification, long-term climate adaptation strategies	NIDM (2012), IPCC AR6 (2021), Dasgupta et al. (2009)
To support coastal disaster preparedness using real-time EO-based early warning	Local to coastal district level	INSAT-3D, Himawari, SCATSAT-1, Oceansat-2/3, SMAP	Real-time SST, rainfall estimates, cloud imagery + data assimilation with weather forecast models (WRF, GFS)	Timely alerts for storm intensity, surge and rainfall prediction accuracy improved	Nayak (2017), Patel et al. (2024)
To assess post-disaster damage and aid recovery planning	Local/ urban scale	Very High Resolution (VHR) imagery (WorldView, QuickBird), Sentinel-1/2	Pre-post image comparison, Damage Proxy Maps (DPMs), OBIA	Infrastructure damage mapped, restoration zones prioritized	UN-SPIDER (2016), Copernicus EMS Reports, Kumar et al. (2022)
Monitor temporal variations in turbidity and suspended sediments in estuaries	Regional scale	MODIS (Terra/Aqua) (250 m–1 km), Sentinel-2 MSI, Landsat-8 OLI	Atmospheric correction, time-series NDWI, turbidity and TSM algorithms using band ratios	Identified seasonal sediment plumes and high-turbidity zones in estuaries	Warrick et al. (2004), Nechad et al. (2010)
Detect oil spills and hydrocarbon pollution in coastal waters	Local to regional	RADARSAT-1, Sentinel-1 SAR, TerraSAR-X, Landsat-8 OLI	SAR backscatter analysis, supervised classification, edge detection for slick boundaries	Mapped oil slick extent with high spatial accuracy during events like Deepwater Horizon and Mumbai offshore spills	Jafarzadeh et al. (2021), Uribe-Martínez et al. (2024)

(continued)

Table 20.4 (continued)

Objectives	Scale	Satellite data used	Methods	Outcomes	References
Assess chlorophyll-a concentration and eutrophication risks	Regional/ global	SeaWiFS, MODIS, Sentinel-3 OLCI, VIIRS, MERIS	Ocean color algorithms (OC4, OC3), empirical calibration with in-situ measurements	Tracked phytoplankton blooms and potential eutrophication hotspots	Gitelson et al. (2008), Blondeau-Patissier et al. (2014), Wang et al. (2022)
Monitor coastal algal bloom occurrences (HABs)	Local to regional	MODIS, Sentinel-2 MSI, Sentinel-3 OLCI	Floating Algae Index (FAI), Fluorescence Line Height (FLH), anomaly detection	Enabled early identification of HABs in sensitive lagoons and fishery zones	Mishra et al. (2021)
Detect thermal pollution from power plants and industries	Local	Landsat-5/7/8 TIR bands, MODIS LST, ASTER	Land Surface Temperature (LST) extraction, hotspot analysis, correlation with outfall locations	Mapped thermal plumes from coastal thermal power plants impacting local marine life	Dash et al. (2010), Huang et al. (2017)
Monitor salinity and dissolved organic matter (CDOM)	Regional scale	MODIS, MERIS, Sentinel-2, Sentinel-3	Ocean color and backscattering coefficients, band ratio indices, in-situ validation	Estimated changes in freshwater discharge zones and pollution inflow from river mouths	Morel and Gentili (2009), Garnesson et al. (2019)
Integrated pollution assessment using multi-sensor fusion	Regional to national	MODIS, Sentinel-1 & -2, Landsat-8, VIIRS, in-situ sensors (Sonde, buoy)	Data fusion, spatiotemporal interpolation, statistical modeling	Holistic pollution load mapping and environmental risk assessment in coastal zones	Chachuła et al. (2022), Middya and Roy (2024)

imagery such as TerraSAR-X and RISAT. Furthermore, advances in Artificial Intelligence and Machine Learning (AI/ML) for remote sensing, including automated shoreline extraction and mangrove health assessment, need to be scaled up for regional planning applications.

- Cross-institutional collaborations between Earth Observation (EO) agencies (such as ISRO, NASA, and ESA), academic institutions, coastal research centers, and civil society organizations can significantly accelerate the development of customized coastal decision-support tools. It is crucial to strengthen initiatives like the Indian National Centre for Coastal Research (NCCR), the Integrated Coastal Zone Management (ICZM) project, and international partnerships through GEOSS, Copernicus, and UN-SPIDER to foster knowledge exchange and capacity transfer.
- Policy recommendations should focus on mandating the use of EO data in Environmental Impact Assessments (EIA), Coastal Regulation Zone (CRZ) demarcation, disaster risk assessments, and marine spatial planning. Furthermore, regulatory mechanisms must be updated to recognize satellite-derived evidence in enforcement and litigation processes officially.

20.6 Discussion and Summary

Coastal zones are dynamic interfaces between land and sea, supporting vital ecosystems, dense human populations, and critical infrastructure. However, they face increasing pressure from urbanization, sea-level rise, storm surges, erosion, and pollution. In response, Earth Observation (EO) tools have emerged as indispensable assets in Integrated Coastal Zone Management (ICZM), offering timely, synoptic, and multi-scale data for monitoring, planning, and decision-making. EO tools—particularly remote sensing satellites such as Landsat, Sentinel-1/2, MODIS, and SAR platforms—have revolutionized the capacity to monitor coastal land use change, wetland degradation, urban sprawl, disaster risk, and water quality over time. For example, Giri et al. (2011) utilized 30 years of Landsat data to map global mangrove distribution and document significant regional declines, thereby reinforcing earlier works, such as Blasco et al. (1998b, a), on tropical mangrove mapping. Hamilton and Casey (2016) further advanced this by presenting high-resolution maps to track the dynamics of mangrove forests.

In urban coastal regions, EO has effectively captured urban sprawl and land use change, with studies like Sudhira et al. (2004) and Seto et al. (2011) highlighting rapid, unplanned urban expansion in Asian megacities. Bhatta et al. (2010) and Taubenböck et al. (2012) employed multi-temporal satellite imagery to assess urban footprint and impervious surface growth, informing sustainable land use policies. EO has been pivotal in disaster risk reduction (DRR), particularly in forecasting and assessing impacts of cyclones, storm surges, and coastal flooding. ISRO's MOSDAC portal, NOAA STAR, and Copernicus Emergency Management Service have delivered real-time products for early warning systems. Notably, Bhatt et al. (2020)

and Alam and Dominey-Howes (2015) emphasized the value of EO-derived data in cyclone impact modeling and DRR policy. Moreover, UN-SPIDER (2016) and Dasgupta et al. (2009) linked EO data with socioeconomic vulnerability to identify coastal hotspots needing adaptation.

Optical and radar sensors have proven effective in detecting various water quality parameters, including chlorophyll concentration, suspended sediments, oil spills, and eutrophication. Research by Nechad et al. (2010), Gitelson et al. (2008), and Blondeau-Patissier et al. (2014) has demonstrated EO's capability to estimate biophysical water parameters through reflectance-based models. Furthermore, Mishra et al. (2021) utilized Sentinel-2 imagery for water turbidity mapping along Indian coasts, a finding supported by earlier studies such as Joshi et al. (2022). From a policy standpoint, insights derived from EO data directly inform Integrated Coastal Zone Management (ICZM) frameworks. These insights facilitate ecosystem-based management, hazard mapping, infrastructure planning, and regulatory enforcement. For instance, Maiti and Bhattacharya (2009) explored vulnerability mapping of Indian coastal zones using integrated EO data, and Appeaning Addo et al. (2011) applied change detection techniques in West Africa to aid shoreline management.

Recent trends also show integration of EO with machine learning (ML), cloud-based platforms (like Google Earth Engine), and IoT-based in-situ sensors, boosting the predictive power and accessibility of coastal monitoring systems (Wang et al. 2022; Garnesson et al. 2019). The transformative role of EO in Coastal Zone Management lies in its ability to provide reliable, repeatable, and scalable data—critical for adaptive, science-based decision-making in ICZM. As climate impacts intensify and coastal population pressures rise, integrating EO tools with socio-economic and ecological models will be vital. Studies across regions and decades (1995–2025) converge on this conclusion, underscoring EO as not merely a monitoring tool but a strategic enabler of sustainable and resilient coastal development.

References

Addo KA (2013) Shoreline morphological changes and the human factor: Case study of Accra Ghana. J Coast Conserv 17:85–91

Alam E, Dominey-Howes D (2015) A new catalogue of tropical cyclones of the Northern Bay of Bengal and the distribution and effects of selected landfalling events in Bangladesh. Int J Climatol 35(6):801–835

Ali A, Kumar A, Bhatnagar A (2018) Coastal ecosystem management: A review. J Coastal Res 34(5):1090–1099. https://doi.org/10.2112/JCOASTRES-D-17-00101.1

Appeaning Addo K, Walkden M, Mills JP (2011) Detection, measurement and prediction of shoreline recession in Accra, Ghana. ISPRS J Photogramm Remote Sens 66(3):259–266

Bagli S, Soille P, Sieber R (2004) Object-oriented classification of IKONOS imagery for mapping coastal erosion. In: Proceedings of 20th ISPRS congress

Bhatt CM et al (2020) Cyclone damage assessment using remote sensing and GIS in Odisha, India. J Indian Soc Remote Sens 48:769–777

Bhatta B, Saraswati S, Bandyopadhyay D (2010) Urban sprawl measurement from remote sensing data. Appl Geogr 30(4):731–740

Bhatti SS, Tripathi NK (2014) Monitoring land use change using remote sensing and GIS techniques: A case study of Kullu-Manali region in Himachal Pradesh, India. Egypt J Remote Sens Space Sci 17(1):111–117

Blasco F, Aizpuru M, Gers C (1998b) Depletion of the mangroves of Continental Asia. Wetlands Ecol Manage 6(1):1–9

Blasco F, Saenger P, Janodet E (1998a) Mangroves as indicators of coastal change. CATENA 27(3–4):167–178

Blondeau-Patissier D et al (2014) A review of ocean color remote sensing and its applications in marine biodiversity research. Remote Sens Environ 140:55–84

Central Intelligence Agency (CIA) (2016) The world factbook. Central Intelligence Agency, Washington, D.C. https://www.cia.gov/the-world-factbook

Chachuła K, Słojewski TM, Nowak R (2022) Multisensor data fusion for localization of pollution sources in wastewater networks. Sensors 22(1):387. https://doi.org/10.3390/s22010387

Chen J, Chang KT (2020) Integration of Landsat and UAS imagery for shoreline change analysis in Taiwan. Remote Sensing 12(22):3805

Chmura GL, Anisfeld SC, Cahoon DR, Lynch JC (2003) Global carbon sequestration in tidal, saline wetland soils. Global Biogeochem Cycles 17(4):1111. https://doi.org/10.1029/2002GB001917

Dasgupta S, Laplante B, Meisner C, Wheeler D, Yan J (2009) The impact of sea level rise on developing countries: a comparative analysis. Clim Chang 93:379–388. https://doi.org/10.1007/s10584-008-9499-5

Dewan AM, Yamaguchi Y (2009) Land use and land cover change in Greater Dhaka, Bangladesh: Using remote sensing to promote sustainable urbanization. Appl Geogr 29(3):390–401

Dibba B, Yaffa S, Sawaneh M, Adzawla W (2025) Land cover transformation and population growth: impacts on coastal environment of The Gambia (1990–2020). Sustainability 17(5):1853. https://doi.org/10.3390/su17051853

Fatoyinbo TE, Simard M, Washington-Allen RA, Shugart HH (2008) Landscape-scale extent, height, biomass, and carbon estimation of Mozambique's mangrove forests with Landsat ETM+ and Shuttle Radar Topography Mission elevation data. J Geophys Res: Biogeosciences 113:G02S06

Garnesson P, Mangin A, Fanton d'Andon O, Demaria J, Bretagnon M (2019) The CMEMS GlobColour chlorophyll a product based on satellite observation: multi-sensor merging and flagging strategies. Ocean Sci 15:819–830. https://doi.org/10.5194/os-15-819-2019

Giri C, Ochieng E, Tieszen LL, Zhu Z, Singh A, Loveland T, Masek J, Duke N (2011) Status and distribution of mangrove forests of the world using earth observation satellite data. Glob Ecol Biogeogr 20(1):154–159. https://doi.org/10.1111/j.1466-8238.2010.00584.x

Gitelson AA et al (2008) A simple semi-analytical model for remote estimation of chlorophyll-a in turbid waters. Remote Sens Environ 112(9):3582–3593

Gornitz V, Daniels RC, White TW, Birdwell KR (1997) The development of a coastal risk assessment database: Vulnerability to sea-level rise in the U.S. southeast. J Coastal Res 14:327–338

Gupta RD, Sharma SVS, Rao VS (2005) Mapping and monitoring of mangroves using remote sensing and GIS. Trop Ecol 46(2):195–204

Hamilton SE, Casey D (2016) Creation of a high spatio-temporal resolution global database of continuous mangrove forest cover for the 21st century. Glob Ecol Biogeogr 25(6):729–738

Huang X, Hu J, Qin F, Quan W, Cao R, Fan M, Wu X (2017) Heavy metal pollution and ecological assessment around the Jinsha coal-fired power plant (China). Int J Environ Res Public Health 14(12):1589. https://doi.org/10.3390/ijerph14121589

Islam MM, Rahman MM, Rahman MR (2015) Urban expansion and environmental degradation: A study on impact of urban sprawl on water bodies in Dhaka city. Environ Dev 13:1–9

Jiang H, Zhang L, Wen Y (2020) Multi-sensor monitoring of coastline changes in China's coastal cities. Remote Sensing 12(1):152

Joshi G, Verma P, Meena B, Goswami P, Peter DM, Jha DK, Vinithkumar NV, Dharani G (2022) Unraveling the potential of bacteria isolated from the equatorial region of Indian Ocean in mercury detoxification. Front Mar Sci 9:986493. https://doi.org/10.3389/fmars.2022.986493

Karim F, Marvanek S, Merrin LE, Nielsen D, Hughes J, Stratford D, Pollino C (2020) Modelling flood-induced wetland connectivity and impacts of climate change and dam. Water 12(5):1278. https://doi.org/10.3390/w12051278

Jafarzadeh H, Mahdianpari M, Homayouni S, Mohammadimanesh F, Dabboor M (2021) Oil spill detection from Synthetic Aperture Radar Earth observations: a meta-analysis and comprehensive review. GIScience Remote Sens 58(7):1022–1051. https://doi.org/10.1080/15481603.2021.1952542

Liu Y, Cui G, Tong S, Wang S, Lu X (2022) Determination of the hydrodynamic characteristics of a typical Inland Saline-Alkali Wetland in Northeast China. Front Ecol Evol 10:939431. https://doi.org/10.3389/fevo.2022.939431

Luijendijk A, Hagenaars G, Ranasinghe R, Baart F, Donchyts G, Aarninkhof S (2018) The state of the world's beaches. Scientific Reports, 8, Article 6641. https://doi.org/10.1038/s41598-018-24630-6

Maiti S, Bhattacharya AK (2009) Shoreline change analysis and its application to prediction: A remote sensing and statistics-based approach. Mar Geol 257(1–4):11–23

Mcleod E, Chmura GL, Bouillon S, Salm R, Björk M, Duarte CM, Lovelock CE, Schlesinger WH, Silliman BR (2011) A blueprint for blue carbon: toward an improved understanding of the role of vegetated coastal habitats in sequestering CO_2. Front Ecol Environ 9:552–560. https://doi.org/10.1890/110004

Middya AI, Roy S (2024) Multiview data fusion technique for missing value imputation in multisensory air pollution dataset. J Ambient Intell Human Comput 15:3173–3191. https://doi.org/10.1007/s12652-024-04816-9

Mishra DR et al (2021) Monitoring coastal water turbidity and sediment plume dynamics using Sentinel-2 MSI. ISPRS J Photogramm Remote Sens 177:276–291

Morel A, Gentili B (2009) The dissolved yellow substance and the shades of blue in the Mediterranean Sea. Biogeosciences 6:2625–2636. https://doi.org/10.5194/bg-6-2625-2009

Mukherjee A, Mitra P (2022) Predicting cyclone landfall using mutual information and dilated recurrent neural network. Environ Data Sci 1:e18. https://doi.org/10.1017/eds.2022.18

Nandy S, Kushwaha SPS (2011) Study on the utility of IRS LISS-III and AWiFS data for characterization and classification of mangrove forests. J Coast Conserv 15:123–137

Nayak S (2017) Coastal zone management in India – present status and future needs. Geo-Spat Inf Sci. https://doi.org/10.1080/10095020.2017.1333715

Nechad B, Ruddick K, Park Y (2010) Calibration and validation of a generic multisensor algorithm for mapping of turbidity in coastal waters. Remote Sens Environ 114(4):854–866

Neumann B, Vafeidis AT, Zimmermann J, Nicholls RJ (2017) Future coastal population growth and exposure to sea-level rise and coastal flooding–A global assessment. PLoS ONE 10(3):e0118571. https://doi.org/10.1371/journal.pone.0118571

Ozyurt G, Ergin A (2010) Improving coastal vulnerability assessments to sea-level rise: A new indicator-based methodology for decision-makers. Ocean Coast Manag 53(10):645–654

Patel A, Yadav SM, Teegavarapu R (2024) Enhancing real-time flood forecasting and warning system by integrating ensemble techniques and hydrologic model simulations. J Water Clim Chang 15(9):4307–4327. https://doi.org/10.2166/wcc.2024.052

Polyakov M, Zhang D (2008) Population growth and land use dynamics along urban–rural gradient. J Agric Appl Econ 40(2):649–666. https://doi.org/10.1017/S1074070800023919

Rahman MA, Rashed T, Khan MH (2011) Urban growth and natural hazards in Dhaka city, Bangladesh: An integrative analysis. Nat Hazards 58:651–675

Rajan KS, Shibasaki R (2013) Urban growth analysis using remote sensing data. Urban Remote Sens: Monit, Synth Model Urban Environ 2:57–72

Ramachandran RM, Jayappa KS, Subramanya K (2020) Coastal vulnerability assessment for climate change using remote sensing and GIS. Environ Monit Assess 192, Article 562. https://doi.org/10.1007/s10661-020-08545-y

Richards DR, Friess DA (2016) Rates and drivers of mangrove deforestation in Southeast Asia, 2000–2012. Proc Nat Acad Sci 113:344–349. https://doi.org/10.1073/pnas.1510272113

Rodriguez JP, Beard TD Jr, Bennett EM, Cumming GS, Cork S, Agard J, Dobson AP, Peterson GD (2005) Ecosystem services and their link to human well-being. Ecol Econ 64(2):441–445. https://doi.org/10.1016/j.ecolecon.2007.01.009

Schneider A, Friedl MA, Potere D (2010) Mapping global urban areas using MODIS 500-m data: New methods and datasets based on 'urban ecoregions.' Remote Sens Environ 114(8):1733–1746

Seto KC, Fragkias M, Guneralp B, Reilly MK (2011) A meta-analysis of global urban land expansion. PLoS ONE 6(8):e23777

Simard M, Fatoyinbo L, Smetanka C, Rivera-Monroy V, Castañeda-Moya E, Thomas N, Van der Stocken T (2019) Mangrove canopy height globally related to precipitation, temperature and cyclone frequency. Nat Geosci 12:40–45

Spalding MD, McIvor AL, Beck MW, Koch EW, Möller I, Reed DJ, Rubinoff P, Spencer T, Tolhurst TJ, Wamsley TV, van Wesenbeeck BK, Wolanski E, Woodroffe CD (2014) The role of ecosystems in coastal protection: Adapting to climate change and coastal hazards. Nat Clim Chang 4:1045–1049. https://doi.org/10.1038/nclimate2271

Sudhira HS, Ramachandra TV, Jagadish KS (2004) Urban sprawl: metrics, dynamics and modelling using GIS. Int J Appl Earth Obs Geoinf 5(1):29–39

Taubenböck H, Wegmann M, Roth A, Mehl H, Dech S (2012) Urbanization in India—spatiotemporal analysis using remote sensing. Comput Environ Urban Syst 36(3):297–306

Thieler ER, Himmelstoss EA, Zichichi JL, Miller TL (2005) Digital shoreline analysis system (DSAS) version 3.0. USGS Open-File Report 2005–1304

United Nations Environment Programme (UNEP) (2021) Adaptation gap report 2021: The gathering storm—adapting to climate change in a post-pandemic world. UNEP, Nairobi. https://www.unep.org/resources/adaptation-gap-report-2021

United Nations Conference on Trade and Development (UNCTAD) (2023) Review of maritime transport 2023. UNCTAD, Geneva. https://unctad.org/webflyer/review-maritime-transport-2023

UN-SPIDER (2016) United Nations international conference on space-based technologies for disaster risk reduction - "Understanding disaster risk". 19 to 21 September 2016. Available at: https://www.un-spider.org/sites/default/files/Information%20note_Beijing%202016_FINAL.pdf

Uribe-Martínez A, Espinoza-Tenorio A, Cruz-Pech JB, Cupido-Santamaría DG, Trujillo-Córdova JA, García-Nava H, Flores-Vidal X, Gudiño-Elizondo N, Herguera JC, Appendini CM, Cuevas E (2024) An affordable operational oil spill monitoring system in action: a diachronic multiplatform analysis of recent incidents in the southern Gulf of Mexico. Environ Monit Assess 196(11):1069. https://doi.org/10.1007/s10661-024-13161-5. PMID: 39419911; PMCID: PMC11486827

Van Beijma S, Comber A, Lamb A (2014) Random forest classification of salt marsh vegetation habitats using seasonal dynamic vegetation indices from time series RapidEye data. Remote Sensing 6(7):6662–6683

Visbeck M, Kronfeld-Goharani U, Neumann B, Rickels W, Schmidt J, van Doorn E, Matz-Lück N (2014) Securing sustainable ocean services. Nat Clim Chang 4:504–508. https://doi.org/10.1038/nclimate2280

Wang M et al (2022) Machine learning-based coastal water quality retrieval using Sentinel-2 data. Remote Sens Environ 269:112786

World Travel & Tourism Council (WTTC) (2022) Global economic impact & trends 2022. WTTC, London. https://wttc.org/research/economic-impact

Chapter 21
Geospatial and Statistical Investigation of Salt Pan Dynamics and Coastal Landscape Changes in the Indian Sundarbans

Swapenendu Lodh, Bipasha Chakraborty, Mousumi Roy, and Ujjwal Bhandari

21.1 Introduction

Coastal ecosystems are composed of salt marshes that originated in low-energy intertidal zones in the interior of estuaries and bays. Several ecosystem services encompass salt marshes, vegetated surfaces due to their location (Costanza et al. 2008; Foster et al. 2013; Möller et al. 2014; Watson et al. 2017). Coastal biodiversity allocates vital roles with salt marshes (Lefeuvre et al. 2003), protecting sea level in the coastal zone (Morris et al., 2002) and contamination (Nelson and Zavaleta 2012; Ahmed et al., 2022), and moderating greenhouse gas emissions (Chmura et al. 2003). Indeed, growing scientific literature and societal interest in storm surge attenuation and nature-based flood defence programs, i.e., storm surge flood risks mitigation and restoration by the marsh and mangrove, highlight the salt marshes (Cheong et al. 2013; Temmerman et al. 2013; Fagherazzi 2020; Sutton-Grier et al. 2015). However, these wetlands are becoming increasingly vulnerable to global climate change and human activities (Erwin 2008; Kirwan and Megonigal 2013; Simas et al. 2001). The global salt marsh coverage has declined dramatically over the past few decades. Currently, less than 50% of the world's original salt marshes remain, and

S. Lodh
Department of Geography, Asutosh College, Kolkata, West Bengal, India

B. Chakraborty
Department of Geography, Visva-Bharati University, Santiniketan, West Bengal, India

M. Roy
Department of Geography, Diamond Harbour Women's University, Sarisha, West Bengal, India

U. Bhandari (✉)
Department of Geography, Syamaprasad College, Kolkata, West Bengal, India
e-mail: ujjwalcosmos@gmail.com

A. Nath et al. (eds.), *Geospatial Application of Coastal Hazards*, Springer Water,
https://doi.org/10.1007/978-3-032-20076-1_21

the loss continues at a rate of approximately 1–2% per year, causing CO_2 emissions around 60 M t yr.$^{-1}$ (Adam 1993; Cope et al. 2008; Olivier et al. 2012; Temmerman et al. 2013; Foster et al. 2013; Leonardi and Fagherazzi 2014, 2015; Bendoni et al. 2016; Leonardi et al. 2016b). Regional to global studies have estimated that sea-level rise (SLR) over the twenty-first century could induce the loss of 30–80% of coastal wetlands, including salt marshes and mangroves (Lovelock et al. 2015; Spencer et al. 2016; Schuerch et al. 2018; Thiéblemont et al. 2019). This would imply the loss of habitat for unique flora and fauna, and the loss of valuable ecosystem services (Brown et al. 2006; Barbier et al. 2011; Mudd et al. 2009; Kirwan and Mudd 2012; Pendleton et al. 2012; Barbier et al. 2013), such as carbon sequestration (McLeod et al. 2011) and flooding and shoreline erosion protection in the hinterland (Gedan et al. 2011; Shepard et al. 2011; Temmerman et al. 2013). Above the mean sea level salt marshes occur, their persistence depends on accretion under SLR (Fagherazzi et al. 2020). Marsh surface elevation results in a vertical rise in the form of the sum of erosion, compaction, sedimentation, shrinkage of mineral and organic sediments, and swelling (Plater et al., 1999; Van de Koppel et al. 2005; Deegan et al. 2012; Fagherazzi et al. 2012; Kirwan et al. 2016; Leonardi et al. 2016a; Ganju et al. 2017). Marsh vegetation die-off is the result of "drowning" of a marsh, and as a result, it turns into salt pans which may lead to flooding as well as if sea level rises, not keep pace with SLR by increasement of accretion (Cooper et al. 2008; Morris et al. 2002; Kirwan and Megonigal 2013; Blankespoor et al. 2014; Schepers et al. 2017). Today, the global mean SLR rate is about 0.25 cm year^{-1} (Nicholls et al. 2021), which is projected to accelerate to values of 1 to 2 cm year^{-1} by 2100 (Rahmstorf 2007; Church et al. 2013). Over recent years, several meta-analyses have compiled published accretion data for salt marshes and mangroves (Lovelock et al. 2015; Kirwan et al. 2016; Crosby et al. 2016; Jankowski et al. 2017).

Salt-rich groundwater with topographic depressions is the criterion for forming a special type of terrestrial wetland, known as salt pans, which are typically found in arid climates (Lowenstein et al. 1985; Shaw et al. 2011). Salt pans are controlled by their formation in a specific environment of fluvial or aeolian processes as well as climate because their morphological, sedimentary, and mineralogical characteristics can be later modified (Bryant 1999). However, at the global scale, their number is declining mainly due to direct human intervention in their hydrology and climate change (Nayak et al. 1989; Williams, 2004; Picado et al. 2009; Wurtsbaugh et al. 2017; Silva et al. 2022; Wasserman et al. 2022). Habitat loss has accelerated in recent years due to processes such as eutrophication, fragmentation, long-term drying, overgrowth with vegetation, and siltation (Horváth et al. 2020). Land use change due to excessive groundwater drainage is largely connected (Zimmermann-Timm and Teubner 2021; Mitter et al., 2021). Marsh vegetation in open water with higher ratios instigate storm surge attenuation (Loder et al. 2009; Temmerman et al. 2012; Zhang et al., 2012; Shen et al. 2012; Hu et al. 2015), implying gradual marsh degradation around the world, which are observed in patchy patterns (Westerink 2010; Wamsley et al. 2009; Zhang et al. 2012; Barbier et al. 2013; Liu et al. 2013; Hu et al. 2015; Schepers et al. 2017b), lead to the loss of the storm protection function of marshes (Temmerman et al. 2012). Surge attenuation influences tidal channels, and numerical simulations show that flood propagation is facilitated through the channels into deeper landward channels, leading to reduced storm surge height (Temmerman et al. 2012; Stark et al. 2016). In comparison to other wetlands, and from a morphological point of view, salt marshes are more resistant to the impact of storms; this has been mainly attributed to the strength of the soil, which is defined by the deeper root systems in the coastal areas with floating marshes and inland wetlands (Morton and Barras 2011). Fagherazzi 2014, stated that very violent storms can be prevented without damage if effective buffering of vegetated shorelines is formed from a morphological standpoint. Moderate storms will be less effective; vice versa, unvegetated surfaces collapse under high energy, but in mild weather conditions, they absorb energy efficiently (Morton and Barras 2011). For salt marshes, ponds generated during storms are generally much smaller and less frequent than brackish and freshwater marsh ponds; they also maintain a more amorphous shape (with no preferential direction) in comparison to the more elongated ponds frequently found in freshwater marshes (Barras et al. 2008). These ponds (Fig. 21.1) are more easily formed with lower terrain, and strong wind-driven currents can erode surface sediments (Morton and Barras 2011). Sediment losses from the marshes change with subsequent storms (Mariotti 2016). Once ponds are formed, they may continue to expand even if the surrounding marsh platform maintains equilibrium with sea-level rise and wave dynamics. As ponds enlarge, they can eventually connect with nearby tidal channels (Mariotti and Fagherazzi 2013; Schepers et al. 2017a). When such hydrological connectivity occurs, a pond can potentially recover if its bed elevation remains above the critical threshold for vegetation growth or if sediment deposition exceeds the rate of sea-level rise. Conversely, when these conditions are not met, wind-induced wave action can lead to sustained erosion, with the eroded sediments either redeposited within the marsh system or exported through tidal channels (Mariotti and Fagherazzi 2013). Consequently, depending on the interplay between biological activity and sedimentation processes, pond formation and expansion can represent either a reversible or irreversible trajectory, with recovery occurring only when the pond bed accretes to an elevation comparable to the surrounding marsh platform (Mariotti and Carr 2014; Leonardi and Fagherazzi 2014, 2015; Mariotti 2016; Leonardi et al. 2016a, b).

Extreme and average storm conditions change coastal areas due to a changing climate (Zhang et al. 2001; Webster et al. 2005; Bacmeister et al. 2016). For example, according to the Intergovernmental Panel on Climate Change (Meehl et al. 2007; Pachauri 2014), it is almost certain (99–100% probability) that cyclone activity has increased in intensity. Regarding extremes, it is likely (66–100% probability) that extreme sea levels, such as those during storm surges, have risen globally since 1970. This trend has mainly been linked to rising mean sea levels, although more research is needed to fully distinguish the effects of global sea level rise from local changes in coastal systems (Pachauri 2014; Leonardi and Plater 2017). Overall, the IPCC (Meehl et al. 2007) suggests that future cyclones will likely feature higher peak wind speeds and increased near-storm rainfall, with more intense storms occurring despite

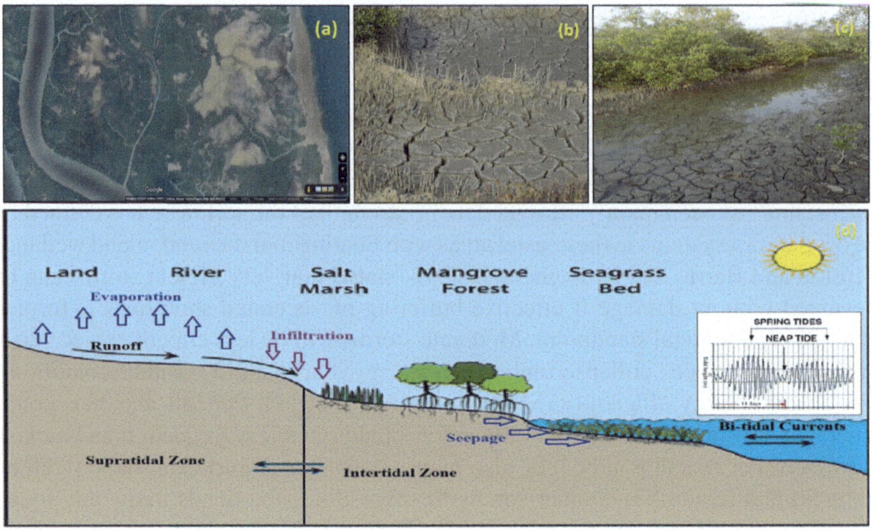

Fig. 21.1 **a** Bird's-eye view of salt marsh and creeks with salt pans, **b** pond-like formation of salt pans, **c** salt pans are formed along the track of the creek, **d** schematic diagram of tidal and marsh coupling with climatic components to generate domains of salt pans

a likely decrease in the total number of storms. On a regional level, it has been shown that storm surge extremes could significantly increase by the end of this century using a barotropic surge model and global conditions of the North Sea coast captured with the IPCC A2 SRES scenarios for 1961–1990 and 2071–2100 (Woth et al. 2006).

However, episodic surface flooding presents considerable challenges for consistent monitoring and analysis of playa environments (Bryant 1993; Ghrefat and Goodell 2011). Remote sensing offers an effective approach to overcome these limitations by providing extensive spatial coverage and multi-temporal observations. Numerous studies have demonstrated the utility of remote sensing for investigating playa surface dynamics, particularly through optical sensors operating in the visible–near infrared and short-wave infrared (VNIR–SWIR) spectral regions (400–2500 nm), which are well suited for analyzing pan surface properties. Previous research has shown that key mineralogical crust types—such as gypsum, carbonate, quartz, and halite—can be successfully identified using broadband multispectral data from platforms such as Landsat 7 Enhanced Thematic Mapper Plus (ETM+) (Epema 1992; Li et al. 2014) and Landsat 8 Operational Land Imager (OLI) (Flahaut et al., 2021). Comparative assessments have further revealed that Landsat imagery can effectively classify broad mineralogical surface types and produce results comparable to those derived from high spectral resolution sensors, including Hyperion and AVIRIS, when integrated with ground-based spectral data (Milewski 2020). The improved temporal resolution of Landsat 8 since its launch in 2013 (up to 22 acquisitions per year) enables detailed examination of seasonal salt-crust development processes at a spatial resolution of 30 m, which serves as a foundation for the present investigation.

21.2 Study Sites

The Sundarban is the largest continuous mangrove forest in the world, extending across the lower deltaic plains of India and Bangladesh. The Indian Sundarban region (Fig. 21.2) occupies the estuarine confluence of the Ganga, Brahmaputra, and Meghna rivers. It is bounded by the Hooghly River (a distributary of the Ganga) to the west, the Kalindi–Raimangal–Hariabhanga River system to the east, the Dampier–Hodges Line to the north and west, and the Bay of Bengal to the south. This deltaic landscape consists of an intricate network of mudflats and islands, formed through the accumulation of sediments transported from the Himalayan hinterland and dissected by anastomosing tidal channels. Geographically, the region extends from 21°40′04″°N to 22°09′21″°N and 88°01′56″°E to 89°06′01″°E, covering an area of approximately 9630 km^2 (Spalding et al. 2010), representing about 3% of the world's total mangrove area (Biswas et al. 2007). The broader Sundarban system spans nearly 10,000 km^2 (Giri et al. 2007), with around 60% in Bangladesh and 40% in India (UNESCO-WHC, 2005). The Indian portion comprises 54 inhabited islands (~5400 km^2) and 48 reserve forest islands (~4200 km^2), distributed across six southeastern blocks of North 24 Parganas and thirteen southern blocks of South 24 Parganas districts in West Bengal.

The Ganga and Brahmaputra rivers discharge mean monthly freshwater flows of approximately 12,000 m^3 s^{-1} and 18,100 m^3 s^{-1}, respectively (Global Runoff Data Centre, 2000), transporting sediment loads of about 729–744 × 10^6 tons yr.$^{-1}$ (Singh et al. 2007) and 595–672 Mt. yr.$^{-1}$ (Darby et al., 2018). Roughly one-third of these sediments are deposited within the delta and floodplain systems (Goodbred and Kuehl 2000), while only ~21% reach the delta front (Allison 1998). The alternating phases of monsoonal (May–November) sediment loading and post-monsoonal unloading induce vertical elastic deformation of up to 6 cm annually across the delta surface (Steckler et al. 2010). The Ganga–Brahmaputra Delta, with a gradient of approximately 5×10^{-5}, is one of the flattest deltas globally, characterized by vast low-lying floodplains. The Sundarban Delta, largely tidally influenced, lies within 2 m above mean sea level (MSL), encompassing more than 6000 km^2 of intertidal land and an additional 2000 km^2 below MSL, protected by embankments (Brakenridge et al. 2013). The region experiences recurrent saline water intrusion, freshwater flooding, and inefficient drainage, leading to persistent natural hazards (Brouwer et al. 2007; Brakenridge et al. 2013). Most sediments are deposited on the subaqueous delta front, while only a small fraction is redistributed by tides and waves into inactive tidal segments, resulting in surface accretion rates of approximately 0–10 mm yr.$^{-1}$, depending on proximity to active channels (Rogers et al. 2013). Thus, the delta-front forests are prograding seaward at ~15 m yr.$^{-1}$ (Michels et al. 1998).

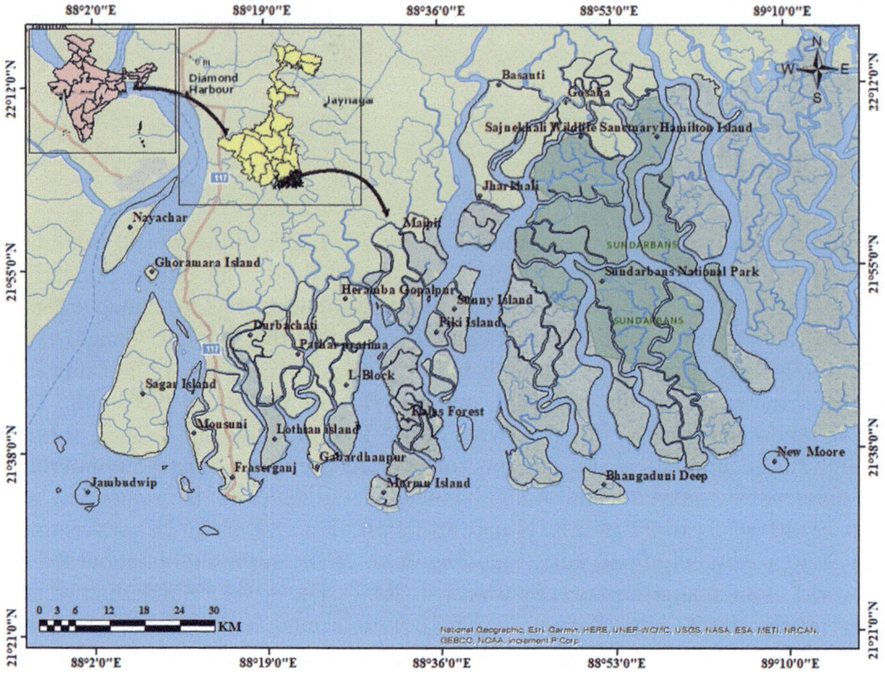

Fig. 21.2 Overview of the Indian Sundarban region

Sundarban consists of 106 islands and supports a good amount of biodiversity with 32 true mangroves and 62 mangrove-associated species (Banerjee, 2013), 127 species of fish and 1287 species of other animals, including 186 species of birds and various types of micro-organisms. Sundarban mangrove forest acts as a barrier for the coastal population from the natural hazards (Das and Vincent 2009). The greater Sundarban region is strongly influenced by tidal processes, with tidal amplitudes within the estuaries ranging from 3.5 to 4 m and exhibiting seasonal variations between 1 and 6 m (Islam et al. 2014). The Hooghly River forms the western boundary of the system, while the Raymangal River defines its eastern limit. The Indian part of the Sundarban supports a population of approximately 4.5 million (Census of India, 2011). Geomorphologically, the delta can be divided into four distinct zones from north to south: the inactive (moribund) delta, mature delta, tidally active delta, and subaqueous delta. Large portions of the first three zones have been reclaimed from the original mangrove forest to support human habitation and agriculture.

Spatial variations in freshwater discharge, coupled with tectonic uplift in the west and subsidence from sediment compaction and human activities in the east, have created a pronounced salinity gradient across the delta. The eastern and western sectors exhibit hyposaline conditions due to substantial freshwater inputs from the Himalayan-fed rivers, whereas the central region remains hypersaline as a result of elevated topography and limited freshwater contribution from the Matla, Bidyadhari, and Harinbhanga rivers (Blasco et al. 1996; Goodbred and Kuehl 2000; Syvitski et al. 2009; Spalding et al. 2010; Hanebuth et al. 2013).

A significant geomorphic feature of the region is the "Swatch of No Ground," a deep submarine canyon located near the western part of the delta, approximately 30 km south of Dublar Char Island at the Raimangal River estuary (21°–21°22′°N) in the Bay of Bengal. This canyon, which incises the continental shelf to within 40 km of the shoreline, serves as an active sediment conduit to the deep sea and plays a crucial role in trapping westward-migrating sediments along the subaqueous delta front (Allison 1998).

In addition to hosting nearly one-third of the world's mangrove tree species, the Sundarban supports exceptionally high biodiversity, encompassing more than 200 additional plant species, over 400 species of fish, more than 300 species of birds, 35 species of reptiles, 42 species of mammals, and numerous benthic invertebrates, fungi, and bacteria (Gopal and Chauhan 2006; IUCN 2001). The Sundarban forest is broadly categorised into two major reserves, viz. Sundarban Biosphere Reserve and Sundarban Tiger Reserve. The tiger reserve was further categorised into two major divisions of National Park East and National Park West. The islands lying on the western bank of the river Matla are under the Sundarban Biosphere Reserve (SBR), while the complete eastern part of the forest is under the Sundarban Tiger Reserve (STR). The Sundarban comprises four key protected areas—Sundarban National Park and the Sundarban West, South, and East Wildlife Sanctuaries—which were designated a UNESCO World Heritage Site in 1987 (Giri et al. 2007) and recognized as a Global Biosphere Reserve in 1989 (Bandyopadhyay, 2012) due to their unique and highly diverse ecosystems. Despite its location near the Tropic of Cancer, the region experiences mean maximum temperatures of approximately 35 °C (Banerjee, 2002), and monsoonal rainfall dominates the climate, with an average annual precipitation of 1920 mm (Gopal and Chauhan 2006). Intense summer storms can produce instantaneous wave surges of up to 7.5 m (Seidensticker 1986), leading to coastal erosion, embankment breaches, and sudden saline water intrusion.

The Sundarban's geographic position along the cyclone-prone Bay of Bengal makes it particularly vulnerable to frequent tropical storms. However, the extensive mangrove forest serves as a natural protective barrier, reducing the impact of such hazards. Nevertheless, this ecologically productive and globally significant ecosystem is increasingly threatened by climate change, sea-level rise, and associated environmental stressors.

21.3 Materials and Methods

The methodology flowchart (Fig. 21.3) is prepared to evaluate coastal salt pan dynamics. The methodology begins with data acquisition, where Landsat satellite imagery data (2000, 2010, and 2024) are utilised to gather high-resolution moisture and vegetation information.

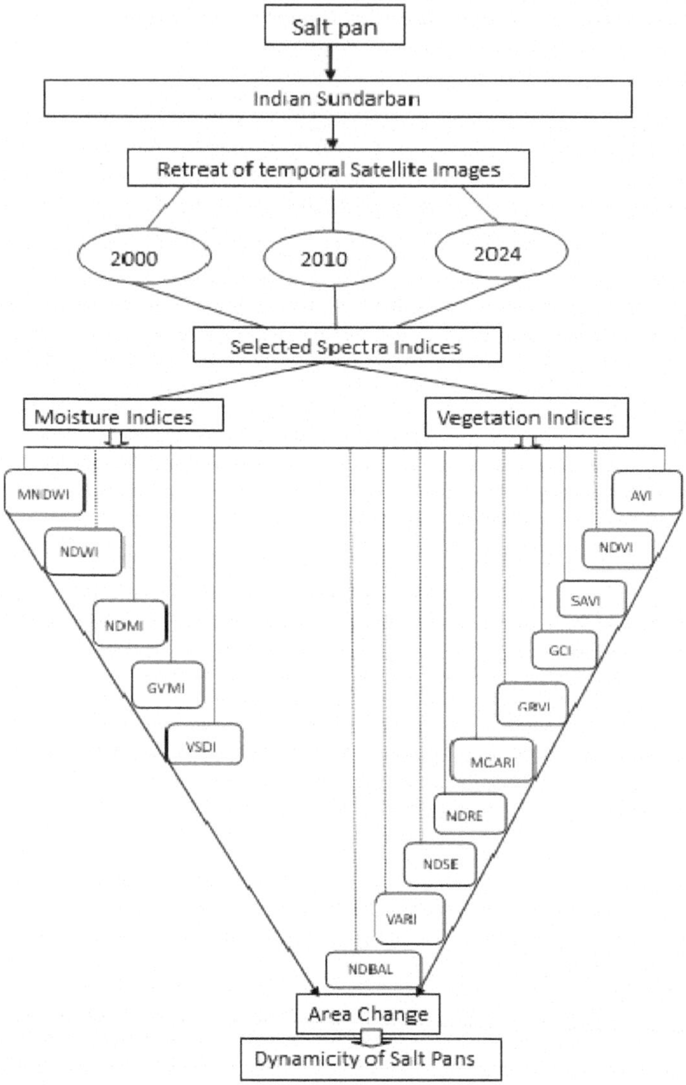

Fig. 21.3 The stepwise methodological framework adopted for this study

These methods have been followed, i.e., preprocessing, removal of cloud cover, and clipping the area of interest (AOI), to ensure data quality and relevance for the Indian Sundarban coastal environment. The next step is the calculation of 15 Spectral Indices, such as AVI, SAVI, NDMI, NDWI, etc., which are derived to assess vegetation health, soil salinity, and moisture dynamics. To predict soil salinity distribution, these methods have been used in various regions (Zhang et al. 2011; Taghadosi and

Hasanlou 2017; MH Yossif 2017; Allbed et al. 2017). Subsequently, temporal analysis of the above-mentioned parameters from 2000 to 2024 is conducted to detect changes over time in the remote sensing techniques. The spatial analysis tool in ArcGIS 10.3 was used to extract those parameters. The processed data is then integrated into class-wise area change detection, which employs geospatial modelling and user-friendly interfaces to provide actionable insights into the dynamicity of salt pans. Finally, monitoring parameters, which integrated the developed system between the salt pan area dynamicity with cyclonic surges.

21.3.1 Data Acquisition and Processing

Based on the atmospheric windows, two different frequency domains can be considered to acquire high-resolution images of salt pans: the optical domain and the microwave domain (Rees 2013). Both optical imagery (Heintzman et al. 2017; Lefebvre et al. 2019; Castañeda et al. 2005; Bowen et al. 2017; Chiloane et al. 2025; Bryant et al. 2003; Li et al. 2019) and microwave imagery (Schlaffer et al. 2022; Hess et al. 2003; Prigent et al. 2007; Reschke and Hüttich 2014; Chew and Small 2020; Schlaffer et al. 2016) have been widely used for water body and soil salinity monitoring. The crucial procedure for deriving the inundation state of salt pans, independent of the wavelength, is pixel-wise recognition of spectral signatures of water and non-water surfaces (Jensen et al. 2007). This is based on the unique physical properties of water, desiccated (saline) surfaces, and the riparian zone surrounding the salt pans (Jensen et al. 2007).

To evaluate coastal salt pan dynamics, we utilised satellite imagery data that covered the study area. Specifically, Landsat 8 Collection 2 Level-2 Surface Reflectance data, Landsat 7 ETM+ data, and ETM+ PAN data with a temporal data accordingly for 2024, 2010, and 2000 with a spatial resolution of 30 m, was obtained from the United States Geological Survey (USGS) Earth Resources Observation and Science (EROS) database through the Google Earth Engine (GEE) platform. Optical remote sensing is a passive technique that relies on solar radiation (Booysen et al. 2021). It is based on electromagnetic radiation frequencies (Fig. 21.4) ranging from 0.3 to 15 m (visible: 0.38 to 0.75 m; near-infrared (NIR): 0.75 to 1.4 m; short-wave infrared (SWIR): 1.4 to 3.0 m) (Kerekes et al. 2008), although there are different spectral subdivisions (Rees 2013). To retrieve the salt pan water surface extent, it is necessary to differentiate between the two main land surface constituents in and around the salt pan basins, i.e., vegetation and (saline) soil from water.

Landsat bands for identifying water pixels from non-water pixels and vegetation to non-vegetation pixels can be used separately (Table 21.1) or in combination. We utilized fifteen indices that are suitable for identifying salt pans.

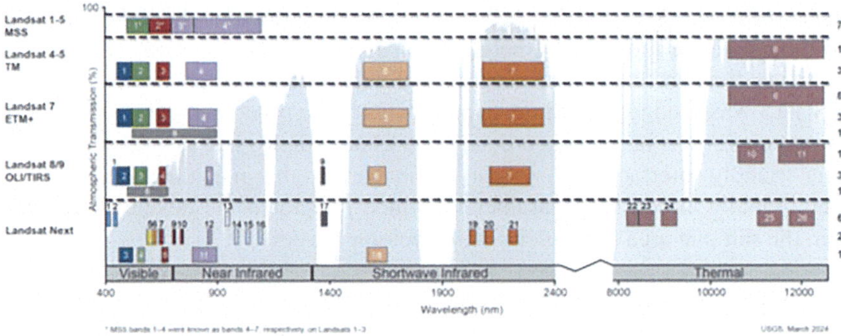

Fig. 21.4 Comparison of landsat spectral bands based on the USGS spectral library (web address: https://www.usgs.gov/landsat-missions/landsat-5 (accessed on 29th July 2025))

21.3.2 Tropical Cyclone-Related Storm Surges and Their Relation With Salt Pans

Tropical cyclones (TCs) are one of the catastrophic natural disasters that affect many countries around the globe, causing considerable loss of human lives, flora and fauna, and destruction to coastal areas' property (Singh et al. 2019). Cyclones are a very common phenomenon in the Bengal delta. It is noticeable that the NIO (North Indian Ocean) basin accounts for 7% of global cyclonic events, which is nearly 5 times higher than the Arabian Sea. Favourable SST conditions with larger vorticity and higher relative humidity change the nature of these cyclones and make the coastal area more vulnerable due to the production of rainfall during their landfall (Wahiduzzaman and Yeasmin 2019; Bhardwaj et al. 2023). Presently, the Indian Sundarban region is facing a critical situation due to such multi-hazard events every year (Table 21.2).

During the last decade, multiple severe cyclones (Fig. 21.5) and their aftermath (other extreme events like salt pans) have made the situation deteriorate. Though in some cases it gives positive feedback, like in the case of a long-time span after a storm surge increases the salinity of the depression zones of the islands, which enhances salt content in these zones, which affects mangroves, and as a result, these salt pans become a biological hazard (Medina-Pérez et al. 2019; Parida et al. 2013; Escapa et al. 2015; Omar 2019). On the other hand, if a cyclonic surge occurs, the overtopping surge water regains its previous salinity and which reduces the salt pan areas (Mclvor et al. 2015; Passeri et al. 2015; Yu et al. 2016). It should also be noted that if an island spatially increases its salt pan area, then the island decreases, and if the salt pan area deteriorates due to flooding water, the island area is accreted with the increasing mangrove succession (McKee et al. 2012; Mitra 2013; Hossain et al., 2023).

Table 21.1 Spectral remote sensing indices for identifying salt pan dynamics

Spectra index bases	Index number	Formula	Source
Moisture indices	Modified normalized difference water index (MNDWI)	(Green − SWIR)/(Green + SWIR)	Du et al. (2016); Tassew et al. (2025)
	Normalized difference water index (NDWI)	(Green − NIR)/(Green + NIR)	Rad et al. (2021)
	Normalized difference moisture index (NDMI)	(NIR − SWIR)/(NIR + SWIR)	Taloor et al. (2021); Xu et al. (2025)
	Global vegetation moisture Index (GVMI)	(NIR + 0.1) − (SWIR +0.02) (NIR + 0.1) + (SWIR +0.02)	Fan et al. (2018); Li et al. (2014)
	Visible and shortwave infrared drought index (VSDI)	1 − [(ρSWIR - ρblue) + (ρred - ρblue)	Xu et al. (2025); Gonnelli et al. (2023)
Vegetation indices	AVI (Advanced vegetation index)	[NIR * (1 − Red) * (NIR − Red)]^(1/3)	Loi et al. (2017), Syakur et al. (2025)
	Normalized difference vegetation index (NDVI)	(NIR − Red)/(NIR + Red)	Tripathi et al. (2014)
	Soil adjusted vegetation index (SAVI)	$((NIR - Red)/(NIR + Red + L))*(1+ L)$	Wilson et al. (2016)
	Green chlorophyll index (GCI)	NIR/Green − 1	Sasagawa et al. (2022); Qiao et al. (2022)
	Green-red vegetation index (GRVI)	(Green − Red)/(Green + Red)	Yin et al. (2022); Li et al. (2023)
	Modified chlorophyll absorption in reflectance index (MCARI)	((ρ700 − ρ670)—0.2 * (ρ700 − ρ550)) * (ρ700/ρ670)	Mielke et al. (2012); Sharifi et al. (2020)
	Normalized difference red edge index (NDRE)	(NIR − RedEdge)/(NIR + RedEdge)	Ljubicic et al. (2024)
	NDSI (normalized difference soil index)	(R − NIR)/(R + NIR)	Piyoosh and Ghosh (2018); Voitik et al. (2023)
	Visible atmospherically resistant index (VARI)	(Green − Red)/(Green + Red − Blue)	Chayhard et al. (2019); Emard (2024)
	Normalized difference bareness index (NDBal)	(Band5 − Band6)/(Band5 + Band6)	Guha and Govil, (2022)

Table 21.2 Cyclone-related chronological surge history of the study region from 2000–2024

Name of the cyclone	Date	Maximum velocity (kmph)	Surge height	Affected areas
Aila	23–5–2009	120	2–3 m, but with astronomical tides, the total water level exceeded 4 m in some areas	West Bengal, Bangladesh
Phailin	9–10–2013 13–10–2013	220	2–2.5 m above the astronomical tide, and the surge reached a maximum of 3.5 m	Andhra Pradesh, Odisha, east east-central Bay of Bengal, Andaman and Nicobar Islands, Gangetic West Bengal, Jharkhand, and Bihar
Hudhud	7–10–2014 14–10–2014	195	1.4 m	Andaman and Nicobar, Andhra Pradesh, Odisha
Komen	30–7–2015	76	(storm surge + astronomical tide) of about 2 m	Nagaland, Manipur, Mizoram, Tripura, Gangetic West Bengal, Myanmar, Bangladesh
Titli	08–10–2018 13–10–2018	126	1.0 m above astronomical tide.	Andhra Pradesh, Odisha, West Bengal, Bangladesh
Fani	26–4–2019 4–5–2019	230	storm surge of 0.5–1 meter, reported a maximum surge height exceeding 5 m in some locations	Odisha, Andhra Pradesh, West Bengal
Bulbul	05–11–2019 11–11–2019	135	1.0 to 2.0 m above the astronomical tide in low-lying areas of South and North 24 Parganas, and 0.5 to 1.0 m in East Medinipur, the Bangladesh coast near the landfall area	West Bengal, Odisha
Amphan	16–5–2020 21–5–2020	260	4–5 meters above the astronomical tide	West Bengal, Odisha
Yaas	23–5–2021 28–5–2021	130	2–4 meters above the astronomical tide inundates low-lying areas	West Bengal, Jharkhand, Odisha

(continued)

Table 21.2 (continued)

Name of the cyclone	Date	Maximum velocity (kmph)	Surge height	Affected areas
Jawad	2–12–2021 6–12–2021	120	4th Dec 2–4 m height is likely along the coast of Odisha and West Bengal, 5th Dec 0.5—1 m height above astronomical tide, low-lying areas of Odisha	Andhra Pradesh, Odisha, Gangetic West Bengal
Sitrang	22–10–2022 25–10–2022	110	storm surge of height about 1 m above the astronomical tide for the West Bengal coast, 2.0 m above the astronomical tide for the Bangladesh coast	Assam, Odisha, West Bengal, Bangladesh
Remal	24–5–2024 28–5–2024	110–120	80 % for the storm surge height exceeding 3 m	Odisha, West Bengal, Bangladesh
Dana	24–10–2024 26–10–2024	120	1–2 m above the astronomical tide to inundate low-lying areas of Kendrapara, Bhadrak, Balasore, and East Medinipur districts during the time of landfall	Odisha, West Bengal

Source Prepared by authors from different IMD datasets

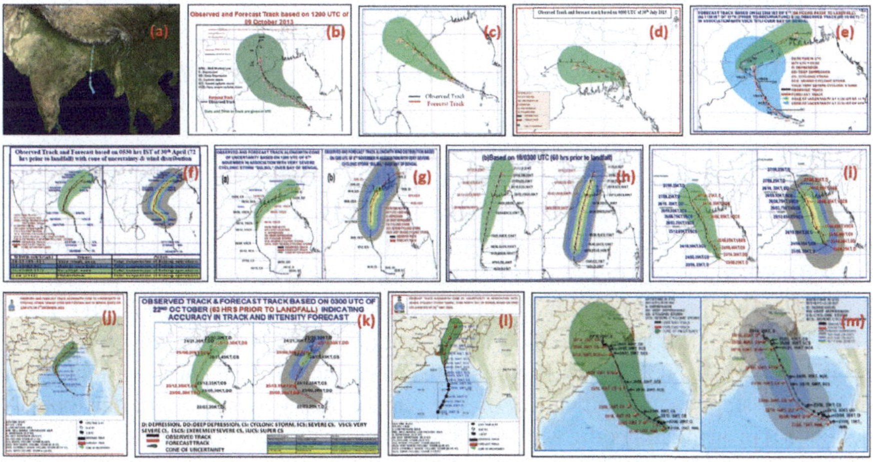

Fig. 21.5 Tropical cyclones of the BoB during 2000 to 2024 and the storm surge-related inundation: **a** Aila, **b** Phailin, **c** Hudhud, **d** Komen, **e** Titli, **f** Fani, **g** Bulbul, **h** Amphan, **i** Yaas, **j** Jawad, **k** Sitrang, **l** Remal, **m** Dana cyclone

21.3.3 Salt Pan Dynamics and Island Area Change

Land degradation, land subsidence, soil salinization, groundwater contamination, sea-level rise, and coastal flooding are serious challenges in India's coastal region, which is home to millions of people. Rising sea levels, ocean tides, and strong waves, with their dynamic location shaping the region by continuous erosion and accretion, resulting in a significant variation from decades ago (Akter et al. 2018; Berus et al. 2017). The morphology of the island has been shaped and reshaped by fluvial and marine processes and influences. The current study focuses on identifying the changing pattern of salt pans with the changing vegetation, which is controlled by climatic and other extreme events. Finally, the area changes for the last 25 years are assembled for all 15 parameters and identify the dynamicity pattern of these salt pans in these reclaimed and non-reclaimed parts of the Indian Sundarban (Bandyopadhyay 2019; Kar et al. 2020; Mandal et al. 2021; Jayanta 2023).

All significant information on geomorphological and geoenvironmental issues should be collected to ensure the long-term development of this vulnerable area. This analysed data can then be used to guide strategic planning and development at all levels to optimize land use, determine appropriateness, increase biological and other resource production, and achieve long-term sustainability. The current research focuses on the change detection and the overall salt marsh environments of the area as an integral part of understanding the area's sustainability and future impacts.

21.4 Results and Analysis

21.4.1 Spectral Indices Change Over The Period 2000–2024

The spatial distribution of moisture and vegetation content was shown in Figs. 21.6, 21.7, and 21.8 to demarcate salt pan distribution for 2000, 2010, and 2024. These indices have given good correlation for the salt-affected areas and salt pan analysis. These indices were chosen for salinity classification and quantification (Poenaru et al., 2015). Recently, these methods have been used worldwide to predict soil salinity distribution (Zhang et al. 2011; Taghadosi and Hasanlou 2017; MH Yossif 2017; Allbed et al. 2017). Sample years computed by the spatial analysis tool in ArcGIS to extract NDVI, SAVI, etc. At first, we included all these variables and conducted a stepwise identification of salt pans in every aspect, and after that, we calculated it in an area-specific way. This predicted model best correlated with coastal area change for this region.

Here, five Moisture indices are selected, like MNDWI, NDWI, NDMI, GVMI, and VSDI (Latha, 2021; Xu et al., 2025). These are derived from remote sensing data, can be valuable tools for identifying and mapping salt pans. These indices, which reflect the amount and distribution of water in the soil and on the surface, can highlight areas with high salinity and specific hydrological conditions characteristic

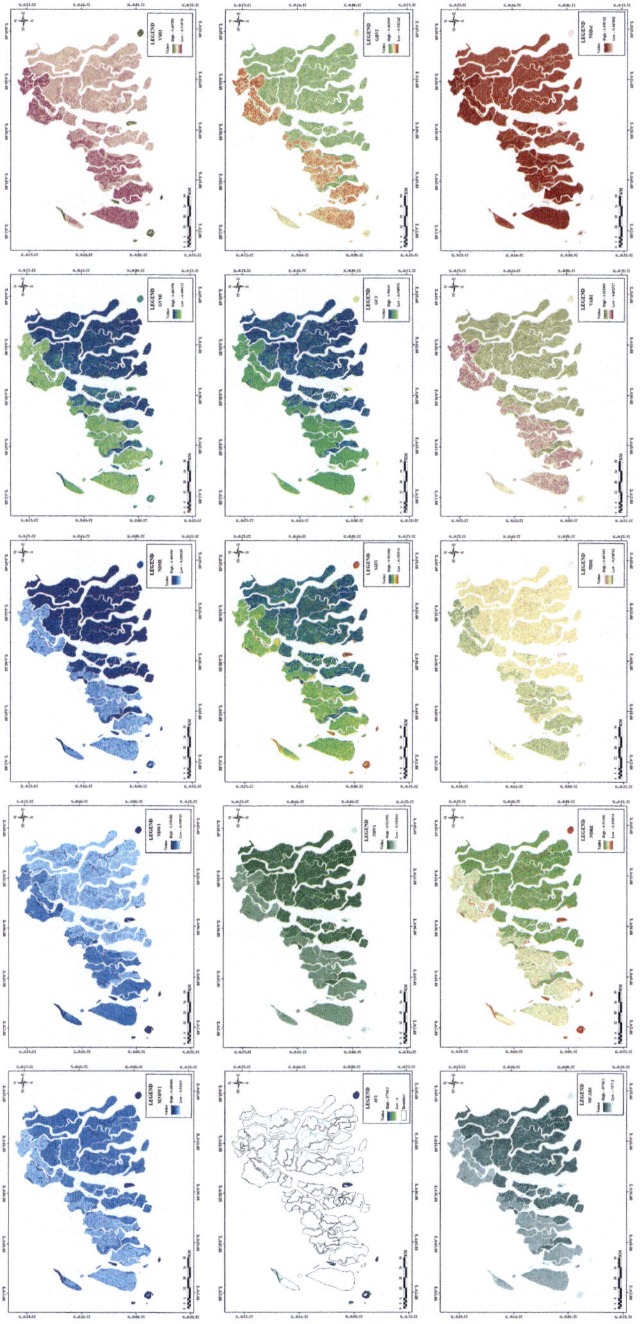

Fig. 21.6 Identifying salt pan dynamism with Spectral indices for 2000 (row-wise indices are from left to right accordingly): MNDWI, NDWI, NDMI, GVMI, VSDI, AVI, NDVI, SAVI, GCI, GRVI, MCARI, NDRE, NDSI, VARI, NDBal

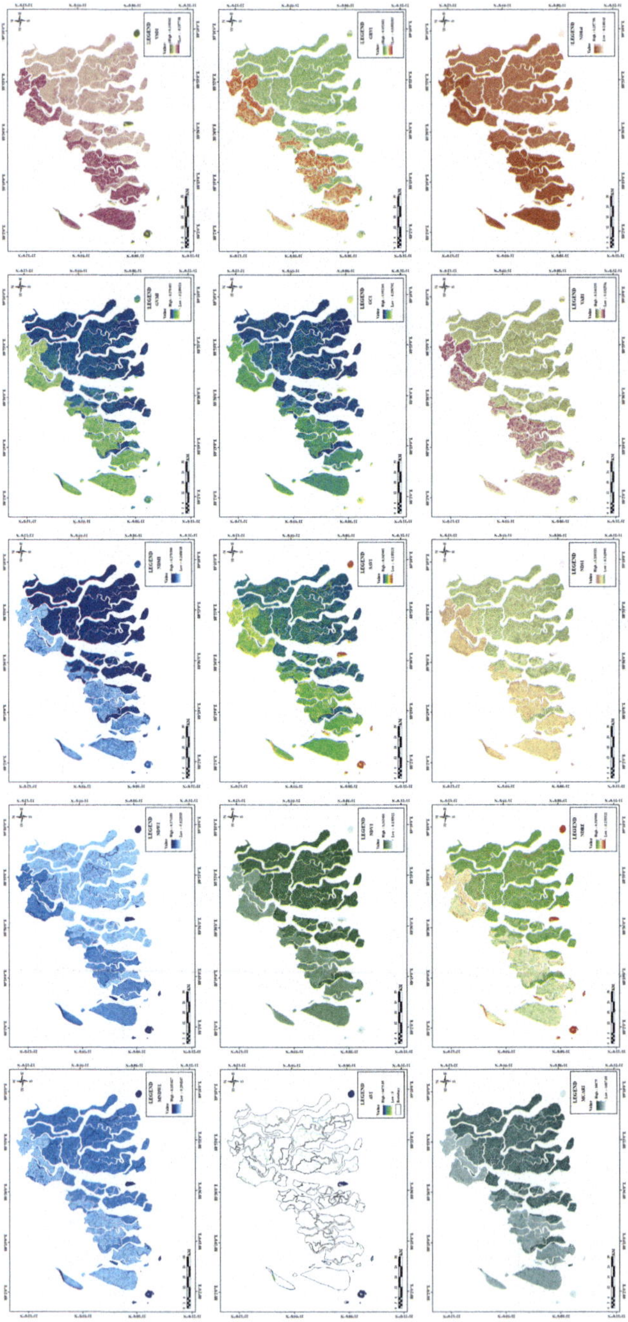

Fig. 21.7 Identifying salt pan dynamism with Spectral indices for 2010 (row-wise indices are from left to right accordingly): MNDWI, NDWI, NDMI, GVMI, VSDI, AVI, NDVI, SAVI, GCI, GRVI, MCARI, NDRE, NDSI, VARI, NDBal

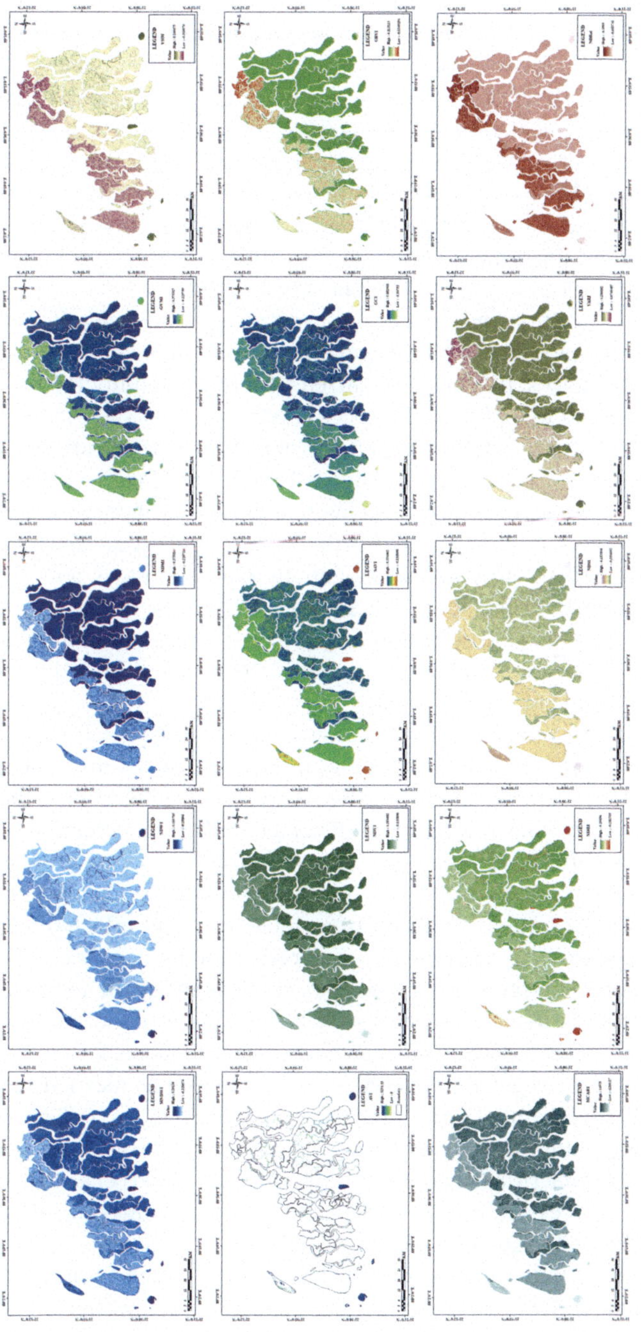

Fig. 21.8 Identifying salt pan dynamism with Spectral indices for 2024 (row-wise indices are from left to right accordingly): MNDWI, NDWI, NDMI, GVMI, VSDI, AVI, NDVI, SAVI, GCI, GRVI, MCARI, NDRE, NDSI, VARI, NDBal

of salt pans. Moisture indices can indicate areas with low soil moisture and high surface dryness, which are common features of salt pans (Wang et al. 2011; Hasab et al. 2020; Rasheed et al. 2022). Salt pans often exhibit sparse or unique vegetation cover due to the harsh environmental conditions. Moisture indices, when combined with vegetation indices (like NDVI), can help identify these areas and differentiate them from surrounding vegetation. NDWI is sensitive to changes in water content and can help identify areas with high surface moisture, which can be indicative of salt pan formation (Milewski et al. 2020; Peplouw 2024). NDMI is sensitive to vegetation water stress and can help identify areas where vegetation is stressed due to high salinity, a common feature of salt pans (Scardino et al. 2023; Pratama and Venugopal 2024; Mishra et al. 2025; Chauhan et al. 2025).

In other cases, Vegetation indices can be used to indirectly identify salt pans by detecting areas of sparse or stressed vegetation, which often correlate with high salinity. Soil salinity indices, like the Normalized Difference Salinity Index (NDSI), are specifically designed to highlight the spectral reflectance of salt-affected areas, particularly those with less vegetation. Vegetation indices like the Normalized Difference Vegetation Index (NDVI) and Soil-Adjusted Vegetation Index (SAVI) show a negative correlation with soil salinity, meaning that as salinity increases, the values of these indices decrease (Nguyen et al. 2020; Teshaev et al. 2020; Salem and Jia 2024; Mehla et al. 2024). Salt pans are essentially bare soil areas with high salt concentrations. Therefore, vegetation indices like NDBal can be used to identify these areas by detecting the lack of healthy vegetation. Therefore, A combination of vegetation indices (5 selected) and salinity indices (10 selected) provides the most effective approach for identifying salt pans.

Changes in moisture indices and vegetation indices can be related to changes in salt pan area. Moisture content significantly influences soil salinity, and vegetation indices like NDVI show moderate negative correlations with salinity. Increased salinity can negatively impact vegetation health and distribution, potentially leading to changes in salt pan areas. Salt-affected soils can restrict vegetation growth due to the toxic effects of high salt concentrations on plant physiology. Changes in NDVI can indicate shifts in the extent or health of vegetation in salt-affected areas. For example, if a salt pan expands, you might see a corresponding decrease in NDVI in that area as vegetation struggles to survive. This study analyses the expansion of a salt marsh using spectral indices as a proxy for the condition of the salt marsh and found that the marsh expanded while some areas experienced dieback, demonstrating the relationship between vegetation health and salt marsh dynamics (Miller et al., 2017; Campbell 2018; Powell 2025). Changes in moisture indices (reflecting water availability) and vegetation indices (reflecting vegetation health) can be indicators of changes in salt pan area (Mandal 2019; Milewski, 2020; Shokri et al. 2024; Wang et al. 2024). These indices can help in understanding the dynamics of salt-affected areas and their impact on vegetation and land use.

Salt pan dynamicity in the Indian Sundarbans refers to the changing patterns of salinity within the mangrove ecosystem (Biswas et al. 2017; Chowdhury et al. 2016; Islam et al. 2023; Samanta et al. 2021). This dynamic is influenced by factors like sea-level rise, river discharge, and human interventions, leading to shifts in salinity levels across different parts of the delta. These changes directly impact mangrove vegetation, favouring salt-tolerant species in some areas while potentially harming others. Rising sea levels contribute to increased salinity intrusion into the Sundarbans, affecting both surface and groundwater. Variations in freshwater input from rivers like the Ganges and Brahmaputra influence salinity patterns (Alam 1996; Hossain 2017; Das et al. 2021; Mukhopadhyay et al. 2024). Siltation of river channels can also alter water flow and salinity distribution. High evaporation rates, especially during the dry season, can concentrate salts in surface waters, creating hypersaline patches, etc. In some areas, particularly on islands like Ajmalmari, Henry's, and Patibunia, dense salt patches develop due to high evaporation and reduced freshwater input (Das 2015). These patches can negatively impact mangrove growth and ecological balance. Effective conservation and management depend on the dynamics of salinity, and it is crucial mangroves of the Sundarbans ecosystem.

Cyclones significantly impact salt pan dynamics in the Sundarbans by causing saline water intrusion, altering salinity levels, and affecting soil properties. These changes can lead to mangrove die-off, reduced blue carbon sequestration, and shifts in mangrove community ecology (Dutta et al. 2019; Das 2023; Mandal et al. 2023; Mandal et al. 2024; Roy et al. 2024). The increased frequency of cyclones, exacerbated by climate change and rising sea levels, further intensifies these effects, posing challenges for the region's ecology and human populations. The altered salinity can lead to a shift in the mangrove community, with more salt-tolerant species potentially replacing less tolerant ones (Liang et al. 2008; Parida and Jha 2010; Rahman 2020). The death of mangroves and changes in vegetation can reduce the Sundarbans' capacity to sequester blue carbon, a crucial ecosystem service. In essence, the Sundarbans experiences a dynamic interplay between cyclones, salinity, and vegetation, with climate change exacerbating these effects. Understanding and addressing these dynamics is crucial for the long-term ecological and socioeconomic well-being of the region.

21.4.2 Spectra Indices Change Statistics Over 2000 to 2024

In the Sundarbans, salt pans form naturally through the buildup of salt deposits on supratidal flats where saline water from tides and storm surges evaporates, leaving behind salt. Climate change impacts, including rising temperatures and declining rainfall, are altering these areas and impacting the mangrove ecosystem (Mitra 2013; Ward et al. 2016; Friess 2025). Human activities like deforestation for agriculture and aquaculture also contribute to increased salinity and changes in the salt pan areas. The Sundarbans, a deltaic region, experiences regular tidal inundation. High tides and storm surges bring saline water onto the land, particularly the supratidal flats. As the water recedes, or when it is trapped in depressions on the forest floor, it evaporates, leaving behind salt deposits (Chhabra 2022; Sharma and Naik 2024). Over time, this process of saline water intrusion and evaporation leads to a gradual accumulation of salt, forming salt pans (Asencio 2013). The Sundarbans are experiencing rising temperatures and declining rainfall, which impacts the delicate balance of the mangrove ecosystem. While mangroves are adapted to saline conditions, significant climate changes can stress these ecosystems and affect their health and distribution, potentially altering salt pan dynamics. Clearing of mangrove forests for agriculture, aquaculture, and other development projects reduces the natural buffering capacity of the mangroves against saltwater intrusion, enhancing the salinity problem and formation of salt pans. Overall, the salt pan areas in the Sundarbans are dynamic and are being impacted by a complex interplay of natural processes and human activities (Neogi et al. 2016; Ghosh et al. 2019; Bandyopadhyay 2019; Mondal et al. 2025). Climate change and human activities are both contributing to changes in salinity and land use patterns in the Sundarbans, leading to both expansion and shifts in salt pan areas, which are calculated in the study (Table 21.3).

The cumulative histograms and Chi-square values presented in the image are highly relevant to analyzing salt pan dynamics when applied to spectral indices derived from remote sensing data (Panuju et al. 2020). This approach allows for the quantification and statistical assessment of changes in salt pan environments over time or across different spatial areas. Once spectral indices are calculated for the pixels within a salt pan, their values form a dataset. Cumulative histograms (Fig. 21.9) can then be generated from these spectral index values to visualize their distribution across the salt pan area or over time.

A chi-square test (Eq. 21.1) can be used to compare cumulative histograms (also known as cumulative distribution functions or CDFs) to assess if two datasets have significantly different distributions. This involves comparing the observed cumulative frequencies from the histograms with the frequencies expected under a null hypothesis (Russell 2018). The chi-square statistic quantifies the discrepancy between observed and expected frequencies, and a p-value is calculated to determine the statistical significance of the difference. From the cumulative histogram, calculate the observed frequency, then make a difference with the expected frequency, square it, and from whole divide it by the expected frequency:

$$\chi^2 = \Sigma \left[(\text{Observed} - \text{Expected})^2 / \text{Expected} \right] \tag{21.1}$$

The graph displays multiple cumulative distribution functions (CDFs) or cumulative histograms, each representing a different distribution. The y-axis, labelled "Cumulative density," indicates the cumulative probability or proportion of data points less than or equal to a given value on the x-axis. The legend below the graph associates each coloured line with a numerical value and a Chi-square distribution, e.g., "0.005587—Chi-square (68.995)". This suggests that each curve represents the CDF of a Chi-square distribution with a specific degree of freedom (the number in parentheses), and the associated numerical value might represent a statistic or a parameter related to that specific Chi-square distribution, possibly a calculated Chi-square statistic or a p-value, depending on the context of the problem.

The cumulative distribution of a specific spectral index (e.g., NDSI) for different areas within the salt pan (e.g., highly saline zones vs. moderately saline zones). The cumulative distribution of a spectral index for the same area at different time points,

Table 21.3 Spatial distribution of different spectra indices for different classes (high to low)

Spectral index bases	Parameters	Indices from high to low (2000)				
		0	1	2	3	4
Moisture Indices	MNDWI	0.005587	722.0731	2657.096	136.1961	0.012876
	NDWI	1.225221	2424.07	980.1247	109.577	0.004832
	NDMI	0.003853	395.0024	1102.059	2017.146	0.777644
	GVMI	0.003853	394.9256	1101.951	2017.328	0.779492
	VSDI	0.005188	569.1256	2812.904	133.3393	0.017911
Vegetation Indices	AVI	20.05606	103.5555	0.025773	0.01417	0.001193
	NDVI	0.011478	120.3942	1199.893	2190.752	3.974717
	SAVI	0.011478	120.3942	1199.893	2190.752	3.974717
	GCI	114.1618	1448.356	1951.598	0.833261	0.018719
	GRVI	0.008596	0.101712	3515.866	0.096597	0.037105
	MCARI	830.376	2668.86	15.83365	0.020547	0.006854
	NDRE	0.011478	120.3942	1199.893	2190.752	3.974717

(continued)

Table 21.3 (continued)

Spectral index bases	Parameters	Indices from high to low (2000)				
		0	1	2	3	4
	NDSI	0.005188	569.1256	2812.904	133.3393	0.017911
	VARI	0.034882	0.118047	3514.075	1.830434	0.046159
	NDBal	0.017911	133.3393	2812.904	569.1256	0.005188
Spectral index bases	Parameters	Indices from high to low (2010)				
		0	1	2	3	4
Moisture indices	MNDWI	92.37969	977.8096	2294.145	83.67038	67.19661
	NDWI	395.2791	2053.071	914.4034	98.58828	53.33697
	NDMI	112.7741	720.1724	590.8734	1589.87	500.4968
	GVMI	112.7742	720.1648	590.8809	1589.871	500.4965
	VSDI	155.5245	2008.593	1229.767	72.51009	47.96639
Vegetation indices	AVI	1.323553	9.805038	26.29903	15.18191	41.76372
	NDVI	53.15897	232.9852	1088.849	1699.933	439.7408
	SAVI	53.15897	232.9852	1088.849	1699.933	439.7408
	GCI	95.89785	453.2371	1421.177	1537.281	7.186556
	GRVI	417.1402	1992.877	1104.27	0.053129	0.0009
	MCARI	643.1863	1199.569	1574.167	97.66168	0.136034
	NDRE	53.15895	232.9853	1088.849	1699.933	439.7405

(continued)

Table 21.3 (continued)

Spectral index bases	Parameters	Indices from high to low (2010)				
		0	1	2	3	4
	NDSI	439.7405	1699.933	1088.849	232.9853	53.15895
	VARI	348.3792	1333.312	1828.473	4.165216	0.0009
	NDBal	47.96642	72.51019	1229.766	2008.593	155.5244
Spectral index bases	Parameters	Indices from high to low (2024)				
		0	1	2	3	4
Moisture indices	MNDWI	6.789023	465.6429	939.1566	1949.112	154.134
	NDWI	951.4591	1867.109	509.4702	109.7149	77.35746
	NDMI	0.154325	346.6419	739.1339	1166.77	1261.6
	GVMI	0.154326	346.639	739.1276	1166.756	1261.624
	VSDI	2.270234	582.0805	2488.158	324.4446	117.8338
Vegetation indices	AVI	1.168367	10.28446	32.08886	29.3787	44.07528
	NDVI	69.03559	136.9847	886.781	1360.567	1061.713
	SAVI	69.03561	136.9847	886.781	1360.567	1061.713
	GCI	130.5377	207.7564	1323.344	1744.138	109.3442
	GRVI	20.97392	444.1805	865.1779	2184.792	0.002342
	MCARI	338.1705	1250.964	1416.834	508.3774	0.589687
	NDRE	77.35746	109.7149	509.4702	1867.111	951.4568
	NDSI	2.270226	582.0805	2488.158	324.4448	117.8338
	VARI	19.9884	376.856	786.9019	2329.75	1.837855
	NDBal	577.1693	2127.395	588.2287	217.1505	4.681645

* All decimal values are represented in sq. km

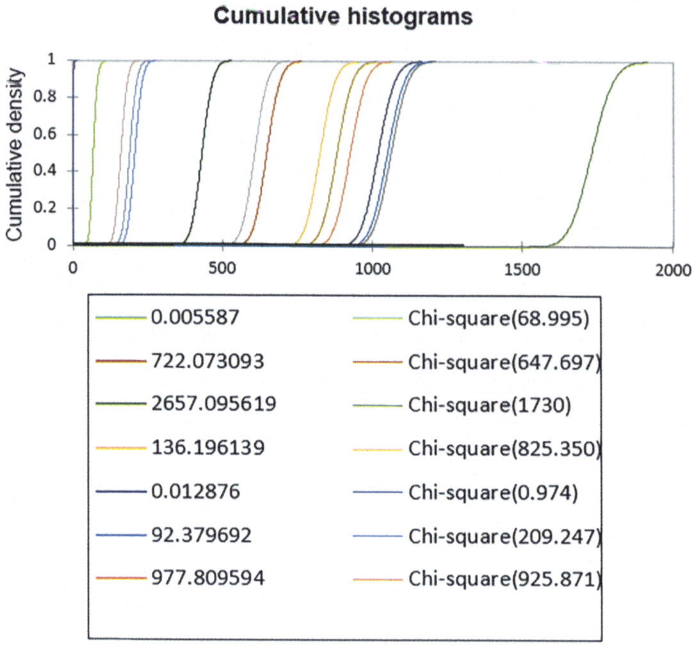

Fig. 21.9 Cumulative histograms of spectral index values over the time span

revealing changes in salinity, vegetation, or water extent over seasons or years (i.e., salt pan dynamics). As the chi-square values are very high in this scenario, this high Chi-square value could indicate a significant departure from an expected state or a previous state, signalling important changes in salt pan dynamics, like increasing salinization or altered water regimes.

Now, to gain more insights into the spectra parameters correlation analysis, a correlation matrix and a coefficient of determination are used. A correlation map (Fig. 21.10a), which visually represents the strength and direction of relationships between different spectra variables. The map uses a colour scheme to indicate the intensity and type of correlation: red for maximum correlation, brown for moderate, blue for low, and white for no correlation. The diagonal cells (where a variable is correlated with itself) typically show a perfect positive correlation (a value of 1), and are often represented by a solid colour for positive correlation, which has high intensity, indicating the changes of variables with respective changes. Whereas, negative correlation is represented by blue or green colours and shows a relationship as one variable increases, the other tends to decrease. The numbers along the bottom and right side of this map (e.g., 0.005587, 136.196139, 977.809594, etc.) represent the different variables or data points being analyzed.

The second image (Fig. 21.10b) displays a heatmap, likely representing a correlation matrix or a similar data visualization, with a colour scale indicating the strength and direction of relationships between different elements. The rows and columns are

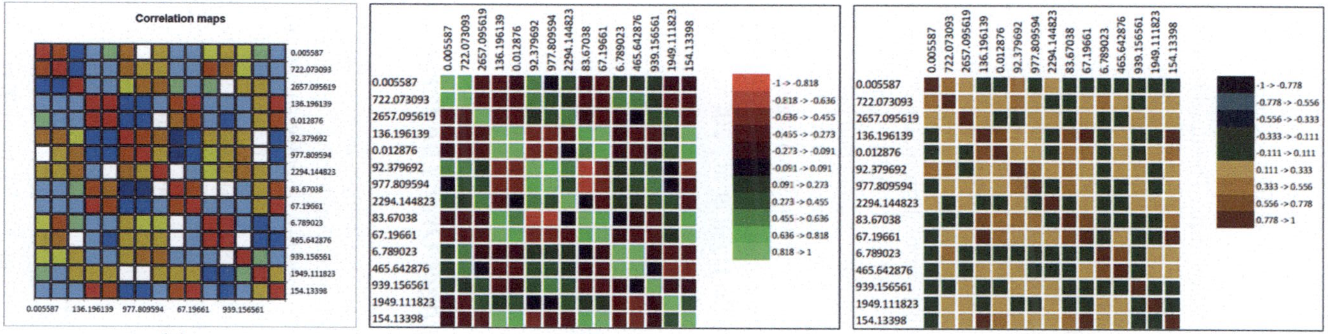

(a) (b) (c)

Fig. 21.10 Correlation analysis of the spectra parameters: **a** correlation matrix heatmap, **b** correlation matrix, **c** coefficient of determination labelled with numerical values, suggesting these represent variables or features in a dataset. The colour legend on the right indicates a range from −1 to 1, with red representing negative values (e.g., negative correlation), green representing positive values (e.g., positive correlation), and black/dark colours representing values close to zero. The intensity of the colour corresponds to the magnitude of the value, with brighter red or green indicating stronger relationships. The third image (Fig. 21.10c) displays a heatmap, likely representing a coefficient of determination matrix, where the intensity and colour of each square indicate a specific value or relationship between the corresponding row and column labels. The legend maps colour ranges to numerical values, with darker colours (browns) representing positive correlations or higher values, and darker greens representing negative correlations or lower values.

Remote sensing techniques are frequently used to study large and dynamic salt pan environments (Singh et al. 2010; Milewski et al. 2017, 2020; Schauer et al. 2023). The data gathered from remote sensing (e.g., spectral characteristics, surface temperature) can be analyzed using correlation matrices, which are then visualized as heatmaps to understand patterns and relationships, aiding in the development and validation of numerical models for predicting salt pan behaviour. The heatmap image, depicting a matrix of coefficients of determination or correlation, can be directly applied to the study of salt pan dynamics by visualizing the relationships between various environmental and physical parameters within a salt pan ecosystem (Safaee and Wang 2020; Zhang et al., 2023; Duan et al., 2025). As most of the observing parameters show the strongest correlations (the most intensely coloured cells), we can assume that most of the parameters significantly influenced the dynamicity within the salt pan environment.

21.5 Discussion

21.5.1 Salt Pan Area Changes Over The Period 2000 to 2024

In this study, a comparative analysis of changes in moisture and vegetation characteristics over 24 years (2000–2024) was conducted based on these spectral indices. While the table (Table 21.3) itself does not directly show changes in area linked to these indices over time, it provides data that can be used to infer such changes if interpreted in the context of remote sensing applications. The "Indices from High to Low" sections for each year (2000, 2010, 2024) likely represent statistics or distributions of the spectral index values across a study area. The numerical values under columns 0, 1, 2, 3, and 4 could correspond to different statistical measures like minimum, maximum, mean, median, or quartiles of the index values within that area, ordered from highest to lowest prevalence or value. Changes in Vegetation Cover (NDVI, SAVI, GCI, etc.) show a decrease in the higher values (e.g., "0" and "1" columns) of a vegetation index like NDVI from 2000 to 2024 could suggest a reduction around healthy, dense vegetation. Fluctuations in the parameters of moisture indices (MNDWI, NDWI, NDMI, GVMI, VSDI) across the years could indicate changes in water body extent, soil moisture levels, or vegetation water content across the study area over time, potentially reflecting drought conditions, or increased wetness. Observing changes across the entire range of parameters (0–4) for a given index would provide insight into shifts in the distribution of that characteristic within the study area. For instance, if the values in column "4" (representing lower values of the index) increase significantly, it might suggest an expansion of less vigorous vegetation or drier conditions in areas previously exhibiting higher moisture content.

21.5.2 Salt Pan Dynamics Predictor Model

By analyzing the changes in these spectral indices over the years, the study aims to determine the "Area Change" of the salt pans. This change in area, coupled with the information from moisture and vegetation indices, helps understand the "Dynamicity of Salt Pans," referring to how these areas are changing and evolving due to natural processes or human activities. Indices like MNDWI and NDWI are specifically designed to enhance water features in satellite images, allowing for the delineation of open water bodies and the estimation of surface wetness in the salt pans. By analyzing the values of these indices across different time points (2000, 2010, 2024), we can track changes in the extent of inundation or desiccation within the salt pans, indicating shifts in water availability. Changes in these indices over time can reveal periods of vegetation growth, stress, or decline, which are often linked to variations in soil salinity and moisture levels, thus providing insights into ecological responses within the salt pan system. This analysis of temporal "Area Change" directly leads to the understanding of the "Dynamicity of Salt Pans," revealing trends of expansion, shrinkage, or shifts in vegetation cover, which can be attributed to natural processes (like flood-desiccation cycles) or anthropogenic impacts (e.g., changes in water management or land use).

21.5.3 Salt Pan Dynamics and Mangrove Distribution

Indian Sundarban, salt pan dynamics, and mangrove distribution are closely linked. It is highly susceptible to salinity changes and is located at the land-sea interface, making it dynamic, influenced by factors like sea-level rise and freshwater flow, which directly affect the types and distribution of mangrove species. Salt-tolerant species like *Avicennia marina* and *Phoenix paludosa* dominate areas with higher salinity, while salinity-sensitive species like Heritiera fomes are limited to areas with lower salinity. Areas with higher salinity, often found in the western and central parts of the Sundarbans, are dominated by salt-tolerant mangrove species. The eastern part of the Sundarbans, influenced by the active river, experiences lower salinity levels, supporting a greater diversity of mangrove species, including some that are less tolerant to high salinity. Sea-level rise and reduced freshwater input (due to upstream dam construction) contribute to increased salinity intrusion, which can negatively impact the growth and survival of certain mangrove species. Understanding the relationship between salt pan dynamics and mangrove distribution is crucial for developing effective conservation strategies for the Sundarbans mangrove forest.

21.6 Conclusions

Recent cyclones and their associated storm surges are significantly impacting the mangrove and salt pan dynamics of the Indian Sundarban, leading to increased salinity, mangrove degradation, and changes in soil properties. The increased frequency and intensity of cyclones, coupled with sea level rise, are causing more saline water to inundate the Sundarban region, affecting mangrove biodiversity and leading to the formation of salt pans in areas previously dominated by mangroves. Salinity-sensitive mangrove species like *Xylocarpus granatum, Xylocarpus moluccensis, Rhizophora mucronata, Bruguiera gymnorrhiza, and Bruguiera cylindrica* are facing local extinction in the affected areas due to increased salinity. The increased salinity can alter the distribution and zonation of mangrove species, with some species potentially retreating from highly saline areas and others expanding into areas previously too fresh for them. While the overall mangrove area may not have significantly changed in recent decades, there is a net loss in core and buffer areas, coinciding with an expansion of the transition zone, potentially due to erosion and increased salinity. Increased salinity can reduce the soil organic blue carbon pool in mangrove areas, as observed in the Sundarban. The Sundarban's coastal communities are heavily reliant on the mangrove ecosystem, and the impacts of cyclones and storm surges on mangroves also affect their livelihoods, displacement, and migration. The destruction of mangroves, which act as natural buffers against cyclones, makes the Sundarban region more vulnerable to future storm surges and other extreme weather events. In conclusion, recent cyclones and associated storm surges are having a profound and detrimental impact on the Indian Sundarban's mangrove and salt pan dynamics, leading to increased salinity, mangrove degradation, changes in soil properties, and increased vulnerability of the region. Knowledge of salt pan dynamics can help in the targeted conservation of specific plant and animal species. It can also inform the development of sustainable resource management practices, such as selective logging and fishing, that minimize negative impacts on the ecosystem. By monitoring salinity fluctuations, we can better understand how the Sundarbans ecosystem responds to environmental changes and adapt management strategies accordingly.

Contribution statement SL: Formal analysis, Map preparation, **BC**: Writing—original draft, **MR**: Design of study, Writing—original draft, **UB**: Conceptualization, Design of study, Formal analysis, Writing—original draft, review & editing.

References

Ahmed S, Sarker SK, Friess DA, Kamruzzaman M, Jacobs M, Islam MA, Pretzsch H (2022) Salinity reduces site quality and mangrove forest functions. From monitoring to understanding. Sci Total Environ 853:158662

Akter N, Islam MZ, Chakrabarty T, Khalequzzaman M (2018) Variability, heritability and diversity analysis for some morphological traits in basmati rice (*Oryza sativa L.*) genotypes. The Agriculturists 16(02):01–14

Alam M (1996) Subsidence of the Ganges—Brahmaputra Delta of Bangladesh and associated drainage, sedimentation and salinity problems. In: Sea-level rise and coastal subsidence: causes, consequences, and strategies. Springer Netherlands, Dordrecht

Allbed A, Kumar L, Shabani F (2017) Climate change impacts on date palm cultivation in Saudi Arabia. J Agric Sci 155(8):1203–1218

Allison MA (1998) Historical changes in the Ganges-Brahmaputra delta front. J Coast Res:1269–1275

Asencio AD (2013) Permanent salt evaporation ponds in a semi-arid Mediterranean region as model systems to study primary production processes under hypersaline conditions. Estuar Coast Shelf Sci 124:24–33

Bandyopadhyay S (2019) Sundarban: a review of evolution and geomorphology. World Bank Group, p 36

Barbier EB, Georgiou IY, Enchelmeyer B, Reed DJ (2013) The value of wetlands in protecting southeast Louisiana from hurricane storm surges. PLoS One 8(3):e58715

Barras JA, Bernier JC, Morton RA (2008) Land area change in coastal Louisiana, a multidecadal perspective (from 1956 to 2006). US Department of the Interior, US Geological Survey, Reston, p 14

Berus T, Halon A, Markiewicz A, Orlowska-Heitzman J, Romanowska-Dixon B, Donizy P (2017) Clinical, histopathological and cytogenetic prognosticators in uveal melanoma–a comprehensive review. Anticancer Res 37(12):6541–6549

Bettencourt P, de Oliveira RP, Fulgêncio C, Canas Â, Wasserman JC (2022) Prospective water balance scenarios (2015–2035) for the management of São Francisco River Basin, Eastern Brazil. Water 14(15):2283

Bhatnagar A, Shukla AD, Ngangom M, Thakkar MG, Chauhan G (2025) Temporal changes in land-sea configuration inferred from post-glacial sedimentation pattern in the western great Rann of Kachchh, India. Jour Geosci Res 10(1):1–6

Biswas H, Mukhopadhyay SK, Sen S, Jana TK (2007) Spatial and temporal patterns of methane dynamics in the tropical mangrove dominated estuary, NE coast of Bay of Bengal, India. J Mar Syst 68(1–2):55–64

Biswas S, Zaman S, Mitra A (2017) Soil characteristics of Indian Sundarbans: the designated world heritage site. Sci J Biomed Eng Biomed Sci 1(2):053–059

Blasco F, Saenger P, Janodet E (1996) Mangroves as indicators of coastal change. Catena 27(3–4):167–178

Brakenridge GR, Syvitski JPM, Overeem I, Higgins SA, Kettner AJ, Stewart-Moore JA, Westerhoff R (2013) Global mapping of storm surges and the assessment of coastal vulnerability. Nat Hazards 66(3):1295–1312

Brouwer R, Akter S, Brander L, Haque E (2007) Socioeconomic vulnerability and adaptation to environmental risk: a case study of climate change and flooding in Bangladesh. Risk Anal: Int J 27(2):313–326

Campbell AD (2018) Monitoring salt marsh condition and change with satellite remote sensing. University of Rhode Island

Castañeda C, Herrero J, Casterad MA (2005) Landsat monitoring of playa-lakes in the Spanish Monegros desert. J Arid Environ 63(2):497–516

Chew C, Small E (2020) Estimating inundation extent using CYGNSS data: a conceptual modeling study. Remote Sens Environ 246:111869

Chhabra R (2022) Nature and origin of salts, classification, area and distribution of salt-affected soils. In: Salt-affected soils and marginal waters: Global perspectives and sustainable management. Springer International Publishing, Cham, pp 1–47
Chiloane CN, Dube T, Sibanda M, Dalu T, Shoko C (2025) A Summary of recent advances in the literature on machine learning techniques for remote sensing of groundwater dependent ecosystems (GDEs) from space. Remote Sens 17(8):1460
Chowdhury A, Sanyal P, Maiti SK (2016) Dynamics of mangrove diversity influenced by climate change and consequent accelerated sea level rise at Indian Sundarbans. Int J Glob Warm 9(4):486–506
Chuang PC, Young MB, Dale AW, Miller LG, Herrera-Silveira JA, Paytan A (2017) Methane fluxes from tropical coastal lagoons surrounded by mangroves, Yucatán Mexico. J Geophys Res: Biogeosciences 122(5):1156–1174
Cuadra S, Picado F, Bengtsson G, Jakobsson K (2009) Application of two probabilistic models, using either environmental monitoring or human biomonitoring data, to estimate human health risk due to mercury exposure in a small scale gold mining region in Central Nicaragua. Epidemiology 20(6):S94
Cui L, Li G, Ouyang N, Mu F, Yan F, Zhang Y, Huang X (2018) Analyzing coastal wetland degradation and its key restoration technologies in the coastal area of Jiangsu China. Wetlands 38(3):525–537
Das B, Bandyopadhyay A (2013) Eco-tourism of Sundarban in Gangetic delta. Int J Sci Eng Res 4(5):459–468
Das GK (2015) Estuarine morphodynamics of the Sunderbans
Das I, Chanda A, Akhand A, Hazra S (2023) Carbon biogeochemistry of the estuaries adjoining the Indian Sundarbans mangrove ecosystem: a review. Life 13(4):863
Das K, Mukherjee A, Malakar P, Das P, Dey U (2021) Impact of global-scale hydroclimatic patterns on surface water-groundwater interactions in the climatically vulnerable Ganges river delta of the Sundarbans. Sci Total Environ 798:149198
Das S, Vincent JR (2009) Mangroves protected villages and reduced the death toll during the Indian super cyclone. Proc Natl Acad Sci 106(18):7357–7360
de Freitas LD, de Moraes JFL, da Costa AM, Martins LL, Silva BM, Avanzi JC, Uezu A (2022) How far can nature-based solutions increase water supply resilience to climate change in one of the most important Brazilian watersheds? Earth 3(3):748–767
Dietrich JC, Bunya S, Westerink JJ, Ebersole BA, Smith JM, Atkinson JH, Roberts HJ (2010) A high-resolution coupled riverine flow, tide, wind, wind wave, and storm surge model for southern Louisiana and Mississippi. Part II: synoptic description and analysis of hurricanes Katrina and Rita. Mon Weather Rev 138(2):378–404
Dunn FE, Nicholls RJ, Darby SE, Cohen S, Zarfl C, Fekete BM (2018) Projections of historical and 21st century fluvial sediment delivery to the Ganges-Brahmaputra-Meghna, Mahanadi, and Volta deltas. Sci Total Environ 642:105–116
Dutta MK, Kumar S, Mukherjee R, Sanyal P, Mukhopadhyay SK (2019) The post-monsoon carbon biogeochemistry of the Hooghly–Sundarbans estuarine system under different levels of anthropogenic impacts. Biogeosciences 16(2):289–307
Emard S (2024) Time series of a forest canopy: detecting changes using the visible atmospherically resistant index and a low-cost drone (Doctoral dissertation, Toronto Metropolitan University)
Epema GF (1992) Atmospheric condition and its influence on reflectance of bare soil surfaces in southern Tunisia. Int J Remote Sens 13(5):853–868
Escapa M, Perillo GM, Iribarne O (2015) Biogeomorphically driven salt pan formation in Sarcocornia-dominated salt-marshes. Geomorphology 228:147–157
Fagherazzi S (2014) Storm-proofing with marshes. Nat Geosci 7(10):701–702
Falkner K, Kropf B, Schmid E, Mitter HA (2021) Bayesian network to support agricultural land and groundwater management under climate change in the Seewinkel region. Strategien für den Agrar-und Ernährungssektor und den ländlichen Raum in Zeiten multipler Krisen 55

Friess DA (2025) The diversity of mangrove forests and geographical biases in their research. Bull Mar Sci

Gao L, Rao MPN, Liu YH, Wang PD, Lian ZH, Abdugheni R, Li WJ (2024) Salinity-induced changes in diversity, stability, and functional profiles of microbial communities in different saline lakes in arid areas. Microb Ecol 87(1):135

Gettelman A, Bresch DN, Chen CC, Truesdale JE, Bacmeister JT (2018) Projections of future tropical cyclone damage with a high-resolution global climate model. Clim Chang 146(3):575–585

Gilvear D, Bryant R (2003) Analysis of aerial photography and other remotely sensed data. Tools Fluv Geomorphol 5:23

Giri C, Long J, Abbas S, Murali RM, Qamer FM, Pengra B, Thau D (2015) Distribution and dynamics of mangrove forests of South Asia. J Environ Manag 148:101–111

Giri C, Pengra B, Zhu Z, Singh A, Tieszen LL (2007) Monitoring mangrove forest dynamics of the Sundarbans in Bangladesh and India using multi-temporal satellite data from 1973 to 2000. Estuar Coast Shelf Sci 73(1–2):91–100

Gonnelli A, Carlà R, Baronti S, Raimondi V (2023) Near-infrared and short-wavelength infrared-based indices to monitor soil moisture from a satellite: a comparative analysis. Eng Proc 51(1):29

Goodbred SL Jr, Kuehl SA (2000) Enormous Ganges-Brahmaputra sediment discharge during strengthened early Holocene monsoon. Geology 28(12):1083–1086

Gopal B, Chauhan M (2006) Biodiversity and its conservation in the Sundarban mangrove ecosystem. Aquat Sci 68(3):338–354

Guo D, Yang D, Zhang H, Song J, Zhang R, Xu R, He Y (2025) Deepseek-r1: incentivizing reasoning capability in llms via reinforcement learning. https://arxiv.org/abs/2501.12948

Hanebuth TJ, Kudrass HR, Linstädter J, Islam B, Zander AM (2013) Rapid coastal subsidence in the central Ganges-Brahmaputra Delta (Bangladesh) since the 17th century deduced from submerged salt-producing kilns. Geology 41(9):987–990

Hasab HA, Dibs H, Dawood AS, Hadi WH, Hussain HM, Al-Ansari N (2020) Monitoring and assessment of salinity and chemicals in agricultural lands by a remote sensing technique and soil moisture with chemical index models. Geosciences 10(6):207

Hess LL, Melack JM, Novo EM, Barbosa CC, Gastil M (2003) Dual-season mapping of wetland inundation and vegetation for the central Amazon basin. Remote Sens Environ 87(4):404–428

Hossain MJ (2017) Relative influences of different physical factors on salinity intrusion in coastal aquifer in Bangladesh

Hou X, Wu T, Hou W, Chen Q, Wang Y, Yu L (2016) Characteristics of coastline changes in mainland China since the early 1940s. Sci China Earth Sci 59(9):1791–1802

Hu K, Chen Q, Wang H (2015) A numerical study of vegetation impact on reducing storm surge by wetlands in a semi-enclosed estuary. Coast Eng 95:66–76

Islam F, Parvin A, Parvin A, Akhtar US, Shaikh MAA, Uddin MN, Hossain MK (2023) Sediment-bound hazardous trace metals (oid) in south-eastern drainage system of Bangladesh: first assessment on human health. Heliyon 9(9)

Islam MS, Bonner JS, Edge BL, Page CA (2014) Hydrodynamic characterization of corpus christi bay through modeling and observation. Environ Monit Assess 186(11):7863–7876

Ito G, Flahaut J, Barthez M, González-Maurel O, Godoy B, Martinot M, Payet V (2021) Strategic mineral mapping in the Altiplano-Puna volcanic complex using orbital multispectral and hyperspectral imageries. In: 2021 11th Workshop on hyperspectral imaging and signal processing: evolution in remote sensing (WHISPERS). IEEE, pp 1–5

Itto I, Mab UI, Tnc L (2001) FAO (food and agricultural organisation), mangrove forest management guidelines', forestry paper, 117. Rome: FAO, 2007. S Afr J Bot:394

Jayanta KM (2023) Biodiversity of the sundarbans: ethics of conservation ecology (1770–2022)

Jensen T, Zhurbenko V, Krozer V, Meincke P (2007) Coupled transmission lines as impedance transformer. IEEE Trans Microw Theory Tech 55(12):2957–2965

Kang P, Pan Y, Ran Y, Li W, Shao M, Zhang Y, Ding X (2023) Soil saprophytic fungi could be used as an important ecological indicator for land management in desert steppe. Ecol Indic 150:110224

Kar NS, Das S, Bandyopadhyay S (2020) Anthropogeomorphology of the lower deltaic west bengal with special reference to the hugli river system. In: Anthropogeomorphology of bhagirathi-hooghly river system in India. CRC Press, pp 83–122

Kerekes JP, Strackerjan KE, Salvaggio C (2008) Spectral reflectance and emissivity of man-made surfaces contaminated with environmental effects. Opt Eng 47(10):106201–106201

Lefebvre G, Davranche A, Willm L, Campagna J, Redmond L, Merle C, Poulin B (2019) Introducing WIW for detecting the presence of water in wetlands with landsat and sentinel satellites. Remote Sens 11(19):2210

Leonardi N, Defne Z, Ganju NK, Fagherazzi S (2016a) Salt marsh erosion rates and boundary features in a shallow Bay. J Geophys Res: Earth Surf 121(10):1861–1875

Leonardi N, Fagherazzi S (2014) How waves shape salt marshes. Geology 42(10):887–890

Leonardi N, Fagherazzi S (2015) Effect of local variability in erosional resistance on large-scale morphodynamic response of salt marshes to wind waves and extreme events. Geophys Res Lett 42(14):5872–5879

Leonardi N, Ganju NK, Fagherazzi S (2016b) A linear relationship between wave power and erosion determines salt-marsh resilience to violent storms and hurricanes. Proc Natl Acad Sci 113(1):64–68

Leonardi N, Plater AJ (2017) Residual flow patterns and morphological changes along a macro-and meso-tidal coastline. Adv Water Resour 109:290–301

Li D, Ju W, Fan W, Gu Z (2014) Estimating the age of deciduous forests in northeast China with enhanced thematic mapper plus data acquired in different phenological seasons. J Appl Remote Sens 8(1):083670–083670

Li H, Li H, Wang J, Hao X (2019) Extending the ability of near-infrared images to monitor small river discharge on the northeastern Tibetan Plateau. Water Resour Res 55(11):8404–8421

Liang S, Zhou R, Dong S, Shi S (2008) Adaptation to salinity in mangroves: implication on the evolution of salt-tolerance. Chin Sci Bull 53(11):1708–1715

Liu X, Dong G, Wang X, Xue Z, Jiang M, Lu X, Zhang Y (2013) Characterizing the spatial pattern of marshlands in the Sanjiang Plain, Northeast China. Ecol Eng 53:335–342

Ljubičić N, Popović V, Kostić M, Vukosavljev M, Buđen M, Stanković N, Stevanović N (2024) The normalized difference red edge index (NDRE) in grain yield and biomass estimation in maize (*Zea mays* L.). In: Proceedings of the XV international scientific agricultural symposium "Agrosym", pp 373–378

Loder NM (2009) An evaluation of the potential of coastal wetlands for hurricane surge and wave energy reduction (Doctoral dissertation)

Loder NM, Irish JL, Cialone MA, Wamsley TV (2009) Sensitivity of hurricane surge to morphological parameters of coastal wetlands. Estuar Coast Shelf Sci 84(4):625–636

Mandal A, Das A, Das M (2024) Assessment of ecosystem service diversity and evenness in Asia's largest freshwater mangrove forest. Assessment 9:1

Mandal A, Ghosh D, Dhaliwal SS, Toor AS, Majumder A (2023) Climate change and its impact on soil carbon storage: an Indian perspective. In: Climate change impacts in India. Springer International Publishing, Cham, pp 211–235

Mandal AK (2019) Modern technologies for diagnosis and prognosis of salt-affected soils and poor-quality waters. In: Research developments in saline agriculture. Springer Singapore, Singapore, pp 95–152

Mandal SK, Ray R, González AG, Pokrovsky OS, Jana TK (2021) Antimony uptake by mangroves and its environmental fate in the Sundarbans, India. Estuar Coast Shelf Sci 248:106923

Mariotti G (2016) Revisiting salt marsh resilience to sea level rise: are ponds responsible for permanent land loss? J Geophys Res: Earth Surf 121(7):1391–1407

Mariotti G, Carr J (2014) Dual role of salt marsh retreat: Long-term loss and short-term resilience. Water Resour Res 50(4):2963–2974

Mauranyapin NP, Madsen LS, Taylor MA, Waleed M, Bowen WP (2017) Evanescent single-molecule biosensing with quantum-limited precision. Nat Photonics 11(8):477–481

McIvor A, Spencer T, Spalding M, Lacambra C, Möller I (2015) Mangroves, tropical cyclones, and coastal hazard risk reduction. In: Coastal and marine hazards, risks, and disasters. Elsevier, pp 403–429

McKee K, Rogers K, Saintilan N (2012) Response of salt marsh and mangrove wetlands to changes in atmospheric CO_2, climate, and sea level. In: Global change and the function and distribution of wetlands. Springer Netherlands, Dordrecht, pp 63–96

Medina-Pérez G, Fernández-Luqueño F, Vazquez-Nuñez E, López-Valdez F, Prieto-Mendez J, Madariaga-Navarrete A, Miranda-Arámbula M (2019) Remediating polluted soils using nanotechnologies: Environmental benefits and risks. Pol J Environ Stud 28(3)

Meehl GA, Covey C, Delworth T, Latif M, McAvaney B, Mitchell JF, Taylor KE (2007) The WCRP CMIP3 multimodel dataset: a new era in climate change research. Bull Am Meteorol Soc 88(9):1383–1394

Mehla MK, Kumar A, Kumar S, Jhorar RK (2024) Soil salinity assessment and mapping using spectral indices and geostatistical techniques-concepts and reviews. In: Remote sensing of soils. Elsevier, pp 429–437

Mercy Latha A, Ghosh SK (2021) A comprehensive review of depressed collectors of slow-wave devices. J Electromagn Waves Appl 35(1):95–137

MH Yossif T (2017) Change detection of land cover and salt affected soils at Siwa Oasis. Egypt Alex Sci Exch J 38(July–September):446–462

Michels KH, Kudrass HR, Hübscher C, Suckow A, Wiedicke M (1998) The submarine delta of the Ganges–Brahmaputra: cyclone-dominated sedimentation patterns. Mar Geol 149(1–4):133–154

Milewski R (2020) Potential of optical remote sensing for the analysis of salt pan environments (Doctoral dissertation, Universität Potsdam)

Milewski R, Chabrillat S, Behling R (2017) Analyses of recent sediment surface dynamic of a Namibian Kalahari salt pan based on multitemporal landsat and hyperspectral hyperion data. Remote Sens 9(2):170

Mishra RN, Kumar MRRA, Marc-Antoni G, Karim Z (2025) Water-in-salt electrolytes: advances and chemistry for sustainable aqueous monovalent-metal-ion batteries. Batteries 11(4):120

Mitra A (2013) Sensitivity of mangrove ecosystem to changing climate, vol 62. Springer, New Delhi, pp 143–157

Mitra A, Sengupta K, Chowdhury MR, Roychowdhury G, Raha AK, Zaman S, Raha AK (2013) Mangroves: a unique gift of nature. In: Sensitivity of mangrove ecosystem to changing climate. Springer, New Delhi, India, pp 33–105

Mohapatra M, Chauhan A, Varshney A, Gurjar S, Bushair MT, Sharma M, KVani G (2023) Short to medium range impact based forecasting of heavy rainfall in India. Mausam 74(2):311–344

Mondal I, Ariful Hossain SK, Bandyopadhyay J, Altuwaijri HA, Das A, Jose F, Juliev M (2025) Potential effects of projected global sea level rise on Sundarbans mangrove wetland ecosystem: insights from SLAMM and hybrid machine learning models. Environ, Dev Sustain:1–34

Morton RA, Barras JA (2011) Hurricane impacts on coastal wetlands: a half-century record of storm-generated features from southern Louisiana. J Coast Res 27(6A):27–43

Mukhopadhyay BP, Chakraborty A, Roy R, Bandyopadhyay A (2024) Assessment of the changing landscape and overall aquifer condition due to saline water intrusion: a review. Hydrol Urban Water Supply:199–213

Natural Resources. Species Survival Commission (2001) IUCN red list categories and criteria. IUCN

Neogi SB, Dey M, Lutful Kabir SM, Masum SJH, Kopprio GA, Yamasaki S, Lara RJ (2016) Sundarban mangroves: diversity, ecosystem services and climate change impacts

Nguyen KA, Liou YA, Tran HP, Hoang PP, Nguyen TH (2020) Soil salinity assessment by using near-infrared channel and vegetation soil salinity index derived from landsat 8 OLI data: a case study in the Tra Vinh Province, Mekong Delta. Vietnam Prog Earth Planet Sci 7(1):1–16

Omer NH (2019) Water quality parameters. Water Qual-Sci, Assess Policy 18:1–34

Pachauri S (2014) Household electricity access a trivial contributor to CO_2 emissions growth in India. Nat Clim Chang 4(12):1073–1076
Panuju DR, Paull DJ, Griffin AL (2020) Change detection techniques based on multispectral images for investigating land cover dynamics. Remote Sens 12(11):1781
Parida AK, Jha B (2010) Salt tolerance mechanisms in mangroves: a review. Trees 24(2):199–217
Parida AK, Tiwari V, Jha B (2013) Impacts of climate change on Asian mangrove forests. In: Mangrove ecosystems of Asia: status, challenges and management Strategies. Springer New York, New York, NY, pp 233–256
Passeri DL, Hagen SC, Medeiros SC, Bilskie MV, Alizad K, Wang D (2015) The dynamic effects of sea level rise on low-gradient coastal landscapes: a review. Earth's Future 3(6):159–181
Peplouw M (2024) Spatiotemporal characteristics of surface water in Sua Pan, Botswana, using Earth Observation data: 1992–2022 Doctoral dissertation, Faculty of Science. University of the Witwatersrand, Johannesburg, South Africa
Pratama M, Venugopal V (2024) Modeling tidal hydrodynamics of the montrose tidal inlet system: ebb jet and eddy formation at a tidal inlet with a high-angle half-delta. J Waterw Port Coast Ocean Eng 150(6):04024016
Prigent C, Papa F, Aires F, Rossow WB, Matthews E (2007) Global inundation dynamics inferred from multiple satellite observations, 1993–2000. J Geophys Res Atmos 112(D12)
Qian W, Zhu Y (2001) Climate change in China from 1880 to 1998 and its impact on the environmental condition. Clim Chang 50(4):419–444
Rasheed A, Raza A, Jie H, Mahmood A, Ma Y, Zhao L, Jie Y (2022) Molecular tools and their applications in developing salt-tolerant soybean (*Glycine max L.*) cultivars. Bioengineering 9(10):495
Rees G (2013) Physical principles of remote sensing. Cambridge University Press
Renssen H, Knoop JM (2000) A global river routing network for use in hydrological modeling. J Hydrol 230(3–4):230–243
Reschke J, Hüttich C (2014) Continuous field mapping of mediterranean wetlands using sub-pixel spectral signatures and multi-temporal landsat data. Int J Appl Earth Obs Geoinf 28:220–229
Rogers K, Saintilan N, Howe AJ, Rodríguez JF (2013) Sedimentation, elevation and marsh evolution in a southeastern Australian estuary during changing climatic conditions. Estuar Coast Shelf Sci 133:172–181
Roy SS, Ghosh T, Hewavithana DK (2024) Transformation of coastal wetlands in the Sundarban Delta (1999–2020). Environ Monit Assess 196(8):758
Rus GM, Ilovan OR, Dulamă ME, Chiş O, Ursu CD, Colcer AM, Horvath C (2020) The anthropic impact on relief and vegetation. Case study: Valea Ierii. European Proceedings of Social and Behavioural Sciences
Russell JA (2018) Statistics in music education research. Oxford University Press
Safaee S, Wang J (2020) Towards global mapping of salt pans and salt playas using Landsat imagery: a case study of western United States. Int J Remote Sens 41(22):8693–8716
Salem OH, Jia Z (2024) Evaluation of different soil salinity indices using remote sensing techniques in Siwa Oasis. Egypt Agron 14(4):723
Samanta S, Hazra S, Mondal PP, Chanda A, Giri S, French JR, Nicholls RJ (2021) Assessment and attribution of mangrove forest changes in the Indian Sundarbans from 2000 to 2020. Remote Sens 13(24):4957
Sawaya WN, Khatchadourian HA, Safi WM, Al-Muhammad HM (1983) Chemical characterization of prickly pear pulp, Opuntia ficus-indica, and the manufacturing of prickly pear jam. Int J Food Sci Technol 18(2):183–193
Scardino G, Mancino S, Romano G, Patella D, Scicchitano G (2023) An integrated approach between multispectral satellite images and geophysical and morpho-topographic surveys for the detection of water stress associated with coastal dune erosion. Remote Sens 15(18):4415
Schauer H, Schlaffer S, Bueechi E, Dorigo W (2023) Inundation–desiccation state prediction for salt pans in the western Pannonian Basin using remote sensing, groundwater, and meteorological data. Remote Sens 15(19):4659

Schepers L, Brennand P, Kirwan M, Guntenspergen G, Temmerman S (2017a) Coastal marsh conversion to ponds induces irreversible elevation loss relative to sea level. Spat Patterns Bio-Geomorphol Eff Veg Loss Submerg Coast Marsh 61

Schepers L, Kirwan M, Guntenspergen G, Temmerman S (2017b) Spatio-temporal development of vegetation die-off in a submerging coastal marsh. Limnol Oceanogr 62(1):137–150

Schlaffer S, Chini M, Dettmering D, Wagner W (2016) Mapping wetlands in Zambia using seasonal backscatter signatures derived from ENVISAT ASAR time series. Remote Sens 8(5):402

Schlaffer S, Chini M, Dorigo W, Plank S (2022) Monitoring surface water dynamics in the prairie pothole region of North Dakota using dual-polarised Sentinel-1 synthetic aperture radar (SAR) time series. Hydrol Earth Syst Sci 26(3):841–860

Seidensticker T (1986) Arabic literary works as a source of documentation for technical terms of the material culture. (Islamkundliche Untersuchungen), vol 98

Sharma LK, Naik R (2024) Saline wetland ecosystems. In: Conservation of saline wetland ecosystems: an initiative towards UN decade of ecological restoration. Singapore, Springer Nature Singapore, pp 33–75

Shokri N, Hassani A, Sahimi M (2024) Multi-scale soil salinization dynamics from global to pore scale: a review. Rev Geophys 62(4):e2023RG000804

Singh D, Ghosh S, Roxy MK, McDermid S (2019) Indian summer monsoon: extreme events, historical changes, and role of anthropogenic forcings. Wiley Interdiscip Rev Clim Chang 10(2):e571

Singh G, Bundela DS, Sethi M, Lal K, Kamra SK (2010) Remote sensing and geographic information system for appraisal of salt-affected soils in India. J Environ Qual 39(1):5–15

Singh M, Singh IB, Müller G (2007) Sediment characteristics and transportation dynamics of the Ganga River. Geomorphology 86(1–2):144–175

Spalding M, Kainuma M, Collins L (2010) World atlas of mangroves. Routledge

Stark J, Plancke Y, Ides S, Meire P, Temmerman S (2016) Coastal flood protection by a combined nature-based and engineering approach: Modeling the effects of marsh geometry and surrounding dikes. Estuar Coast Shelf Sci 175:34–45

Steckler MS, Nooner SL, Akhter SH, Chowdhury SK, Bettadpur S, Seeber L, Kogan MG (2010) Modeling Earth deformation from monsoonal flooding in Bangladesh using hydrographic, GPS, and gravity recovery and climate experiment (GRACE) data. J Geophys Res Solid Earth 115(B8)

Sultana S, Paul SC, Parveen S, Alam S, Rahman N, Jannat B, Karim MM (2020) Isolation and identification of salt-tolerant plant-growth-promoting rhizobacteria and their application for rice cultivation under salt stress. Can J Microbiol 66(2):144–160

Syakur S, Kausar AN, Fazlina YD (2025) Analysis of mangrove vegetation distribution and density in East Aceh Regency. In: IOP conference series: earth and environmental science, vol. 1477, no. 1, p 012011. IOP Publishing

Syvitski JP, Kettner AJ, Overeem I, Hutton EW, Hannon MT, Brakenridge GR, Nicholls RJ (2009) Sinking deltas due to human activities. Nat Geosci 2(10):681–686

Taghadosi MM, Hasanlou M (2017) Trend analysis of soil salinity in different land cover types using landsat time series data (case study Bakhtegan salt lake). Int Arch Photogramm Remote Sens Spat Inf Sci 42:251–257

Tassew GT, Mitiku AB, Mekonnen TM, Geleta FB, Woldemichael AT (2025) Evaluating flood extent using synthetic aperture radar (SAR) and modified normalized difference water index (MNDWI) methods. Remote Sens Earth Sys Sci 8(2):673–695

Temmerman S, De Vries MB, Bouma TJ (2012) Coastal marsh die-off and reduced attenuation of coastal floods: a model analysis. Glob Planet Chang 92:267–274

Teshaev N, Mamadaliyev B, Ibragimov A, Khasanov S (2020) The soil-adjusted vegetation index for soil salinity assessment in Uzbekistan. ICIGIS 26:324–333

Thomas CD, Cameron A, Green RE, Bakkenes M, Beaumont LJ, Collingham YC, Williams SE (2004) Extinction risk from climate change. Nature 427(6970):145–148

Tripathi P, Patel NR, Kushwaha SPS, Dadhwal VK (2014) Upscaling of leaf area index in Terai forest plantations using fine-and moderate-resolution satellite data. Int J Remote Sens 35:7749–7762. https://doi.org/10.1080/01431161.2014.976886

Upadhyay VP, Ranjan R, Singh JS (2002) Human–mangrove conflicts: the way out. Curr Sci 83(11):1328–1336

Voitik A, Kravchenko V, Pushka O, Kutkovetska T, Shchur T, Kocira S (2023) Comparison of NDVI, NDRE, MSAVI and NDSI indices for early diagnosis of crop problems. Agric Eng 27

Wahiduzzaman M, Yeasmin A (2019) Statistical forecasting of tropical cyclone landfall activities over the North Indian Ocean rim countries. Atmos Res 227:89–100

Wang W, Wang S, Ma X, Gong J (2011) Recent advances in catalytic hydrogenation of carbon dioxide. Chem Soc Rev 40(7):3703–3727

Ward RD, Friess DA, Day RH, Mackenzie RA (2016) Impacts of climate change on mangrove ecosystems: a region by region overview. Ecosyst Health Sustain 2(4):e01211

Webster PJ, Holland GJ, Curry JA, Chang HR (2005) Changes in tropical cyclone number, duration, and intensity in a warming environment. Science 309(5742):1844–1846

Wen L, Ryan S, Powell M, Ling JE (2025) From clusters to communities: enhancing wetland vegetation mapping using unsupervised and supervised synergy. Remote Sens 17(13):2279

Woth K, Weisse R, Von Storch H (2006) Climate change and North Sea storm surge extremes: an ensemble study of storm surge extremes expected in a changed climate projected by four different regional climate models. Ocean Dyn 56(1):3–15

Wurtsbaugh WA, Miller C, Null SE, DeRose RJ, Wilcock P, Hahnenberger M, Moore J (2017) Decline of the world's saline lakes. Nat Geosci 10(11):816–821

Xu H, Sun H, Xu Z, Wang Y, Zhang T, Wu D, Gao J (2025) kNDMI: a kernel normalized difference moisture index for remote sensing of soil and vegetation moisture. Remote Sens Environ 319:114621

Yusuf A, Lei D, Sun Y, Duan S, Zhang Y (2025) Nitrogen transformation and microbial community interactions in hydrodynamic heterogeneous hyporheic zone sediment: insights for ecosystem sustainability. Limnologica 112:126252

Zhang F, Pant D, Logan BE (2011) Long-term performance of activated carbon air cathodes with different diffusion layer porosities in microbial fuel cells. Biosens Bioelectron 30(1):49–55

Zhang K, Liu H, Li Y, Xu H, Shen J, Rhome J, Smith TJ III (2012) The role of mangroves in attenuating storm surges. Estuar Coast Shelf Sci 102:11–23

Zimmermann-Timm H, Teubner K (2021) Folgen der Grundwassersenkung am Beispiel des Neusiedlers See Seewinkel (Burgenland, Österreich). In: Lozán JLS-W, Breckle HG, Kasang D, (Eds) Warnsignal Klima: Boden und Landnutzung, pp 142–149

Chapter 22
Navigating Towards Improved Caribbean Coastal Resilience: Solutions and Challenges

Anton Ali, Kailas Sekhar Banerjee, and Deborah Villaroel-Lamb

22.1 Introduction

The Caribbean region (Fig. 22.1) spanning roughly between latitudes 9°to 22°N and longitudes 89°to 60°W boast of diverse coastlines, inclusive of volcanic and raised limestone terrace-type landforms (Banton et al. 2017). The region has long been hailed as the world's premier tourist destination, linked primarily to the region's climate and its coastal regions (Duval 2004). Consequently, many of the economies of the Caribbean islands are heavily dependent on tourism (Pentelow and Scott 2011) with few exceptions such as Trinidad and Tobago and quite possibly Guyana with its newly found oil reserves (CAF 2025). The region can be broken into two divisions: (1) the Greater Antilles, comprising Cuba, Jamaica, Hispaniola and Puerto Rico; (2) the Lesser Antilles comprising the remaining smaller island arc chain south and east of the Caribbean Sea.

Based upon an analysis of the 2025 World Resource Institute (WRI) and Central Intelligence Agency (CIA) World Factbook data, the combined coastline length of all the Caribbean countries is approximately 39,000 km. The coastal geomorphology across this total length ranges beyond the protected bay beaches associated with the region. Tectonic, volcanic, riverine, and coastal processes (which vary spatiotemporally) are responsible for this variability (Quesada-Román and Pérez-Briceño, 2019).

A. Ali · K. S. Banerjee · D. Villaroel-Lamb (✉)
Department of Civil and Environmental Engineering, Faculty of Engineering, The University of the West Indies, St. Augustine, Trinidad and Tobago
e-mail: deborah.villarroel-lamb@uwi.edu

A. Ali
e-mail: anton.ali@uwi.edu

K. S. Banerjee
e-mail: kailas.banerjee@uwi.edu

A. Nath et al. (eds.), *Geospatial Application of Coastal Hazards*, Springer Water,
https://doi.org/10.1007/978-3-032-20076-1_22

Fig. 22.1 Map of Caribbean (adapted from Darsan et al. 2012)

Noteworthy coastal geomorphological landforms evident in the region are the estuarine Nariva mangrove swamp in Trinidad as detailed by Darsan et al. (2012) and Jobos Bay National Estuarine Research Reserve in Puerto Rico which has been the subject of many investigations such as the pollution monitoring by Martínez-Colón et al. (2021). Various Coral reef systems, such as sections of the fringing Mesoamerican reef found in Belize (Breton et al. 2006b, a) or Buccoo Reef/Bon Accord Lagoon Complex in Tobago. Sea grass beds, such as those found in the Dominican Republic (Breton et al. 2006b, a) and across much of the Lesser Antilles. Sand spit and tombolo formations such as the Cocal sand spit along the east coast of Trinidad as detailed by Darsan et al. (2012) or the Palisadoes at the Kingston harbour in Jamaica noted by Ramirez and Townsend (2006). Caves, stacks, stumps, blowholes and wave cut platforms such as those observed along the north coast of Trinidad (Darsan et al. 2012).

The various costal geomorphological landforms span a range of ecological, scientific and socio-economic benefits to the region. The estuaries in the region provide the benefits of nursery and reproductive growth of marine species, carbon sequestration, cycling of nutrients and mitigation against the impacts of climate change (Haddout et al. 2024). The coral reef, mangrove and seagrass bed formations in the tropics often coexist (Carlson et al. 2021). Research specific to this interconnected complex in the Caribbean is ongoing with noteworthy contributions being Kjerfve (1998), Harborne et al. (2006), Dorenbosch et al. (2007) and Carlson et al. (2021). Economically, the value of mangroves, coral reefs and other coastal ecosystems in the Caribbean is estimated at approximately 15 billion annually, ranging across fisheries, carbon sequestration and tourism (Resilient Islands 2019).

The vulnerability of Caribbean coastal geomorphological formations originates from both man-made and natural hazards. The main man-made hazards include: unplanned (and by extension, planned) coastal development as detailed by Scott et al. (2012); land and marine based pollution as detailed by Sarwat (2024); sediment budget alterations either via sand mining (IMA 2022) or sediment starvation due to river damming (Zhao et al. 2021); alterations to coastal hydrodynamics by river damming or implemented coastal structures. Examples outlining beach mining and coastal development in the Caribbean region are detailed in WRA (2001), Smith Warner Team (2022) and Mycoo and Chadwick (2012).

The main natural hazards in the region are volcanic eruptions, earthquakes, tsunamis, and tropical cyclones, of which primarily the latter two pose significant threats to coastal regions. Whilst past tsunami events in the Caribbean Sea have been recorded, an analysis of the documented data formulated by Harbitz et al. (2012) highlights that only 51 tsunamis affected Caribbean islands between 1498 and 2006. Comparatively, from the beginning of the official Atlantic tropical cyclone record in 1851 up until 2024, there have been 1,370 tropical cyclones reaching tropical storm strength in the North Atlantic area (which includes the Caribbean) (McAdie et al. 2009). Thus, tropical cyclones pose the most frequent natural hazard in the region.

The region's tropical cyclone susceptibility coincides with that of the Atlantic basin. Specifically, an average of 14 named storms, 7 hurricanes and 3 major hurricanes are expected yearly, based upon the climatology from 1991–2021 (NOAA 2022). Noteworthy examples that have severely impacted the region are the 2017 hurricanes Irma, Maria and Harvey which have left many countries with increased debt following the storms' occurrences (Cavallo et al. 2024). Coupled with the dilemma increasing sea levels due to climate change, an increase in the number of intense storms and hurricanes is envisioned. Projections and correlations such as Patricola and Wehner (2018), Murakami et al. (2018), Knutson et al. (2020), Reed et al. (2021) and Reed et al. (2022) have established this link.

The region exhibits coastal zone vulnerability due to man-made and natural hazards, and these Caribbean coasts are also vulnerable from the context of land elevations and socio-economic circumstances. Flat low-lying lands are typical of the Caribbean basin (Straus and Kulp 2018). Countries such as the Bahamas, Barbuda, and parts of Jamaica are noted as being within 10 m of mean sea level (Rhiney 2015) and thus, particularly vulnerable to coastal hazards and Sea Level Rise (SLR). This vulnerability is further exacerbated by the fact that the population concentration in the Caribbean is mainly along coastal regions (Cavallo et al. 2024). In fact, the majority of Caribbean Island capitals are situated along the coast. Except for the Greater Antilles, over 70% of the population in these islands resides within 5 km of the coast (Cashman and Nagdee 2017). In many Caribbean territories, critical infrastructure—including schools, hospitals, transportation networks, and utilities—is concentrated along coastal zones. Even in islands characterized by significant topographic variation, substantial capital investments have been made in nearshore areas that are highly vulnerable to environmental hazards (Lewsey et al. 2004). Notable examples include Jamaica, as examined by Richards (2008), and Trinidad.

Consequently, a pathway towards enhanced coastal resilience in the region is paramount. Recently, Masselink and Lazarus (2019) provided the formal definition of coastal resilience as the ability of both socio-economic and natural systems in the coastal environment to adapt to varying natural, climatic and anthropogenic disturbances without compromising their respective functions. The Caribbean must adopt philosophies congruent to this definition, especially against the backdrop of global climate change and accompanying SLR effects. This article presents the typical coastal resiliency approaches used in the Caribbean in Sect. 22.2, followed by a discussion of future pathways for improved coastal resiliency in the region in Sect. 22.3.

22.2 Coastal Resiliency Approaches in the Caribbean

As a region, the Caribbean currently adopts a range of approaches in the pursuit of coastal resilience. These approaches broadly fall into the classifications of Traditional Infrastructure and Engineered Solutions, Nature-Based Solutions (NBS), National and Regional Initiatives: policy and planning, as well as International Collaboration, Capacity Building Initiatives, and Stakeholder Involvement.

22.2.1 Traditional Infrastructure and Engineered Solutions

Traditional infrastructure and engineered solutions have been the main approach to coastal protection in the region. These solutions fall into hard/grey engineering (sea walls, revetments, breakwaters, rock groynes, etc.), soft/natural engineering (dune stabilization, beach nourishment, cliff stabilization, etc.) and hybrid engineering (some combination of the latter two) classifications. Historically, hard engineering has been commonly used (Villaroel-Lamb et al. 2024), accounting for 51% of documented coastal adaptation interventions in the region (Laurent and Duvat 2024). In the Caribbean such solutions are ubiquitous, examples being the offshore breakwater in Port St. Charles, Barbados; riprap seawalls in Orange Bay and Buff Bay, Jamaica; and the rock revetment in Cocos Bay, Trinidad. Despite their popularity, numerous studies have detailed the adverse impacts of hard infrastructures on marine and coastal ecosystems (Perricone et al. 2023). Additionally, whilst hard engineered solutions may be the most feasible and cost-effective solution in built environment contexts such as harbours and ports (Schueler 2017), they are generally costly to build and to maintain (Narayan et al. 2016; Schueler 2017).

Traditional soft and hybrid engineering solutions can also be found in the Caribbean. These solutions incorporate strategies/techniques that mimic natural systems as closely as possible to achieve natural coastal defence. An example of the hybrid solution, beach nourishment with an accompanying submerged breakwater in

Fig. 22.2 Hybrid submerged breakwater and beach nourishment solution on the Bahamar, Bahamas (adapted from Banton et al. 2017)

the region, is at the Bahamar in the Bahamas (Fig. 22.2). Beach nourishment is infrequently applied in the Caribbean, largely due to limited institutional capacity and low governmental prioritization of coastal management (Neufville 2015; Spencer et al. 2022). Other examples of hybrid coastal solutions in the region include the managed retreat at Rocky Point, Jamaica and beach nourishment—groyne implementation at Junakoo Beach, Bahamas.

22.2.2 Nature-Based Solutions (NBS)

In a general context, nature-based solutions can be applied across both terrestrial and coastal settings and be defined as all actions inspired, supported or copied from nature to sustainably combat various environmental, social and economic challenges (European Commission 2015). Thus, via the European Commission (2015) definition, a case can be argued for the inclusion of traditional soft or hybrid engineering approaches under the umbrella of nature-based solutions. In fact, whilst detailing the historical development of coastal nature-based solutions, Pontee and Bassetti (2024) cited traditional soft and hybrid examples as far back as 1900 as past milestones towards the adoption of nature-based solutions in coastal settings. However, against the backdrop of intended design philosophies and/or ethos, a distinction between traditional soft/hybrid solutions and nature-based solutions can be made.

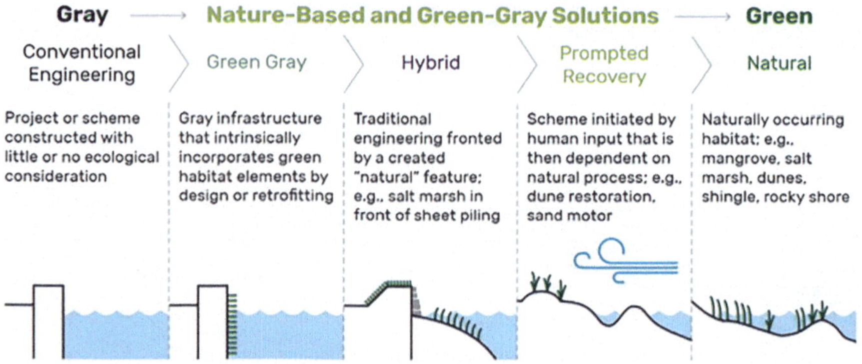

Fig. 22.3 Spectrum range of possible nature inclusion in nature-based solutions (adapted from Suedel et al. 2021)

Specifically, traditional soft or hybrid engineered solutions which have primarily focused on attaining coastal defence as the main design goal (via the means of replicating natural features/process) are arguably different from nature-based solutions whose design goals span beyond solely coastal defence to include environmental and socio-economic systems/features/processes from initial design conception.

Thus, a more accurate definition of coastal nature-based solutions can be given as solutions that utilize natural features to attain reduced coastal flooding and erosion risk whilst also promoting additional environmental and socioeconomic benefits (Pontee and Bassetti 2024). These types of solutions encompass many terms such as "ecosystem engineering", "engineering with nature" and "living shorelines" as detailed by Perricone et al. (2023). They span a range of interventions depending on the specific role of nature incorporated in the solution, as detailed by Suedel et al. (2021) (Fig. 22.3).

The coastal defence functionality of nature-based solutions is well recognized. For instance, coral reefs are known to be an effective natural coastal defence system via wave attenuation, sand production and sand retention facilitated by these structures (Ferrario et al. 2014; Pascal et al. 2016; Elliff and Silva 2017; Reguero et al. 2018). Oyster reefs can also facilitate wave attenuation resulting in sediment stabilization and shoreline protection (Meyer et al. 1997). Mangrove forests/swamps are hailed as one of the most effective natural costal defence systems due to their ability to facilitate wave attenuation, settle sediments, reduce wind passage and restrict flow energy (Perricone et al. 2023). Saltmarshes and seagrass beds are also recognized for their ability to facilitate wave energy reduction and stabilize sediments via their root structure systems as detailed by many studies such as Lara et al. (2016), Garzon et al. (2019a, b), Ondiviela et al. (2014) (and many other literature cited within).

In the Caribbean, nature-based solutions have surfaced as a modern, next generation approach to coastal protection. Several islands have employed such coastal solution strategies. Key examples being mangrove and coral restoration activities in Jamaica as detailed by Lee et al. (2022), ecosystem-based adaptation regarding the

Fig. 22.4 Mangrove replanting in Guyana (adapted from Banton et al 2017)

health of the Mesoamerican reef in Belize as detailed by Mercer et al. (2012), beach nourishment, revegetation, mangrove restoration and coral restoration at Grenville Bay, Grenada as detailed by Suedel et al. (2023) and mangrove restoration in Guyana due to national and international partnerships (Fig. 22.4).

Despite the above-mentioned examples, generally the use of coastal nature-based solution techniques in Latin America and the Caribbean remains relatively low (Marsters et al. 2021) linked to engineering performance limitations and knowledge gaps surrounding these solutions (Seddon et al. 2020; Vogelsang et al. 2023; Villaroel-Lamb et al. 2024).

22.2.3 National and Regional Initiatives: Policy and Planning

Policy and planning approaches which include land-use and marine-use planning, management strategies, development codes/legislation and disaster response and preparedness measures are also essential to coastal resilience in the Caribbean. Beavers et al. (2016) outlined the need for policy and planning in the pursuit of coastal resilience especially against the context of adaptation and climate change.

In the Caribbean these approaches are practiced at both regional and national levels. Nationally, most Caribbean countries possess some form of sectoral or Integrated Coastal Zone Management (ICZM) policy (Table 22.1). Examples of Caribbean countries which have established national ICZM policies and resulting national instruments are Barbados (The Barbados Policy Framework and Integrated

Coastal Zone Management Plan), Belize (The National Integrated Coastal Zone Management Strategy for Belize and Belize Integrated Coastal Zone Management Plan) and Trinidad and Tobago (Integrated Coastal Zone Management Policy Framework and Integrated Coastal Zone Management Plan and Marine Spatial Plan) (Muñoz 2020).

Comparatively between sectoral management and ICZM approaches, the latter is superior, being better able to facilitate national action in response to the many problems and conflicts that affect the coastal zone in the region (Munoz 2020). Moreover, typical sectoral management is inefficient in remedying the complex cross-sectoral management issues in the zone which often yield conflicts between varying sectors (Thia-Eng 1993; Scura et al. 1992; Fallon and Chua 1990). Previously, hinderances

Table 22.1 National coastal management policies and ICZM regulations and instruments in the Caribbean (adapted from Muñoz 2020)

Country	Coastal management policies/strategies	ICZM regulations	Main instrument/s for ICZM
Bahamas	–	Coastal Protection Act (1968)	Notification of and Request for Coastline Activity and projects
Barbados	The Barbados Policy Framework	Coastal Zone Management Act (1998)	Integrated Coastal Zone Management Plan
Belize	The National Integrated Coastal Zone Management Strategy for Belize (2000)	Coastal Zone Management Act (1998, revised 2000)	Belize Integrated Coastal Zone Management Plan (2016)
Cuba	–	Management of the Coastal Zone (2000)	Decleration of Areas under the Integrated Coastal Zone Management Regime (2007)
Jamaica	Ocean and Coastal Zone Management Policy (2002)	*Multiple laws, no single ICZM law*	National Spatial Plan Vision 2030
Puerto Rico	Adoption of the United States Coastal Zone Management Policy	Coastal Zone Management Act (1972)	Coastal Zone Management Program of Puerto Rico (1978, revised in 2009)
The Dominican Republic	–	(Draft) Sectoral Law of Marine Coastal Resources (2008)	Coastal and Marine Resources Management Program
Trinidad and Tobago	Approved Integrated Coastal Zone Management Policy Framework (2023)	*To be developed*	Integrated Coastal Zone Management Plan (ICZM) and Marine Spatial Plan

against greater adoption of ICZM in the region stemmed from a combination of national legislative, regulatory, structural and behavioural origins as detailed by Tompkins et al. (2002) who examined Trinidad and Tobago as a study area. However, recent advancement in the uptake of ICZM in Latin America and the Caribbean are noted, though changes and improvements are slow in most countries (Muñoz 2020).

Via a regional vantage point, the Caribbean community, through the intergovernmental organization, CARICOM, adopts a range of regional policies all geared towards the region's resilience. The recent Caribbean Community Environmental and Natural Resources Policy Framework (CENRPF) produced by CARICOM is one such example. CENRPF (2022–2032) clearly outlines the management of coastal resources as a key objective and the need for special attention on issues pertaining to the coastal environment, climate change, and natural disasters impact and management amongst other key considerations (Griffith 2025). Complementing the various regional policies adopted by CARICOM are established regional institutions, of which, the Caribbean Disaster Emergency Management Agency (CDEMA) and the Caribbean Community Climate Change Centre (CCCCC) are noteworthy. The former, for regional disaster management, where one of the institution's outlined priorities in the 2014–2024 strategy is the "Building of Disaster Resilience" (CDEMA 2025). The latter for the coordination of the region's response to climate change and being a central regional hub for information, management and adaptation on climate change in the Caribbean (CCCCC 2024).

Noting the mentioned, policy and planning approaches may not be effective at building coastal resilience in the region due to scope limitations and lack of enforcement. Robinson (2018), via semi-structured interviews with policymakers in the region, revealed that the scope of national policies, legislation and regulations along with deficiencies with monitoring, evaluation and enforcement capacity are indeed limitations to climate change adaptation in the Caribbean. Additionally, the existence of ICZM policies and regulations does not guarantee effective outcomes, though it can be presumed more effective in countries with sufficient institutional capacities (Muñoz 2020). In-actioning of developed policies is also one of the barriers to institutional climate resilience Palma-Oliveira et al. (2018). Still, policy planning approaches are essential to Caribbean coastal resilience. Indeed, against the reality of climate change, Caribbean policymakers recognize the need to build resilience into natural and human systems for effective adaptation (Robinson 2018).

22.2.4 International Collaboration, Capacity Building Initiatives and Stakeholder Involvement

Along with regional interventions, the Caribbean has also been the subject of international collaboration initiatives centred around coastal resilience and management (Davy and Brenton 2006; Chakalall et al. 1998; Garcia 2002). These interventions have proven pivotal in the advancement of ICZM in the region. For example, the

ICZM progress made in Jamaica was only possible via the United Nations Environmental Programme (UNEP) collaboration within the framework of the National Spatial Plan Vision 2030; Whilst the ICZM progress made in Barbados and Belize are the result of successful international collaboration (Muñoz 2020). Similarly, the ICZM policy framework developed by Trinidad and Tobago benefitted from technical studies funded by the Inter-American Development Bank (IADB) (ICZM Inter-Ministerial Committee 2023).

International collaboration has also assisted in furthering climate change adaptation in the region. One such example is the cooperation between the European Union (EU), UNEP and the Planning Institute of Jamaica (PIOJ) to forward climate change adaptation, disaster risk reduction including improved coastal ecosystem management in Jamaica (Mercer et al. 2012). International financing for climate change adaptation is also noteworthy in the region. Fifteen Caribbean countries received a total of US$824 million in commitments from the Organization for Economic Co-operation and Development (OECD) members during 2010–2014 (Robinson 2018). According to Robinson (2018), the principal donors to the Caribbean—namely France, the European Union (EU), Norway, Spain, and the United States—collectively accounted for 79% of total aid commitments to the region.

Despite international support, implementing collaborative and financial initiatives in the region remains challenging. Breton and Davy (2006) postulated that many international collaborations in the region are established based on the political, geographical, and linguistic diversity of the region. Breton and Davy (2006) further stated the result of such types of collaborations are sporadic exchanges with reduced effectiveness and inefficiencies, ultimately yielding individual products that cannot be compared and analyses lacking for development in the region. Additionally, although commitments and international pledges have been made, Caribbean countries continue to encounter challenges in acquiring funding for climate change (Mohan 2023). International financing is often unequal across the region and typically accompanied by eligibility requirements and specific conditions related to use of provided funds (Robinson 2018).

Capacity building via enhancing the technical capabilities in the region is also pivotal to Caribbean coastal resilience. In fact, technical capacity and resources are noted by regional policymakers as the second most significant hinderance to climate change adaptation in the Caribbean (Robinson 2018). Regionally, CARICOM has implemented a series of projects which outlined the building of capacity as one of its main pursuits (Medeiros 2011). Resultantly, several institutions in the region offer training and capacity building programs linked to coastal resilience and management. CDEMA's Regional Training Center (RTC) is one such example. CDEMA RTC was established in 2015 and follows the mandate of building capacity for comprehensive disaster management amongst regional members (CDEMA 2018). Localized knowledge transfer and skill training initiatives in the region are also essential to capacity building in the region. Polar (2025) detailed the example of a two-day workshop facilitated by the Coastal Hazard Vulnerability Project in Dominica. The workshop focused on GIS and field training, which brought together 19 participants across varying sectors to build Dominica's local capacity towards coastal protection (Polar 2025).

Equally significant are the exemplary roles regional universities play in the context of building technical capacity via education, inducting coastal management graduate programs over only a few decades (Breton and Davy 2006). The University of the West Indies (UWI) is particularly noteworthy in this regard, having introduced an optional level III coastal engineering course within the undergraduate B.Sc. in Civil Engineering degree in 2002 and the development of two graduate offerings: Post Graduate Diploma in Coastal Zone Processes and M.Sc. In Coastal Zone Engineering and Management (Ibrahim and Charles 2006). Currently, the UWI is the only regional tertiary institution offering an accredited M.Sc. in Coastal Engineering and Management programme spanning a range of core courses delivered by a blended modality (Table 22.2). Still, regional capacity differences are noted in terms of individual countries' research organization capacities (Breton and Davy 2006).

Stakeholder participation in the planning and management of coastal zones is also essential in attaining coastal resiliency in the Caribbean region. Public stakeholder

Table 22.2 UWI M.Sc. in coastal engineering and management course outline (adapted from UWI 2025)

Module	Course name	Credits	Assessment	Notes
Semester 1				
CZEM 6100	Coastal Processes and Hazards	4	100% Coursework	
CZEM 6130	Principles of Coastal Defence, Management and Environmental Assessment	4	100% Coursework	Intensive group project
CZEM 6101	Coastal Geomorphology	4	100% Coursework	Virtual field trips
COEM 6016	Natural Hazards Management	4	100% Coursework	
Semester 2				
CZEM 6102	Coastal Zone Metrics	4	100% Coursework	Use of GIS software. Intensive one week field activity
CZEM 6106	Design of Coastal Structures	4	100% Coursework	Problem-based learning. Oral presentation to group
CZEM 6112	Coastal Zone Modelling	4	100% Coursework	Use of Numerical Models
COEM 6020	Research methods	3	100% Coursework	Oral presentation to group
Semester 3				
CZEM 6108	Research Project	9	100% Coursework	

participation at local levels allows communities to develop a sense of ownership as stewards when they are purposefully involved in the protection of the coastal environment (Lee et al. 2022). Caribbean coastal management should begin and continually develop at the community level, as evidence continually suggests that local level stakeholders critically contribute to the knowledge base for effective management of the zone (Davy and Breton 2006). Indeed, Community-based Disaster Risk Reduction (CBDRR) and Community-based Adaptation (CBA) relies significantly upon the action of local communities and the local knowledge (Shaw et al. 2009; Kelman and Gailard 2010; UNDP 2025). Additionally, planning and policy developed upon an understanding of the individual climate change challenges faced by local communities are more effective in the protection of human infrastructure and systems (Robinson and Wren 2020).

At a regional and national level, Robinson (2018) noted public awareness activities and stakeholder consultations as a climate change adaptation action being undertaken by policymakers in the Caribbean. In fact, stakeholder involvement can also address many governmental challenges linked to social conflicts, adaptation planning uncertainties and capability limitations when pursuing the complexities that may arise with Sea Level Rise (SLR) (IPCC 2022). National and regional ICZM endeavours inherently necessitates stakeholder participation; as successful ICZM programs recognizes the need for stakeholder involvement such that public approval and stakeholder requirements are met (Thai-Cheng 1993; UNDP 2025). A noteworthy example of such consultations is observed in the development of the Trinidad and Tobago ICZM Policy Framework where governmental agencies, business and energy sectors, fisherfolk and the general public were engaged through thematic working groups and public consultations (ICZM Inter-Ministerial Committee 2023).

Relevant Non-Governmental Organizations (NGOs) in the region also form a critical part of the stakeholder body, as they reinforce the role of civil society and are prevalent institutions concerned with the management approaches adopted (Davy and Breton 2006). The Future Centre Trust in Barbados is one such NGO which is actively involved with the education, awareness and beach clean-up actions on the island, with the goal of better informing tourist and locals on the significance of Barbados's ecosystems for sustainability and climate change adaptation (Mercer et al. 2012).

Although there are notable benefits of stakeholder involvement spanning various levels, observable challenges have been identified within the region within the specific context of coastal zone management and climate change adaptation. Specifically, coastal zone management in the Caribbean has focused primarily on the regional and national levels with little recognition of the importance of local community-based knowledge or the need to mobilize communities at the local level (Breton and Davy 2006). Additionally, there is the perceived notion amongst the private sector and other stakeholders that the government holds the sole responsibility for climate change action (Robinson 2018).

22.3 Discussion

Individually, the afore-mentioned four broadly outlined coastal resilience approaches (namely, 1. Traditional Infrastructure and Engineered Solutions; 2. NBS; 3. National and Regional Initiatives: Policy and Planning; 4. International Collaboration, Capacity Building Initiatives and Stakeholder Involvement) when considered individually, presents specific drawbacks and challenges within the Caribbean context. However, an overlap of these approaches can permit the drawing on the strengths of one approach as a means of mitigating or overriding the shortcoming of another. For example, NbS can encompass ecosystem restoration or area-based conservation along with an ICZM Policy (Lee et al. 2022). In such a consolidation of approaches, the NbS engineering limitations and knowledge gaps mentioned by Seddon et al. (2020) and Volgesang et al. (2023) can be observed, documented and possibly addressed by a suitable ICZM Policy. Similarly, the apparent lack of interest by certain non-governmental stakeholders in the region regarding climate change adaptation strategies (Robinson 2018), may be overcome by the inclusion of NbS in such efforts, as research has shown a propensity of stakeholders to support NbS as compared to grey solutions (Deely and Hynes 2020). The purposeful design inclusion of natural ecosystems alongside traditional hard engineered solutions permits for the incorporation of sustainability, adaptation and self-sustaining characteristics not typical of traditional hard/grey coastal defence structures (Perricone et al. 2023). Conversely, the time-dependent protection level of the natural elements of NbS, along with the vulnerability of these natural NbS components to anthropogenic and climate change (Boström-Einarsson et al. 2020; Hughes et al. 2017; Lovelock et al. 2022) can be mitigated by the use of grey-green configurations over designated time periods throughout the solution's design life. Likewise, the national-level funding competitions in the region which have resulted in the highly unsuccessful ICZM policies in the region (Muñoz 2020) can be addressed by the incorporation of international collaborations and capacity building initiatives outlined in Sect. 22.2.

The specific combination/s and/or ordering of approaches to be used within the region must be examined on a project specific basis, as Klöck and Nunn (2019) outlined the need to better understand where adaptations may be possible and the suitability of interventions in such cases. This inherently implies the need for adequate information and data records prior to decision-making. In fact, quality data information, data sharing, along with simplicity of access and transfer, are critical to building climate resilience across the region (CDB 2018). Regionally, however, the lack of data and records are noted by policymakers as the third major limitation towards climate change adaptation (Robinson 2018). This limitation can negatively impact upon the region's resilience even at the planning and policy level. Specifically, in developing resiliency policies, information and knowledge-based data is needed to shape the specifics of policy with respect to design, prioritizations and resource allocation (Galaitsi et al. 2024). Additionally, whilst several tools specific to the coastal zone have been developed in recent years (e.g. Montaño et al. 2020; Vos et al. 2019; Almeida et al. 2021), limited data and data integration remain a challenge for both short- and long-term coastal monitoring programmes in developing countries (Oyedotun and Burningham 2021). Also noteworthy is the "Lack of data" being listed as a NbS implementation barrier in the region (Bailey et al. 2022), leading to the need for comprehensive data collection as an intervention in support of wider adoption of NbS in the region (Villaroel-Lamb et al. 2024).

The outlook of Caribbean's coastal resilience is also dependent on the region's ability to transcend beyond the multifaceted challenges which plague the region. Issues relating to geopolitical and environmental characteristics, geographic distribution, socio-economic inequalities, and cultural diversity are unique hinderances against the sustainable development of the marine and coastal resources in the region (Gómez et al. 2006). Limited economy diversification, societal hierarchy inequalities, and ecological degradation are notable vulnerability factors which stem from the colonial history of the region (Laurent and Duvat 2024; Baclay et al. 2019; Popke and Rhiney 2019; Stancioff et al. 2018). Additionally, the issue of existing high economic debt in the region limits the financial capabilities needed for climate change adaptation and adaptation-related governance; thus, the region's economic and human vulnerability further increases against the backdrop of climate change (Laurent and Duvat 2024). Robinson (2018) noted governance and politics as areas of concern regarding climate change adaptation and the reduced vulnerability of the region. The role of politics in promoting vulnerability must be overcome for the depoliticizing of climate change (Robinson, 2018; Smith and Rhiney 2016). Simultaneously, Robinson (2018) noted the region's subpar performance on the corruption perception index as a basis for improvements in reducing the region's vulnerability linked to governance. Societal legacies of colonialism and the holdover of past mindsets are noteworthy factors which limit the progression of coastal resilience in the region. For example, the "seawall mindset" imposed by developed countries unto developing ones over decades have resulted in a propensity of the region to primarily adopt hard costal defence structures (Banton et al., 2017; Jackson et al. 2012; Klock et al. 2022), since their use provides an immediate sense of protection to local communities (Betzold and Mohamed 2017; Logan et al. 2018; Nunn et al. 2021; Ratter et al. 2019).

However, it is apparent from the foregoing that regionally there is the need for the adoption of transdisciplinary (including the collaborative methods of co-creation and co-design), cross-sectoral, multilevel institutional approaches to address these intractable areas within the Caribbean region to achieve coastal resilience. Resilient coastal communities can only emanate from solutions that focus on the people who use those areas and abandoning solutions that destroy or degrade valuable ecosystems (Villarroel-Lamb et al. 2024). Resource constraints, both human and financial, that limit achieving resilience in coastal communities can be overcome by judicious projects through transdisciplinary teams that involve a wide-range of stakeholder engagement activities (Bailey et al. 2022).

Notwithstanding the significance of all the coastal resiliency approaches detailed in Sect. 22.2, the implementation of ICZM and NbS techniques appear markedly noteworthy in the Caribbean. ICZM progress in Latin America and the Caribbean has significantly improved since 2001, where as many as nine countries progressed beyond "pre-initial" stage grouping in terms of ICZM progress (Muñoz 2020). NbS use in the region has also increased, accounting for 22%, the second highest climate change adaptation action taken in the region (Laurent and Duvat 2024). Coastal NbS also appears particularly attractive due to its ability to sustain or promote natural ecosystems whilst also support community-based livelihoods and sustainable tourism in the region; key components given the restricted land size, high coastal population and limited resources typical of small island developing states of the Caribbean (Villaroel-Lamb et al. 2024). Still, the region can continually pursue further advancements towards improved coastal resilience. Most evident is the need to advance the mechanisms by which coastal NbS can be implemented regionally through development of design guidelines and policies and legislative instruments that prioritize these types of solutions (Cohen-Shacham et al. 2019). Two aligning developments are the recent adoption of Artificial Intelligence (A.I.) in coastal engineering and management (e.g. Masria and Abouelsaad 2025; Juan et al. 2025) and the Allen et al. (2019) definition of resilience from three distinct vantage points: (1) resilience as a rate—the normative view, based on engineering perspectives, (2) resilience as a process—another normative view, the primary framing for community resilience and natural hazard management, and (3) resilience as an emergent property—accounting for configurations of differing adaptive and transformative social-ecological systems spanning varying scales (Clement et al. 2024).

22.4 Conclusion

This chapter pursued the primary goal of highlighting the challenges and pathways towards an augmented level of coastal resilience in the Caribbean region. Firstly, a non-exhaustive overview of the Caribbean was presented, detailing the key range of coastal geomorphological features, vulnerability aspects and prominent risks and hazards prevalent in the region. Secondly, the main coastal resilience approach strategies currently adopted in the region were outlined. Namely, i. Traditional Infrastructure and Engineered Solutions, ii. Nature-based Solutions (NbS), iii. National and Regional Initiatives: policy and planning, as well as iv. International Collaboration, Capacity Building Initiatives and Stakeholder Involvement. Brief details of each approach were presented, with notable examples of their use in Caribbean settings. It was observed that each named approach possessed strengths and weaknesses with no single approach emerging as the most superior. However, the most used approach in the region is the Traditional Infrastructure and Engineered Solutions encompassing conventional hard, soft and hybrid interventions. Finally, some stimulating observations were presented that outlined the potential combinations of suitable interventions with potential benefits for the region. Also succinctly included are region specific challenges which hinder the effective implementation of various approaches, either implemented individually or as a combination of multiple methods. Particularly significant are the social, economic and political challenges that hinder tangible progress towards coastal resilience in the Caribbean region. Finally, this pathway towards improved coastal resilience is indeed a multifaceted problem, necessitating the need for a transdisciplinary approach spanning multiple sectors, at differing levels, and encompassing a wide range of affected and influential stakeholders.

AI Disclosure The authors have not employed the use of AI in the research.

Competing Interests The authors have no conflicts of interest to declare that are relevant to the content of this chapter.

Plagiarism The authors declare that the chapter contains a similarity index of _11%__.

References

Allen CR et al (2019) Resilience reconciled. Nat Sustain 2:898–900. https://doi.org/10.1038/s4189301904014

Almeida LP et al (2021) Coastal analyst system from space imagery engine (CASSIE): Shoreline management module. Environ Model Softw 140:105033. https://doi.org/10.1016/j.envsoft.2021.105033

Query ID="Q4" Text="" Bailey J et al (2022) Building a more resilient and low-carbon Caribbean: report 4: infrastructure resilience in the Caribbean through nature-based solutions. Inter-American Development Bank, Washington, DC, USA

Banton JD et al (2017) Selection of appropriate coastal protection strategies for Caribbean coastlines. In: Coastal structures and solutions to coastal disasters 2015

Barclay J et al (2019) Historical trajectories of disaster risk in Dominica. Int J Disaster Risk Sci 10:149–165. https://doi.org/10.1007/s137530190215z

Beavers R, Babson A, Schupp C (eds) (2016) Coastal adaptation strategies handbook. NPS 999/134090 National Park Service, Washington DC

Betzold C, Mohamed I (2017) Seawalls as a response to coastal erosion and flooding: a case study from Grande Comore, Comoros (West Indian Ocean). Reg Environ Change 17:1077–1087. https://doi.org/10.1007/s101130161044x

Boström-Einarsson L et al (2020) Coral restoration—a systematic review of current methods, successes, failures and future directions. PLoS ONE 15:e0226631. https://doi.org/10.1371/journal.pone.0226631

Breton Y, Davy B (2006a) Analytical insights, lessons learnt, and recommendations. In: Breton Y, Brown D, Davy B et al (eds) Coastal resource management in the wider caribbean: resilience, adaptation, and community diversity. Ian Randle Publishers, Kingston Jamaica, pp 223–254

Breton Y et al (2006b) Social sciences and the diversity of caribbean communities. In: Breton Y, Brown D, Davy B et al (eds) Coastal resource management in the wider caribbean: resilience, adaptation, and community diversity. Ian Randle Publishers, Kingston Jamaica, pp 17–49

CAF (Development Bank of Latin America and the Caribbean) (2025) The Caribbean: development that demands resilience. In: Caf.com. https://www.caf.com/en/currently/news/the-caribbean-development-that-demands-resilience. Last accessed 24 Jul 2025.

Caribbean Development Bank (CDB) (2018) CDB announces USD1.5mn to build greater climate resilience through technology|Caribbean Development Bank. In: Caribank.org. https://www.caribank.org/newsroom/news-and-events/cdb-announces-usd15mn-build-greater-climate-resilience-through-technology. Last accessed 24 Jul 2025

Caribbean Disaster Emergency Management Agency (CDEMA) (2025) About CDM. In: CDEMA—Caribbean disaster emergency management Agency. https://www.cdema.org/index.php/cdm. Last accessed 24 Jul 2025

Carlson RR et al (2021) Synergistic benefits of conserving land-sea ecosystems. Glob Ecol Conserv 28:e01684. https://doi.org/10.1016/j.gecco.2021.e01684
Cashman A, Nagdee M (2017) Impacts of climate change on settlements and infrastructure in the coastal and marine environments of Caribbean small island developing states (SIDS), pp 155–173
Cavallo EA et al (2024) Climate change, hurricanes, and sovereign debt in the Caribbean basin. Inter-American Development Bank (IADB), Washington DC
Caribbean Community Climate Change Centre (CCCCC) (2024) About Us. https://caribbeanclimate.org/about/?_gl=1. Last accessed 31 Jul 2025
Caribbean Disaster Emergency Management Agency (CDEMA) (2018) Regional training centre. In: Regional training centre. https://cdema.org/rtc/index.php?option=com_sppagebuilder&view=page&id=6. Last accessed 29 Jun 2025
Chakalall B, Mahon R, McConney P (1998) Current issues in fisheries governance in the Caribbean community (CARICOM). Mar Policy 22:29–44. https://doi.org/10.1016/S0308597X(97)800024
Clement S et al (2024) How resilience is framed matters for governance of coastal social ecological systems. Environ Policy Gov 34:65–76. https://doi.org/10.1002/eet.2056
Cohen-Shacham E et al (2019) Core principles for successfully implementing and upscaling Nature based solutions. Environ Sci Policy 98:20–29. https://doi.org/10.1016/j.envsci.2019.04.014
Darsan J et al (2012) Coastal conservation project: status of beaches and bays in Trinidad (2004–2008). Institute of Marine Affairs, Trinidad
Davy B, Breton Y (2006) Introduction. In: Breton Y, Brown D, Davy B et al (eds) Coastal resource management in the wider Caribbean: resilience, adaptation, and community diversity. Ian Randle Publishers, Kingston Jamaica, pp 1–16
Deely J, Hynes S (2020) Bluegreen or grey, how much is the public willing to pay? Landsc Urban Plan 203:103909. https://doi.org/10.1016/j.landurbplan.2020.103909
Dorenbosch M et al (2007) Influence of habitat configuration on connectivity between fish assemblages of Caribbean seagrass beds, mangroves and coral reefs. Mar Ecol Prog Ser 334:103–116. https://doi.org/10.3354/meps334103
Duval DT (2004) Tourism in the caribbean: Trends, development, prospects, 1st edn. Routledge, London
Elliff CI, Silva IR (2017) Coral reefs as the first line of defense: Shoreline protection in face of climate change. Mar Environ Res 127:148–154. https://doi.org/10.1016/j.marenvres.2017.03.007
European Commision (2015) Towards an EU research and innovation policy agenda for nature-based solutions and re-naturing cities. https://doi.org/10.2777/479582
Fallon L, Chua T (1990) Towards strengthening policy and strategies orientation for fisheries resource management: the role of coastal area management. Trop Coast Area Manag 5:3–5
Ferrario F et al (2014) The Effectiveness of coral reefs for coastal hazard risk reduction and adaptation. Nat Commun 5:1–9. https://doi.org/10.1038/ncomms4794
Galaitsi SE et al (2024) Balancing climate resilience and adaptation for Caribbean small island developing states (SIDS): building institutional capacity. Integr Environ Assess Manag 20:1237–1255. https://doi.org/10.1002/ieam.4860
Garcia M (2002) The Caribbean: main experiences and regularities in capacity building for the management of coastal areas. Capacity Buildingpapers from the Global Conference on Oceans and Coasts at Rio+10: toward the 2002 world summit on sustainable development 45:677–693. https://doi.org/10.1016/S09645691(02)000935
Garzon JL et al (2019a) Wave Attenuation by Spartina saltmarshes in the Chesapeake Bay under storm surge conditions. J Geophys Res Oceans 124:5220–5243. https://doi.org/10.1029/2018JC014865
Garzon JL et al (2019b) Fieldbased numerical model investigation of wave propagation across marshes in the Chesapeake Bay under storm conditions. Coast Eng 146:32–46. https://doi.org/10.1016/j.coastaleng.2018.11.001

Gómez LM et al (2006) Community mobilisation and education in contaminated coastal ecosystems. In: Breton Y, Brown D, Davy B et al (eds) Coastal resource management in the wider caribbean: resilience, adaptation, and community diversity. Ian Randle Publishers, Kingston Jamaica, pp 137–171

Griffith M (2025) Caribbean community environmental and natural resources policy framework 2022–2032. CARICOM

Haddout S et al (2024) Estuarine management and technologies: a way forward for estuarine research. Estuar Manag Technol 1:1–6. https://doi.org/10.3897/emt.1.116778

Harbitz CB et al (2012) Tsunami hazard in the Caribbean: regional exposure derived from credible worst case scenarios. Cont Shelf Res 38:1–23. https://doi.org/10.1016/j.csr.2012.02.006

Harborne AR et al (2006) The functional value of Caribbean Coral Reef, seagrass and Mangrove habitats to ecosystem processes. Adv Mar Biol 50:57–189. https://doi.org/10.1016/s0065-2881(05)50002-6

Hughes TP et al (2017) Coral reefs in the Anthropocene. Nature 546:82–90. https://doi.org/10.1038/nature22901

Ibrahim JC, Charles R (2006) Coastal zone engineering and management: medium and long-term solutions for coastal disasters in the Caribbean. In: Fourth LACCEI International Latin American and Caribbean conference for engineering and technology: breaking frontiers in engineering: education, research and practice

ICZM Inter-Ministerial Committee (2023) ICZM Policy Framework—ICZM-TT. In: ICZM-TT. https://www.iczm.gov.tt/policy/draft-policy-document/

Institute of Marine Affair (IMA) (2022) Coastal issues. In: IMA. https://www.ima.gov.tt/2022/10/10/coastal-issues/

Intergovernmental Panel on Climate Change (IPCC) (2022) Sea level rise and implications for lowlying islands, coasts and communities. In: The ocean and cryosphere in a changing climate: special report of the intergovernmental panel on climate change. Cambridge University Press, Cambridge, pp 321–446

Jackson CW et al (2012) Documenting beach loss in front of seawalls in Puerto Rico: pitfalls of engineering a small island nation shore. In: Andrew, Pilkey OH (eds) Pitfalls of shoreline stabilization: selected case studies. Springer Netherlands, Dordrecht, pp 53–71

Juan NP, Rodríguez JO, Campo D, Valdecantos VN (2025) Artificial intelligence in coastal engineering: application to climate change studies. J Coast Res 113:180–184. https://doi.org/10.2112/JCRSI113036.1

Kelman I, Gaillard JC (2010) Embedding climate change adaptation within disaster risk reduction. In: Shaw R, Pulhin JM, Pereira J (eds) Climate change adaptation and disaster risk reduction: issues and challenges. Emerald Group Publishing Limited

Kjerfve B (1998) CARICOMP: Caribbean coral reef, seagrass and mangrove sites. UNESCO

Klöck C, Nunn PD (2019) Adaptation to climate change in small island developing states: a systematic literature review of academic research. J Environ Dev 28:196–218. https://doi.org/10.1177/1070496519835895

Klöck C et al (2022) Maladaptive diffusion? The spread of hard protection to adapt to coastal erosion and flooding along island coasts in the Pacific and Indian ocean. Reg Environ Change 22:136. https://doi.org/10.1007/s1011302201989x

Knutson T et al (2020) Tropical cyclones and climate change assessment: part ii: projected response to anthropogenic warming. Bull Am Meteor Soc 101:E303–E322. https://doi.org/10.1175/BAMSD180194.1

Lara JL et al (2016) Largescale 3D experiments of wave and current interaction with real vegetation. Part 1: guidelines for physical modeling. Coast Eng 107:70–83. https://doi.org/10.1016/j.coastaleng.2015.09.012

Laurent C, Duvat VKE (2024) Addressing the climate adaptation tracking gap: an assessment method and its application to the Caribbean region. Reg Environ Change 24:147. https://doi.org/10.1007/s10113024023019

Lee S et al (2022) The role of Nature based Solutions in disaster resilience in coastal Jamaica: current and potential applications for "building back better." Disasters 46:S78–S100. https://doi.org/10.1111/disa.12539

Lewsey C et al (2004) Assessing climate change impacts on coastal infrastructure in the Eastern Caribbean. Mar Policy 28:393–409. https://doi.org/10.1016/j.marpol.2003.10.016

Logan TM et al (2018) Hardadaptive measures can increase vulnerability to storm surge and tsunami hazards over time. Nat Sustain 1:526–530. https://doi.org/10.1038/s4189301801376

Lovelock CE et al (2022) Tackling the mangrove restoration challenge. PLoS Biol 20:e3001836. https://doi.org/10.1371/journal.pbio.3001836

Marsters L et al (2021) Nature-based solutions in Latin America and the Caribbean: financing mechanisms for regional replication. Inter-American Development Bank (IADB) and World Resources Institute, Washington DC

Martínez-Colón M et al (2021) Bioaccumulation and biomagnification of potential toxic elements (PTEs): an Avicennia germinans–Uca rapax trophic transfer story from Jobos Bay, Puerto Rico. Ecol Indic 121:107038. https://doi.org/10.1016/j.ecolind.2020.107038

Masria A, Abouelsaad O (2025) Artificial intelligence applications in coastal engineering and its challenges—a review. Cont Shelf Res 286:105425. https://doi.org/10.1016/j.csr.2025.105425

Masselink G, Lazarus ED (2019) Defining Coastal Resilience. Water 11. https://doi.org/10.3390/w11122587

McAdie C et al (2009) Tropical cyclones of the North Atlantic Ocean, 1851–2006: With 2007 and 2008 track maps included. US Department of Commerce, National Oceanic and Atmospheric Administration (NOAA)

Medeiros D et al (2011) Review of current and planned adaptation action: the Caribbean. International Institute for Sustainable Development

Mercer J et al (2012) Ecosystem based adaptation to climate change in Caribbean small island developing states: integrating local and external knowledge. Sustainability 4:1908–1932. https://doi.org/10.3390/su4081908

Meyer DL et al (1997) Stabilization and erosion control value of Oyster cultch for intertidal marsh. Restor Ecol 5:93–99. https://doi.org/10.1046/j.1526100X.1997.09710.x

Mohan PS (2023) Financing climate change mitigation and adaptation in Caribbean SIDS. PLOS Clim 2:e0000167. https://doi.org/10.1371/journal.pclm.0000167

Montaño J et al (2020) Blind testing of shoreline evolution models. Sci Rep 10:2137. https://doi.org/10.1038/s41598020590l8y

Muñoz JMB (2020) Progress of coastal management in Latin America and the Caribbean. Ocean Coast Manag 184:105009. https://doi.org/10.1016/j.ocecoaman.2019.105009

Murakami H et al (2018) Dominant effect of relative tropical Atlantic warming on major hurricane occurrence. Science 362:794–799. https://doi.org/10.1126/science.aat6711

Mycoo M, Chadwick A (2012) Adaptation to climate change: the coastal zone of Barbados. Marit Eng 165:159–168. https://doi.org/10.1680/maen.2011.19

Narayan S et al (2016) The effectiveness, costs and coastal protection benefits of natural and nature based defences. PLOS ONE 11. https://doi.org/10.1371/journal.pone.0154735

National Oceanic and Atmospheric Administration (NOAA) (2022) Climate prediction center—Atlantic Hurricane outlook. In: Noaa.gov. https://www.cpc.ncep.noaa.gov/products/outlooks/hurricane2022/August/hurricane.shtml. Last accessed 30 Jul 2025

Neufville Z (2015) Row Erupts over Jamaica's Bid to slow beach erosion. In: Inter press service. https://www.ipsnews.net/2015/02/row-erupts-over-jamaicas-bid-to-slow-beach-erosion/. Last accessed 30 Jul 2025

Nunn PD, Klöck C, Duvat V (2021) Seawalls as maladaptations along island coasts. Ocean Coast Manag 205:105554. https://doi.org/10.1016/j.ocecoaman.2021.105554

Ondiviela B et al (2014) The role of seagrasses in coastal protection in a changing climate. Coasts@Risks: THESEUS, a new wave in coastal protection 87:158–168. https://doi.org/10.1016/j.coastaleng.2013.11.005

Oyedotun TDT, Burningham H (2021) The need for data integration to address the challenges of climate change on the Guyana coast. Geogr Sustain 2:288–297. https://doi.org/10.1016/j.geosus.2021.11.003

Palma-Oliveira et al (2018) Community driven hypothesis testing: a solution for the tragedy of the Anticommons. Risk Anal 38:620–634. https://doi.org/10.1111/risa.12860

Pascal N et al (2016) Economic valuation of coral reef ecosystem service of coastal protection: a pragmatic approach. Ecosyst Serv 21:72–80. https://doi.org/10.1016/j.ecoser.2016.07.005

Patricola CM, Wehner MF (2018) Anthropogenic influences on major tropical cyclone events. Nature 563:339–346. https://doi.org/10.1038/s4158601806732

Pentelow L, Scott DJ (2011) Aviation's inclusion in international climate policy regimes: implications for the Caribbean tourism industry. J Air Transp Manag 17:199–205. https://doi.org/10.1016/j.jairtraman.2010.12.010

Perricone V et al (2023) Naturebased and bioinspired solutions for coastal protection: an overview among key ecosystems and a promising pathway for new functional and sustainable designs. ICES J Mar Sci 80:1218–1239. https://doi.org/10.1093/icesjms/fsad080

Polar P (2025) Building local capacity for coastal resilience in Dominica. In: Hitresetcaribbean.org. https://hitresetcaribbean.org/building-local-capacity-for-coastal-resilience-in-dominica/. Last accessed 14 Jul 2025

Pontee N, Bassetti L (2024) Using naturebased solutions to improve coastal resilience. Proc Inst Civ Eng 177:54–66. https://doi.org/10.1680/jcien.24.00946

Popke J, Rhiney K (2019) Introduction to special issue on the Caribbean after Irma and Maria: climate, development & the PostHurricane context. J of Extr Even 06:1902001. https://doi.org/10.1142/S2345737619020019

Quesada-Román A, Pérez-Briceño PM (2019) Geomorphology of the Caribbean coast of Costa Rica. J Maps 15:363–371. https://doi.org/10.1080/17445647.2019.1600592

Ramirez DR, Townsend W (2006) Seaweed and Mangroves: Improving Environmental Practices in Coastal Communities to Secure Sustainable Livelihoods. In: Breton Y, Brown D, Davy B et al (eds) Coastal resource management in the wider Caribbean: resilience, adaptation, and community diversity. Ian Randle Publishers, Kingston Jamaica, pp 172–191

Ratter B et al (2019) Challenges for shared responsibility—political and social framing of coastal protection transformation in the Maldives. DIE ERDE—J Geogr Soc Berl 150:169–183. https://doi.org/10.12854/erde-2019-426

Reed KA et al (2021) Anthropogenic influence on Hurricane Dorian's extreme rainfall. Bull Am Meteor Soc 102:S9–S15. https://doi.org/10.1175/BAMSD200160.1

Reed KA et al (2022) Attribution of 2020 hurricane season extreme rainfall to humaninduced climate change. Nat Commun 13:1905. https://doi.org/10.1038/s41467022293791

Reguero BG et al (2018) Coral reefs for coastal protection: a new methodological approach and engineering case study in Grenada. J Environ Manag 210:146–161. https://doi.org/10.1016/j.jenvman.2018.01.024

Resilient Islands (2019) Valuing benefits of mangroves and coral reefs in the Caribbean. Bvearmbdo. https://bvearmb.do/handle/123456789/5664

Rhiney K (2015) Geographies of Caribbean vulnerability in a changing climate: issues and trends. Geogr Compass 9:97–114. https://doi.org/10.1111/gec3.12199

Richards A (2008) Development trends in Jamaica's coastal areas and the implications for climate change. Planning Institute of Jamaica, Kingston

Robinson S, Wren C (2020) Geographies of vulnerability: a research note on human system adaptations to climate change in the Caribbean. Geogr TidsskriftDanish J Geogr 120:79–86. https://doi.org/10.1080/00167223.2020.1733432

Robinson S (2018) Adapting to climate change at the national level in Caribbean small island developing states. Isl Stud J 13:79–100. https://doi.org/10.24043/isj.59

Sarwat MS (2024) Impacts and effects of pollution on marine and coastal resources. Uttar Pradesh J Zool 45:403–409. https://doi.org/10.56557/upjoz/2024/v45i164322

Schueler K (2017) Naturebased solutions to enhance coastal resilience. Inter-American Development Bank (IADB), Bahamas

Scott D et al (2012) The vulnerability of Caribbean coastal tourism to scenarios of climate change related sea level rise. J Sustain Tour 20:883–898. https://doi.org/10.1080/09669582.2012.699063

Scura LF et al (1992) Lessons from integrated coastal zone management: the ASEAN experience. In: Integrative Framework and Methods for Coastal Area Management, pp 1–68

Seddon N et al (2020) Understanding the value and limits of naturebased solutions to climate change and other global challenges. Philos. Trans. R. Soc. B: Biol. Sci. 375:20190120. https://doi.org/10.1098/rstb.2019.0120

Shaw R et al (2009) Indigenous knowledge and disaster risk reduction: from practice to policy. Nova Science Publishers Inc., New York, USA

Smith RJ, Rhiney K (2016) Climate (in)justice, vulnerability and livelihoods in the Caribbean: the case of the indigenous Caribs in northeastern St. Vincent. Geoforum 73:22–31. https://doi.org/10.1016/j.geoforum.2015.11.008

Smith Warner Team (2022) Every grain counts. In: Smith Warner International Ltd. https://www.smithwarner.com/every-grain-counts/. Last accessed 30 Jul 2025

Spencer N et al (2022) Sea level rise under climate change: Implications for beach tourism in the Caribbean. Ocean. Coast. Manag. 225. https://doi.org/10.1016/j.ocecoaman.2022.106207

Stancioff CE et al (2018) Local perceptions of climate change impacts in St. Kitts (Caribbean Sea) and Malé, Maldives (Indian Ocean). Atmosphere 9. https://doi.org/10.3390/atmos9120459

Strauss B, Kulp S (2018) Sea-level rise threats in the Caribbean: data, tools, and analysis for a more resilient future. Inter-American Development Bank (IADB), Princeton NJ

Suedel BC et al (2021) Enhancing structural measures for environmental, social, and engineering benefits. In: Mohan RK (ed) International guidelines on natural and nature-based features for flood risk mitigation. U.S. Army Engineer Research and Development Center, Vicksburg, MS, USA, pp 701–771

Suedel BC et al (2023) Naturebased solutions for building resilience in small island developing states in the Caribbean. In: D MC, Bennett ER (eds) Building resilience to climate change in small island developing states (SIDS) in the Caribbean. Springer Nature Switzerland, Cham, pp 35–48

Thia-Eng C (1993) Essential elements of integrated coastal zone management. Integr Coast Manag 21:81–108. https://doi.org/10.1016/09645691(93)90021P

Tompkins E, Adger WN, Brown K (2002) Institutional networks for inclusive coastal management in Trinidad and Tobago. Environ Plan A 34:1095–1111. https://doi.org/10.1068/a34213

United Nations Development Programme (UNDP) (2025) Community-Based Resilience-Building: A UNDP Guidance Note. https://www.undp.org/sites/g/files/zskgke326/files/2025-05/undp-community-based-resilience-building-guidance-note.pdf. Last accessed 29 Jul 2025

Villarroel-Lamb D et al (2024) Transcending disciplinary boundaries: NbS for Caribbean coastal resilience. Proc Inst Civ EngineersCivil Eng 177:76–88. https://doi.org/10.1680/jcien.24.00939

Vogelsang LG et al (2023) Assessing the costeffectiveness of Naturebased Solutions under climate change uncertainty and learning. Water Resour Econ 43:100224. https://doi.org/10.1016/j.wre.2023.100224

Vos K et al (2019) CoastSat: a google earth Engineenabled Python toolkit to extract shorelines from publicly available satellite imagery. Environ Model Softw 122:104528. https://doi.org/10.1016/j.envsoft.2019.104528

Water Resources Agency (WRA) (2001) Integrating the management of watersheds and coastal areas in Trinidad and Tobago

University of the West Indies (2025) Welcome to Coastal Zone Engineering. Faculty of Engineering: Coastal Zone Engineering. https://sta.uwi.edu/eng/coastal/. Last accessed 19 Jul 2025

Zhao B et al (2021) Effects of river damming and delta erosion on organic carbon burial in the Changjiang Estuary and adjacent East China Sea inner shelf. Sci Total Environ 793:148610. https://doi.org/10.1016/j.scitotenv.2021.148610

Chapter 23
Developing an Approach to Measure Social Sustainability and Livelihood Vulnerability to Coastal Hazards Under Anthropogenic Impacts in the Indian Sundarbans

Bireswar Patra, Ujjwal Bhandari, Mousumi Roy, and Bappaditya Koley

23.1 Introduction

Coastal regions worldwide are increasingly vulnerable to climate-induced hazards, with severe implications for human well-being, livelihoods, and ecosystems. The Indian Sundarbans, a UNESCO World Heritage Site and the world's largest contiguous mangrove forest, epitomizes this fragility (Newton et al. 2020; Sarkar 2022; Cons 2025; Mallick 2025). The delta is highly dynamic, shaped by tidal processes and riverine sedimentation, yet it faces mounting threats from sea-level rise, saline water intrusion, and recurrent cyclonic events. Anthropogenic stressors, including population pressure, land-use changes, and infrastructural deficiencies, further compound the system's vulnerability.

Hydrometeorological extremes such as cyclones, storm surges, and tidal flooding have intensified in frequency and severity over the past two decades (He et al. 2018; Mandal et al. 2019; Zhang et al. 2021; Mohammed et al. 2024). Events like Cyclones

B. Patra
Department of Geography, Asutosh College, Kolkata, West Bengal, India

U. Bhandari (✉)
Department of Geography, Syamaprasad College, Kolkata, West Bengal, India
e-mail: ujjwalcosmos@gmail.com

M. Roy
Department of Geography, Diamond Harbour Women's University, Diamond Harbour, West Bengal, India

B. Koley
Department of Geography, Bankim Sardar College, South 24 Parganas, Tangrakhali , West Bengal, India

A. Nath et al. (eds.), *Geospatial Application of Coastal Hazards*, Springer Water,
https://doi.org/10.1007/978-3-032-20075-4_23

Aila (2009), Amphan (2020), and Yaas (2021) inflicted widespread devastation—submerging farmlands, damaging homes, and displacing thousands. Inhabitants, predominantly dependent on climate-sensitive sectors like agriculture, aquaculture, and forest resource extraction, have limited adaptive capacity. This reliance on natural resources, combined with structural poverty and weak institutional support, renders their livelihoods exceptionally sensitive to environmental shocks.

Globally, vulnerability assessment has evolved toward multidimensional frameworks integrating social, economic, and ecological parameters (Depietri 2020; Gunaratne et al. 2023). Composite indices, such as the Human Development Index (HDI) and Multidimensional Poverty Index (MPI), have advanced poverty and well-being research (Santos and Seth 2023; James 2025). However, micro-level assessments of livelihood vulnerability and social sustainability within climate-stressed coastal systems remain scarce. The Sundarbans context demands such an approach, as conventional macro-level analyses fail to capture localized disparities crucial for targeted adaptation planning.

This study addresses these gaps by developing a composite methodological framework to measure social sustainability and livelihood vulnerability at the block level in the Indian Sundarbans. The specific objectives are to: (i) construct a Sustainable Livelihood Index (SLI) and Composite Vulnerability Index (CVI) using socio-economic and infrastructural indicators; (ii) map spatial variations in vulnerability and sustainability across 25 C.D. blocks; and (iii) derive insights for climate-resilient development and livelihood planning.

The proposed approach integrates statistical techniques, geospatial mapping, and livelihood frameworks, contributing to vulnerability research and informing evidence-based policy interventions for one of the most threatened coastal ecosystems globally.

23.1.1 Sustainable Livelihoods as an Analytic Lens

The sustainable livelihoods (SL) approach positions people's **capabilities, assets, and activities** at the center of analysis. It evaluates how they cope with shocks while maintaining or enhancing these assets without degrading the resource base. The conceptual roots are widely traced to Chambers & Conway's classic work and subsequent formalizations that shaped the now-standard pentagon of assets (human, physical, natural, social, financial) and their interactions with institutions and vulnerability contexts. These foundations were consolidated and operationalized in DFID's *Sustainable Livelihoods Guidance Sheets*, which also popularized graphics and diagnostic steps still used in applied research and policy analyses. Scoones' framework further clarified typical livelihood strategy bundles (intensification, diversification, migration) and the trade-offs households face when recombining assets under stress. Together, these contributions provide the conceptual platform for our asset-grouped indicators and for interpreting sustainability and vulnerability in a dynamic, hazard-prone setting.

23.1.2 Composite Measures of Development, Poverty, and Vulnerability

Composite indices allow multi-dimensional phenomena to be summarized for comparison across places and time. The **Human Development Index (HDI)** established a durable template by normalizing components and aggregating across dimensions; its strengths and limitations (e.g., reductionism, sensitivity to indicator choice and weighting) are now well documented. In climate and livelihoods research, the **Livelihood Vulnerability Index (LVI)** popularized by Hahn et al. (2009) provides a pragmatic, indicator-based way to characterize exposure, sensitivity, and adaptive capacity (Zainab and Shah 2024), typically using **min–max normalization** and dimension-wise aggregation before a final composite score. Numerous recent applications refine the LVI structure, adopt alternative weighting schemes, or link it more explicitly to IPCC vulnerability components; many continue to report values on the 0–0.5 scale described in methodological notes. Our study follows these well-established index-construction practices (min–max scaling, asset-grouped dimensions, equal weights) while tailoring indicators to the Sundarbans context.

23.1.3 Coastal Hazards, Climate Stressors, and Livelihood Risks in the Sundarbans

The Indian Sundarbans—part of the world's largest mangrove delta—has long been documented as a **multi-hazard, high-exposure** environment where cyclones, storm surges, tidal flooding, and salinity intrusion interact with historical land use and infrastructural legacies (Ehsan et al. 2022). Reviews highlight centuries of ecological change and direct human impacts, while remote-sensing assessments track mangrove dynamics and shoreline change (Samanta et al. 2020; Thakur et al. 2021; Biswas et al. 2024). More recent, village- and block-level studies triangulate **physical exposure** (elevation, coastline proximity, cyclone tracks, surge zones) with **social sensitivity and adaptive capacity** (poverty, housing quality, health and education access, connectivity) using multi-criteria or composite-index approaches—patterns that consistently identify southern island blocks and eastern margins as risk hotspots.

The last decade's extreme events—**Aila (2009), Amphan (2020), Yaas (2021)**—catalyzed a surge of impact and risk research (Zaid 2021; Rahman 2022; Arif et al. 2023; Das 2023; Biswas and Sen 2024; Pathak et al. 2025). Studies report extensive inundation, embankment breaches, salinity-driven agricultural loss, and damage to housing and public services; satellite assessments around Amphan show large flooded extents and vegetation impacts, while ground-based analyses document prolonged soil salinization and recovery challenges. Work on specific islands (e.g., Mousuni) and cross-sector reviews (water, food, etc.) emphasize how recurrent cyclones amplify pre-existing vulnerabilities and force everyday adaptations that are costly and uneven across social groups. These findings underscore the need for

micro-spatial diagnostics (e.g., at the block level) to support targeted adaptation, which our study provides through SLI/CVI mapping.

23.1.4 Methodological Advances and Gaps

Methodologically, Sundarbans studies increasingly integrate **GIS-based exposure layers** with **socio-economic indicators** and apply multi-criteria decision methods or composite indices (Bera and Maiti 2021; Mondal et al. 2024). Yet three gaps remain. **First**, indicator sets often replicate generic vulnerability checklists rather than **asset-aligned** measures tied to the livelihoods literature; aligning indicators to the SL asset pentagon helps with interpretation and policy relevance. **Second**, many assessments optimize for hazard exposure but comparatively under-specify **health, food security, and social capital**—dimensions central to livelihood sustainability in deltaic ecologies. **Third**, while normalization and aggregation are standard, **transparency in weighting, thresholding, and classification** varies; clearer reporting of scaling choices and class breaks (and their sensitivity) would aid comparability across studies. The present work addresses these gaps by (i) structuring indicators explicitly within the SL asset framework, (ii) including **health and food-security** outcomes alongside asset stocks, and (iii) documenting the normalization, aggregation, and classification scheme to enhance reproducibility.

23.1.5 Synthesis: Positioning the Present Study

Taken together, the foundational SL literature legitimizes an **asset-grouped, people-centred** approach; composite-index traditions (HDI; LVI) provide **operational tools** for normalization and aggregation; and the Sundarbans hazard literature establishes the **spatial signature of risk** and the salience of embankments, salinity, and access to services. Building on these strands, our contribution is a **micro-level, multi-dimensional** assessment across 25 C.D. blocks that (a) jointly measures **social sustainability** (SLI) and **livelihood vulnerability** (CVI), (b) integrates indicators spanning assets and outcomes (including **food security** and **health**), and (c) renders results as **decision-ready maps** to support targeted, climate-resilient planning in one of the world's most exposed coastal socio-ecological systems.

23.2 Study Area

The Indian Sundarbans, forming the western segment of the Ganges–Brahmaputra-Meghna delta, spans approximately 10,000 km^2, of which 38% lies in India and 62% in Bangladesh (Ariful 2022; Bandyopadhyay 2019; Samanta et al. 2021; Uddin et al. 2024). The Indian share extends between **21°30′N to 22°50′N latitude and 88°01′E to 89°02′E longitude**. Administratively, it comprises parts of the **South 24 Parganas and North 24 Parganas districts** in West Bengal (Table 23.1). The study area includes 25 Community Development (C.D.) blocks (Fig. 23.1): 17 from South 24 Parganas (e.g., Kakdwip, Sagar, Namkhana, Patharpratima, Basanti, Gosaba, Kultali, Joynagar-I & II) and 8 from North 24 Parganas (e.g., Hingalganj, Hasnabad, Sandeshkhali-I & II).

Table 23.1 Census data-wise comparison of demographic data between two districts of the Indian part of Sundarban

District	Total area (km^2)	Total population (2011)	Population density (/ km^2)	Literacy rate (%)	Sex ratio (♀ per 1,000 ♂)	Children 0–6 (no.)	Children 0–6 (% of pop.)
North 24 Parganas	4,094	10,009,781	2,445	84.06	955	957,973	9.57
South 24 Parganas	9,960	8,161,961	819	77.51	956	1,025,679	12.57

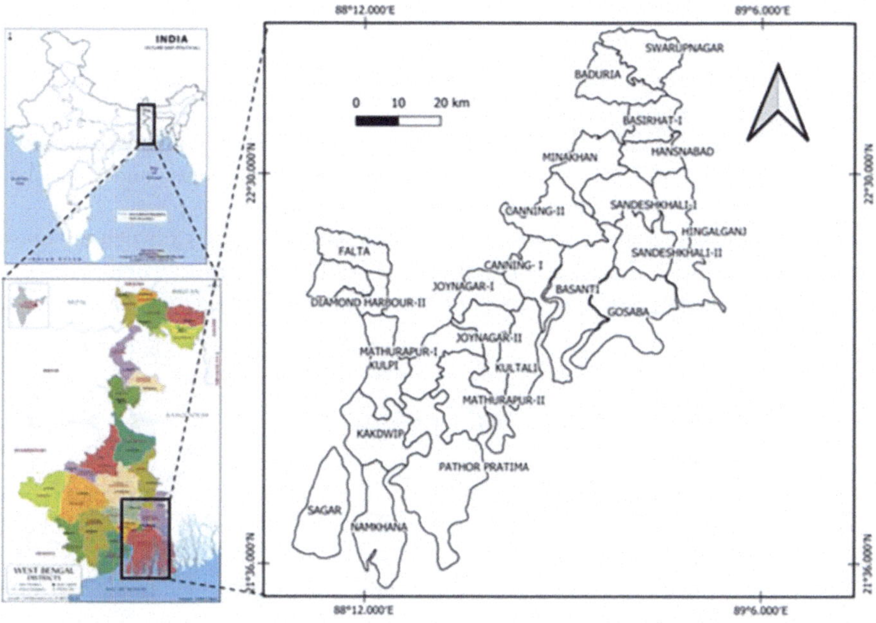

Fig. 23.1 Location map of the study area with block boundaries and major rivers

23.2.1 Physiographic and Socio-Economic Characteristics

The region is characterized by an intricate network of tidal rivers, creeks, and estuaries, interspersed with numerous islands. Soils are alluvial and saline-prone, limiting agricultural productivity. The climate is tropical monsoonal, with annual rainfall exceeding 1,700 mm. Cyclones and storm surges frequently breach embankments, inundating croplands and settlements (Mehzabin 2019; Giri et al. 2022; Miller 2023). Infrastructure—roads, healthcare, drinking water supply—remains underdeveloped.

Population density is extremely high, exceeding **1,000 persons per km²** in several blocks. Livelihoods depend on paddy cultivation, shrimp farming, fishing, honey collection, and casual labour. High poverty incidence, gender inequities, and limited access to education and healthcare exacerbate vulnerability. These socio-environmental stresses, compounded by climate change, necessitate an integrated sustainability assessment (Ghosh 2018a, b; Chaturvedi, 2022; Ghosh and Dutta 2024; Parthasarathy et al. 2025).

23.3 Data and Methods

23.3.1 Data Sources

The study employed a combination of **primary and secondary data** (Table 23.2). Secondary data were extracted from:

District Census Handbooks (2011) for demographic and housing indicators; Human Development Reports for South and North 24 Parganas; and Published literature and government reports for contextual information.

Primary data were collected through **household surveys**, using a structured questionnaire aligned with the **DFID Sustainable Livelihoods Framework** (1999). A stratified sampling approach was adopted to capture variability across blocks.

23.3.2 Indicator Selection and Classification

A total of **15 socio-economic indicators** were selected, categorized into seven dimensions (Table 23.3). Each indicator reflects critical aspects of livelihood resilience and social sustainability.

The selection of these 15 socio-economic indicators is grounded in the **Sustainable Livelihood Framework** (DFID 1999), which conceptualizes livelihoods as an interplay of **five core asset types—human, natural, physical, financial, and social capital**—supplemented here by **food security and health outcomes** due to their critical role in climate-vulnerable coastal contexts. This expanded framework (Fig. 23.2)

Table 23.2 Data sources and software used in the study

Category	Data/tool	Description/purpose	Source
Demographic data	Population, literacy, housing, workforce	Block-level socio-economic and demographic indicators	Census of India, District Census Handbook (2011)
Human development	HDI-related indicators	Education, health, and income-related attributes	District human development reports (South & North 24 Parganas) (2009 & 2010)
Livelihood parameters	Household survey data	Primary data on assets, income diversification, and food security	Field survey, structured questionnaire (2023)
Geospatial layers	Block boundaries, river networks	Spatial reference for mapping and analysis	Survey of India toposheets, NRSC bhuvan portal
Hazard context data	Cyclone and flood-prone zones	Historical hazard exposure and vulnerability context	IMD reports, published literature
Software	**ArcGIS 10.8**	Spatial analysis and choropleth mapping	ESRI
	QGIS (v3.x)	Open-source spatial analysis and visualization	QGIS.org
	MS Excel	Data preprocessing, normalization, and statistical calculations	Microsoft corporation
	SPSS/R (optional)	Statistical validation, index correlation analysis	IBM SPSS, R foundation

ensures a **comprehensive and multidimensional assessment** of sustainability and vulnerability rather than relying solely on income-based measures.

Human assets reflect education and skill levels, which influence the capacity to adopt adaptive strategies, access information, and diversify income sources during shocks. **Physical assets**, such as housing quality and energy access, provide structural security and facilitate livelihood diversification, reducing sensitivity to climate extremes. **Natural assets**, particularly landholding size and irrigation coverage, are essential in an agrarian economy like the Sundarbans, where productive resources determine food availability and income stability.

Financial assets (income diversification and savings) capture the ability to withstand and recover from shocks, acting as buffers against economic disruption. **Social assets**, represented by institutional participation and community networks, enhance **collective action and informal support systems**, which are vital in disaster-prone regions. The inclusion of **food security indicators** acknowledges the close link between agricultural performance and nutritional well-being in subsistence

Table 23.3 Selected socio-economic indicators and their justification

Dimension	Indicator	Rationale for selection	Supporting references
Human assets	Literacy rate (%)	Higher literacy improves access to information, disaster preparedness, and adaptive capacity	Chambers and Conway (1992); DFID (1999)
	School attendance (%)	Indicates educational attainment and intergenerational resilience capacity	Scoones (1998); Alkire (2011)
	Skilled labor ratio (%)	Reflects human capital quality and the ability to diversify livelihoods	Carney (1998); Bebbington (1999)
Physical assets	Housing quality (PUCCA%)	Strong housing reduces disaster impacts and represents physical security	Hahn et al. (2009); Dempsey et al. (2011)
	Access to electricity (%)	Proxy for infrastructure and modern energy access linked to livelihood diversification	UNDP (1990); FAO (2017)
Natural assets	Average landholding size	Indicates agricultural resource availability and food security	DFID (1999); Ellis (2000)
	Irrigation coverage (%)	Enhances agricultural productivity and drought resilience	Scoones (1998); Saha (2017)
Financial assets	Income diversification index	Multiple income sources reduce vulnerability to shocks	Ashley and Carney (1999); Lindenberg (2002)
	Household savings (%)	Indicates financial resilience and ability to invest in recovery or adaptation	Roslina (2018); Alkire and Foster (2011)
Social assets	Participation in local institutions (%)	Strong social networks and governance participation enhance collective resilience	Narayan (2002); Szabo et al. (2016)
	Social support networks (%)	Informal support mitigates post-disaster livelihood shocks	Bebbington (1999); Brocklesby and Fisher (2003)
Food security	Per capita food availability (kg/year)	Adequate food supply is fundamental for household resilience and well-being	FAO (2017); Hahn et al. (2009)
	Agricultural productivity (kg/ha)	Core determinant of food security and income stability	Ellis (2000); Lindenberg (2002)
Health outcomes	Access to healthcare facilities (%)	Access to health services reduces morbidity during and after disasters	UNDP (1990); Das and Misri (2013)
	Child malnutrition (%)	Reflects the chronic vulnerability and overall well-being of households	UNICEF (2016); Alkire (2011)

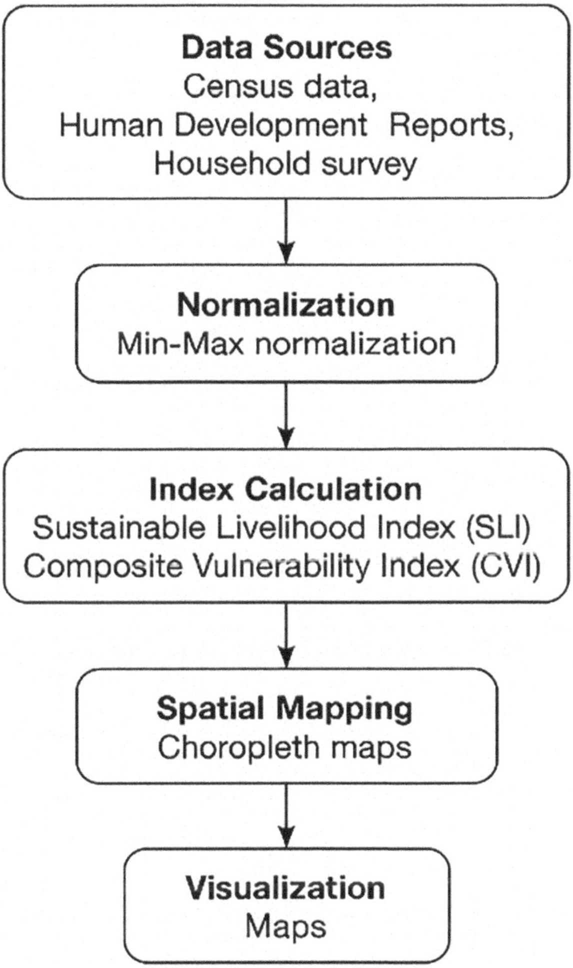

Fig. 23.2 Methodological Flowchart showing steps

economies. At the same time, **health outcomes** (access to healthcare and child nutrition) reflect chronic vulnerability dimensions that exacerbate disaster impacts.

Together, these seven dimensions encompass the **core determinants of adaptive capacity, resilience, and well-being** in a region repeatedly exposed to cyclones, salinity intrusion, and embankment failures. This structure also aligns with **previous empirical studies** on vulnerability assessment in coastal and deltaic regions (Hahn et al. 2009; Scoones 1998; Carney 1998), while extending their scope to include health and food outcomes—two domains often underrepresented in vulnerability indices. By adopting this comprehensive set of indicators, the study ensures a **holistic representation of both structural capacities and functional outcomes** necessary for sustainable livelihoods in the Sundarbans.

23.3.3 Normalization and Index Construction

Indicators were standardized using **Min–Max normalization** (Eq. 23.1) to transform values into a 0–1 scale:

$$S_d = \frac{X_d - S_{\min}}{S_{\max} - S_{\min}} \tag{23.1}$$

where X_d = observed value, S_{min} and S_{max} = minimum and maximum values across blocks.

Composite indices (Eq. 23.2) were computed as follows:

- **Sustainable Livelihood Index (SLI)**: average of seven-dimension scores with equal weights.
- **Composite Vulnerability Index (CVI)**: derived from ranked scores of all indicators.

$$\text{C.I} = \sum \text{Xi} \tag{23.2}$$

23.3.4 Threshold Calculation and Classification

Blocks were categorized (Eq. 23.3) into **low**, **moderate**, and **high vulnerability/ sustainability** using a threshold value:

$$V_t = \frac{\sum n \times \sum d}{k} \tag{23.3}$$

where n = number of blocks, d = number of parameters, and k = 2.

23.3.5 Spatial Analysis

Choropleth maps and composite score visualizations were generated using **ArcGIS 10.8 and QGIS**. Violin plots were used to depict statistical distribution and inter-block variability.

Table 23.4 Statistical summary of selected socio-economic indicators

Indicator	Range	Mean	Std. Dev	Classification scheme
Overall literacy rate (%)	65.2–4.9	74.6	4.2	Very Low (<68), Low (68–72), Moderate (72–76), High (76–80), Very High (>80)
Health service index	0.15–0.62	0.38	0.12	Very Low (<0.25), Low (0.25–0.35), Moderate (0.35–0.45), High (>0.45)
Crop cultivation index	0.35–0.87	0.56	0.14	Low (<0.40), Moderate (0.40–0.60), High (>0.60)
Infrastructure index	10–43	26.8	7.6	Low (<20), Moderate (20–30), High (>30)
Sustainable livelihood index	0.24–0.78	0.52	0.11	Low (<0.40), Moderate (0.40–0.60), High (>0.60)
Poverty index	0.42–0.89	0.63	0.10	Low (<0.50), Moderate (0.50–0.70), High (>0.70)

23.4 Results

23.4.1 Literacy Rate

The literacy rate in the study area varies from **65.2% (Basanti)** to **84.9% (Falta)**, with an overall mean of **74.6%** (Table 23.4). Higher literacy is concentrated in the **western and northern inland blocks**, such as Falta, Diamond Harbour-II, and Hasnabad, which benefit from better accessibility to urban centers and educational infrastructure. Conversely, **southern island blocks**—Gosaba, Kultali, and Sagar—record lower literacy levels (<70%), reflecting their geographic isolation, recurrent cyclone damage to schools, and limited educational outreach (Fig. 23.3a).

23.4.2 Health Service Index

The Health Service Index exhibits a **range of 0.15 to 0.62**, with a mean value of **0.38**. The index is relatively high in blocks adjacent to district headquarters and major roads, such as Diamond Harbour-II and Kakdwip, indicating better availability of health facilities. However, **peripheral island blocks**—Basanti, Gosaba, and Kultali—register the lowest scores (<0.25), highlighting severe healthcare deficits

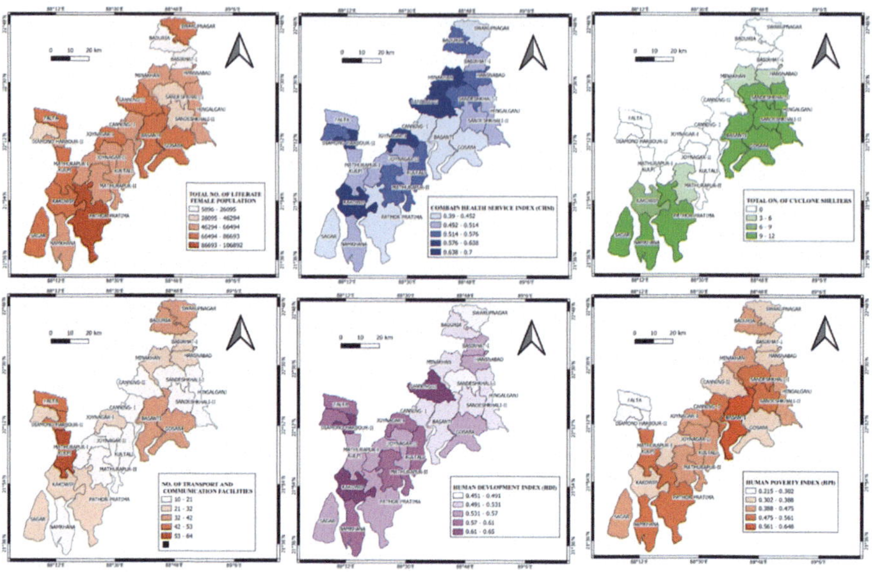

Fig. 23.3 Spatial distribution of socio-economic indicators across 25 community development blocks in the Indian Sundarbans: **a** Literacy rate, **b** Health service index, **c** Crop cultivation index, **d** Infrastructure index, **e** Sustainable livelihood index, **f** Poverty index. *Source* Census of India (2011), District Human Development Reports, and Field Survey (2023)

in disaster-prone zones. This inadequacy compounds vulnerability during cyclonic events, where evacuation and emergency medical support are critical (Fig. 23.3b).

23.4.3 Crop Cultivation Index

Agricultural sustainability, represented by the Crop Cultivation Index, demonstrates substantial spatial variation, with values ranging from **0.35 to 0.87** (mean = 0.56). Inland blocks such as Joynagar I and II show higher scores (>0.60) due to better irrigation and soil fertility, while **salinity-prone southern islands**—Sagar and Patharpratima—record lower productivity (<0.40) (Fig. 3c). This reflects the combined effect of **salinity intrusion, tidal flooding, and monocropping**, which constrains crop diversity and yields.

23.4.4 Infrastructure Index

Infrastructure development exhibits marked disparity, with index values ranging from **10 to 43** (mean = 26.8). Blocks like **Falta and Diamond Harbour-II**, located closer to urban peripheries and major transport corridors, exhibit high infrastructure scores (>30), supported by road connectivity, electricity, and housing quality. In contrast, **eastern and southern peripheral blocks** such as Gosaba, Hingalganj, and Kultali exhibit poor infrastructure (<20), impeding disaster preparedness and limiting access to markets and services (Fig. 3d).

23.4.5 Sustainable Livelihood Index (SLI)

The Sustainable Livelihood Index, constructed by aggregating seven asset-based dimensions, ranges from **0.24 (Basanti)** to **0.78 (Falta)**, with a mean of **0.52**. High-SLI blocks, notably **Falta, Kakdwip, and Diamond Harbour-II**, demonstrate diversified income sources, stronger institutional support, and better physical assets. Conversely, **Basanti, Kultali, and Sagar** exhibit low SLI values (<0.40), indicating limited adaptive capacity and high dependence on climate-sensitive activities (Fig. 3e).

23.4.6 Poverty Index

The Poverty Index highlights persistent socio-economic deprivation across the study area, with values ranging from **0.42 to 0.89** (mean = 0.63). High-poverty zones (>0.70) cluster in **southern deltaic islands**, such as Gosaba, Kultali, and Sagar, which also exhibit low infrastructure and livelihood diversification. Inland blocks like Falta and Diamond Harbour-II show relatively lower poverty levels (<0.50), consistent with their higher SLI and better service provision (Fig. 3f). The spatial correlation between poverty and vulnerability underscores the **reinforcing loop of socio-economic fragility and environmental stress** in marginal areas.

23.4.7 Composite Vulnerability and Spatial Patterns

The cumulative analysis reveals a **clear spatial gradient of vulnerability**, intensifying toward the **coastal and island blocks**. These zones exhibit **low literacy, poor healthcare, fragile infrastructure, and high poverty**, compounded by frequent hazard exposure. Conversely, inland blocks proximal to **urban nodes and transport networks** demonstrate higher sustainability and resilience, driven by infrastructural advantages and diversified livelihoods.

The **overlap of low SLI and high poverty** in southern and eastern Sundarbans underscores **policy-critical hotspots** requiring targeted interventions in health, education, infrastructure, and livelihood diversification.

This section presents the spatial distribution and variability of socio-economic indicators across 25 Community Development (C.D.) blocks in the Indian Sundarbans. The selected indicators are categorized under seven dimensions of the Sustainable Livelihood Framework and normalized to create composite indices. The results reveal pronounced **intra-regional disparities** in literacy, health services, crop cultivation, infrastructure, poverty levels, and sustainable livelihood conditions. Figure 23.3 illustrates the block-level variations for six critical indicators.

SLI values revealed substantial inter-block disparities. Inland blocks such as **Falta, Kakdwip, and Diamond Harbour-II** recorded higher sustainability scores (>0.60), attributed to better infrastructure, diversified livelihoods, and greater access to services. In contrast, **Basanti, Gosaba, Kultali, and Sagar** exhibited low sustainability (<0.40), reflecting limited access to healthcare, high dependency on mono-cropping, and frequent disaster exposure.

The statistical analysis of livelihood vulnerability (Table 23.5) indicators across the Sundarbans C.D. blocks reveals distinct regional disparities that directly correspond to livelihood insecurity and infrastructural inequalities. Blocks such as Gosaba, Basanti, Hingalganj, and Sandeshkhali-II register high values in the livelihood assets index (Ai), reflecting significant dependence on fragile natural resources and high exposure to hazards. However, their socio-economic status (Bi) and infrastructure/service access (Ci) scores remain considerably low, underscoring a structural vulnerability where livelihood dependency is not matched by institutional or infrastructural support. By contrast, blocks like Diamond Harbour-II, Falta, and Baduria score relatively higher in Ci, suggesting better connectivity and services, but their overall vulnerability persists due to lower livelihood asset resilience and weaker socio-economic structures.

These patterns underscore that vulnerability is multidimensional, shaped not only by ecological exposure but also by disparities in infrastructure and human development. Policy priorities must therefore focus on strengthening critical infrastructure (transport, health, education, banking, and communication) in blocks with high livelihood dependency but weak service bases, such as Gosaba and Hingalganj. Simultaneously, socio-economic diversification through skill development, alternative livelihoods, and financial inclusion is crucial for reducing dependency on climate-sensitive occupations. Targeting these block-level disparities through place-based strategies can significantly enhance resilience and support sustainable development in the Sundarbans, where hazard exposure and socio-economic fragility intersect.

23.4.8 Infrastructure and Service Accessibility

The spatial assessment of socio-economic and infrastructural indicators across 25 Community Development (C.D.) blocks of the Indian Sundarbans reveals significant disparities in livelihood assets and service accessibility. Figures 23.3 and 23.4 depict the block-level variations across education, health, livelihood, poverty, and

Table 23.5 Statistical summary of composite livelihood vulnerability indicators across C.D. blocks in the Indian Sundarbans

Sub-division	C.D. Block	Total Households	Ai (Livelihood Assets)	Bi (Socio-Economic Status)	Ci (Infrastructure/ Services)	Composite Index (L)
Kakdwip	Kakdwip	60,201	0.367	0.369	0.264	0.562
	Sagar	43,716	0.371	0.449	0.163	0.596
	Namkhana	41,433	0.356	0.410	0.231	0.579
	Patharpratima	69,641	0.456	0.385	0.152	0.622
Canning	Basanti	70,818	0.668	0.223	0.109	0.665
	Gosaba	58,197	0.751	0.172	0.067	0.675
	Canning-I	64,041	0.407	0.329	0.257	0.552
	Canning-II	49,711	0.492	0.324	0.174	0.601
Diamond Harbour	Diamond Harbour-I	33,206	0.070	0.421	0.498	0.383
	Diamond Harbour-II	42,522	0.043	0.303	0.653	0.301
	Kulpi	59,078	0.318	0.319	0.343	0.468
	Mathurapur-I	40,602	0.438	0.209	0.349	0.497
	Mathurapur-II	45,888	0.527	0.210	0.251	0.547
	Falta	57,153	0.049	0.294	0.657	0.299
Baruipur	Kultali	45,099	0.620	0.243	0.139	0.650
	Joynagar-I	55,734	0.228	0.305	0.454	0.411
	Joynagar-II	50,413	0.503	0.185	0.297	0.511
Basirhat	Hingalganj	46,048	0.731	0.133	0.153	0.672
	Sandeshkhali-I	37,344	0.477	0.317	0.207	0.603

(continued)

Table 23.5 (continued)

Sub-division	C.D. Block	Total Households	Ai (Livelihood Assets)	Bi (Socio-Economic Status)	Ci (Infrastructure/ Services)	Composite Index (L)
	Sandeshkhali-II	37,771	0.662	0.252	0.080	0.673
	Hasnabad	47,739	0.327	0.256	0.409	0.451
	Minakhan	43,756	0.248	0.406	0.339	0.498
	Basirhat-I	38,033	0.094	0.269	0.632	0.308
	Baduria	67,671	0.041	0.319	0.633	0.486
	Swarupnagar	61,611	0.037	0.442	0.520	0.390

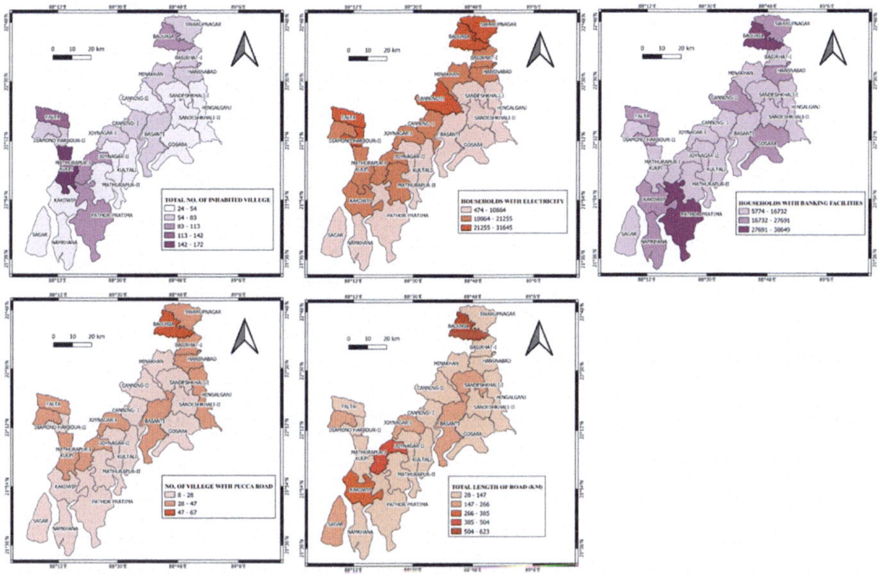

Fig. 23.4 Spatial distribution of infrastructure and service accessibility indicators across the Indian Sundarbans: **a** Percentage of inhabited villages, **b** Villages with electricity, **c** Villages with banking facilities, **d** Villages with Pucca Roads, **e** Length of roads (km). *Source* Census of India (2011), District Statistical Handbooks

infrastructure-related indicators. These patterns highlight the persistence of structural vulnerabilities in ecologically fragile zones.

Infrastructure indicators demonstrate marked inter-block disparities (Fig. 23.4):

- **Percentage of inhabited villages** varies between 60 and 112 per block, with denser settlements in **northern and north-eastern blocks** (Hasnabad, Minakhan), while southern islands exhibit sparser distribution due to geomorphic instability and recurrent erosion (Fig. 4a).
- **Electricity coverage** remains above 90% in most blocks, but peripheral island blocks face issues of reliability during extreme weather events (Fig. 4b).
- **Banking facility coverage** is critically low (<15%) in several blocks, notably Gosaba and Kultali, limiting financial inclusion and adaptive investment opportunities (Fig. 4c).
- **Pucca road connectivity** ranges from 40 to 87%. Falta and Diamond Harbour-II exhibit better connectivity (>70%), while Gosaba, Hingalganj, and Kultali remain poorly served (<50%) (Fig. 4d).
- **Road length** varies significantly from 152 km (Kultali) to over 500 km (Falta), highlighting a persistent accessibility gap in remote island blocks (Fig. 4e).

These infrastructural deficits exacerbate isolation during disasters and restrict access to markets, education, and health services.

Table 23.6 Statistical summary table

Indicator	Range	Mean	Std. Dev	Classification scheme
% of inhabited villages	60–112	85	12	Low (<70), Moderate (70–90), High (>90)
% of villages with electricity	82–100	95	5	Low (<85), Moderate (85–95), High (>95)
% of villages with banking facilities	5–36	18	8	Low (<15), Moderate (15–25), High (>25)
% of villages with Pucca roads	40–87	62	12	Low (<50), Moderate (50–70), High (>70)
Length of roads (km)	152–506	320	88	Low (<250), Moderate (250–400), High (>400)

Infrastructure and service accessibility indicators exhibit significant intra-regional disparities across the Sundarbans blocks (Fig. 23.4). The proportion of inhabited villages ranges between 60 and 112 per block, with higher concentrations in northern inland blocks such as Hasnabad and Baduria, reflecting dense rural settlement patterns. In contrast, southern islands (Sagar, Namkhana) exhibit lower village counts due to geomorphic constraints and recurrent erosion.

Electricity coverage exceeds 95% in most inland blocks but drops marginally in remote island blocks (Table 23.6), indicating uneven access to a reliable energy supply. Banking facilities remain a critical gap, with several blocks reporting coverage below 15%, constraining financial inclusion and adaptive investments. Road connectivity follows a similar pattern, with Diamond Harbour-II and Falta recording the highest share of pucca roads (>70%) and road length (>400 km), while peripheral blocks such as Gosaba and Kultali remain poorly connected (<50% pucca road coverage). These deficits in basic services and infrastructure amplify livelihood vulnerability and slow post-disaster recovery, reinforcing structural marginalization in geographically isolated blocks.

23.4.9 Composite Vulnerability Gradient

The integrated analysis reveals a **clear north–south vulnerability gradient**, with inland and semi-urban fringe blocks exhibiting stronger resilience due to **higher literacy, diversified livelihoods, better infrastructure, and institutional access**. Conversely, **southern, and eastern coastal islands** demonstrate **systemic vulnerability**, marked by **low educational attainment, poor health and banking services, fragile connectivity, high poverty, and reliance on agriculture**. This pattern underscores the **intersection of socio-economic fragility and environmental risk**, necessitating targeted interventions.

23.4.10 Composite Development Index: Spatial Categorization

This map (Fig. 23.5) represents the **Composite Development Index** for the Indian Sundarbans blocks, classifying them into two categories:

- **Developed (Green)**: Composite Index values between **107–187**
- **Backward (Red)**: Composite Index values between **187–299**

This visualization summarizes the overall status based on multiple socio-economic and infrastructure indicators, offering a consolidated view of regional disparities.

The Composite Development Index (CDI), constructed from 15 socio-economic and infrastructure indicators, reveals stark regional inequalities across the Indian

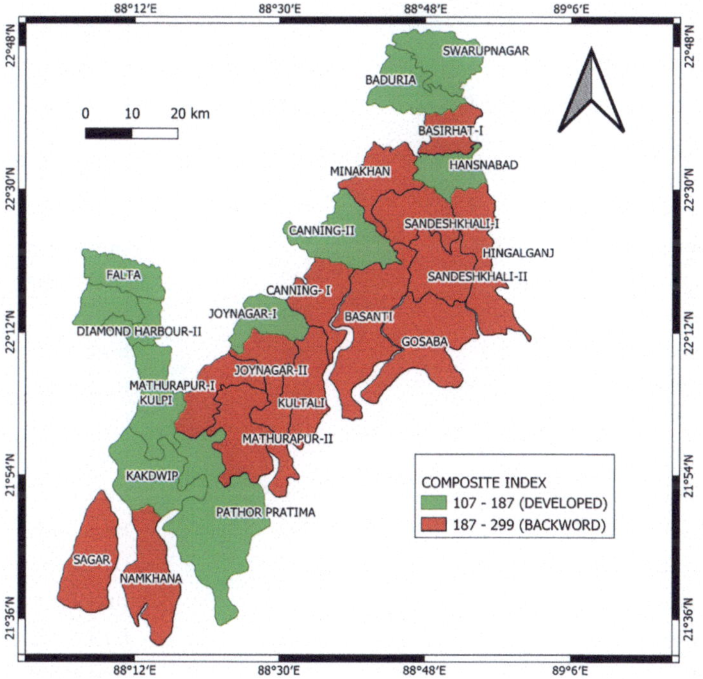

Fig. 23.5 Composite development Index classification for the Indian Sundarbans C.D. blocks: Developed (green, 107–187) and Backward (red, 187–299). *Source* Computed from Census (2011), District Handbooks, and Primary Survey (2023)

Sundarbans (Fig. 23.5). Out of the 25 C.D. blocks, **10 blocks (40%) fall under the 'Developed' category** (Table 23.7), with index values ranging from **107 to 187**, while **15 blocks (60%) are categorized as 'Backward'**, exhibiting higher composite scores (187–299) indicative of multidimensional deprivation.

Developed blocks, such as **Falta, Diamond Harbour-II, Kakdwip, Kulpi, Patharpratima, Joynagar-I, Joynagar-II, and Swarupnagar** (Table 23.8), benefit from better infrastructure, higher literacy, improved connectivity, and diversified livelihoods. Conversely, **backward blocks**, including **Basanti, Kultali, Gosaba, Namkhana, Sagar, Hingalganj, and Sandeshkhali-II**, are geographically isolated and exhibit poor service access, high poverty, and low adaptive capacity.

Table 23.7 Composite development indices

ID	C.D. Blocks	Total area in Sq. Km	Composite Index	Status of development (Threshold = 187.5)
1	Kakdwip	252.73	117	Developing
2	Sagar	282.11	222	Backward
3	Namkhana	370.62	229	Backward
4	Pathorpratima	484.48	162	Developing
5	Kultali	306.18	235	Backward
6	Basanti	404.2	199.5	Backward
7	Gosaba	296.72	219	Backward
8	Kulpi	210.84	155.5	Developing
9	Canning-I	187.86	152.5	Developing
10	Canning-Ii	214.93	287.5	Backward
11	Diamond Harbour-I	70.25	178	Developing
12	Diamond Harbour-II	94.46	133	Developing
13	Joynagar-I	131.01	132	Developing
14	Joynagar-Ii	186.25	252	Backward
15	Mathurapur-I	147.3	198	Backward
16	Mathurapur-Ii	227.45	210.5	Backward
17	Falta	130.67	107	Developing
18	Hingalganj	283.8	236.5	Backward
19	Sandeshkhali-I	182.3	258.5	Backward
20	Sandeshkhali-Ii	197.22	299	Backward
21	Hasnabad	153.07	173	Developing
22	Minakhan	158.73	223	Backward
23	Basirhat-I	111.87	223	Backward
24	Baduria	179.73	120	Developing
25	Swarupnagar	215.13	169.5	Developing

Table 23.8 Categorization of blocks based on composite development index

Category	index range	Number of blocks	Blocks
Developed	107–187	10	Falta, Diamond Harbour-II, Kakdwip, Kulpi, Patharpratima, Joynagar-I & II, Swarupnagar, Canning-I & II
Backward	187–299	15	Basanti, Kultali, Gosaba, Namkhana, Sagar, Hingalganj, Sandeshkhali-I & II, Minakhan, Hansnabad, Baduria, Mathurapur-I & II

This dichotomy highlights a **north-west vs. south-east gradient**, with developed blocks clustered along transport corridors and semi-urban fringes, while backward blocks dominate the **southern deltaic islands**, where recurrent cyclones and salinity intrusion amplify socio-economic vulnerabilities.

23.4.11 Vulnerability Patterns

The analysis of vulnerability patterns across the Sundarbans blocks reveals a distinct **northwest-to-southeast gradient**, with vulnerability increasing significantly toward the remote tidal islands. The **southern and southeastern blocks**, including **Basanti, Kultali, Gosaba, Namkhana, Sagar, Hingalganj, and Sandeshkhali I and II**, exhibit the **highest levels of vulnerability** due to compounded socio-economic fragility and environmental exposure. These areas are characterized by **low literacy rates (often below 70%)**, limited educational infrastructure, and **poor access to health services**, as reflected in their low health service index scores (<0.25). Infrastructural deficits are acute, with **less than 50% of villages connected by pucca roads**, minimal banking coverage (<15%), and inadequate road length. These weaknesses coincide with **high poverty levels (above 0.70)**, low livelihood sustainability (SLI < 0.40), and heavy dependence on agriculture that is increasingly undermined by **salinity intrusion and tidal flooding**. The frequent occurrence of cyclones such as **Aila, Amphan, and Yaas** has further aggravated these vulnerabilities through embankment breaches and saline water ingress.

A **moderate vulnerability zone** occupies the **central part of the Sundarbans**, covering blocks such as **Mathurapur-I and II, Patharpratima, Canning-I and II, Minakhan, and Hansnabad**. These blocks present **mixed performance**—with moderate literacy (70–75%) and partial access to services—but continue to struggle with **limited healthcare facilities, banking infrastructure, and climate-resilient transport networks**. While poverty levels are somewhat lower than in the high-vulnerability zone, these blocks remain exposed to flood and cyclone risks, and their agricultural systems show moderate sensitivity to salinity intrusion.

In contrast, the **northwestern and inland fringe blocks**, including **Falta, Diamond Harbour-II, Kulpi, Kakdwip, Joynagar-I and II, Swarupnagar, and**

Baduria, form the **low-vulnerability zone**, benefiting from **better connectivity, diversified livelihoods, and institutional reach**. These blocks exhibit **higher literacy rates (above 78%)**, improved access to electricity and banking services, and relatively low poverty (<0.50). Proximity to urban centers and transport corridors provides greater market access and employment opportunities, enhancing adaptive capacity. However, despite their relative advantage, these areas remain **ecologically sensitive** and exposed to cyclone-induced disruptions, underlining the need for continued investment in resilient infrastructure.

Overall, this pattern underscores the **spatial concentration of multidimensional vulnerability in the southern tidal islands**, where socio-economic deprivation intersects with environmental risks. Addressing this disparity requires **targeted, place-specific interventions** aimed at strengthening infrastructure, diversifying livelihoods, and expanding institutional services in the most affected blocks, while consolidating resilience gains in the relatively developed zones.

23.4.12 Block-Level Multi-dimensional Profile Analysis

These radar (spider) charts represent **multi-dimensional socio-economic and infrastructure indicators** for each block in the Sundarbans, likely normalized on a 0–1 scale. They visually highlight **strengths and weaknesses** across key domains like education, health, roads, banking, communication, and housing.

Radar diagrams for all 25 blocks (Fig. 23.6) provide a comparative visualization of the normalized scores across socio-economic and infrastructure indicators. The diagrams reveal distinct patterns of **intra-regional disparity**. Blocks such as **Falta, Diamond Harbour-II, and Kakdwip** exhibit broader and more symmetrical radar shapes, reflecting balanced development across most indicators, including **education, health, road infrastructure, and banking services**. Conversely, blocks like **Basanti, Gosaba, Kultali, and Hingalganj** display highly irregular profiles, with sharp deficits in multiple dimensions such as **health facilities, permanent housing, and transport connectivity**, indicating acute livelihood vulnerability.

Central blocks like **Mathurapur-I and II, Canning-I and II, and Patharpratima** demonstrate intermediate profiles—moderate performance in education and housing but pronounced weaknesses in health and communication infrastructure. The **southern island blocks** (Sagar, Namkhana) show the most compressed radar shapes, signifying severe infrastructure gaps coupled with high sensitivity to environmental hazards. In contrast, **northern fringe blocks** (Baduria, Swarupnagar) achieve relatively higher scores in educational and communication services but still lag in livelihood diversification. This visualization reinforces the **multi-scalar nature of vulnerability**, where physical exposure overlaps with structural underdevelopment, creating composite risk landscapes.

The comparative analysis of the highest and lowest indicators across blocks highlights the **spatial unevenness in development and service delivery** within the Sundarbans region (Table 23.9). Blocks such as **Falta, Diamond Harbour-II,**

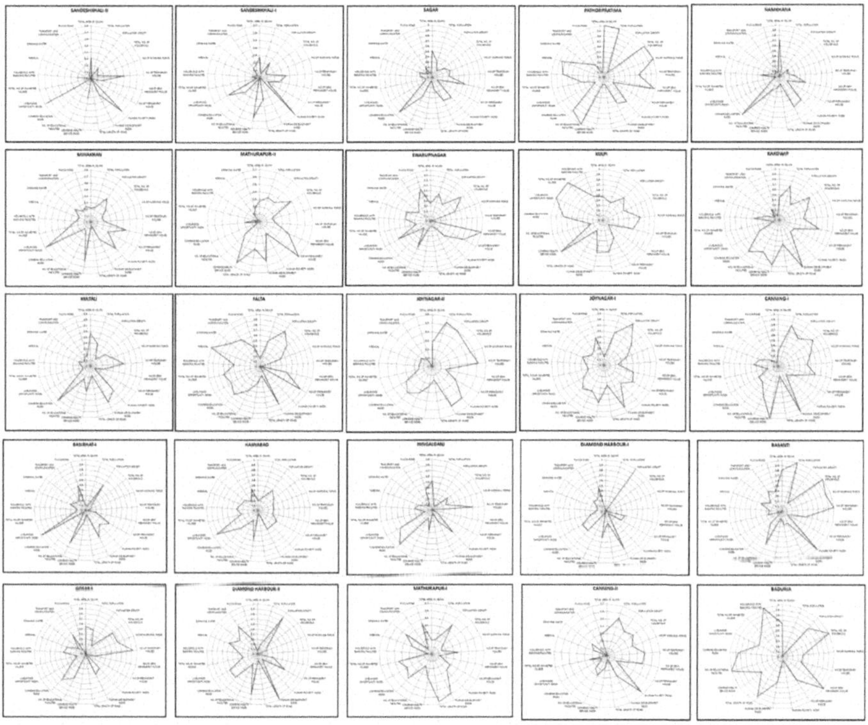

Fig. 23.6 Radar diagrams showing block-wise condition of each parameter

Kakdwip, and Swarupnagar show strong performance in educational facilities, communication infrastructure, and road connectivity, suggesting a higher degree of institutional presence and relatively better physical accessibility. However, even in these relatively developed blocks, **financial inclusion (banking access)** and **healthcare availability** consistently emerge as weak points, reflecting a systemic service gap across the entire region.

Conversely, blocks such as **Basanti, Gosaba, Kultali, Namkhana, and Hingalganj** demonstrate pronounced deficiencies in **health facilities, banking access, and livelihood diversification**, compounding their exposure to environmental hazards. These blocks also exhibit lower scores for drinking water availability and permanent housing, intensifying household-level vulnerability during extreme events such as cyclones and tidal floods. The **critical gap analysis** further underscores that **healthcare infrastructure and financial services are the most universally underperforming dimensions**, requiring targeted interventions across all vulnerability categories. In contrast, **road connectivity and educational access show spatial clustering of advantage**, mostly in blocks proximate to urban corridors.

For infrastructure planning, these patterns emphasize the need for a **multi-sectoral intervention strategy** that prioritizes health and financial access in high-risk blocks,

Table 23.9 Highest, lowest scoring indicators and critical gaps per block

Block	Highest indicator(s)	Lowest indicator(s)	Critical gaps
Sandeshkhali-II	Total area, population density	Banking access, medical facilities	Health, banking, pucca roads
Sandeshkhali-I	Population density, communication facilities	Pucca road, drinking water	Pucca roads, health, drinking water
Sagar	Population density	Health facilities, permanent housing	Health, permanent housing, livelihood opportunities
Patharpratima	Population density, total length of roads	Banking access, drinking water	Banking, health, drinking water
Namkhana	Population density	Medical facilities, livelihood opportunity index	Health, livelihood opportunities, banking
Minakhan	Education facilities	Banking access, permanent housing	Banking, permanent housing, drinking water
Mathurapur-II	Road length, population density	Health facilities, banking access	Health, banking, medical facilities
Swarupnagar	Education, communication	Permanent housing, drinking water	Permanent housing, health, drinking water
Kulpi	Road connectivity, pucca road	Banking access, medical facilities	Banking, health, drinking water
Kakdwip	Road connectivity, education facilities	Banking access, drinking water	Banking, health, drinking water
Kultali	Population density	Health facilities, livelihood opportunity index	Health, livelihood opportunities, banking
Falta	Road connectivity, education, communication	Banking access, medical facilities	Banking, health, drinking water
Joynagar-II	Road connectivity	Banking access, medical facilities	Banking, health, drinking water
Joynagar-I	Education, road connectivity	Livelihood opportunity index, drinking water	Health, livelihood opportunities, banking
Canning-I	Population density	Banking access, health facilities	Banking, health, drinking water
Basirhat-I	Education, communication facilities	Permanent housing, banking access	Banking, permanent housing, health
Hasnabad	Communication Facilities	Drinking water, health facilities	Health, drinking water, banking

(continued)

Table 23.9 (continued)

Block	Highest indicator(s)	Lowest indicator(s)	Critical gaps
Hingalganj	Population density	Banking access, health facilities	Banking, health, livelihood opportunities
Diamond Harbour-II	Road connectivity, education facilities	Health facilities, drinking water	Health, drinking water, banking
Gosaba	Population density	Banking access, medical facilities	Health, banking, permanent housing
Diamond Harbour-I	Education, road connectivity	Banking access, health facilities	Banking, health, drinking water
Baduria	Education, communication	Permanent housing, health facilities	Health, permanent housing, banking
Basanti	Population Density	Medical facilities, drinking water	Health, drinking water, banking

while strengthening **climate-resilient transport and drinking water systems** in moderately vulnerable areas. Furthermore, livelihood support programs, including skill development and diversification, should be concentrated in **southern island blocks and central transitional zones**, where economic dependence on climate-sensitive sectors like agriculture remains high. This approach will facilitate **balanced development and risk reduction**, contributing to long-term social sustainability in the Sundarbans.

The violin plot (Fig. 23.7) illustrates the distribution of MMNI scores across the 25 Community Development blocks of the Indian Sundarbans, sorted by median values. The visualization reveals pronounced inter-block disparities in multi-modal network integration. Blocks such as **Sagar, Namkhana, Basanti, and Kultali** exhibit lower median MMNI scores with wide dispersion, indicating poor accessibility coupled with uneven distribution of basic services. Conversely, inland blocks like **Diamond Harbour-II, Falta, and Kakdwip** show higher medians and relatively narrow spreads, reflecting better infrastructural integration and connectivity. The heterogeneity within several blocks underscores persistent infrastructural and service gaps, reinforcing spatial inequality in accessibility.

Prioritizing investments in road connectivity, transport networks, and public utilities is essential for low-MMNI blocks such as Sagar, Basanti, and Kultali to enhance accessibility. For high-variance blocks, balanced infrastructure development must be implemented to address uneven service concentration and ensure equitable resource distribution. In contrast, high-performing blocks like Falta and Diamond Harbour-II should focus on service resilience and climate-proof infrastructure to maintain accessibility during extreme weather events. Additionally, southern tidal island blocks require the introduction of ferry-based emergency routes and all-weather connectivity to mitigate isolation risks during cyclones and floods.

The box plot (Fig. 23.8) illustrates the distribution of MMNI scores across 25 blocks of the Indian Sundarbans, arranged in ascending order of their median

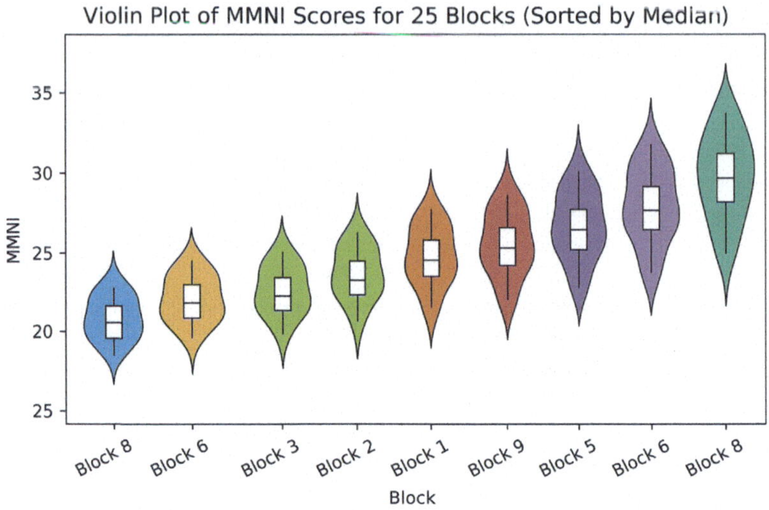

Fig. 23.7 The violin plot shows MMNI (Min–Max normalization score) for all median blocks

values. The visualization highlights substantial inter-block disparities in multi-modal network integration. Blocks at the lower end, such as Sagar, Namkhana, Basanti, and Kultali, exhibit lower medians with relatively shorter interquartile ranges, reflecting persistent infrastructural constraints and limited accessibility. In contrast, blocks positioned toward the higher end, including Diamond Harbour-II, Falta, and Canning I, display higher median values and wider variability, signifying better connectivity but with uneven infrastructural distribution within these areas. The gradual increase in medians from left to right suggests a spatial gradient of accessibility, with central and northern blocks outperforming southern tidal island blocks. This uneven distribution of connectivity not only reinforces socio-economic disparities but also heightens vulnerability to cyclones and flooding in low-MMNI blocks, where limited access to emergency services and evacuation routes poses significant challenges for disaster preparedness and response.

The observed variation in MMNI scores underscores significant spatial disparities in accessibility across the Sundarbans, with southern tidal island blocks exhibiting the lowest medians, reflecting poor connectivity and infrastructural isolation. Such gaps exacerbate vulnerability during extreme events, as inadequate transport networks and service concentration hinder timely evacuation and emergency response. Addressing these challenges requires **targeted infrastructure investment in low-MMNI blocks, the introduction of climate-resilient and all-weather connectivity options, and ferry-based emergency systems**. For high-variance blocks, **balanced development strategies** should ensure equitable service distribution, while high-performing blocks must focus on **resilience upgrades** to sustain accessibility under escalating climate risks.

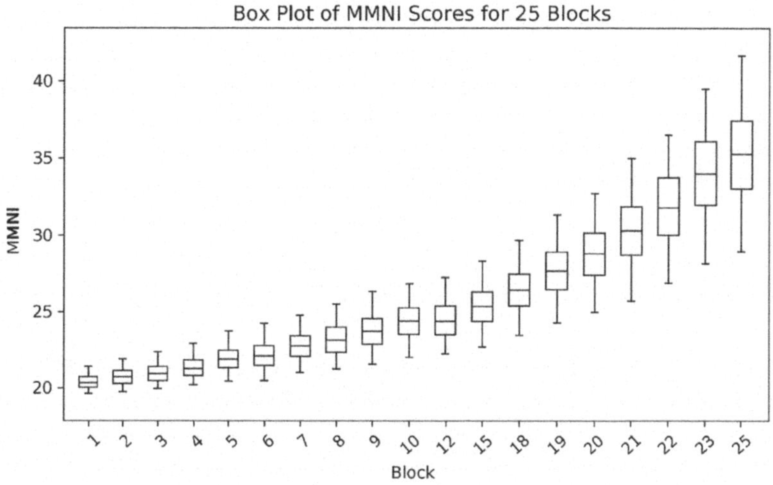

Fig. 23.8 Box plot of MMNI scores for 25 blocks (sorted by median)

23.5 Discussion

The present study provides a comprehensive assessment of **livelihood vulnerability and social sustainability across the Indian Sundarbans blocks**, integrating fifteen socio-economic indicators categorized into seven dimensions. The results reveal a striking **northwest–southeast gradient of vulnerability**, with blocks closer to urban fringes and transport corridors showing stronger performance in education, road connectivity, and communication, while the southern and southeastern tidal island blocks remain severely underdeveloped and disproportionately exposed to climate-related hazards. These findings not only corroborate earlier studies that highlighted the entrenched socio-economic deprivation of Sundarbans Island communities (Hazra et al. 2010; Bhattacharya and Sarkar 2019) but also expand upon them by offering a **multi-dimensional, indicator-driven framework** that captures localized disparities with higher granularity.

23.5.1 Spatial Inequalities and Structural Vulnerabilities

The block-level radar profiles demonstrate that **health services and financial access are the weakest domains across nearly all blocks**, even in those categorized as relatively developed. This indicates a **systemic service deficit** that extends beyond geographic peripherality, pointing toward institutional neglect and underinvestment in critical social infrastructure. Similar concerns were reported by Dasgupta et al. (2019), who noted that despite improvements in literacy and electricity coverage in parts of the Sundarbans, **human development outcomes stagnate**

where health and financial services remain absent. In contrast, indicators such as literacy and pucca road connectivity show relatively higher scores in northwestern blocks like Falta, Diamond Harbour-II, and Kakdwip, reflecting their **proximity to metropolitan Kolkata and enhanced market integration**. However, these advantages are not sufficient to fully mitigate vulnerabilities to climate events, which cut across developed and backward blocks alike.

23.5.2 Environmental Hazards and Livelihood Fragility

The results underscore that **southern island blocks such as Sagar, Namkhana, Basanti, Gosaba, and Kultali are epicenters of livelihood fragility**, given their heavy dependence on agriculture and fisheries, sectors highly sensitive to salinity intrusion, tidal flooding, and cyclonic disturbances. The compounded impact of recurrent cyclones (Aila 2009, Amphan 2020, Yaas 2021) has not only eroded physical infrastructure but also **undermined social resilience**, pushing vulnerable households into cycles of poverty and migration. Previous works (Ghosh 2018a, b; Mukhopadhyay et al. 2021) have similarly emphasized how repeated embankment breaches and saline water ingress exacerbate social vulnerability, but the present study adds to this literature by quantitatively linking these hazard impacts with **multi-dimensional deprivation patterns** at the block level.

23.5.3 Composite Index and Development Disparities

The **Composite Development Index (CDI)** analysis further confirms that **60% of Sundarbans blocks are classified as backward**, revealing a concentration of underdevelopment in the southeastern delta. These findings are consistent with national-level vulnerability assessments (Planning Commission 2013), which identified Sundarbans blocks as lagging in health, infrastructure, and livelihood diversification. Importantly, this study demonstrates that even among developed blocks, **gaps in critical sectors such as healthcare and financial services persist**, which calls for a more nuanced classification beyond a binary developed–backward divide. Such evidence supports arguments made by Adger (2006) and Cutter and Finch (2008) on the necessity of adopting **multidimensional frameworks of vulnerability** that integrate social, economic, and institutional determinants.

23.5.4 Policy Implications

From a planning perspective, these results highlight several **priority intervention areas**. First, strengthening **healthcare systems** in the Sundarbans is urgent, as almost all blocks scored poorly on health service indices. This requires both infrastructural investment (primary health centers, cyclone-resilient facilities) and human resource deployment (doctors, trained paramedics). Second, the persistent gap in **banking and financial services** needs immediate attention, as financial inclusion can enhance household adaptive capacity by enabling access to credit, savings, and insurance schemes. Third, **climate-resilient infrastructure**, including embankments, all-weather roads, and sustainable drinking water systems, is critical in the high-vulnerability southern blocks. Finally, **livelihood diversification programs**, such as skill development, ecotourism, and value-added fishery enterprises, must be promoted to reduce dependence on climate-sensitive agriculture.

23.5.5 Comparative Insights and Contribution

While earlier vulnerability assessments in the Sundarbans often focused on either **biophysical hazards** or **single socio-economic indicators**, this study contributes by developing a **holistic, indicator-based framework** that integrates exposure, sensitivity, and adaptive capacity across multiple dimensions. By combining spatial analysis, statistical summaries, and radar chart visualizations, it provides a block-level diagnostic tool that can guide **targeted and differentiated policy responses**. The approach is scalable and can be applied in other coastal and deltaic regions globally, making it relevant to international debates on **socio-ecological resilience and sustainable development** under anthropogenic stressors.

23.6 Conclusion

The present study developed a **multi-dimensional approach to assess social sustainability and livelihood vulnerability to coastal hazards in the Indian Sundarbans**, integrating 15 socio-economic indicators under seven key dimensions. The analysis revealed **persistent spatial disparities in development**, with southern tidal island blocks such as **Gosaba, Basanti, Kultali, and Namkhana** emerging as the most vulnerable due to poor infrastructure, limited health facilities, inadequate financial access, and high exposure to climate extremes. In contrast, northern and western blocks such as **Falta, Diamond Harbour-II, and Kakdwip** demonstrated relatively better socio-economic conditions but still exhibited critical weaknesses in **healthcare and banking services**, underscoring the systemic nature of these gaps.

The composite index and radar diagram analyses highlighted the **unequal distribution of adaptive capacity**, with well-connected blocks performing better across education, road connectivity, and communication, while remote and hazard-prone blocks remain development-deficient. These findings reaffirm the argument that **vulnerability in deltaic regions is not only shaped by biophysical hazards but is**

deeply rooted in social and infrastructural inequities. The methodological framework presented here, which combines spatial indicators with composite vulnerability measures, provides a **scalable and replicable model for integrated vulnerability assessment**, particularly in other climate-sensitive coastal and deltaic ecosystems.

In essence, **achieving social sustainability and resilience in the Sundarbans requires a multidimensional strategy that addresses infrastructural deficits, livelihood insecurity, and institutional limitations**, while also incorporating climate change adaptation into regional development planning.

23.7 Recommendations

Based on the findings, the following strategic recommendations are proposed to reduce vulnerability and enhance resilience in the Indian Sundarbans:

- **Strengthening Health Infrastructure:**

Healthcare emerged as the most critical gap across all blocks. Establishing cyclone-resilient health centers, improving mobile health units, and deploying trained medical staff should be prioritized. Telemedicine services can bridge accessibility gaps in remote islands.

- **Enhancing Financial Inclusion:**

Financial services such as banking, microcredit, and insurance should be expanded, particularly in southern and eastern blocks. Financial inclusion will enable households to access credit for recovery and invest in livelihood diversification.

- **Climate-Resilient Infrastructure Development:**

Investments in **all-weather pucca roads, embankment reinforcement, and sustainable drinking water systems** are essential for reducing exposure and improving post-disaster recovery capacity. These measures must integrate climate adaptation considerations.

- **Livelihood Diversification and Skill Development:**

Reducing dependence on climate-sensitive agriculture and fisheries through alternative livelihoods, such as eco-tourism, handicrafts, and aquaculture with salinity-tolerant species, will enhance adaptive capacity. Skill-building programs should be block-specific.

- **Education and Awareness Programs:**

Enhancing community knowledge on disaster preparedness, sustainable resource management, and financial literacy will strengthen social resilience. Schools can serve as hubs for climate adaptation awareness.

- **Integrated Governance and Policy Coordination:**

A **multi-level governance mechanism** involving local governments, NGOs, and state authorities is needed to implement development plans effectively. Participatory planning and community engagement should be central to these initiatives.

- **Monitoring and Data Integration:**

Establishing a real-time socio-economic and environmental monitoring system will help track vulnerability dynamics and support evidence-based policy-making. Integration of **GIS-based vulnerability mapping** into planning tools is essential.

Acknowledgements The authors express their sincere gratitude to the **Department of Geography, Syamaprasad College,** for providing logistical support during the course of this research and facilitating access to secondary data and necessary research infrastructure. Special thanks are due to the **Office of the Census of India** and the **Government of West Bengal** for making block-level demographic and socio-economic datasets available. We also acknowledge the support of **local administrative bodies and community stakeholders** in the Sundarbans region, whose insights and cooperation were invaluable in contextualizing the vulnerability assessment. Finally, the authors thank the **peer reviewers** for their constructive feedback, which helped improve the quality and clarity of this work.

References

Adger WN (2006) Vulnerability. Glob Environ Chang 16(3):268–281

Alkire S (2011) Multidimensional poverty measurement and analysis. Oxford University Press

Alkire S, Foster J (2011) Counting and multidimensional poverty measurement. J Public Econ 95(7–8):476–487

Arif MSI, Mahdi I, Rafi MA, Khan SJ, Rahman MM (2023) Cyclone exposure mapping in coastal Bangladesh: a multi-criteria decision analysis. Heliyon 9(10)

Ariful I (2022) Assessment of vegetation damage of the Sundarbans and inundation in the southwest coastal region of Bangladesh for three recent cyclones using satellite images

Ashley C, Carney D (1999) Sustainable livelihoods: lessons from early experience. DFID

Bandyopadhyay S (2019) Sundarban: a review of evolution and geomorphology. World Bank Group, 36

Bebbington A (1999) Capitals and capabilities: a framework for analyzing peasant viability, rural livelihoods and poverty. World Dev 27(12):2021–2044

Bera R, Maiti R (2021) Multi hazards risk assessment of Indian Sundarbans using GIS based analytic hierarchy process (AHP). Reg Stud Mar Sci 44:101766

Bhattacharya S, Sarkar S (2019) Livelihood vulnerability and adaptation in the Indian Sundarbans: a block-level assessment. Environ Dev Sustain 21(5):2301–2323

Biswas D, Sen A (2024) An analysis of local political forces around the management of climate change-led disasters in the Indian Sundarbans. Asian J Soc Sci 52(2):63–75

Biswas S, Sarkar K, Das TK (2024) Decadal changes (1980–2021) of shoreline and mangrove cover in Sundarban Delta, India, using remote sensing and GIS. J Phys Oceanogr 54(8):1655–1674

Brocklesby MA, Fisher E (2003) Community development in sustainable livelihoods approaches—an introduction. Community Dev J 38(3):185–198

Carney D (ed) (1998) Sustainable rural livelihoods: what contribution can we make? Department for International Development

Census of India (2011) District census handbooks: South 24 Parganas & North 24 Parganas. Office of the Registrar General & Census Commissioner, India
Chambers R, Conway G (1992) Sustainable rural livelihoods: practical concepts for the 21st century (IDS Discussion Paper No. 296). Institute of Development Studies
Chaturvedi S (2022) Adaptation governance. In: Handbook on climate change and disasters. Edward Elgar Publishing, pp 183–198
Cons J (2025) Delta futures: time, territory, and capture on a climate frontier. University of California Press, p 212
Cutter SL, Finch C (2008) Temporal and spatial changes in social vulnerability to natural hazards. Proc Natl Acad Sci 105(7):2301–2306
Das & Misra (2013) Health infrastructure and disaster vulnerability in coastal West Bengal. Indian J Public Health 57(3):177–183
Das GK (2023) Climatic Hazards of the Coast. Coastal environments of India: a coastal West Bengal perspective. Springer International Publishing, Cham, pp 135–173
Dasgupta S, Huq M, Khan ZH, Ahmed MMZ, Mukherjee N, Khan MF, Pandey K (2019) Coping with climate change in the Sundarbans: lessons from multidimensional vulnerability analysis. World Dev 120:204–214
Dempsey N, Bramley G, Power S, Brown C (2011) The social dimension of sustainable development: defining urban social sustainability. Sustain Dev 19(5):289–300
Depietri Y (2020) The social–ecological dimension of vulnerability and risk to natural hazards. Sustain Sci 15(2):587–604
Development & Planning Department, Government of West Bengal (2009) District human development report: South 24 Parganas. Government of West Bengal, Kolkata. http://www.wbpspm.gov.in/publications/DHDR_South24Pgs.pdf
Development & Planning Department, Government of West Bengal (2010) District human development report: North 24 Parganas. Government of West Bengal, Kolkata. http://www.wbpspm.gov.in/publications/DHDR_North24Pgs.pdf
DFID (1999) Sustainable livelihoods guidance sheets. Department for International Development
District census handbook: [South 24 Paraganas & North 24 Paraganas], census of India 2011, West Bengal, series 20, part XII-A/B. Directorate of Census Operations, West Bengal
Ehsan S, Begum RA, Maulud KNA (2022) Household external vulnerability due to climate change in Selangor coast of Malaysia. Clim Risk Manag 35:1–32
Ellis F (2000) Rural livelihoods and diversity in developing countries. Oxford University Press
FAO (2017) The state of food security and nutrition in the world 2017. Food and Agriculture Organization of the United Nations
Ghosh A (2018a) Human–nature interactions in the Indian Sundarbans: ecological vulnerability, resilience and sustainability. Ambio 47(7):725–738
Ghosh A (2018) Sustainability conflicts in coastal India. Springer Nature, Gewerbestrasse
Ghosh A, Dutta K (2024) Health threats of climate change: from intersectional analysis to justice-based radicalism. Ecol Soc 29(2)
Giri S, Daw TM, Hazra S, Troell M, Samanta S, Basu O, Chanda A et al (2022) Economic incentives drive the conversion of agriculture to aquaculture in the Indian Sundarbans: livelihood and environmental implications of different aquaculture types. Ambio 51(9):1963–1977
Gunaratne MS, Firdaus RBR, Sosai A (2023) The importance of multidimensional vulnerability analysis in sustainable development. J Sustain Sci Manage 18(9):38–57
Hahn MB, Riederer AM, Foster SO (2009) The livelihood vulnerability index: a pragmatic approach to assessing risks from climate variability and change—a case study in Mozambique. Clim Change 94(1–2):71–110
Hazra S, Samanta K, Mukhopadhyay A, Akhand A (2010) Temporal change detection (2001–2008) study of Sundarban. School of Oceanographic Studies, Jadavpur University
He L, Shen J, Zhang Y (2018) Ecological vulnerability assessment for ecological conservation and environmental management. J Environ Manage 206:1115–1125

India Meteorological Department (2020) Very severe cyclonic storm "AMPHAN" (16–21 May 2020): a report. Government of India.
IPCC (2014) Climate change 2014: impacts, adaptation, and vulnerability (WGII AR5). Cambridge University Press
IPCC (2022) Climate change 2022: impacts, adaptation and vulnerability (WGII AR6). Cambridge University Press
James P (2025) The human development index as a multidimensional alternative to income-based measures of development
Lindenberg M (2002) Measuring household livelihood security at the family and community level in the developing world. World Dev 30(2):301–318
Mallick JK (2025) Mangrove flora and fauna in Sundarbans and their significance in a global perspective
Mandal UK, Maji B, Mullick S, Nayak DB, Mahanta KK, Raut S (2019) Global climate change and human interferences as risk factors, and their impacts on geomorphological features as well as on farming practices in Sundarbans eco-region. The Sundarbans: a disaster-prone eco-region: increasing livelihood security. Springer International Publishing, Cham, pp 405–437
Mehzabin S (2019) Assessment of livelihood adaptation strategies to climate variability in a coastal area of Southwest Bangladesh
Miller SR (2023) Over the seawall: tsunamis, cyclones, drought, and the delusion of controlling nature. Island Press
Mohammed, Sultana F, Khan A, Ahammed S, Saimun MSR, Bhuiyan MS, Arfin-Khan MA et al (2024) Assessing vulnerability to cyclone hazards in the world's largest mangrove forest, The Sundarbans: a geospatial analysis. Forests 15(10):1722
Mondal BK, Mahata S, Basu T, Das R, Patra R, Abdelrahman K, Praharaj S et al (2024) Analysis of the post-cyclonic physical flood susceptibility and changes of mangrove forest area using multi-criteria decision-making process and geospatial analysis in Indian Sundarbans. Atmosphere 15(4):432
Mukhopadhyay A, Das D, Ghosh T, Hazra S, Mitra D (2021) Anthropogenic influence on the geomorphology of the Indian Sundarbans. Environ Dev 37:100582
Narayan D (2002) Empowerment and poverty reduction: a sourcebook. World Bank
Newton A, Icely J, Cristina S, Perillo GM, Turner RE, Ashan D, Kuenzer C et al (2020) Anthropogenic, direct pressures on coastal wetlands. Front Ecol Evol 8:144
NRSC (n.d.) Bhuvan geoportal. National Remote Sensing Centre, ISRO
Parthasarathy D, Srivastava S, Mehta L, Bose S, Movik S (2025) Transformation in the context of uncertainty and compounding effects: insights from marginal environments in India and Bangladesh. Glob Environ Chang 93:103025
Pathak S, Mukherjee J, Sen A, Choudry A (2025) Whose habitat? Exploring human-tiger conflict in the riskscapes of the Indian Sundarbans. J Polit Ecol 32(1)
Planning Commission, Government of India (2013) Twelfth five year plan (2012–2017), volume i: faster, more inclusive and sustainable growth. Oxford University Press
Rahman R (2022) Community resilience to climate change through disaster resilient housing practice in coastal area: a case study of dacope upazila in Khulna
Roslina K (2018) Contribution of Brackish and freshwater aquaculture to livelihood of small-scale rural aquaculture farmers in Kedah, Malaysia. Pertanika J Soc Sci Humanit 26(3)
Saha S (2017) Groundwater potential mapping using analytical hierarchical process: a study on Md. Bazar block of Birbhum District, West Bengal. Spat Inf Res 25(4):615–626
Samanta S, Hazra S, Mondal PP, Chanda A, Giri S, French JR, Nicholls RJ (2021) Assessment and attribution of mangrove forest changes in the Indian Sundarbans from 2000 to 2020. Remote Sens 13(24):4957
Santos ME, Seth S (2023) Measurement and analysis of well-being, human development and poverty with multiple dimensions
Sarkar SK (2022) Sundarban Mangrove wetland (a UNESCO world heritage site): a comprehensive global treatise. Elsevier

Scoones I (1998) Sustainable rural livelihoods: a framework for analysis (IDS Working Paper No. 72). Institute of Development Studies
Survey of India (various years) Topographical sheets (1:50,000 & 1:250,000). Survey of India
Szabo S et al (2016). Population dynamics, delta vulnerability and environmental change: comparison of the Mekong, Ganges–Brahmaputra and Amazon delta regions. Sustain Sci 11(4):539–554 (for comparative delta framing)
Thakur S, Mondal I, Bar S, Nandi S, Ghosh PB, Das P, De TK (2021) Shoreline changes and its impact on the mangrove ecosystems of some islands of Indian Sundarbans, North-East coast of India. J Clean Prod 284:124764
Uddin SA, He L, Hossain MJ, Nusrat N, Debi M (2024) Ganges-Brahmaputra-Meghna River Delta. In: Delta sustainability: a report to the mega-delta programme of the UN Ocean Decade. Springer Nature, Singapore, pp 89–116
UNDP (1990) Human development report 1990. United Nations Development Programme
UNICEF, UNICEF (2016) Strategy for water, sanitation and hygiene 2016–2030
UNISDR (now UNDRR) (2015) Sendai framework for disaster risk reduction 2015–2030. United Nations Office for Disaster Risk Reduction
World Bank (2014) Building resilience for sustainable development of the Sundarbans: strategy report. The World Bank
Zaid Al F (2021) Technologies of water control and social management in flooded coastal Bangladesh
Zainab A, Shah KU (2024) Taking stock of recent progress in livelihood vulnerability assessments to climate change in the developing world. Climate 12(7):100
Zhang W, Luo M, Gao S, Chen W, Hari V, Khouakhi A (2021) Compound hydrometeorological extremes: drivers, mechanisms and methods. Front Earth Sci 9:673495

Chapter 24
Living on the Edge: Urban Resilience in Dynamic Coastal Environments

Subhankar Roy and Gouri Sankar Bhunia

24.1 Introduction

Urbanization in coastal areas is accelerating at an unprecedented pace, driven by economic opportunities, port accessibility, tourism, and population growth. Over 40% of the world's population resides within 100 km of a coastline, and this number continues to rise (UN-Habitat 2020). Coastal areas have historically been centers of trade, migration, and culture. As noted by Neumann et al. (2015), the majority of global megacities are located along coasts due to strategic economic advantages. However, rapid urbanization often occurs in an unregulated manner, leading to the encroachment on ecologically sensitive zones such as mangroves, wetlands, and dunes (Seto et al. 2011). Urban expansion exacerbates coastal risks by altering natural drainage patterns and reducing the ability of ecosystems to buffer against these risks. Coastal zones are inherently dynamic—shaped by natural processes such as tides, wave action, erosion, sediment transport, and increasingly, climate change-related phenomena like sea-level rise and extreme weather events. This juxtaposition of intensive urban development with natural coastal volatility presents complex risks and vulnerabilities.

According to the IPCC (2021), global mean sea level has risen by approximately 20 cm since 1900, with the rate of increase accelerating in recent decades. Coastal erosion, storm surges, and saltwater intrusion threaten infrastructure, freshwater resources, and agriculture (Nicholls and Cazenave 2010). Urban development's built without consideration of these dynamics become highly vulnerable to disasters. Meerow et al. (2016) argue that resilience must be viewed through multiple lenses—ecological, infrastructural, social, and institutional. In coastal areas, this includes

S. Roy
Department of Geography, Seacom Skills University, Bolpur, West Bengal, India

G. S. Bhunia (✉)
Kharagpur, West Bengal, India
e-mail: rsgis_gouri@rediff.com

A. Nath et al. (eds.), *Geospatial Application of Coastal Hazards*, Springer Water,
https://doi.org/10.1007/978-3-032-20076-1_24

not only hard infrastructure like sea walls but also soft measures like ecosystem restoration, early warning systems, and inclusive governance. Studies such as Aerts et al. (2014) and Anguelovski et al. (2016) highlight how vulnerability is unevenly distributed. In many coastal cities, informal settlements are located in the most hazard-prone areas, lacking access to basic services or disaster insurance.

The concept of urban resilience is defined as the capacity of urban systems to absorb disturbances, adapt to changing conditions, and maintain essential functions—has become central to the planning and sustainability discourse in these environments (Meerow, Newell and Stults, 2016). Coastal urban centers, such as Jakarta, Mumbai, and New York City, are grappling with flooding, subsidence, groundwater salinization, and infrastructure vulnerabilities. These risks are not only environmental but also social, with marginalized communities often being disproportionately affected. Based on the above discussion, this study aims to critically examine the evolving landscape of urban resilience in dynamic coastal environments by analyzing global scholarly contributions, research networks, and thematic trends from 2000 to 2025. It seeks to identify key interventions—both technological and nature-based—that enhance coastal resilience, while highlighting existing research gaps, influential contributors, and policy-relevant insights to guide future urban planning and adaptive strategies.

24.2 Methods

This analysis focuses on the period from 2000 to 2025. The primary data source for this analysis is the Dimensions AI database, chosen for its broad coverage of peer-reviewed articles, conference proceedings, policy documents, and technical reports in the English language. A total of 13,362 relevant publications were extracted using a structured keyword search encompassing terms such as "urbanization," "dynamic coast," "climate change," "remote sensing," "GIS," and "cutting-edge technology." Data preprocessing involved normalization of author names and institutional affiliations to ensure accuracy in network mapping. The analysis leveraged VOSviewer, a widely used tool for constructing and visualizing bibliometric networks (Fig. 24.1).

Two primary analytical techniques were employed: co-citation analysis to identify influential authors, journals, and documents within the knowledge domain, and co-authorship analysis to assess collaborative networks among researchers and institutions. Additionally, advanced citation metrics including Field Citation Ratio (FCR) and Relative Citation Ratio (RCR) were used to evaluate the scientific influence and contextual relevance of the publications. The share of publications with at least one citation ($\geq$1 citations) was computed to assess the citation visibility and research impact across the dataset. Network visualizations highlighted thematic clusters, revealing interdisciplinary linkages across domains such as coastal engineering, urban planning, environmental science, and geoinformatics. This methodological framework not only illuminates the evolution of research focus over the 25 years but also identifies influential research nodes and emerging areas, offering strategic

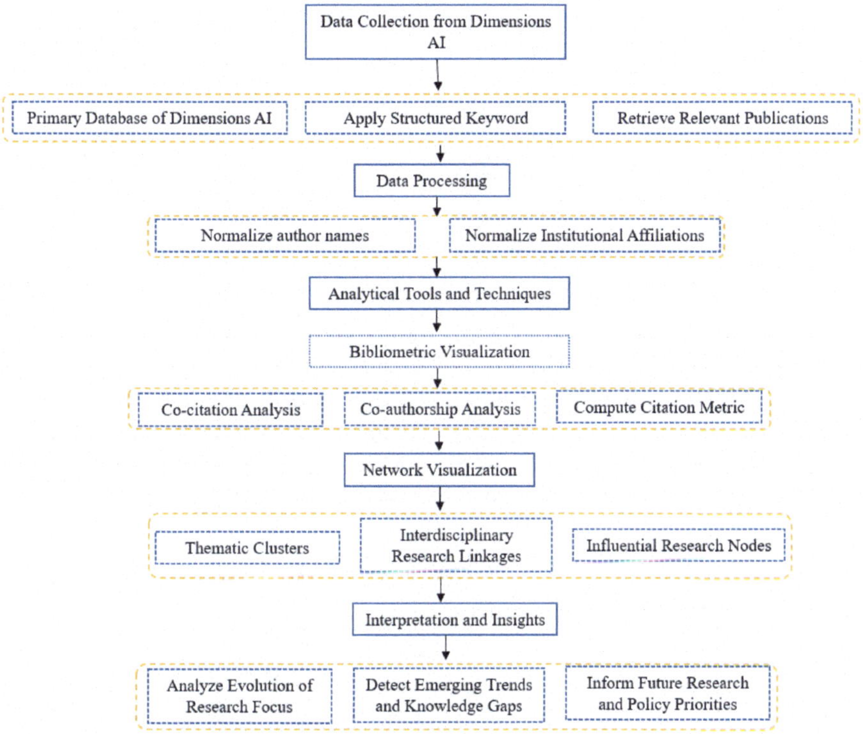

Fig. 24.1 Flow chart map of the overall methodology

insights for future scholarship and policy formulation in the context of coastal urban resilience.

24.3 Results and Discussion

24.3.1 Research Publications

The majority of these publications (46.56%, or 6222) are chapters, indicating that much of the discourse is embedded within larger compendiums, handbooks, or edited volumes. Edited books closely follow, accounting for 34.26% (4578 publications), further underscoring the importance of comprehensive, multi-author volumes in shaping this research area. Journal articles, often considered the gold standard for scientific communication, constitute only 10.93% (1461) of the total publications. Proceedings papers account for 4.30% (575), indicating a reasonable presence of the

Table 24.1 Number and percentage of publications as per the research categories

Research categories	Number	Percent
Chapter	6222	46.56
Edited book	4578	34.26
Article	1461	10.93
Proceeding	575	4.30
Monograph	515	3.85
Preprint	11	0.08

topic in conferences and symposia, where emerging ideas, pilot studies, and innovative methodologies are shared (Table 24.1). Overall, the publication pattern underscores a field that is both academically rich and practically grounded, with a preference for collective knowledge production and thematic consolidation through books and chapters, while maintaining a steady stream of peer-reviewed and conference contributions for methodological and empirical advancement.

24.3.2 Author-Wise Publications

Among the top authors, Yun-Chul Hong from Seoul National University, South Korea, leads with 38 publications. These have received a relatively modest 11 citations, yielding a mean citation rate of 0.29, which may indicate a broader engagement in producing knowledge or policy-driven outputs rather than high-impact academic studies. Mi Na Ha, affiliated with Dankook University, South Korea, follows with 25 publications and 19 citations, with a mean citation of 0.76, suggesting a slightly higher impact per publication. A notable cluster of prolific authors hails from the University of Maribor, Slovenia, including Jurij Krope, Darko Goričanec, and Stanislav Božičnik, each contributing 24 publications with low citation counts (3 citations each, mean 0.13). This could reflect collaborative institutional research efforts or localized studies with limited global reach (Table 24.2). Dorian Q. Fuller from University College London, UK, stands out with 22 publications and a remarkable 185 citations, resulting in a mean citation of 8.41—the highest among the group. Krystyna Konstankiewicz from the Polish Academy of Sciences and William Joseph Edmonds exhibit relatively strong citation performances with mean citations of 2.86 and 2.94 respectively, reflecting a higher resonance of their work in the academic community. Abdul Ghani Olabi of the University of Sharjah, UAE, with 24 publications and 6 citations, and I. P. Abrol, with 17 publications and 62 citations (mean 3.65), make notable contributions to the body of literature, indicating a balance between output volume and academic impact.

Table 24.2 Top ten authors of publication of coastal dynamics

Sr. No.	Author Name	Organization, Country	Publications	Citations	Mean
1	Yun-Chul Hong	Seoul National University, South Korea	38	11	0.29
2	Mi Na Ha	Dankook University, South Korea	25	19	0.76
3	Jurij Krope	University of Maribor, Slovenia	24	3	0.13
4	Abdul Ghani Olabi	University of Sharjah, United Arab Emirates	24	6	0.25
5	Darko Goričanec	University of Maribor, Slovenia	24	3	0.13
6	Stanislav Božičnik	University of Maribor, Slovenia	24	3	0.13
7	Dorian Q Fuller	University College London, United Kingdom	22	185	8.41
8	Krystyna Konstankiewicz	Polish Academy of Sciences, Poland	21	60	2.86
9	William Joseph Edmonds		18	53	2.94
10	I P Abrol		17	62	3.65

24.3.3 *Journal-Wise Publication*

The data, extracted from Dimensions AI for the period 2000–2025, shows that "Sustainable Cities and Communities" emerges as the most prolific SDG-linked journal category, with 566 publications contributing to the field, accumulating 6853 citations and a mean citation score of 12.11. This highlights the increasing importance of sustainability-driven urban planning in the context of coastal vulnerabilities. Notably, "Life on Land" emerges as highly influential with 453 publications but a striking 19,439 citations, averaging 42.91 citations per publication, likely due to its overlap with biodiversity and ecosystem-based solutions in coastal zones. Journals such as "Zero Hunger" and "Good Health and Well Being" also show high citation counts (26.62 and 14.69 average, respectively), indicating the integrative view of food security and public health in vulnerable coastal areas. Meanwhile, "Clean Water and Sanitation" and "Life Below Water"—directly tied to hydrological and marine challenges—register strong engagement with mean citations of 18.19 and 30.59 respectively (Table 24.3). Technical and policy-focused themes like "Industry, Innovation and Infrastructure" and "Responsible Consumption and Production" reflect moderate scholarly attention with citation means of 7.22 and 7.57.

Table 24.3 Top ten publications in coastal dynamics

Sr. No	Name	Publications	Citations	Mean
1	Sustainable cities and communities	566	6853	12.11
2	Climate action	526	10,008	19.03
3	Life on land	453	19,439	42.91
4	Affordable and clean energy	437	4292	9.82
5	Zero hunger	333	8866	26.62
6	Good health and well being	329	4833	14.69
7	Responsible consumption and production	211	1598	7.57
8	Industry, innovation and infrastructure	195	1408	7.22
9	Quality education	177	2924	16.52
10	Life below water	123	3763	30.59

24.3.4 Temporal Trends Publication Versus Citation

This research reflects a steadily growing scholarly interest, punctuated by sharp spikes that correlate with the intensification of climate-related coastal events and global policy attention. In the early 2000s, the number of publications remained relatively low, ranging from 27 to 45 annually. However, a substantial surge in publication output began post-2008, when contributions rose sharply to 369 in 2008 and peaked at 1439 in 2020. This trend was likely influenced by global climate accords (e.g., the Paris Agreement in 2015), widespread urban flooding events, and an increasing focus on climate-smart urbanism (Fig. 24.2).

The upward trajectory in publication output is reflected in the citation counts, which increased from a modest 363 in 2000 to a peak of 17,954 in 2024, followed by a slight decline to 11,872 in 2025. This slight drop is potentially due to the lag in

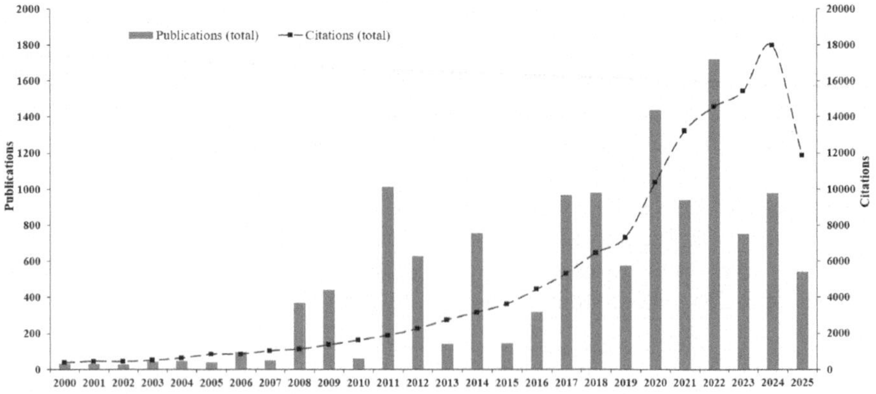

Fig. 24.2 Publication versus citation during the period between 2000 and 2025

citation accumulation for recent publications. This citation momentum reflects the increasing scientific and policy relevance of the topic. Meanwhile, the percentage of publications with at least one citation shows variability. Early years (2000–2005) reported high percentages (85–97%), indicating high influence despite fewer publications. However, between 2008 and 2019, the percentage fluctuated widely (30–67%), possibly due to a broader spectrum of publications, including grey literature and conference outputs. Notably, in the years 2011 and 2017–2018, the percentage of cited papers dipped below 40%, suggesting a need for quality refinement or delayed impact recognition. Furthermore, recent years (2020–2024) have seen a shift towards interdisciplinary research that merges geospatial analysis, policy assessment, climate modeling, and socio-economic vulnerability mapping, thereby enhancing citation potential and policy applicability.

This analysis, using key metrics such as Field Citation Ratio (FCR) and Relative Citation Ratio (RCR) from 2000 to 2025, reveals critical insights into the scholarly impact and research dynamics of this domain. Notably high FCR values were observed in 2005 (12.68) and 2014 (4.64), suggesting landmark studies or review articles that gained significant scholarly attention. The year 2005, in particular, stands out with a sharp peak, possibly corresponding to growing global concerns post the 2004 Indian Ocean Tsunami, which sparked new paradigms in coastal resilience planning and disaster preparedness (Fig. 24.3).

Years like 2006 (FCR: 0.87) and 2008–2009 (FCR: 0.51 each) show lower geometric means, indicating either a proliferation of low-cited papers or that publications from these years had not yet matured in citation accumulation by the data cut-off point. However, from 2011 onwards, the RCR begins to reflect more citation impact, albeit sporadically. Years like 2014 (RCR: 3.05) and 2023 (RCR: 11.12) mark significant spikes, underscoring the production of highly influential articles. The exceptional RCR of 2023 suggests breakthrough publications, likely reflecting the synergy

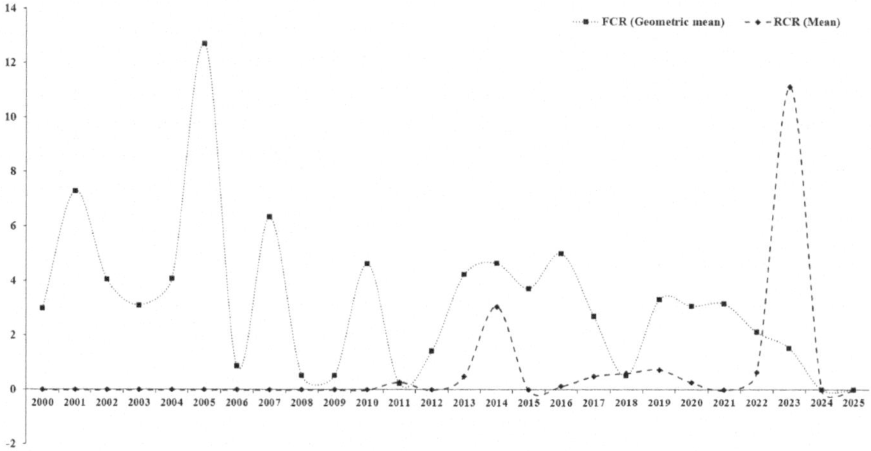

Fig. 24.3 Temporal variation of RCR and FCR from 2000 and 2025

between AI-based forecasting, climate modeling, and resilience infrastructure in coastal megacities.

24.3.5 Co-citation Analysis

Based on a dataset comprising 500 researchers, the analysis identified only one citation link, a surprisingly low figure given the total of five citations across the network. The presence of 499 distinct clusters—nearly one per researcher—further reinforces this fragmentation, indicating that most researchers operate in silos, engaging with different conceptual frameworks, geographies, or methodological approaches without significant overlap (Fig. 24.4). Researchers from other domains may be addressing resilience through domain-specific lenses, thereby limiting mutual citations and collaboration. Moreover, the relatively low citation density hints at either a nascent or highly specialized body of work that has not yet consolidated into a cohesive knowledge base with widely recognized seminal works or shared theoretical anchors.

24.3.6 Co-authorship Analysis

A dataset comprising 176 researchers reveals that they form 521 co-authorship links, resulting in a total of 1463 citations. This relatively dense web of collaborations indicates that many researchers in this field are engaging in joint investigations, often crossing disciplinary or institutional boundaries to co-produce knowledge on coastal resilience. The formation of 15 distinct clusters within the co-authorship network signifies the presence of several well-defined research groups or thematic alliances. The clustering also suggests that while the field is collaborative, it still retains a degree of specialization, with researchers tending to co-author within topical or regional silos rather than across them. Furthermore, the relatively high citation count (1463) in conjunction with these co-authorship patterns underscores the impact and visibility of collaborative work in this domain (Fig. 24.5). It suggests that studies resulting from cooperative research efforts are not only being published but also being referenced more frequently, possibly due to their interdisciplinary depth, policy relevance, or methodological robustness.

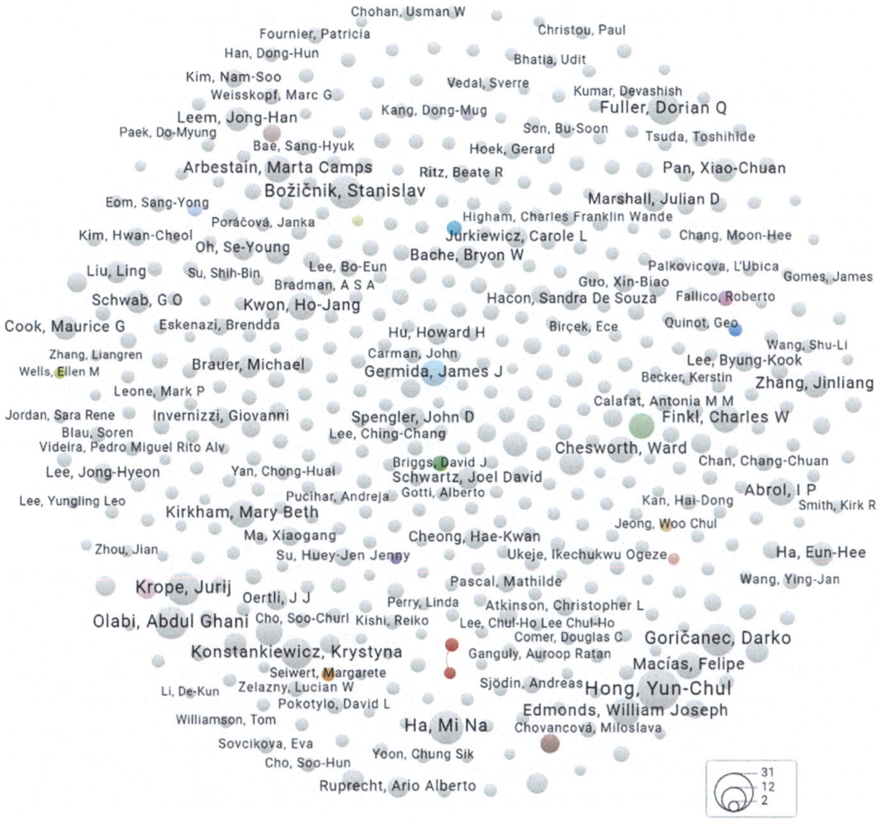

Fig. 24.4 Co-citation analysis of urbanization in coastal areas

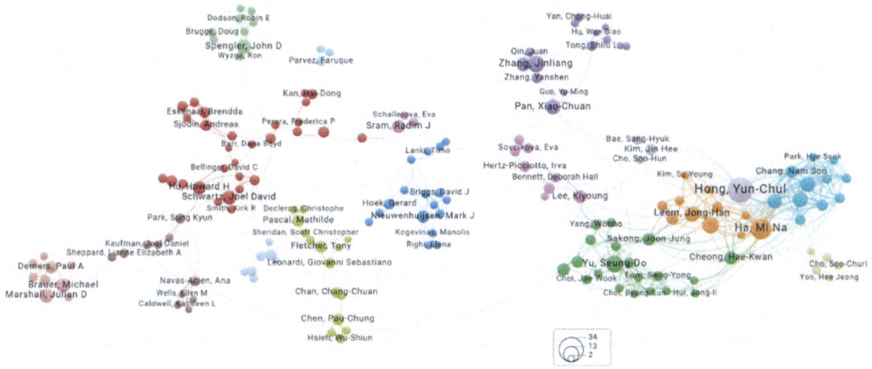

Fig. 24.5 Co-authorship analysis urbanization in coastal areas

24.3.7 *Understanding Urbanization in Coastal Areas*

24.3.7.1 Global Trends and Drivers of Urban Growth in Coastal Urbanization

Urbanization in coastal areas has become a defining feature of global demographic and spatial transformation. More than 40% of the world's population resides within 100 km of the coast, and this figure is expected to continue growing as coastal cities become increasingly central to economic activity, migration, and infrastructure development (Neumann et al. 2015; UN-Habitat 2020). Global trends indicate that urban expansion is occurring at a faster pace in coastal zones compared to inland regions, particularly in Asia and Africa. According to Seto et al. (2011), urban land cover is expected to increase by more than threefold between 2000 and 2030, with the majority of this expansion occurring in coastal zones of the Global South. These areas are not only experiencing rapid growth but are also increasingly exposed to hazards such as sea-level rise, storm surges, coastal erosion, and flooding, driven by both anthropogenic pressures and climate change (IPCC 2019).

Economic opportunities such as trade, tourism, port-based industries, and the blue economy attract people and investments to coastal cities. Urban centers like Shanghai, Mumbai, and Lagos have proliferated due to their strategic geographic positioning along key maritime trade routes. Natural advantages, including access to flat land, fertile deltas, and navigable waterways, also make coastal areas attractive for settlement (Small and Nicholls 2003). Furthermore, demographic factors such as high birth rates and rural-to-urban migration, particularly from climate-affected inland areas, continue to push populations toward coastal urban centers. For instance, in Bangladesh, people displaced by riverbank erosion and saline intrusion in rural areas frequently migrate to coastal cities like Chittagong, often settling in hazard-prone informal settlements (Rana and Islam 2022). In many developing contexts, weak governance, speculative real estate development, and the lack of enforceable land-use regulations further fuel unplanned urban sprawl in coastal zones (Kuffer et al. 2018).

24.3.7.2 Coastal Urbanization Patterns in Developing Versus Developed Regions

In developed countries, coastal urban growth is generally more regulated and supported by advanced infrastructure, zoning laws, and disaster preparedness systems. Countries like the Netherlands exemplify best practices in coastal management through integrated urban planning, flood-resilient infrastructure, and nature-based solutions. Similarly, in the United States, coastal cities have implemented adaptive strategies such as managed retreat and green infrastructure, especially following disasters like Hurricane Sandy (Siders 2019). However, even in these regions, events like Hurricane Katrina and sea-level rise in Florida highlight persistent vulnerabilities. In contrast, developing countries face more severe challenges. Coastal urbanization is often rapid, informal, and under-resourced, with cities like Jakarta experiencing subsidence due to groundwater extraction and inadequate stormwater management. In Lagos, urban growth into wetlands and floodplains has led to recurring urban floods, affecting millions of low-income residents (Ajibade et al. 2020). In these regions, the expansion of informal settlements into ecologically fragile and flood-prone areas underscores the urgent need for inclusive urban planning and resilience-building measures. Understanding the dynamics of coastal urbanization is crucial for designing policies that promote sustainable development and resilience. The convergence of urban expansion, environmental degradation, and climate risks in coastal zones presents a multidimensional governance challenge. Addressing it requires an integrated approach that combines spatial planning, climate adaptation, social equity, and infrastructure investment, particularly in vulnerable developing countries.

24.3.8 Impacts of Urbanization on Coastal Dynamics

24.3.8.1 Land Use Change, Habitat Fragmentation, Encroachment and Coastal Erosion

One of the primary impacts is land use change and habitat fragmentation. Rapid urban expansion in coastal areas leads to the conversion of natural land covers—such as mangroves, wetlands, dunes, and estuaries—into built-up areas including residential, industrial, and commercial zones. Mangroves and coastal wetlands, which act as natural buffers against storms and provide critical breeding grounds for marine biodiversity, are particularly vulnerable to encroachment. According to Seto et al. (2012), urban land expansion often occurs at the expense of high-biodiversity zones, resulting in fragmentation of ecological habitats and loss of ecological connectivity. Valiela et al. (2001) also highlight how fragmented wetlands lose their hydrological integrity, decreasing their capacity to absorb floodwaters and sequester nutrients.

Coastal erosion is a significant consequence, often intensified by unregulated development and human encroachment on the shoreline. Construction of buildings, roads, ports, and sea walls in close proximity to the waterline interferes with sediment transport processes and destabilizes beach profiles. In places such as Chennai, India, and Lagos, Nigeria, unplanned coastal development has led to the erosion of beaches and an intensification of coastal erosion (Vousdoukas et al. 2020). The problem is exacerbated by the removal of vegetation and sand mining, which further weakens the coastal buffer zones. Additionally, the reduction in natural sediment supplies due to damming of rivers upstream limits beach nourishment and promotes shoreline retreat.

24.3.8.2 Pollution, Water Quality Degradation, Infrastructure Development and Hydrodynamic Changes

Pollution and water quality degradation are serious consequences of coastal urbanization. Urban runoff, sewage discharge, solid waste, and industrial effluents are frequently released untreated into coastal water bodies. This influx of nutrients and contaminants leads to eutrophication, harmful algal blooms, and oxygen depletion in coastal waters, negatively affecting marine life and fisheries. In Jakarta, for instance, over 75% of wastewater is discharged untreated, contributing to widespread coastal pollution (ADB 2016). Plastic pollution from coastal cities is also a growing concern, with coastal megacities such as Manila and Mumbai contributing significantly to marine debris due to inadequate waste management systems (Jambeck et al. 2015).

The construction of roads, embankments, ports, drainage systems, and real estate developments interferes with natural water flow, tidal exchange, and sediment dynamics. Hard engineering structures, such as sea walls, breakwaters, and jetties, while intended to protect urban infrastructure, often cause unintended consequences such as downdrift erosion and changes in current patterns. Urban drainage systems also frequently channel runoff at high velocity into the sea, leading to coastal scouring and saline intrusion into inland aquifers. In Shanghai and Bangkok, excessive groundwater withdrawal to meet urban demand has led to land subsidence, which in turn worsens coastal flooding and increases vulnerability to sea-level rise (Nicholls et al. 2008).

24.3.9 Impacts of Coastal Dynamics on Urban Settlements

24.3.9.1 Sea Level Rise, Urban Flooding, Saltwater Intrusion and Water Scarcity

Coastal dynamics—comprising sea level rise, tidal variations, storm surges, and sediment transport—pose significant threats to the resilience of urban settlements situated in low-lying coastal zones. One of the most pressing issues is sea level rise, which exacerbates urban flooding, particularly during extreme weather events or high tides. According to the IPCC Sixth Assessment Report (2021), global mean sea levels have risen by approximately 20 cm since 1900, and projections indicate a further rise of 0.29–1.10 m by 2100 under various emission scenarios. This gradual inundation increases the frequency of urban flooding, affecting infrastructure, displacing populations, and disrupting essential services. For instance, cities like Jakarta, Indonesia, are experiencing chronic flooding, partly due to sea-level rise and land subsidence, which has led to the government's decision to relocate the national capital (Abidin et al. 2011; BBC News 2019).

Saltwater intrusion, another consequence of coastal dynamics, poses a direct threat to urban water security, particularly in deltaic and estuarine regions. This phenomenon occurs when rising sea levels cause saline water to be pushed into freshwater aquifers and rivers, contaminating drinking water sources and reducing agricultural productivity. In the Mekong Delta of Vietnam, increased salinity levels have been recorded several kilometers inland, impacting millions of residents (Smajgl et al. 2015). Similarly, Chennai, India, has faced acute drinking water shortages partially due to saline intrusion into its groundwater sources (Tiwari et al. 2020).

24.3.9.2 Climate Change and Disaster Risk Amplification

Climate change has intensified the frequency and severity of coastal disasters, including cyclones, storm surges, and heatwaves, thereby increasing disaster risks for coastal urban populations. The increased energy in ocean systems due to rising temperatures results in stronger tropical storms, as observed in the increasing intensity of cyclones in the Bay of Bengal (Emanuel 2005). The devastating impact of Cyclone Amphan (2020) on the eastern Indian city of Kolkata and coastal Bangladesh demonstrated how densely populated urban areas in coastal zones are at extreme risk of infrastructure damage, displacement, and health emergencies (UNDRR 2020). Several case studies highlight the vulnerability of coastal cities to dynamic coastal processes. For example, New Orleans, USA, has faced repeated flooding crises due to its geographic location, subsidence, and weakened levee systems, culminating in the catastrophic effects of Hurricane Katrina (2005) (Colten 2006). In Lagos, Nigeria, rapid coastal urbanization coupled with poor drainage infrastructure has left the city susceptible to flash floods and coastal erosion (Adelekan 2010). These examples underscore the urgent need for integrated coastal zone management, resilient infrastructure planning, and proactive climate adaptation strategies in coastal urban areas.

24.3.10 Governance and Policy Framework

24.3.10.1 Coastal Zone Regulation, Urban Planning and Integrated Coastal Zone Management (ICZM)

Governance plays a crucial role in managing the interface between urban development and coastal ecosystems. Coastal Zone Regulation (CZR) frameworks aim to regulate and monitor land-use activities in ecologically sensitive coastal areas. For example, in India, the Coastal Regulation Zone (CRZ) Notification, 2011 under the Environment Protection Act restricts construction and other activities within a specific distance from the High Tide Line (MoEFCC 2011). However, enforcement gaps, overlapping jurisdictions, and real estate pressures often lead to violations, compromising ecosystem integrity and increasing urban vulnerability (Menon and Kohli 2016). Urban planning in coastal areas needs to integrate dynamic coastal processes, sea-level projections, and risk-sensitive land-use zoning, yet many coastal cities still rely on static master plans that fail to address long-term coastal hazards (Brown et al. 2018).

ICZM represents a holistic governance approach that seeks to balance ecological conservation with economic development and social equity across the land-sea interface. It promotes stakeholder participation, inter-sectoral coordination, and science-based policymaking. The World Bank-supported ICZM project in India, launched in 2010, exemplifies such an approach, combining environmental monitoring, livelihood support, and infrastructure development in states like Odisha and Gujarat (World Bank 2019). ICZM facilitates the mapping of hazard-prone zones, the identification of ecosystem services, and the mainstreaming of climate adaptation in planning. Yet, challenges persist in data availability, institutional capacities, and cross-jurisdictional coordination (Cormier et al. 2015).

24.3.10.2 Local Governance, Institutional and International Best Practices and Conventions

Local governments are the front-line actors in managing coastal urban resilience but often face resource constraints, limited technical knowledge, and fragmented mandates. In many countries, decentralization has not been accompanied by sufficient capacity-building or financial devolution. Successful examples—such as the Municipality of Semarang, Indonesia, which developed a participatory climate-resilient action plan—highlight the importance of empowering local bodies to engage communities, enforce building codes, and integrate disaster preparedness (Leitmann 2015). Institutional synergies between urban development authorities, port trusts, and coastal management agencies are essential for effective implementation.

International conventions and frameworks guide national policies on coastal governance. The United Nations Convention on the Law of the Sea (UNCLOS), the Convention on Biological Diversity (CBD), and the Sendai Framework for Disaster Risk Reduction (2015–2030) underscore the need for ecosystem-based approaches and disaster-resilient infrastructure. The European Union's ICZM Recommendation (2002/413/EC) and Marine Spatial Planning Directive (2014/89/EU) offer frameworks for harmonized coastal governance integrating marine and terrestrial planning. Additionally, global best practices include the Living Shorelines initiative in the U.S., which blends natural and engineered coastal defenses, and the Netherlands' Room for the River project, which realigns infrastructure with river and coastal dynamics to reduce flood risk (van Herk et al. 2014).

24.4 Technological and Nature-Based Solutions

24.4.1 Geostpatial Based Modeling and Nature-Based Solutions (Mangroves, Wetlands, Dunes)

Technological and nature-based solutions have become critical components in addressing the challenges posed by urbanization in dynamic coastal environments. Geographic Information Systems (GIS), remote sensing, and hydrodynamic modeling tools offer precise spatial and temporal analysis capabilities for monitoring coastline changes, assessing flood risks, and planning urban development. These tools are vital for early warning systems, vulnerability mapping, and data-driven decision-making (Mansour et al. 2020; Klemas 2011). For instance, GIS-based coastal vulnerability indexing has been effectively implemented in regions like the Bay of Bengal and the Mississippi Delta to prioritize risk-prone zones (Dasgupta et al. 2011).

Nature-based solutions, such as the restoration of mangroves, coastal wetlands, and dune systems, offer sustainable buffers against storm surges, erosion, and sea-level rise. Mangroves, for example, reduce wave energy and protect inland areas, while enhancing biodiversity and carbon sequestration (Temmerman et al. 2013; Alongi 2008). Wetlands play a critical role in water filtration and flood retention, while dunes offer natural barriers against high tides and wind erosion.

24.4.2 Coastal Engineering Approaches and Smart Urban Planning for Coastal Resilience

Coastal infrastructure and engineering approaches, including seawalls, breakwaters, and tidal gates, remain important but are increasingly seen as complementary to soft, nature-based strategies. Hybrid infrastructure systems that integrate grey and green components are gaining traction as cost-effective and environmentally sustainable alternatives (Narayan et al. 2016a, 2016b).

Smart urban planning for coastal resilience emphasizes compact urban form, strategic zoning, managed retreat, and green-blue infrastructure networks that integrate stormwater management and ecological corridors (Meerow et al. 2016). Urban resilience frameworks now increasingly rely on predictive analytics, participatory planning, and adaptive infrastructure design, supported by global initiatives like the UN-Habitat's City Resilience Profiling Tool and the Rockefeller Foundation's 100 Resilient Cities.

24.5 Future Outlook and Sustainable Strategies

24.5.1 Urban Resilience Adaptation Measures and Community Participation and Livelihood Protection

Modern responses increasingly rely on a synergy of technological innovation and nature-based strategies to enhance resilience. Geographic Information Systems (GIS), Remote Sensing, and coastal modeling tools (e.g., MIKE21, Delft3D) have revolutionized the way coastal cities monitor land-use changes, predict flood risk zones, and analyze shoreline dynamics. These tools support evidence-based decision-making in areas like vulnerability mapping, hydrological modeling, and real-time disaster forecasting (Dewan et al. 2012; Dasgupta et al. 2009). GIS-based tools also enable urban planners to simulate various sea-level rise scenarios, providing insight into future urban exposure and guiding relocation or adaptation strategies.

In parallel, nature-based solutions (NbS), such as the restoration and conservation of mangroves, wetlands, coral reefs, and coastal dunes, have gained global recognition for their dual role in buffering storm surges and supporting biodiversity. These ecological assets naturally dissipate wave energy, reduce erosion, and serve as carbon sinks while enhancing the social-ecological resilience of coastal communities (Narayan et al. 2016a, 2016b; Barbier et al. 2011). For instance, post-2004 Indian Ocean tsunami research highlighted how intact mangrove belts in Tamil Nadu mitigated destruction, showcasing their protective value.

24.5.2 Green Infrastructure with Blue Economy Approaches and Monitoring, Evaluation, and Adaptive Management

Coastal infrastructure such as seawalls, breakwaters, and surge barriers is engineered to reduce flooding and prevent shoreline retreat. While effective in short-term protection, these "gray" solutions often alter sediment transport and ecological processes, necessitating careful integration with ecosystem-based approaches (Temmerman et al. 2013).

Smart urban planning in coastal zones entails adaptive zoning regulations, green buffer zones, and resilient housing typologies that incorporate climate risks. Urban designs now increasingly embed resilience principles such as permeable pavements, elevated infrastructure, and early warning systems. For example, the "Room for the River" program in the Netherlands exemplifies how urban and hydrological planning can be harmonized for safety and sustainability (Rijke et al. 2012).

24.6 Summary

The research indicates a rich yet fragmented knowledge base, with content primarily spanning chapters (46.56%), edited books (34.26%), and journal articles (10.93%). The knowledge ecosystem shows a significant academic focus on theoretical frameworks, interdisciplinary syntheses, and emerging technological interventions such as GIS, remote sensing, and AI-driven modeling for fostering coastal urban resilience. Our co-authorship and co-citation analysis, conducted using VOSviewer, highlights diverse scholarly collaborations across 176 researchers within 15 clusters. We observed notable contributions from institutions in South Korea, Slovenia, the United Kingdom, and the United Arab Emirates. Specifically, Dorian Q. Fuller (UCL, UK) and I. P. Abrol have emerged as influential thought leaders, demonstrated by their high citation-to-publication ratio, signifying foundational contributions to this domain. However, the co-citation network appears weakly connected, with 500 researchers forming 499 discrete clusters. This suggests a fragmented intellectual structure and indicates a need for deeper academic integration and discourse convergence within the field.

This study corroborates earlier research by Ahern (2011), which emphasized the need for resilience-based planning in urban ecosystems prone to environmental shocks. It also aligns with Adger et al. (2005), who framed resilience as a function of both ecological thresholds and socio-institutional capacity. Furthermore, Pelling (2011) argued that resilience is not merely infrastructural but also social, contingent upon governance, risk perception, and community adaptation. In our bibliometric synthesis, we found a strong representation in categories such as chapters (46.56%) and edited books (34.26%), signaling the dominance of foundational or policy-oriented discourse in this domain.

The presence of globally distributed top authors such as Dorian Q. Fuller (UCL) and Krystyna Konstankiewicz (Polish Academy of Sciences), who have high mean citation rates, suggests that impactful work often emerges from interdisciplinary approaches combining environmental archaeology, coastal ecology, and systems thinking. This aligns with findings from Meerow et al. (2016), who advocated for a systems-based redefinition of urban resilience, especially in complex and fluid coastal geographies. By employing tools like VOSviewer, co-citation and co-authorship analysis, and bibliometric indicators such as FCR, RCR, and citation percentages, this study offers a structured understanding of research evolution, collaboration patterns, and influential scholarly contributions. Essentially, the findings not only validate prior conceptual models but also extend them by revealing how technological interventions (e.g., GIS, remote sensing, smart infrastructure) are increasingly featured in resilience discourse. This mirrors recent trends noted by Leitner et al. (2018) on geospatial intelligence in urban risk management. This growing integration of digital tools and collaborative authorship marks a promising shift toward anticipatory and data-informed urban resilience strategies in the world's most at-risk coastal zones.

To strengthen the urban resilience paradigm in dynamic coastal environments, future research must prioritize transdisciplinary integration, combining urban planning, environmental science, socio-technical systems, and digital technologies. There is also a pressing need to mainstream equity, community-based adaptation, and ecosystem-based solutions in both policy frameworks and practical interventions. Expanding the focus toward data-driven adaptive governance, machine learning–based early warning systems, and nature-based infrastructure solutions could bridge the existing theory-practice gap. Furthermore, fostering international research networks, improving open-access coastal datasets, and supporting South-South academic collaborations will enhance knowledge equity and contextual relevance. The temporal bibliometric trend (2000–2025) clearly suggests a steep rise in scholarly interest post-2015, aligned with the Sendai Framework, SDG 11, and the Paris Agreement, providing a policy-backed momentum for impactful research.

References

Abidin HZ et al (2011) Land subsidence in Jakarta (Indonesia): characteristics, impacts and causes. Environ Geol 28(7):899–910

ADB (Asian Development Bank) (2016) Revitalizing Jakarta's Waterways. https://www.adb.org/publications/revitalizing-jakarta-waterways

Adelekan IO (2010) Vulnerability of poor urban coastal communities to climate change in Lagos, Nigeria. Environ Urban 22(2):433–450

Aerts JCJH, Botzen WJW, Emanuel K, Lin N, de Moel H, Michel-Kerjan EO (2014) Evaluating flood resilience strategies for coastal megacities. Science 344(6183):473–475. https://doi.org/10.1126/science.1248222

Ajibade I, Barth JM, Okafor VO (2020) Climate change, urban flooding, and the rights of the urban poor in Africa. Environ Urban 32(1):175–194

Anguelovski I, Shi L, Chu E, Gallagher D, Goh K, Lamb Z, Teicher H (2016) Equity impacts of urban land use planning for climate adaptation: critical perspectives from the global south. J Plan Educ Res 36(3):333–348

Alongi DM (2008) Mangrove forests: resilience, protection from tsunamis, and responses to global climate change. Estuar Coast Shelf Sci 76(1):1–13. https://doi.org/10.1016/j.ecss.2007.08.024

Barbier EB, Hacker SD, Kennedy C, Koch EW, Stier AC, Silliman BR (2011) The value of estuarine and coastal ecosystem services. Ecol Monogr 81(2):169–193

BBC News (2019) Why is Indonesia moving its capital city?. https://www.bbc.com/news/world-asia-49457273

Brown S, Nicholls RJ, Woodroffe CD, Hanson S, Hinkel J, Kebede AS, Neumann B (2018) Land use and urban planning for coastal resilience. Environ Sci Policy 82:57–62

Cicin-Sain B, Knecht RW (1998) Integrated coastal and ocean management: concepts and practices. Island Press, Washington DC

Colten CE (2006) Vulnerability and place: flat land and uneven risk in New Orleans. Am Anthropol 108(4):731–734

Cormier R, Kannen A, Elliott M, Hall P, Davies I (2015) Marine spatial planning quality management system. Mar Policy 51:334–341

Dasgupta S, Laplante B, Meisner C, Wheeler D, Yan J (2009) The impact of sea level rise on developing countries: a comparative analysis. Clim Chang 93(3):379–388

Dasgupta S, Laplante B, Murray S, Wheeler D (2011) Exposure of developing countries to sea-level rise and storm surges. Clim Chang 106(4):567–579. https://doi.org/10.1007/s10584-010-9959-6

Dewan AM, Yamaguchi Y, Rahman MZ (2012) Dynamics of land use/cover changes and the analysis of landscape fragmentation in Dhaka metropolitan, Bangladesh. GeoJournal 77(3):315–330

Emanuel K (2005) Increasing destructiveness of tropical cyclones over the past 30 years. Nature 436(7051):686–688

Government of India (2011) Coastal Regulation Zone (CRZ) Notification. Ministry of Environment, Forest and Climate Change (MoEFCC)

IPCC (2019) Special report on the ocean and cryosphere in a changing climate. https://www.ipcc.ch/srocc/

IPCC (2021) Climate change 2021: the physical science basis. Intergovernmental Panel on Climate Change. https://www.ipcc.ch/report/ar6/wg1/

Jambeck JR, Geyer R, Wilcox C, Siegler TR, Perryman M, Andrady A, Law KL (2015) Plastic waste inputs from land into the ocean. Science 347(6223):768–771. https://doi.org/10.1126/science.1260352

Klemas V (2011) Remote sensing techniques for studying coastal ecosystems: an overview. J Coast Res 27(1):2–17. https://doi.org/10.2112/JCOASTRES-D-10-00103.1

Kuffer M, Pfeffer K, Sliuzas R (2018) Slums from space—fifteen years of slum mapping using remote sensing. Remote Sens 10(6):855

Leitmann J (2015) Building urban resilience in East Asia. World Bank

Mansour S, Al-Belushi M, Al-Awadhi T (2020) Monitoring urban expansion and land use/land cover changes using remote sensing and GIS techniques. Remote Sens Appl Soc Environ 17:100297. https://doi.org/10.1016/j.rsase.2020.100297

Meerow S, Newell JP, Stults M (2016) Defining urban resilience: a review. Landsc Urban Plan 147:38–49. https://doi.org/10.1016/j.landurbplan.2015.11.011

Menon M, Kohli K (2016) Chronicles of coastal land conversion in India: the shifting contours of the CRZ notification. CPR-Namati Environmental Justice Program

Narayan S, Beck MW, Wilson P, Thomas CJ, Guerrero A, Shepard CC, Trespalacios D (2016a) Coastal wetlands and flood damage reduction: using risk industry-based models to assess natural defenses in the northeastern USA. Lloyd's Tercentenary Research Foundation

Narayan S, Beck MW, Wilson P, Thomas CJ, Guerrero A, Shepard CC, Reguero BG (2016b) The value of coastal wetlands for flood damage reduction in the northeastern USA. Sci Rep 7(1):9463

Neumann B, Vafeidis AT, Zimmermann J, Nicholls RJ (2015) Future coastal population growth and exposure to sea-level rise and coastal flooding–a global assessment. PLoS One 10(3):e0118571. https://doi.org/10.1371/journal.pone.0118571

Nicholls RJ, Cazenave A (2010) Sea-level rise and its impact on coastal zones. Science 328(5985):1517–1520

Nicholls RJ, Wong PP, Burkett VR, Codignotto JO, Hay JE, McLean RF, Woodroffe CD (2008) Coastal systems and low-lying areas. In: Parry ML et al (eds) Climate change 2007: impacts, adaptation and vulnerability. IPCC WG II. Cambridge University Press

Rana MMP, Islam MS (2022) Vulnerability of urban poor to climate-induced flooding in Dhaka and Chittagong, Bangladesh. Clim Dev 14(4):314–327

Rijke J, van Herk S, Zevenbergen C, Ashley R (2012) Room for the river: delivering integrated river basin management in The Netherlands. Int J River Basin Manag 10(4):369–382

Seto KC, Güneralp B, Hutyra LR (2011) Global forecasts of urban expansion to 2030 and direct impacts on biodiversity and carbon pools. PNAS 109(40):16083–16088

Seto KC, Fragkias M, Güneralp B, Reilly MK (2012) A meta-analysis of global urban land expansion. PLoS One 6(8):e23777. https://doi.org/10.1371/journal.pone.0023777

Siders AR (2019) Managed retreat in the United States. One Earth 1(2):216–225. https://doi.org/10.1016/j.oneear.2019.09.008

Smajgl A et al (2015) Responding to rising sea levels in the Mekong Delta. Nat Clim Chang 5(2):167–174

Small C, Nicholls RJ (2003) A global analysis of human settlement in coastal zones. J Coast Res 19(3):584–599

Temmerman S, Meire P, Bouma TJ, Herman PMJ, Ysebaert T, De Vriend HJ (2013) Ecosystem-based coastal defence in the face of global change. Nature 504:79–83. https://doi.org/10.1038/nature12859

Tiwari R et al (2020) Assessment of salinity intrusion in groundwater in coastal Tamil Nadu, India. Groundw Sustain Dev 11:100427

Tyler S, Moench M (2012) A framework for urban climate resilience. Clim Dev 4(4):311–326. https://doi.org/10.1080/17565529.2012.745389

UNDRR (2020) Report on cyclone Amphan: recovery and reconstruction. United Nations Office for Disaster Risk Reduction

UN-Habitat (2020) World cities report 2020: the value of sustainable urbanization. https://unhabitat.org

Valiela I, Bowen JL, York JK (2001) Mangrove forests: one of the world's threatened major tropical environments. Bioscience 51(10):807–815. https://doi.org/10.1641/0006-3568(2001)051[0807:MFOOTW]2.0.CO;2

van Herk S, Rijke J, Zevenbergen C, Ashley R (2014) Collaborative research to support transition to flood resilience in urban areas. J Flood Risk Manag 7(4):339–350

Vousdoukas MI, Mentaschi L, Hinkel J, Ward PJ, Feyen L (2020) Climatic and socioeconomic controls of future coastal flood risk in Europe. Nat Clim Chang 10:485–490. https://doi.org/10.1038/s41558-020-0769-1

World Bank (2019) India–integrated coastal zone management project: implementation completion and results report

Chapter 25
Evaluating the Effectiveness of Ecosystem-Based Risk Reduction: Insights from Pichavaram's Policy Framework

S. Dhivya, K. Kirubagaran, N. Karthi, and Devaraj Asir Ramesh

25.1 Introduction

Coastal zones worldwide are increasingly susceptible to Hazards include tsunamis, storm surges, cyclones, and rising sea levels. In response, nature-based solutions like mangrove ecosystems are acknowledged as effective, sustainable tools for reducing disaster risk and enhancing resilience (Alongi 2020). In India, Pichavaram Mangrove Forest, located between Vellar and Coleroon Estuaries in Tamil Nadu stands out as a critical site for studying NbS applications (Pichavaram 2025). This mangrove system, covering approximately 1478 ha (Fig. 25.1) with 51 interconnected islets, supports rich biodiversity, including *Avicennia* and *Rhizophora* species, and provides critical habitat for migratory birds (Mangroves in India 2025). Empirical evidence indicates that Pichavaram's mangrove patches significantly lower coastal exposure a quantitative index of hazard risk—reducing values from 3.47 (very high) to 2.80 on a 5-point scale (Singha et al. 2024). Furthermore, economic valuation studies using the Contingent Valuation Method (CVM) reveal that local communities particularly in MGR Thittu village recognize these benefits, with 73% of surveyed households willing to pay annually for conservation, reporting mean contributions of ₹ 686 for women

S. Dhivya (✉) · K. Kirubagaran
Department of Humanities and Social Sciences, Saveetha School of Law, SIMATS, Chennai, India
e-mail: dhivya.meet@gmail.com

K. Kirubagaran
e-mail: kirubagarank.ssl@saveetha.com

N. Karthi
Chennai Metropolitan Development Authority, Chennai, India

D. A. Ramesh
National Center for Sustainable Coastal Management, Ministry of Environment, Forest and Climate Change, Chennai, India

A. Nath et al. (eds.), *Geospatial Application of Coastal Hazards*, Springer Water,
https://doi.org/10.1007/978-3-032-20076-1_25

and ₹ 916 for men (Debroy et al. 2014). These findings underscore the ecological, economic, and socio-cultural importance of Pichavaram's mangrove ecosystem. The System of Environmental Economic Accounting along with Ecosystem Accounting (SEEA-EA) framework, provides a robust methodology to integrate biophysical data with economic valuation, enabling informed policy and investment decisions (United Nations 2021). While SEEA-EA applications to Pichavaram remain limited, case studies in other Indian mangrove regions demonstrate its potential for quantifying ecosystem benefits and supporting sustainable management (NSO India 2020). India's National Environment Policy (NEP, 2006) reinforces the need for integrated valuation. While Principle V stresses economic efficiency in conservation and necessitates the pricing of ecosystem products and services, Objective IV urges the integration of environmental issues into economic and social growth (Government of India 2006). Strategy 5.1.3 further advocates reversing the perception of environmental resources as "free goods" and internalizing externalities. These goals align with UN Sustainable Development Goal (SDG) 15.9, which promotes integrating ecosystem and biodiversity values into planning and accounting, and SDG 13 (Climate Action), which recognizes the role of carbon-rich ecosystems like mangroves in climate mitigation (United Nations 2015). Statutory frameworks such as The Environment Protection Act's Coastal Regulation Zone (CRZ) notification was initially published in 1991. (1986), categorize Pichavaram as a CRZ-I 'Ecologically Sensitive Area,' highlighting its importance for biodiversity conservation and the need for stringent protection (Government of India 1991). Despite these legal protections, many of Pichavaram's ecosystem services remain undervalued or unquantified. Applying the SEEA-EA framework—linking ecosystem condition and service flows to economic measures—can bridge this gap, enhancing evidence-based policy and fostering community co-management approaches (Debroy et al. 2014; United Nations 2021).

25.1.1 Ecological Services of Pichavaram–Coastal Protection

Pichavaram's 1478 hectares (Ramsar Convention Secretariat, 2024) along with 982 hectares of bird nesting grounds, 52 hectares of mudflats, and 12 hectares of salt marshes, collectively act as a natural barrier. These ecosystems reduce the exposure to extreme hazards e.g., lowering coastal vulnerability indices from 3.47 to 2.80 during tsunami events (Fig. 25.3). Their presence significantly reduces mortality and asset loss, as witnessed during the 2004 tsunami in Parangipettai Block, where hamlets shielded by mangroves saw fewer casualties and damage (Tables 25.1, 25.2, 25.4).

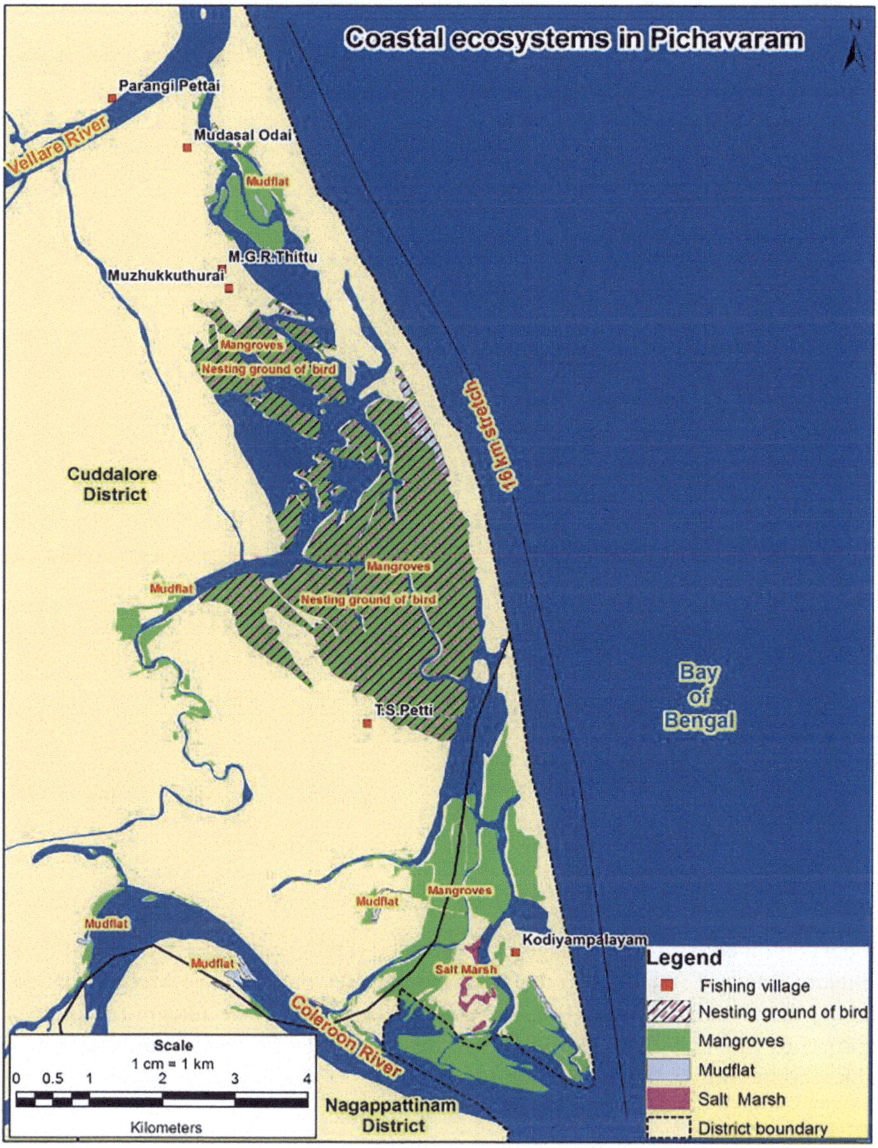

Fig. 25.1 Pichavaram coastal ecosystem Tamilnadu

25.2 Data and Methodology

This study employs four complementary valuation methods to quantify the Total Economic Value (TEV) of the coastal protection service provided by the Pichavaram mangrove ecosystem: (1) the InVEST Coastal Vulnerability and Coastal Protection

Table 25.1 Villages near mangroves ecosystem in Pichavaram mangroves

S.No.	Name of the villages	Households	Population	Sample size
1.	Mudasal odai*			
2.	MGR Thittu*			
3.	Muzhukkuthurai*			
4.	Sengalankuppum*			
5.	Chinnavaikal*			
6.	Kannagi Nagar*			
7.	Pillumedu*			
8.	MGR Nagar*			
9.	Meenavar colony*			
10.	Thirunalthoppu*			
11.	Thaikkal*			
12.	Kuchchipalayam*			
13.	Ponnanthittu*			
14.	Pattancherri*			
15.	Killai	3359	13,608	250
16.	Thandavarayan cholangan pettai (T.S. Pettai)	306	1243	41
17.	Radhavilagam	388	1387	44
18.	Pichavaram	873	3463	42
19.	Illanthaimedu	–	–	–
20.	Kodiyampalayam	–	–	–
	Total	**4926**	**19,701**	**377**

Source 2011 Census of India, Government of India, Ministry of Home Affairs
*Those hamlets are come under Killai

model combined with the Damage Cost Avoided (DCA) method, (2) the Contingent Valuation Method (CVM), (3) the Replacement/Restoration Cost Method (RCM), and (4) the Benefit Transfer Method (BTM) (Fig. 25.2). The integration of these approaches provides a comprehensive assessment of both market and non-market values, addressing both use and non-use benefits.

25.2.1 InVEST Model and Damage Cost Avoided Method

The InVEST Coastal Vulnerability Model was applied to estimate the reduction in exposure of people, property, and infrastructure due to the presence of mangroves. Input datasets included: The model generates a relative exposure index

Table 25.2 Fisher folk villages near mangrove ecosystem in Pichavaram

S. No	Name of the village	Fishermen families	Fisher folk population	Sample size
1.	Chinna Vaikal	83	264	12
2.	Kalaignar Nagar Killai	180	655	31
3.	Killai	346	1156	54
4.	Killai M.G.R. Nagar	120	453	23
5.	M.G.R. Thittu	211	771	36
6.	Mudasal Odai	253	829	39
7.	Muzhukkuthurai	162	556	26
8.	Nadumudasal Odai	30	138	10
9.	Kodiyampalayam	378	1278	60
10.	Pillumedu	85	359	17
11.	T.S.Pettai	216	761	36
12.	Vadakku Mudasal Odai	137	463	22
	Total fishermen families	**2201**	**7683**	**366**

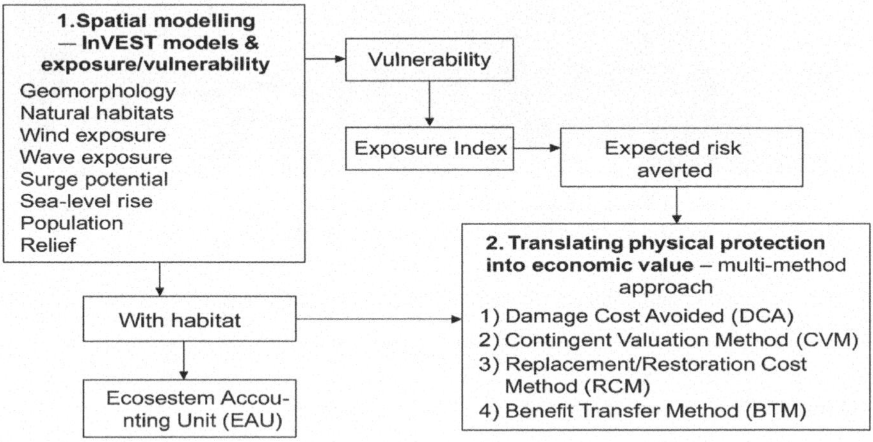

Fig. 25.2 Methodology of Pichavaram mangrove coastal protection services

for each coastal segment under two scenarios: (a) **with mangroves** and (b) **without mangroves**. The DCA method was then used to estimate avoided damages by multiplying the difference in exposure values by the replacement value of assets at risk and the probability of hazard occurrence (Barbier 2016). Annual values of avoided damage were computed in monetary terms, reflecting lives, livelihoods, and infrastructure preserved (Fig. 25.3).

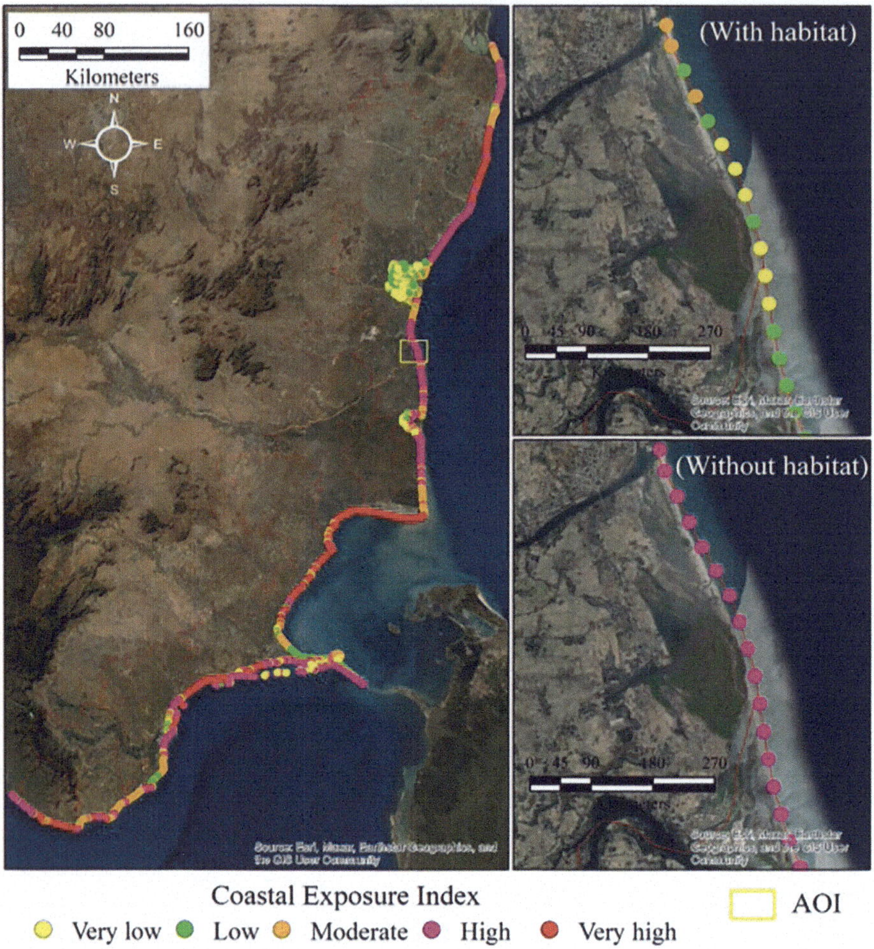

Fig. 25.3 Pichavaram and Muthupet mangrove coastal exposure index. *Source* Adapted from Singha et al. 2024

Damage Cost Avoided (DCA)

$$DCA = Damage_{Without} - Damage_{With}$$

where:

DCA = Damage Cost Avoided (₹).

$Damage_{Without}$ = Estimated damages from the **without mangroves** InVEST run.

$Damage_{With}$ = Estimated damages from the **with mangroves** InVEST run.

Total Avoided Cost (TAC)

$$TAC = L_h + D_b + L_C + H_S + R_P + G_i$$

where:

L_h = Loss of human life avoided (₹).
D_b = Damage to boats avoided (₹).
L_C = Loss of cattle avoided (₹).
H_S = Household structural damage avoided (₹).
R_P = Relief household packages avoided (₹).
G_i = Government infrastructure projects avoided (₹).
Per Hectare Avoided Cost (PHaAC).

$$\text{PHaAC} = \frac{TAC}{A}$$

where A = Mangrove area (ha)

25.2.2 *Contingent Valuation Method*

The Willingness to Pay (WTP) for mangrove-based coastal protection was estimated using the Contingent Valuation Method (CVM). Survey responses were screened to exclude protest bids and zero WTP responses, as these do not reflect the respondents' true economic valuation.

The mean WTP was calculated as:

$$WTP = \frac{\sum_{i=1}^{n} WTP_i}{n}$$

where:

- WTP_i = annual willingness to pay stated by household i
- n = total number of valid responses (after exclusions)

The aggregate annual WTP for the entire study area was then computed as:

$$Total\ WTP = WTP \times H$$

where:

H = total number of households in the study area (4926).
To standardize the value per unit area, the per-hectare WTP was estimated as:

$$WTP\ per\ ha/year = \frac{Total\ WTP}{A}$$

where:

A = total mangrove area (1478 ha).

25.2.3 Annualized Restoration Cost Method

Over a predetermined programme period, the yearly cost of maintaining and restoring mangroves is estimated using the Replacement Cost Method. The annualized cost is calculated using the Capital Recovery Factor (CRF), which converts a total multi-year cost into an equivalent uniform annual cost.

$$\text{Annual Restoration Cost (Total)} = C_{9y} \times \frac{r(1+r)^n}{(1+r)^n - 1} \times A$$

where:

C_{9y} = Total restoration and maintenance cost per hectare over 9 years (₹/ha).

r = Real social discount rate (in decimal form).

n = Project duration in years.

A = Total mangrove area (hectares).

The term $\frac{r(1+r)^n}{(1+r)^n-1}$ is the Capital Recovery Factor (CRF).

25.2.4 Benefit Transfer Method

The BTM was employed to estimate coastal protection values by transferring economic estimates from peer-reviewed studies of similar mangrove ecosystems, adjusted for local socio-economic and ecological conditions.

Standardizing Ecosystem Service Values to 2024 Constant USD.

$$PHV_{2024} = PHV_{Si} \times \frac{D_P}{D_s} \times E \times \prod_{t=i+1}^{2024} 1 + D_t$$

where:

PHV_{2024} = Per-hectare per-year value of the ES at the policy site in 2019 constant USD.

PHV_{Si} = Per-hectare per-year value of the ES at the study site in its base year i.

D_p = GDP deflator for the policy site year (local currency).

D_s = GDP deflator for the study site year (local currency).

E = Exchange rate (USD per local currency unit).

D_t = Annual GDP deflator growth rate for each year from i + 1 to 2024.

This formula is to standardize 1. Price-level differences between the study site and the policy site. 2. Currency conversion to USD. 3. Inflation adjustment to express all values in 2019 constant USD.

25.3 Results

To quantify Pichavaram's coastal protection value, we apply four valuation methods: (1) a geospatial risk model (InVEST) coupled with **damage-cost avoided**, (2) **contingent valuation (CV)** (survey-based willingness-to-pay), (3) **replacement/restoration cost**, and (4) **benefit transfer**. Each method uses different data and yields complementary estimates of economic value. The sections below detail the methodology of each approach and compare their outcomes (Table 25.3).

25.3.1 *InVEST Model and Damage Cost Avoided Method*

The **InVEST Coastal Vulnerability** model (Natural Capital Project) combines coastal geophysical data and habitat maps to assess flood/erosion exposure (Arkema et al. 2013). It produces an index (1–5) of flood risk for each shoreline segment under storm scenarios. To value the mangroves' benefit, we run InVEST in two scenarios: *with mangroves* (current habitat) and *without mangroves*. By overlaying these outputs with population and asset data (extent of households, infrastructure, crop area), we estimate **expected damages** in each scenario.

Table 25.3 Various methods of coastal protection service estimation

Method	Ecosystem benefit captured	Data and inputs	Typical output and notes
InVEST + damage avoided	Lives and assets protected	GIS layers (bathymetry, coastline, mangroves, land use), storm surge data, (Singha et al. 2024) population/ infrastructure values (Department of Environment and Forests, 2005)	Avoided damage in Rs per year of people or houses protected with/ without mangroves to compute the difference
Contingent valuation (CVM)	Total WTP for mangrove protection (non-market values)	Survey of residents (20 villages), sample size (377)	Annual WTP (Rs/ yr) aggregated over the local population. Captures intangible values
Replacement/ restoration	Cost to restoration	Restoration project costs, lifespans (MoEFCC 2023–2024)	Annualised cost (Rs/yr) or present value (Rs) of substitute solution
Benefit transfer	Value from analogous studies (unit values)	Published value estimates (Rs/ha/ yr)	Total estimated value (Rs). Quick estimate but context-sensitive

Table 25.4 Physical flow of coastal protection service of Pichavaram mangroves ecosystem

S.No.	Name of the villages	Households	Population	No of deaths/ missing	Boats	Cattle death	Projects
1	Mudasal odai*						Bridge construction at 11/4 km of Killai road CWSS Renovation of health centers repairs to road
2	MGR Thittu*						
3	Muzhukkuthurai*						
4	Sengalankuppum*						
5	Chinnavaikal*						
6	Kannagi Nagar*						
7	Pillumedu*						
8	MGR Nagar*						
9	Meenavar colony*						
10	Thirunalthoppu*						
11	Thaikkal*						
12	Kuchchipalayam*						
13	Ponnanthittu*						
14	Pattancherri*						
15	Killai	3359	13,608	130	1027	87	
16	Thandavarayan cholangan pettai (T.S. Pettai)	306	1243	4	24	54	
17	Radhavilagam	388	1387	0	0	0	
18	Pichavaram	873	3463	0	42	0	
19	Illanthaimedu						
20	Kodiyampalayam						
Total		4926	19,701	134	1093	141	

The **total avoided cost** due to the protective role of Pichavaram mangroves is approximately **₹ 5059.27 lakhs** (or ₹ 50.59 crores) annually (Table 25.5). This represents the estimated value of damages prevented by the mangroves during coastal hazard events such as storms and floods.

Breaking down the total avoided cost into categories:

- **Household structural damage avoided (₹ 3201.90 lakhs)** is the largest component, indicating that mangroves substantially reduce the need for expensive repairs or reconstruction of homes in the coastal communities.
- **Government infrastructure projects avoided (₹ 981.40 lakhs)** highlights savings from prevented damage to public assets like roads, bridges, and health centers.
- **Loss of human life avoided (₹ 536.00 lakhs)** quantifies the value assigned to lives protected, using government compensation or insurance rates.

Table 25.5 Monetary benefit of coastal protection service of Pichavaram mangroves (Rs in lakhs/yr)

S.No.	Name of the villages	No of deaths/ missing	Boats	Cattle death	Household damage cost	Releif houseld package	Projects
0.00	Mudasal odai*						
0.00	MGR Thittu*						
0.00	Muzhukkuthurai*						
0.00	Sengalankuppum*						
0.00	Chinnavaikal*						
0.00	Kannagi Nagar*						
0.00	Pillumedu*						
0.00	MGR Nagar*						
0.00	Meenavar colony*						
0.00	Thirunalthoppu*						
0.00	Thaikkal*						
0.00	Kuchchipalayam*						900.00
0.00	Ponnanthittu*						62.00
0.00	Pattancherri*						15.00
0.00	Killai	520.00	154.05	32.62	2183.35	83.97	4.40
0.00	Thandavarayan cholangan pettai (T.S. Pettai)	16.00	3.60	20.25	198.90	7.65	
0.00	Radhavilagam	0.00	0.00	0.00	252.20	9.70	
0.00	Pichavaram	0.00	6.30	0.00	567.45	21.82	
0.00	Illanthaimedu						
0.00	Kodiyampalayam						
Total		536.00	163.95	52.88	3201.90	123.15	981.40
Total		5059.27					

- **Damage to boats avoided (₹ 163.95 lakhs)** and **Relief household packages avoided (₹ 123.15 lakhs)** reflect savings to both private livelihoods and government disaster relief efforts.
- **Loss of cattle avoided (₹ 52.88 lakhs)** shows protection of livestock assets (Table 25.6).

The **Per Hectare Avoided Cost** (approximately ₹ 3.42 lakhs per hectare per year) provides an economic valuation of each hectare of mangrove forest in terms of avoided damages, useful for cost-benefit analyses and restoration prioritization.

Table 25.6 Damage cost avoided (DCA) valuation for Pichavaram mangroves

Metric/category	(in ₹ crore)	Notes
L_h: Loss of human life	5.36	Valued using compensation rates per life lost/missing
D_b: Damage to boats avoided	1.64	Replacement value of fishing boats
L_C: Loss of cattle avoided	0.53	Market value of livestock
H_S: Household damage avoided	32.02	Repair/reconstruction cost
R_P: Relief fund household avoided	1.23	Government relief fund savings
G_i: Govt. infrastructure damage avoided	9.81	Roads, bridges, health centres
TAC (Total avoided cost)	**50.59**	Annualised from event-based data
Damage cost avoided (DCA)	Formula	Value (₹ lakhs)
Total avoided cost (TAC)	$L_h + D_b + L_C + H_S + R_P + G_i$	5059.27
Per hectare avoided cost (PHaAC)	$\frac{TAC}{A}$	3.423

Table 25.7 Descriptive statistics and analytical summary of WTP survey data

Parameter	Value
Total respondents	377
Mean WTP (₹/year)	457.85
Median WTP (₹/year)	490.00
Minimum WTP (₹/year)	0.00
Maximum WTP (₹/year)	1000.00
Standard deviation (₹)	232.47
Number of zero WTP responses	38 (10.08%)
Number of protest responses	18 (4.77%)
Interquartile range (₹)	300.00

Table 25.8 WTP estimates for coastal protection by Pichavaram mangroves

Parameter	Value
Mean WTP per household/year (survey)	₹507.63
Total households in study area	4926
Total annual WTP for the area	₹2,500,589.17
Mangrove area	1478 ha
WTP per hectare/year	₹1691.87

Table 25.9 Affected villages and non-affected villages by tsunami

S.No.	Name of the villages	Households	Population
Affected villages by tsunami			
1.	Chinnavaikal	–	–
2.	Kannagi Nagar	–	–
3.	Pillumedu	–	–
4.	MGR Thittu	–	–
5.	Mudasal odai	–	–
Non-affected villages by tsunami			
5.	**Killai**	–	–
6.	Muzhukkuthurai	–	–
7.	Sengalankuppum	–	–
8.	Kuchchipalayam	–	–
9.	Pattancherri	–	–
10.	**Thandavarayan cholangan pettai (T.S. Pettai)**	**306**	**1243**
11.	**Radhavilagam**	**388**	**1387**
12.	**Pichavaram**	**873**	**3463**
13.	Vadukku Pichavaram	–	–

Source 2011 Census of India, Government of India, Ministry of Home Affairs, Field visit identification

25.3.2 Contingent Valuation Method (CVM)

The **Contingent Valuation Method (CVM)** was applied to estimate the economic value of coastal protection services provided by the Pichavaram mangroves. This stated preference approach uses direct survey responses to hypothetical scenarios, allowing valuation of non-market environmental goods.

A total of **377 households** from 15 coastal villages were surveyed (Table 25.7). Respondents were presented with a hypothetical scenario describing the role of mangroves in mitigating coastal hazards and were asked their maximum annual willingness to pay (WTP) for conservation and management of these mangroves.Responses were screened to exclude:

- **Protest bids** (4.77%), where respondents objected to the payment vehicle rather than valuing the service at zero.
- **Zero WTP** responses (10.08%), where respondents expressed no willingness to pay, either due to non-use or inability to contribute.

The contingent valuation survey conducted among 377 households across the Pichavaram coastal villages yielded an average willingness to pay (WTP) of ₹507.63 per household per year for mangrove-based coastal protection, after excluding protest and zero responses. Using the total household count in the study area

Table 25.10 Valuation of coastal protection and risk reduction services of mangroves

Valuation method	Physical flow basis	Original value	Study	Country
Avoided mortality method	0.0148 lives saved/ha; 11.7 M Rs/life saved	173,333 Rs	Verma/Das et al. (2009)	India
Replacement cost	Erosion/storm-control structures per ha	300 USD	Gunawardena et al. (2005)	Sri Lanka
Replacement cost	75 m mangroves, 20 yrs. annualized benefit	1800 USD	Sathirathai (1998)	Thailand
Replacement cost	75 m mangroves, 20 yrs. @10% discount	3000 USD	Sathirathai/ Barbier (2001)	Thailand
Damage cost	Change in 1 km^2 mangrove area/year	3679 USD	Barbier (2007)	Thailand
Damage cost	NPV/ha @10–15% discount	11,000 USD	Barbier (2009)	Thailand
InVEST model (damage cost)	Vulnerability index with/ without mangroves	742–756 USD	UNEP (2015)	Vietnam
Benefit transfer	From Das (2007) Orissa cyclone	653.98 USD	Kavi Kumar et al. (2016)	India
Damage cost	990 casualties over 20 yrs.; sea dike cost	391,816 USD	Bann (1997)	Vietnam
Damage cost	1999 cyclone: Protected vs unprotected villages	116.28 USD	Ruchi Badola (2003)	India
Damage cost	Wind, surge, population, socio-econ. Factors	51,168 Rs	Saudamini Das (2009)	India (Orissa)
Replacement cost	Cost of tsunami wall construction	2,181,818 Rs	Piyashi DebRoy (2012)	India
Damage cost	Kenya coastline severe weather risk 5%	91.7 USD	Janis Hoberg (2011)	Kenya
Damage cost	Crop loss, injuries, income loss	3300–9500 USD	Ranasinghe et al./IUCN (2006)	Sri Lanka
Replacement cost	Erosion prevention/soil accretion agricultural output from local production	240 USD	Ruitenbeek (1992)	Indonesia
Benefit transfer	From Tridoyo (1998)	672 USD	Giselle/Alan (2007)	Philippines
Replacement cost	Breakwater construction cost	476 USD	Abdul Malik et al. (2015)	Indonesia

(4926), the aggregate annual WTP is estimated at ₹2.50 million. Given the total mangrove extent of 1478 ha, this corresponds to an annual value of ₹1691.87 per hectare (Table 25.8). These results highlight the substantial economic importance of Pichavaram mangroves in providing coastal protection services, supporting the case for conservation and sustainable management efforts.

Table 25.11 Meta-analysis study value and descriptive study

Study	*2024 standardized value ha/yr*
Janis Hoberg	4260
Batagoda	4414
UoN and UNEP (1)	6597
IUCN	19,885
UoN and UNEP (2)	20,582
UNEP	32,661
Gunawardena et.al	34,473
Saudamini Das	60,232
Barbier et al.	92,641
Costanza	110,761
Sathirathai	182,588
Salem and Mercer	183,703
Verma and Das et.al	204,656
Sathirathai and Barbier	210,835
Ranasinghe et.al and IUCN	324,012
Metric	*Value*
Count	15
Total (meta-analysis sum)	1,492,300
Mean	99486.67
Median	60,232
Geometric mean	48004.61
Trimmed mean (10%)	89540.62
Q1	20233.5
Q3	183145.5
IQR	162,912
Outlier threshold (>Q3 + 1.5*IQR)	427513.5

From the survey of 377 households, the mean WTP was found to be ₹507.63/year. Using H = 4926 and A = 1478 ha, the results are:

= ×

Total WTP = 507.63 × 4926 = ₹2,500,589.17 per year.

$$WTP\ per\ ha/year = \frac{Total\ WTP}{A}$$

WTP per ha/year = 2,500,589.17/1478 ha = ₹1691.87

The CVM results suggest that the households in the Pichavaram region place a significant monetary value on mangrove conservation for coastal protection. The derived aggregate annual WTP underscores the economic importance of these ecosystems.

25.3.3 Restoration Cost

Using the MISHTI programme restoration cost of ₹162,037 per hectare over nine years, and annualizing this amount with a 3% real discount rate (capital recovery factor), we estimate an annual restoration cost of ₹20,805 per ha per year. For the Pichavaram mangrove extent (1478 ha) this implies an annualized restoration value of ≈ ₹30.74 million (₹3.07 crore) per year, which we use as a lower-bound proxy for the annual value of the coastal protection service provided by these mangroves. We report sensitivity to discounting and note that this figure should be scaled by an attribution/survival factor if only a portion of the restoration cost is assigned to coastal protection.

Given Values

C_9y = ₹162,037 per hectare, $r = 3\% = 0.03$, $n = 9$ years, $A = 1478$ hectares.

Capital recovery factor $\frac{r(1+r)^n}{(1+r)^n-1}(1 + r)^n = (1 + 0.03)^9 = 1.304773.$

$r \times (1 + r)^n = 0.03 \times 1.304773 = 0.039143.$

$(1 + r)^n - 1 = 1.304773–1 = 0.304773.$

$CRF = 0.039143/0.304773 = 0.12845.$

Annual Cost per ha = ₹ 162,037 × 0.12845 = ₹20,805 per ha per year.

Annual Total Cost = ₹20,805 × 1478 ha = ₹30,743,000 per year (₹3.07 crore/ year).

These values represent the annualized equivalent of the 9-year restoration and maintenance cost based on the MISHTI programme estimate and using a 3% real discount rate (Table 25.9).

25.3.4 Benefit (Value) Transfer Method

A synthesis of global and regional valuation studies (Table 25.10) highlights the diverse methodological approaches and economic significance of mangroves in providing coastal protection and risk reduction services. The avoided mortality method, as demonstrated by Verma and Das et al. in India, quantified the value of lives saved during extreme weather events, estimating an economic benefit of approximately ₹173,333 ha^{-1} based on 0.0148 lives saved per hectare and an average value of ₹11.7 million per life saved. The replacement cost method—which assesses the cost of man-made infrastructure required to replicate the same protective benefits was extensively applied in Sri Lanka, Thailand, and India, with estimated values ranging from 300 USD ha^{-1} to ₹2.18 million ha^{-1}. These estimates reflect the high capital investment needed for artificial coastal defences such as sea walls and breakwaters, thereby emphasizing the cost-effectiveness of natural mangrove buffers.

Similarly, the *damage cost avoided method*—which quantifies economic losses prevented by the presence of mangroves yielded considerable variation across regions. Barbier reported net present values between 3679 USD ha^{-1} and 11,000 USD ha^{-1} for Thailand, while Bann estimated an exceptionally high benefit of 391,816 USD ha^{-1} in Vietnam, reflecting avoided dike construction costs and human casualties. In India, Ruchi Badola and Saudamini Das (2009) estimated 116.28 USD ha^{-1} and ₹51,168 ha^{-1}, respectively, linking mangrove extent with reduced cyclone damage and mortality risk. Comparable studies in Sri Lanka and Kenya provided valuations ranging between 91.7 USD ha^{-1} and 9500 USD ha^{-1}. Furthermore, model-based evaluations such as the UNEP (2015) InVEST framework quantified vulnerability differences with and without mangroves, producing values between 742 and 756 USD ha^{-1} in Vietnam.

The benefit transfer method has also been adopted for large-scale ecosystem accounting and policy applications, as demonstrated by Kavi Kumar et al. and Giselle and Alan, who derived per-hectare protective values of 653.98 USD and 672 USD, respectively, by adapting estimates from earlier cyclone-based assessments. Overall, the reviewed literature indicates that valuation outcomes vary widely from 91.7 USD ha^{-1} to over ₹2.18 million ha^{-1} depending on the method employed, site characteristics, and hazard exposure. Notably, replacement cost and damage cost avoided approaches consistently produce higher valuations, underscoring the immense economic worth of mangroves as natural infrastructure for coastal protection. These findings provide a robust empirical basis for benefit transfer applications in the Pichavaram mangrove context, supporting the integration of ecosystem-based risk reduction values into coastal management and climate adaptation planning (Table 25.11).

BTM Value = Median Value per ha × Total Mangrove Area (ha)where: **Median Value per ha** = ₹60,232 (from your meta-analysis) and **Total Mangrove Area** = 1478 ha

BTM Value = ₹ 60,232 × 1478 ha = ₹88,993,096. That's ₹ 8.899 crore as the estimated total value.

25.4 Conclusion and Policy Relevance

This study applied the InVEST Coastal Vulnerability (CV) model to quantify the *physical flow* of coastal protection services provided by the Pichavaram mangroves and converted these flows into monetary terms using the Damage Cost Avoided (DCA) method. The results indicate a per hectare avoided cost of ₹ 3.423 lakh and a total value of ₹ 505.93 crore for the area, representing tangible, market-linked benefits in avoided damages to life, property, and infrastructure. As this valuation is grounded in spatially explicit physical modelling, it offers a *holistic, empirical measure* of regulating ecosystem services. Complementary non-market valuation methods — the Contingent Valuation Method (CVM) (₹ 0.0169 lakh/ha; ₹ 0.25 crore total) and the Benefit Transfer Method (BTM) (₹ 0.602 lakh/ha; ₹ 8.90 crore total) — provide insights into society's willingness to pay and comparable service values from other contexts. While these capture important cultural and indirect benefits, they do not link services to physical processes as directly as the DCA–InVEST approach.

From an environmental economics perspective, integrating these findings under the SEEA-EA (System of Environmental-Economic Accounting—Ecosystem Accounting) framework allows for a *Total Economic Value (TEV)* assessment that is both comprehensive and directly comparable with national accounts. The DCA–InVEST results from the market-linked core of the TEV, while CVM and BTM represent non-market extensions. From an environmental law perspective, such integrated accounting operationalises India's constitutional mandate under Article 48A, supports international commitments under the Paris Agreement and Convention on Biological Diversity, and strengthens application of the *polluter pays* and *public trust* doctrines in justifying public investment in mangrove conservation.

In the context of the Sustainable Development Goals (SDGs), this valuation directly supports:

- SDG 13 (*Climate Action*) — through quantification of climate adaptation benefits;
- SDG 14 (*Life Below Water*) and SDG 15 (*Life on Land*) — via conservation of coastal and terrestrial biodiversity;
- SDG 1 (*No Poverty*) and SDG 11 (*Sustainable Cities and Communities*) — by safeguarding livelihoods and reducing disaster risk; and
- SDG 17 (*Partnerships for the Goals*) — by providing a standardized accounting framework for cross-sectoral policy integration.

The evidence is conclusive: the annual economic benefits of conserving Pichavaram's mangroves far outweigh the costs of management. A SEEA-integrated, physically modelled DCA approach provides a legally defensible, transparent, and repeatable framework for embedding ecosystem service values into Tamil Nadu's Vision 2047 strategy, State Land Use Plans, and climate resilience financing.

Acknowledgments The authors would like to cordially thank the anonymous reviewers, the editor for their insightful comments and suggestions. The authors wish to thank the colleagues at various departments of National center for sustainable coastal management (NCSCM) especially, Dr. Ramesh Ramachandran, Dr. Purvaja Ramachandran and Dr. Rajakumari and Dr. Muruganandam. R. The views presented in this article are entirely of the authors, not of the organizations they are and were affiliated to. Any remaining errors in the article are entirely of the authors.

References

Alongi DM (2020) Mangroves: nature's defence against tsunamis, storms, and erosion. Estuar Coast Shelf Sci 248:106796. https://doi.org/10.1016/j.ecss.2020.106796
Arkema KK, Guannel G, Verutes GM, Wood SA, Guerry A, Ruckelshaus M, Kareiva P, Lacayo M, Silver JM (2013) Coastal habitats shield people and property from sea-level rise and storms. Nat Clim Chang 3(10):913–918. https://doi.org/10.1038/nclimate1944
Barbier EB (2016) The protective service of mangrove ecosystems: a review of valuation methods. Mar Pollut Bull 109(2):676–681. https://doi.org/10.1016/j.marpolbul.2016.01.033
Das S, Vincent JR (2009) Mangroves protected villages and reduced death toll during Indian super cyclone. Proc Natl Acad Sci 106(18):7357–7360. https://doi.org/10.1073/pnas.0810440106
Debroy P, Jayaraman R, Krishnan M, Criddle KR (2014) Importance of mangrove conservation and valuation to women and men—a case study of Pichavaram mangroves in India. Asian Fish Sci Spec Issue 27S:79–90
Department of Environment and Forests, Government of Tamil Nadu (2005) Post-tsunami assessment report: Cuddalore district
Government of India (1991) Coastal regulation zone notification. Ministry of Environment and Forests, environment protection act, 1986
Government of India (2006) National Environment Policy. Ministry of Environment and Forests
Mangroves in India (2025) In Wikipedia. https://en.wikipedia.org/wiki/Mangroves_in_India
Ministry of Environment, Forest and Climate Change (2023–24) Operational guidelines: MISHTI programme. Government of India. https://nationalcampa.nic.in/dashboard/GuidelinesPDF/665052fa8dd59.pdf
NSO India (2020) Natural capital accounting and valuation of ecosystem services–India. National Statistical Office and United Nations Statistics Division
Pichavaram (2025) In Wikipedia. https://en.wikipedia.org/wiki/Pichavaram
Singha P, Dasgupta R, Hashimoto S, Mitra BK (2024) An integrated coastal exposure modelling approach to assist mangrove ecosystem-based disaster risk reduction (eco-DRR) in Tamil Nadu, India. Prog Disaster Sci 24:100388. https://doi.org/10.1016/j.pdisas.2024.100388
United Nations (2015) Transforming our world: the 2030 agenda for sustainable development. United Nations
United Nations (2021) System of environmental-economic accounting—ecosystem accounting: technical report on valuation. https://seea.un.org

The manufacturer's authorised representative in the EU is Springer Nature Customer Service Centre GmbH, Europaplatz 3, 69115 Heidelberg, Germany. If you have any concerns regarding our products, please contact ProductSafety@springernature.com

Printed and bound by CPI Group (UK) Ltd, Croydon, CR0 4YY
07/10/2026
02229357-0018